A2-Level Biology

The Revision Guide

Contents

How Science Works

The Scientific Process ... 2

Section 1 — Photosynthesis and Respiration

Photosynthesis, Respiration and ATP ... 4
Photosynthesis ... 6
Limiting Factors in Photosynthesis ... 10
Aerobic Respiration ... 14
Respiration Experiments ... 18
Aerobic and Anaerobic Respiration ... 20

Section 2 — Populations

Niches and Adaptations ... 22
Investigating Populations and Abiotic Conditions ... 24
Investigating Populations and Analysing Data ... 27
Variation in Population Size ... 30
Human Populations ... 32

Section 3 — Energy Flow and Nutrient Cycles

Energy Transfer and Productivity ... 34
Pyramid Diagrams and Energy Transfer ... 36
Farming Practices and Productivity ... 38
The Nitrogen Cycle and Eutrophication ... 40
The Carbon Cycle ... 42

Section 4 — Global Warming

Introduction to Global Warming ... 44
Causes of Global Warming ... 46
Effects of Global Warming ... 48
Reducing Global Warming ... 52

Section 5 — Succession and Conservation

Succession ... 54
Conservation ... 56
Conservation Evidence and Data ... 58
Conservation of Ecosystems ... 60

Section 6 — Inheritance

Inheritance ... 62
Phenotypic Ratios and Epistasis ... 66
The Chi-Squared Test ... 68
Meiosis ... 70

Section 7 — Variation, Natural Selection and Evolution

Variation ... 72
Natural Selection and Genetic Drift ... 74
The Hardy-Weinberg Principle ... 76
Speciation ... 78
Evolution Evidence and Artificial Selection ... 80

Section 8 — Responding to the Environment

Nervous and Hormonal Communication ... 82
Receptors ... 85
Nervous System — Neurones ... 88
Nervous System — Synaptic Transmission ... 92
Effectors — Muscle Contraction ... 96
Effectors — Glands ... 102
Responses in Animals ... 104
Responses in Plants ... 106
Plant Hormones ... 108
Investigating Responses in Plants ... 110

Section 9 — Homeostasis

Homeostasis Basics ... 112
Control of Body Temperature ... 114
Control of Blood Glucose Concentration ... 116
Control of the Menstrual Cycle ... 120

Section 10 — Genetics

DNA and RNA ... 122
Protein Synthesis ... 124
The Genetic Code and Nucleic Acids ... 126
Regulation of Transcription and Translation ... 128
Control of Protein Synthesis ... 130
Protein Activation and Gene Mutation ... 132
Mutations, Genetic Disorders and Cancer ... 134
Diagnosing and Treating Cancer and Genetic Disorders ... 136
Stem Cells ... 138
Stem Cells in Medicine ... 140

Section 11 — Using Gene Technology

Making DNA Fragments ... 142
Common Techniques ... 144
Gene Cloning ... 146
Genetic Engineering ... 148
Genetic Fingerprinting ... 154
Sequencing Genomes and Restriction Mapping ... 156
DNA Probes in Medical Diagnosis ... 158
Gene Therapy ... 160

Section 12 — Excretion

OCR only

The Liver and Excretion 162
The Kidneys and Excretion 164
Controlling Water Content 166
Kidney Failure and Detecting Hormones 168

Section 13 — Cloning and Biotechnology

OCR only

Cloning 170
Biotechnology 173

Section 14 — The Brain and Behaviour

Brain Structure and Function 176
Brain Development and Habituation 179
Development of the Visual Cortex 182
Behaviour 184

Section 15 — Microorganisms and Immunity

Edexcel only

Viral and Bacterial Infections 188
Infection and The Non-Specific Immune Response ... 190
The Specific Immune Response 192
Developing Immunity 194
Antibiotics 196
Microbial Decomposition and Time of Death 198

Section 16 — Heart Rate, Ventilation and Exercise

Edexcel only

Electrical Activity in the Heart 200
Variations in Heart Rate and Breathing Rate 202
Investigating Ventilation 204
Exercise and Health 206

Section 17 — Drugs

Edexcel only

Drugs and Disease 210
Producing Drugs Using GMOs 212

Answering Data Questions

How to Interpret Experiment and Study Data 214

Answers 217
Acknowledgements 234
Index 235

Published by Coordination Group Publications Ltd.

Editors:
Ellen Bowness, Katie Braid, Joe Brazier, Charlotte Burrows, Katherine Craig,
Rosie Gillham, Murray Hamilton, Jane Towle.
Contributors:
Gloria Barnett, Jessica Egan, Mark Ellingham, James Foster, Julian Hardwick, Derek Harvey,
Adrian Schmit, Sophie Watkins.

Proofreaders:
Sue Hocking, Glenn Rogers.

With thanks to Laura Stoney for copyright research.

This book covers:

AQA
OCR
Edexcel

There are notes on the pages to tell you which bits you need for your syllabus.

ISBN: 978 1 84762 262 4

Groovy website: www.cgpbooks.co.uk
Jolly bits of clipart from CorelDRAW®
Printed by Elanders Ltd, Newcastle upon Tyne.

Based on the classic CGP style created by Richard Parsons.

The Scientific Process

This stuff may look similar to what you learnt at AS, but that's because you need to understand How Science Works for A2 as well. 'How Science Works' is all about the scientific process — how we develop and test scientific ideas. It's what scientists do all day, every day (well, except at coffee time — never come between a scientist and their coffee).

Scientists Come Up with **Theories** — Then **Test Them**...

Science tries to explain **how** and **why** things happen — it **answers questions**. It's all about seeking and gaining **knowledge** about the world around us. Scientists do this by **asking** questions and **suggesting** answers and then **testing** them, to see if they're correct — this is the **scientific process**.

1) **Ask** a question — make an **observation** and ask **why or how** it happens. E.g. why do plants grow faster in glasshouses than outside?
2) **Suggest** an answer, or part of an answer, by forming a **theory** (a possible **explanation** of the observations), e.g. glasshouses are warmer than outside and plants grow faster when it's warmer because the rate of photosynthesis is higher. (Scientists also sometimes form a **model** too — a **simplified picture** of what's physically going on.)
3) Make a **prediction** or **hypothesis** — a **specific testable statement**, based on the theory, about what will happen in a test situation. E.g. the rate of photosynthesis will be faster at 20 °C than at 10 °C.
4) Carry out a **test** — to provide **evidence** that will support the prediction (or help to disprove it). E.g. measure the rate of photosynthesis at various temperatures.

Simone predicted her hair would be worse on date night, based on the theory of sod's law.

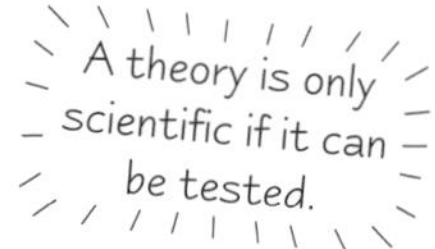
A theory is only scientific if it can be tested.

...Then They **Tell** Everyone About Their **Results**...

The results are **published** — scientists need to let others know about their work. Scientists publish their results in **scientific journals**. These are just like normal magazines, only they contain **scientific reports** (called papers) instead of the latest celebrity gossip.

1) Scientific reports are similar to the **lab write-ups** you do in school. And just as a lab write-up is **reviewed** (marked) by your teacher, reports in scientific journals undergo **peer review** before they're published.
2) The report is sent out to **peers** — other scientists who are experts in the **same area**. They examine the data and results, and if they think that the conclusion is reasonable it's **published**. This makes sure that work published in scientific journals is of a **good standard**.
3) But peer review **can't guarantee** the science is **correct** — other scientists still need to **reproduce** it.
4) Sometimes **mistakes** are made and flawed work is published. Peer review **isn't perfect** but it's probably the best way for scientists to self-regulate their work and to publish **quality reports**.

...Then **Other Scientists** Will **Test** the Theory Too

Other scientists read the published theories and results, and try to **test the theory** themselves. This involves:

- Repeating the **exact same experiments**.
- Using the theory to make **new predictions** and then testing them with **new experiments**.

If the **Evidence** Supports a Theory, It's **Accepted — for Now**

1) If all the experiments in all the world provide good evidence to back it up, the theory is thought of as **scientific 'fact'** (for now).
2) But it will never become **totally indisputable** fact. Scientific **breakthroughs or advances** could provide new ways to question and test the theory, which could lead to **new evidence** that **conflicts** with the current evidence. Then the testing starts all over again...

And this, my friend, is the **tentative nature of scientific knowledge** — it's always **changing** and **evolving**.

The Scientific Process

So scientists need evidence to back up their theories. They get it by carrying out experiments, and when that's not possible they carry out studies. But why bother with science at all? We want to know as much as possible so we can use it to try and improve our lives (and because we're nosy).

Evidence Comes from Lab Experiments...

1) Results from **controlled experiments** in **laboratories** are **great**.
2) A lab is the easiest place to **control variables** so that they're all **kept constant** (except for the one you're investigating).
3) This means you can draw meaningful **conclusions**.

For example, if you're investigating how light intensity affects the rate of photosynthesis you need to keep everything but the light intensity constant, e.g. the temperature, the concentration of carbon dioxide etc.

...and Well-Designed Studies

1) There are things you **can't** investigate in a lab, e.g. whether using a pesticide on farmland affects the number of non-pest species. You have to do a study instead.
2) You still need to try and make the study as controlled as possible to make it **more reliable**. But in reality it's **very hard** to control **all the variables** that **might** be having an effect.
3) You can do things to help, like having a **control** — e.g. an area of similar farmland nearby where the pesticide isn't applied. But you can't easily rule out every possibility.

Having a control reduced the effect of exercise on the study.

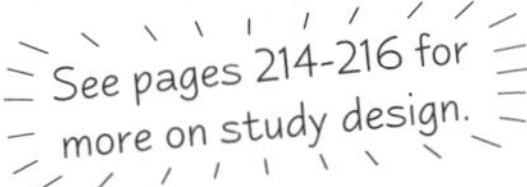

See pages 214-216 for more on study design.

Society Makes Decisions Based on Scientific Evidence

1) Lots of scientific work eventually leads to **important discoveries** or breakthroughs that could **benefit humankind**.
2) These results are **used by society** (that's you, me and everyone else) to **make decisions** — about the way we live, what we eat, what we drive, etc.
3) All sections of society use scientific evidence to make decisions, e.g. politicians use it to devise policies and individuals use science to make decisions about their own lives.

Other factors can **influence** decisions about science or the way science is used:

Economic factors

- Society has to consider the **cost** of implementing changes based on scientific conclusions — e.g. the **NHS** can't afford the most expensive drugs without **sacrificing** something else.
- Scientific research is **expensive** so companies won't always develop new ideas — e.g. developing new drugs is costly, so pharmaceutical companies often only invest in drugs that are likely to make them **money**.

Social factors

- **Decisions** affect **people's lives** — E.g. scientists may suggest **banning smoking** and **alcohol** to prevent health problems, but shouldn't **we** be able to **choose** whether **we** want to smoke and drink or not?

Environmental factors

- Scientists believe **unexplored regions** like remote parts of rainforests might contain **untapped drug** resources. But some people think we shouldn't **exploit** these regions because any interesting finds may lead to **deforestation** and **reduced biodiversity** in these areas.

So there you have it — how science works...

Hopefully these pages have given you a nice intro to how science works, e.g. what scientists do to provide you with 'facts'. You need to understand this, as you're expected to know how science works — for the exam and for life.

Photosynthesis, Respiration and ATP

These pages are for AQA Unit 4, OCR Unit 4 and Edexcel Unit 4.

OK, this isn't the easiest topic to start a book on, but 'cos I'm feeling nice today we'll take it slowly, one bit at a time...

Biological Processes Need Energy

Plant and animal cells **need energy** for biological processes to occur:

- **Plants** need energy for things like **photosynthesis**, **active transport** (e.g. to take in minerals via their roots), **DNA replication**, **cell division** and **protein synthesis**.
- **Animals** need energy for things like **muscle contraction**, maintenance of **body temperature**, **active transport**, **DNA replication**, **cell division** and **protein synthesis**.

Without energy, these biological processes would stop and the plant or animal would die.

Photosynthesis Stores Energy in Glucose

1) **Plants** are **autotrophs** — they can **make** their **own food** (**glucose**). They do this using **photosynthesis**.
2) **Photosynthesis** is the process where **energy** from **light** is used to **make glucose** from H_2O and CO_2 (the light energy is **converted** to **chemical energy** in the form of glucose).
3) Photosynthesis occurs in a **series** of **reactions**, but the overall equation is:

$$6CO_2 + 6H_2O + \text{Energy} \longrightarrow C_6H_{12}O_6 \text{ (glucose)} + 6O_2$$

4) So, energy is **stored** in the **glucose** until the plants **release** it by **respiration**.
5) **Animals** are **heterotrophs** — they **can't make** their **own food**. So, they obtain glucose by **eating plants** (or **other animals**), then respire the glucose to release energy.

Cells Release Energy from Glucose by Respiration

1) **Plant** and **animal** cells **release energy** from **glucose** — this process is called **respiration**.
2) This energy is used to power all the **biological processes** in a cell.
3) There are two types of respiration:
   - **Aerobic respiration** — respiration **using oxygen**.
   - **Anaerobic respiration** — respiration **without oxygen**.
4) Aerobic respiration produces **carbon dioxide** and **water**, and releases **energy**. The overall equation is:

$$C_6H_{12}O_6 \text{ (glucose)} + 6O_2 \longrightarrow 6CO_2 + 6H_2O + \text{Energy}$$

Respiration includes many steps, each of which is controlled by a specific enzyme.

ATP is the Immediate Source of Energy in a Cell

1) A cell **can't** get its energy **directly** from glucose.
2) So, in respiration, the **energy released** from glucose is used to **make ATP** (adenosine triphosphate). ATP is made from the nucleotide base **adenine**, combined with a **ribose sugar** and **three phosphate groups**.
3) It **carries energy** around the cell to where it's **needed**.
4) **ATP** is **synthesised** from **ADP** and **inorganic phosphate** (P_i) using energy from an **energy-releasing** reaction, e.g. the **breakdown** of **glucose** in **respiration**. The energy is stored as **chemical energy** in the **phosphate bond**. The enzyme **ATP synthase** catalyses this reaction.

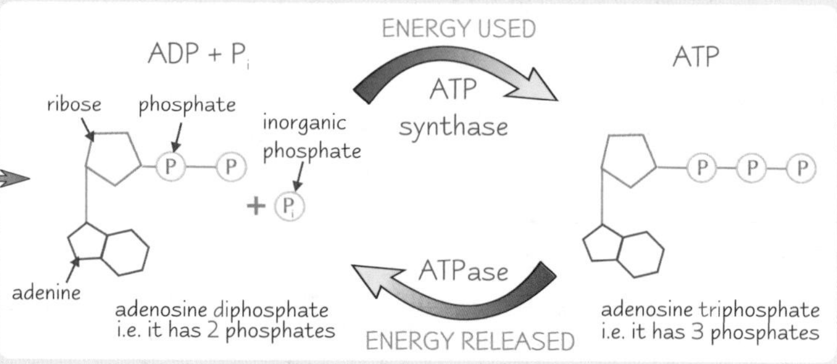

5) ATP **diffuses** to the part of the cell that **needs** energy.
6) Here, it's **broken down** back into **ADP** and **inorganic phosphate** (P_i). Chemical **energy** is **released** from the phosphate bond and used by the cell. **ATPase** catalyses this reaction.
7) The ADP and inorganic phosphate are **recycled** and the process starts again.

Inorganic phosphate (P_i) is just the fancy name for a single phosphate.

Photosynthesis, Respiration and ATP

ATP has *Specific Properties* that Make it a *Good Energy Source*

1) ATP stores or releases only a **small**, **manageable amount** of energy at a time, so **no** energy is **wasted**.
2) It's a **small**, **soluble** molecule so it can be **easily transported** around the cell.
3) It's **easily broken down**, so energy can be **easily released**.
4) It can **transfer energy** to another molecule by transferring one of its **phosphate groups**.
5) ATP **can't pass out** of the **cell**, so the cell **always** has an immediate supply of energy.

Karen needed a lot of energy just to keep her headdress on...

You Need to *Know Some Basics* Before You Start

There are some pretty confusing technical terms in this section that you need to get your head around:

- **Metabolic pathway** — a **series** of **small reactions** controlled by **enzymes**, e.g. **respiration** and **photosynthesis**.
- **Phosphorylation** — **adding phosphate** to a molecule, e.g. **ADP** is phosphorylated to **ATP** (see previous page).
- **Photophosphorylation** — **adding phosphate** to a molecule using **light**.
- **Photolysis** — the **splitting** (lysis) of a molecule using **light** (photo) energy.
- **Hydrolysis** — the **splitting** (lysis) of a molecule using **water** (hydro).
- **Decarboxylation** — the **removal** of **carbon dioxide** from a molecule.
- **Dehydrogenation** — the **removal** of **hydrogen** from a molecule.
- **Redox reactions** — reactions that involve **oxidation** and **reduction**.

Remember redox reactions:

1) If something is **reduced** it has **gained electrons** (e^-), and may have **gained hydrogen** or lost oxygen.
2) If something is **oxidised** it has **lost electrons**, and may have **lost hydrogen** or gained oxygen.
3) Oxidation of one molecule **always** involves reduction of another molecule.

One way to remember electron and hydrogen movement is OILRIG. Oxidation Is Loss, Reduction Is Gain.

Photosynthesis and *Respiration* Involve *Coenzymes*

1) A **coenzyme** is a molecule that **aids** the **function** of an **enzyme**.
2) They work by **transferring** a **chemical group** from one molecule to another.
3) A coenzyme used in **photosynthesis** is **NADP**. NADP transfers **hydrogen** from one molecule to another — this means it can **reduce** (give hydrogen to) or **oxidise** (take hydrogen from) a molecule.
4) Examples of coenzymes used in **respiration** are: **NAD**, **coenzyme A** and **FAD**.
   - NAD and FAD transfer **hydrogen** from one molecule to another — this means they can **reduce** (give hydrogen to) or **oxidise** (take hydrogen from) a molecule.
   - **Coenzyme A** transfers **acetate** between molecules (see pages 15-16).

When hydrogen is transferred between molecules, electrons are transferred too.

Practice Questions

Q1 Write down three biological processes in animals that need energy.

Q2 What is photosynthesis?

Q3 What is the overall equation for aerobic respiration?

Q4 How many phosphate groups does ATP have?

Q5 Give the name of a coenzyme involved in photosynthesis.

Exam Question

Q1 ATP is the immediate source of energy inside a cell.
Describe how the synthesis and breakdown of ATP meets the energy needs of a cell. [6 marks]

Oh dear, I've used up all my ATP on these two pages...

Well, I won't beat about the bush, this stuff is pretty tricky... nearly as hard as a cross between Mr T, Hulk Hogan and Arnie. But, with a little patience and perseverance (and plenty of [chocolate] [coffee] [marshmallows] — delete as you wish), you'll get there. Once you've got these pages straight in your head, the next ones will be easier to understand.

Photosynthesis

***These pages are for** AQA Unit 4, OCR Unit 4 **and** Edexcel Unit 4.*

Right, pen at the ready. Check. Brain switched on. Check. Cuppa piping hot. Check. Sweets on standby. Check. Okay, I think you're all sorted to start photosynthesis. Finally, take a deep breath and here we go...

Photosynthesis Takes Place in the **Chloroplasts** of **Plant Cells**

1) **Chloroplasts** are **small, flattened organelles** found in **plant cells**.
2) They have a **double membrane** called the **chloroplast envelope**.
3) **Thylakoids** (fluid-filled sacs) are **stacked up** in the chloroplast into structures called **grana** (singular = **granum**). The grana are **linked** together by bits of thylakoid membrane called **lamellae** (singular = **lamella**).
4) Chloroplasts contain **photosynthetic pigments** (e.g. **chlorophyll a**, **chlorophyll b** and **carotene**). These are **coloured substances** that **absorb** the **light energy** needed for photosynthesis. The pigments are found in the **thylakoid membranes** — they're attached to **proteins**. The protein and pigment is called a **photosystem**.
5) There are **two** photosystems used by plants to capture light energy. **Photosystem I** (or PSI) absorbs light best at a wavelength of **700 nm** and **photosystem II** (PSII) absorbs light best at **680 nm**.
6) Contained within the inner membrane of the chloroplast and **surrounding** the thylakoids is a gel-like substance called the **stroma**. It contains **enzymes**, **sugars** and **organic acids**.
7) Carbohydrates produced by photosynthesis and not used straight away are stored as **starch grains** in the **stroma**.

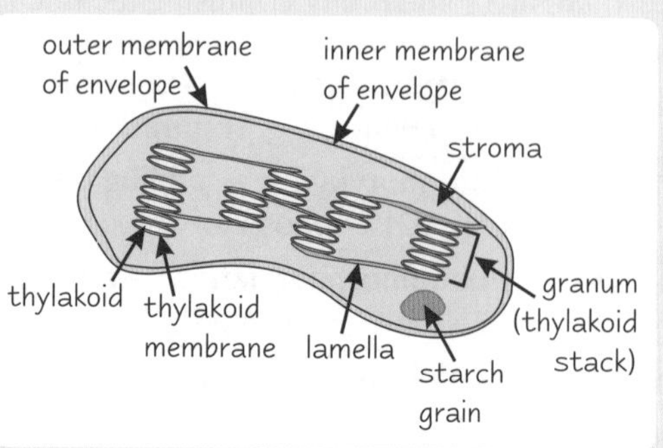

Photosynthesis can be Split into **Two Stages**

There are actually **two stages** that make up **photosynthesis**:

See p. 8 for loads more information on the Calvin cycle.

1 The Light-Dependent Reaction

1) As the name suggests, this reaction **needs light energy**.
2) It takes place in the **thylakoid membranes** of the chloroplasts.
3) Here, light energy is absorbed by **photosynthetic pigments** in the **photosystems** and converted to **chemical energy**.
4) The light energy is used to add a phosphate group to ADP to form **ATP**, and to reduce NADP to form **reduced NADP**. **ATP transfers energy** and reduced **NADP transfers hydrogen** to the light-independent reaction.
5) During the process H_2O is **oxidised** to O_2.

2 The Light-Independent Reaction

1) This is also called the **Calvin cycle** and as the name suggests it **doesn't use light energy** directly. (But it does **rely** on the **products** of the light-dependent reaction.)
2) It takes place in the **stroma** of the chloroplasts.
3) Here, the **ATP** and **reduced NADP** from the light-dependent reaction supply the **energy** and **hydrogen** to make **glucose** from CO_2.

This diagram shows how the two reactions link together in the chloroplast:

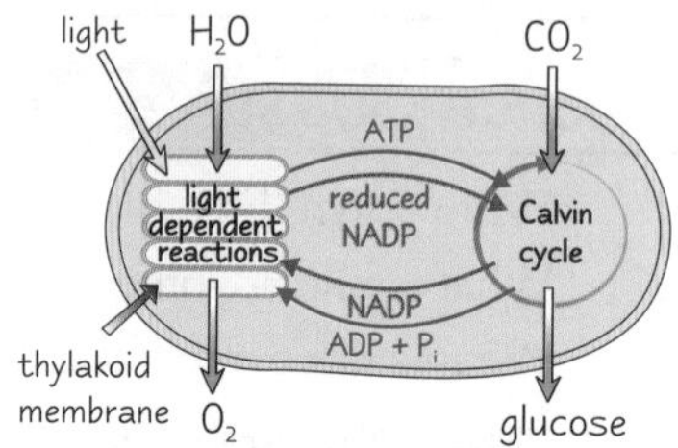

In the **Light-Dependent Reaction ATP** is Made by **Photophosphorylation**

In the light-dependent reaction, the **light energy** absorbed by the photosystems is used for **three** things:

1) Making **ATP** from **ADP** and **inorganic phosphate**. This reaction is called **photophosphorylation** (see p. 5).
2) Making **reduced NADP** from **NADP**.
3) Splitting **water** into **protons** (H^+ **ions**), **electrons** and **oxygen**. This is called **photolysis** (see p. 5).

The light-dependent reaction actually includes **two types** of **photophosphorylation** — **non-cyclic** and **cyclic**. Each of these processes has **different products**.

Photosynthesis

Non-cyclic Photophosphorylation Produces ATP, Reduced NADP and O_2

To understand the process you need to know that the photosystems (in the thylakoid membranes) are **linked** by **electron carriers**. Electron carriers are **proteins** that **transfer electrons**. The photosystems and electron carriers form an **electron transport chain** — a **chain** of **proteins** through which **excited electrons flow**. All the processes in the diagrams are happening together — I've just split them up to make it easier to understand.

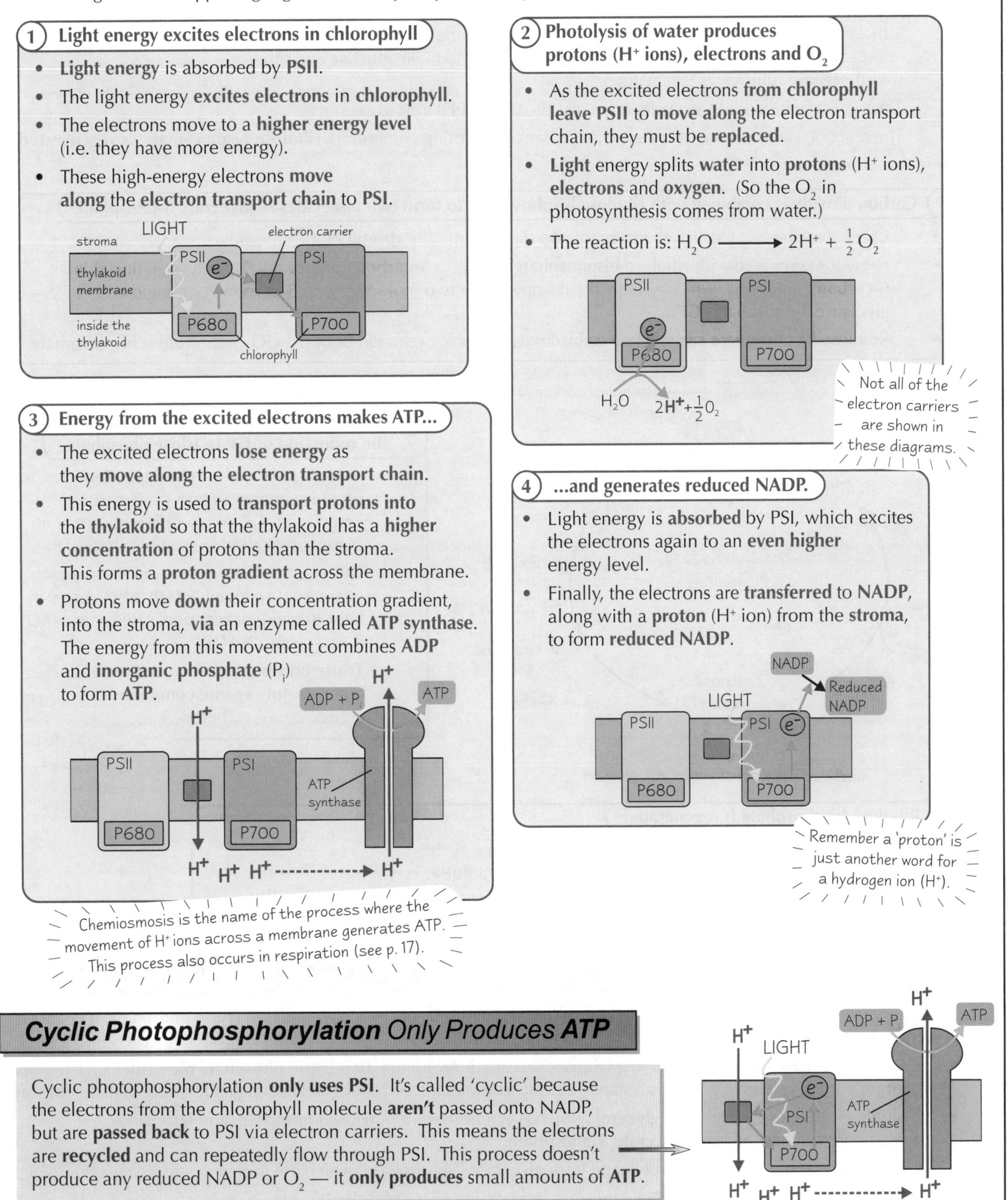

1) Light energy excites electrons in chlorophyll

- **Light energy** is absorbed by **PSII**.
- The light energy **excites electrons** in **chlorophyll**.
- The electrons move to a **higher energy level** (i.e. they have more energy).
- These high-energy electrons **move along** the **electron transport chain** to **PSI**.

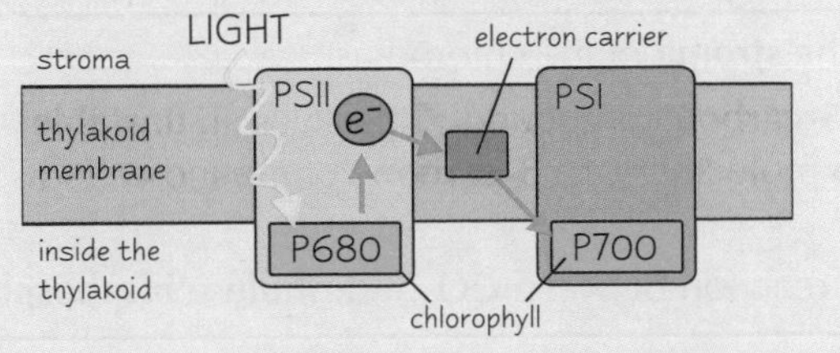

2) Photolysis of water produces protons (H^+ ions), electrons and O_2

- As the excited electrons **from chlorophyll leave PSII** to **move along** the electron transport chain, they must be **replaced**.
- **Light** energy splits **water** into **protons** (H^+ ions), **electrons** and **oxygen**. (So the O_2 in photosynthesis comes from water.)
- The reaction is: $H_2O \longrightarrow 2H^+ + \frac{1}{2}O_2$

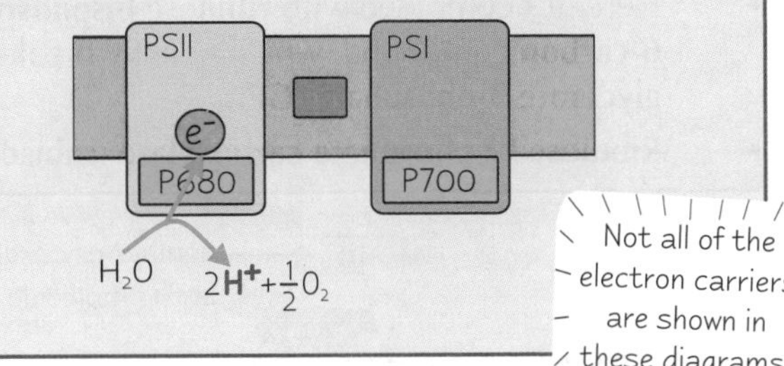

Not all of the electron carriers are shown in these diagrams.

3) Energy from the excited electrons makes ATP...

- The excited electrons **lose energy** as they **move along** the **electron transport chain**.
- This energy is used to **transport protons into** the **thylakoid** so that the thylakoid has a **higher concentration** of protons than the stroma. This forms a **proton gradient** across the membrane.
- Protons move **down** their concentration gradient, into the stroma, **via** an enzyme called **ATP synthase**. The energy from this movement combines **ADP** and **inorganic phosphate** (P_i) to form **ATP**.

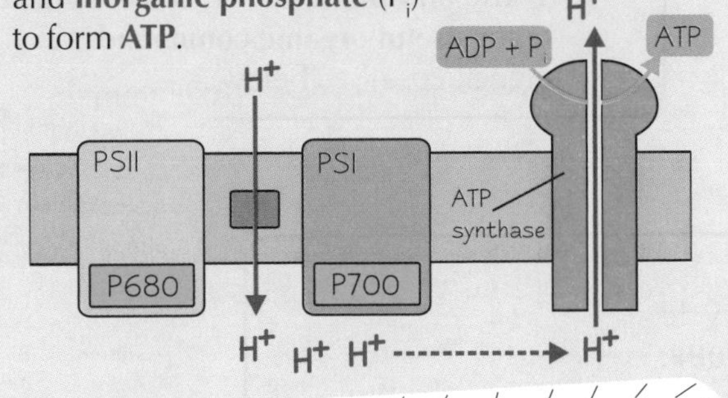

Chemiosmosis is the name of the process where the movement of H^+ ions across a membrane generates ATP. This process also occurs in respiration (see p. 17).

4) ...and generates reduced NADP.

- Light energy is **absorbed** by PSI, which excites the electrons again to an **even higher** energy level.
- Finally, the electrons are **transferred** to **NADP**, along with a **proton** (H^+ ion) from the **stroma**, to form **reduced NADP**.

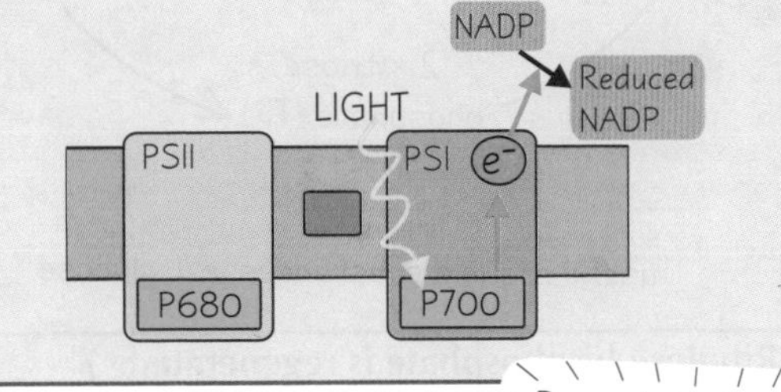

Remember a 'proton' is just another word for a hydrogen ion (H^+).

Cyclic Photophosphorylation Only Produces ATP

Cyclic photophosphorylation **only uses PSI**. It's called 'cyclic' because the electrons from the chlorophyll molecule **aren't** passed onto NADP, but are **passed back** to PSI via electron carriers. This means the electrons are **recycled** and can repeatedly flow through PSI. This process doesn't produce any reduced NADP or O_2 — it **only produces** small amounts of **ATP**.

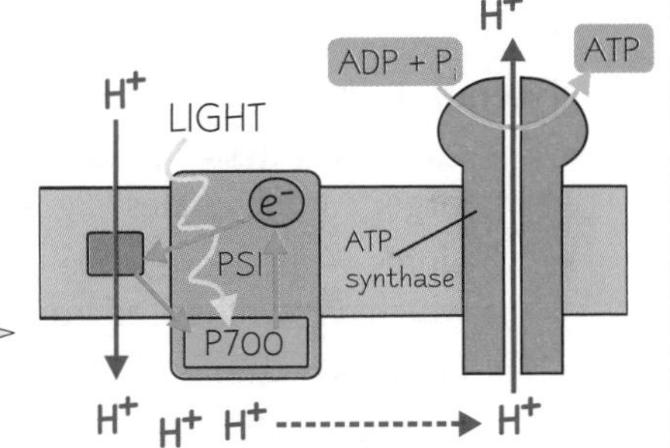

Photosynthesis

These pages are for AQA Unit 4, OCR Unit 4 and Edexcel Unit 4.

Don't worry, you're over the worst of photosynthesis now. Instead of electrons flying around, there's a nice cycle of reactions to learn. What more could you want from life? Money, fast cars and nice clothes have nothing on this...

The **Light-Independent** Reaction is also called the **Calvin Cycle**

1) The Calvin cycle takes place in the **stroma** of the chloroplasts.
2) It makes a molecule called **triose phosphate** from CO_2 and **ribulose bisphosphate** (a 5-carbon compound). Triose phosphate can be used to make **glucose** and other **useful organic substances** (see below).
3) There are a few steps in the cycle, and it needs **ATP** and **H^+ ions** to keep it going.
4) The reactions are linked in a **cycle**, which means the starting compound, **ribulose bisphosphate**, is **regenerated**.

The Calvin cycle is also called carbon fixation, because carbon from CO_2 is 'fixed' into an organic molecule.

Here's what happens at each stage in the cycle:

1) Carbon dioxide is combined with ribulose bisphosphate to form two molecules of glycerate 3-phosphate

- CO_2 enters the leaf through the **stomata** and diffuses into the **stroma** of the chloroplast.
- Here, it's combined with **ribulose bisphosphate (RuBP)**, a **5-carbon** compound. This gives an **unstable 6-carbon** compound, which quickly breaks down into **two** molecules of a **3-carbon** compound called **glycerate 3-phosphate (GP)**.
- **Ribulose bisphosphate carboxylase (rubisco)** catalyses the reaction between CO_2 and **ribulose bisphosphate**.

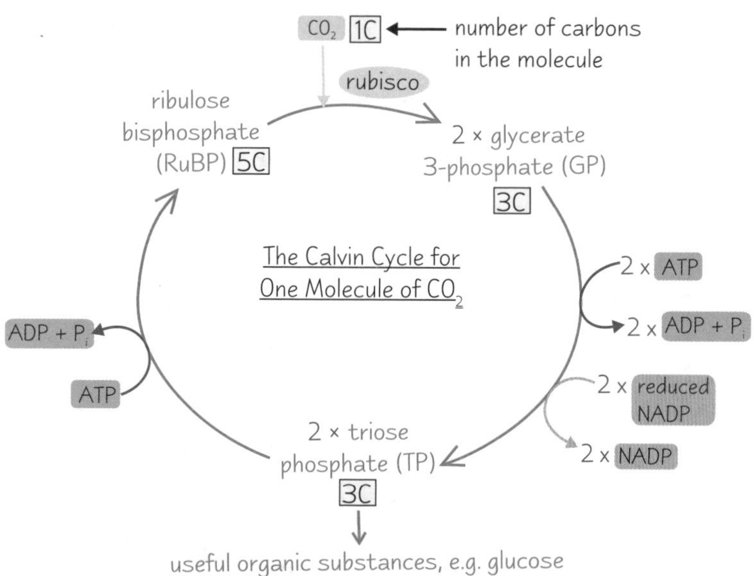

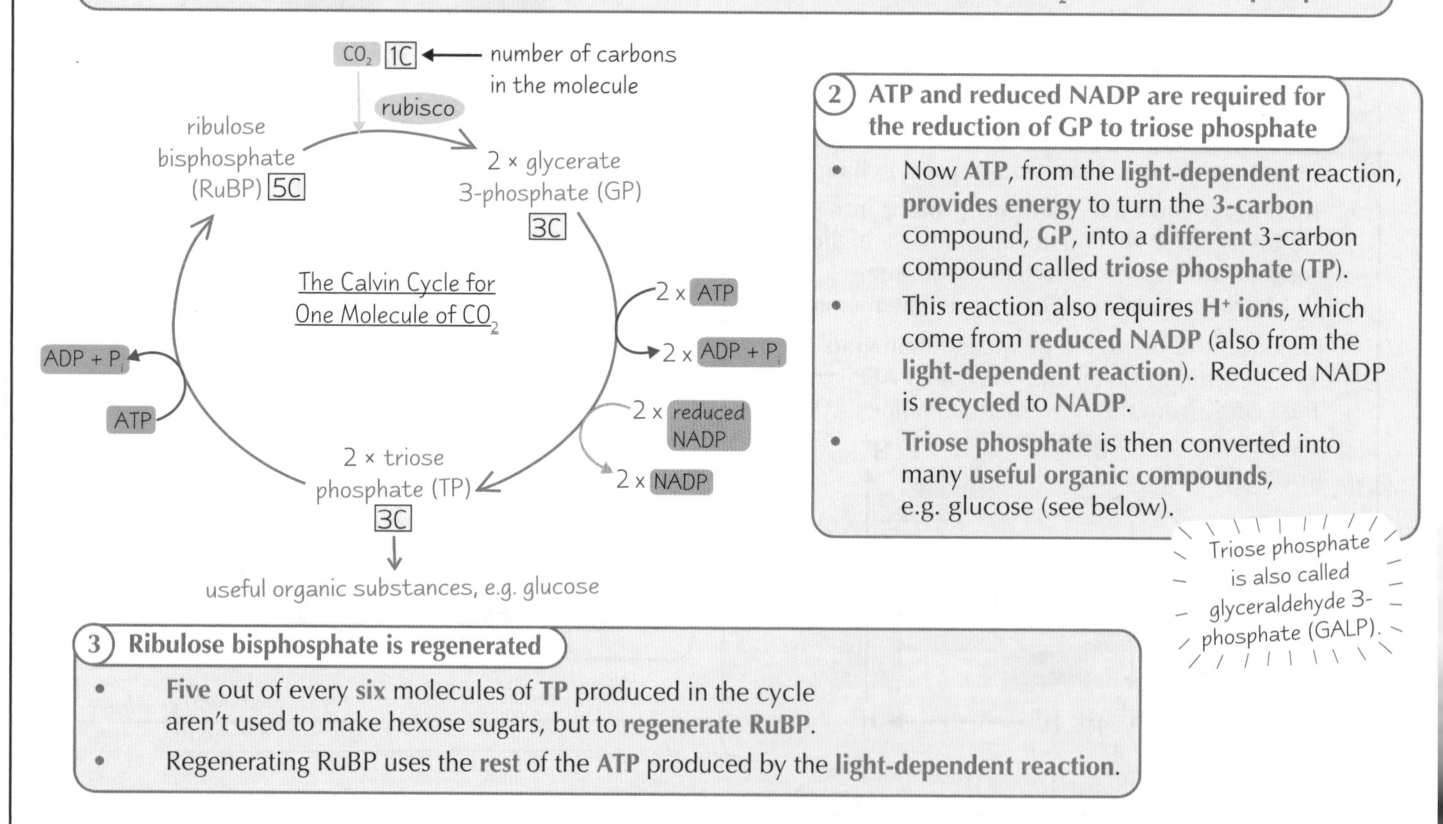

2) ATP and reduced NADP are required for the reduction of GP to triose phosphate

- Now **ATP**, from the **light-dependent** reaction, **provides energy** to turn the **3-carbon** compound, **GP**, into a **different** 3-carbon compound called **triose phosphate (TP)**.
- This reaction also requires **H^+ ions**, which come from **reduced NADP** (also from the **light-dependent reaction**). Reduced NADP is **recycled to NADP**.
- **Triose phosphate** is then converted into many **useful organic compounds**, e.g. glucose (see below).

Triose phosphate is also called glyceraldehyde 3-phosphate (GALP).

3) Ribulose bisphosphate is regenerated

- **Five** out of every **six** molecules of **TP** produced in the cycle aren't used to make hexose sugars, but to **regenerate RuBP**.
- Regenerating RuBP uses the **rest** of the **ATP** produced by the **light-dependent reaction**.

TP and ***GP*** are ***Converted*** into ***Useful Organic Substances*** like ***Glucose***

The Calvin cycle is the starting point for making **all** the organic substances a plant needs. **Triose phosphate** (TP) and **glycerate 3-phosphate** (GP) molecules are used to make **carbohydrates**, **lipids**, **proteins** and **nucleic acids**:

- **Carbohydrates** — **hexose sugars** (e.g. glucose) are made by joining **two triose phosphate molecules** together and **larger** carbohydrates (e.g. sucrose, starch, cellulose) are made by joining **hexose sugars** together in **different ways**.
- **Lipids** — these are made using **glycerol**, which is synthesised from **triose phosphate**, and **fatty acids**, which are synthesised from **glycerate 3-phosphate**.
- **Proteins** — some **amino acids** are made from **glycerate 3-phosphate**, which are joined together to make proteins.
- **Nucleic acids** — the sugar in **RNA** (**ribose**) is made using **triose phosphate**.

Photosynthesis

The Calvin Cycle Needs to Turn Six Times to Make One Hexose Sugar

1) **Three turns** of the cycle produces **six** molecules of **triose phosphate** (TP), because two molecules of TP are made for every one CO_2 molecule used.
2) **Five** out of **six** of these TP molecules are used to **regenerate ribulose bisphosphate** (RuBP).
3) This means that for **three turns** of the cycle only **one TP** is produced that's used to make a **hexose sugar**.
4) A hexose sugar has **six carbons** though, so **two TP** molecules are needed to form one hexose sugar.
5) This means the cycle must turn **six times** to produce **two molecules** of **TP** that can be used to make **one hexose sugar**.
6) Six turns of the cycle need **18 ATP** and **12 reduced NADP** from the light-dependent reaction.

The Structure of a Chloroplast is Adapted for Photosynthesis

OCR and Edexcel only

1) The **chloroplast envelope** keeps the **reactants** for photosynthesis **close** to their **reaction sites**.
2) The **thylakoids** have a **large surface area** to allow as much **light energy** to be **absorbed** as possible.
3) **Lots of ATP synthase** molecules are present in the thylakoid membranes to **produce ATP** in the light-dependent reaction.
4) The **stroma** contains all the **enzymes**, **sugars** and **organic acids** for the light-independent reaction to take place.

Practice Questions

Q1 Name two photosynthetic pigments in the chloroplasts of plants.

Q2 At what wavelength does photosystem I absorb light best?

Q3 What three substances does non-cyclic photophosphorylation produce?

Q4 Which photosystem is involved in cyclic photophosphorylation?

Q5 Where in the chloroplasts does the light-independent reaction occur?

Q6 Name two organic substances made from triose phosphate.

Q7 How many CO_2 molecules need to enter the Calvin cycle to make one hexose sugar?

Q8 Describe two ways in which a chloroplast is adapted for photosynthesis.

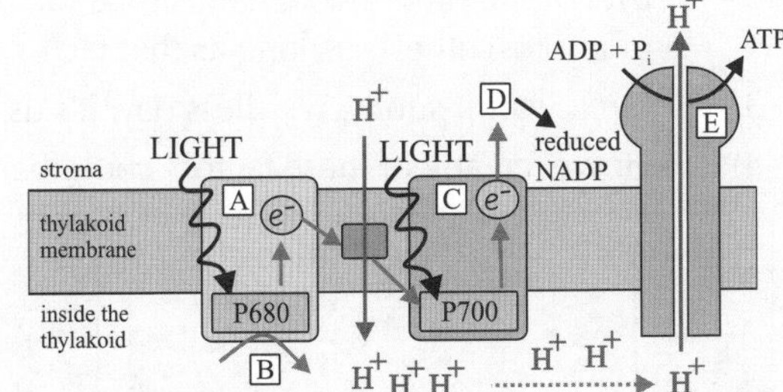

Exam Questions

Q1 The diagram above shows the light-dependent reaction of photosynthesis.

a) Where precisely in a plant does the light-dependent reaction of photosynthesis occur? [1 mark]

b) What is A? [1 mark]

c) Describe process B and explain its purpose. [4 marks]

d) What is reactant D? [1 mark]

Q2 Rubisco is an enzyme that catalyses the first reaction of the Calvin cycle.
CA1P is an inhibitor of rubisco.

a) Describe how triose phosphate is produced in the Calvin cycle. [6 marks]

b) Briefly explain how ribulose bisphosphate (RuBP) is regenerated in the Calvin cycle. [2 marks]

c) Explain the effect that CA1P would have on glucose production. [3 marks]

Calvin cycles — bikes made by people that normally make pants...

Next thing we know there'll be male models swanning about in their pants riding highly fashionable bikes. Sounds awful I know, but let's face it, anything would look better than cycling shorts. Anyway, it would be a good idea to go over these pages a couple of times — you might not feel as if you can fit any more information in your head, but you can, I promise.

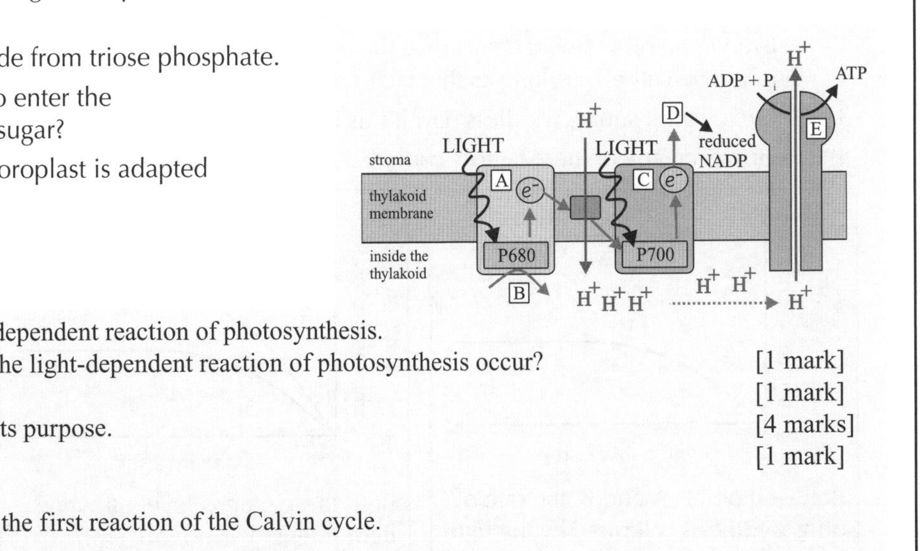

Limiting Factors in Photosynthesis

These pages are for AQA Unit 4 and OCR Unit 4. If you're doing Edexcel you can skip to page 14.

Now you know what photosynthesis is it's time to find out what conditions make it speedy and what slows it down.

There are **Optimum Conditions** for Photosynthesis

The **ideal conditions** for photosynthesis vary from one plant species to another, but the conditions below would be ideal for **most** plant species in temperate climates like the UK.

1. **High light intensity** of a certain **wavelength**

- Light is needed to provide the **energy** for the **light-dependent reaction** — the **higher** the **intensity** of the light, the **more energy** it provides.
- Only certain **wavelengths** of light are used for photosynthesis. The photosynthetic pigments chlorophyll a, chlorophyll b and carotene only **absorb** the **red** and **blue** light in sunlight. (**Green** light is **reflected**, which is why plants look green.)

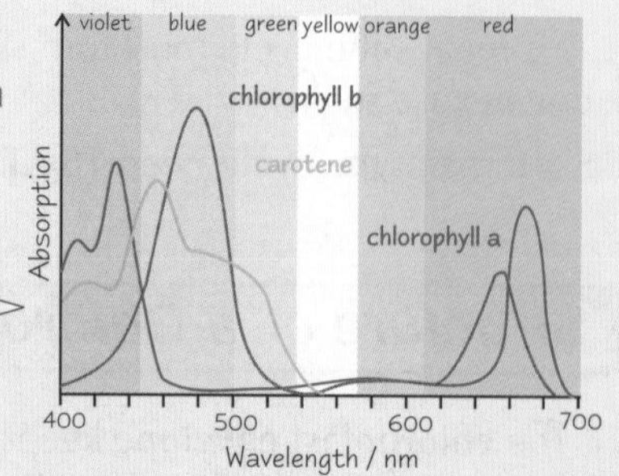

2. **Temperature** around **25 °C**

- Photosynthesis involves **enzymes** (e.g. ATP synthase, rubisco). If the temperature falls **below 10 °C** the enzymes become **inactive**, but if the temperature is **more than 45 °C** they may start to **denature**.
- Also, at **high** temperatures **stomata close** to avoid losing too much water. This causes photosynthesis to slow down because **less CO_2** enters the leaf when the stomata are closed.

3. **Carbon dioxide at 0.4%**

- Carbon dioxide makes up **0.04%** of the gases in the atmosphere.
- Increasing this to **0.4%** gives a **higher rate** of photosynthesis, but any higher and the stomata start to **close**.

Plants also need a constant supply of water — too little and photosynthesis has to stop but too much and the soil becomes waterlogged (reducing the uptake of magnesium for chlorophyll a).

Light, **Temperature** and **CO_2** can all **Limit Photosynthesis**

1) **All three** of these things need to be at the **right level** to allow a plant to photosynthesise as quickly as possible.
2) If any **one** of these factors is **too low** or **too high**, it will **limit photosynthesis** (slow it down). Even if the other two factors are at the perfect level, it won't make **any difference** to the speed of photosynthesis as long as that factor is at the wrong level.
3) On a warm, sunny, windless day, it's usually **CO_2** that's the limiting factor, and at night it's the **light intensity**.
4) However, **any** of these factors could become the limiting factor, depending on the **environmental conditions**.

Light intensity

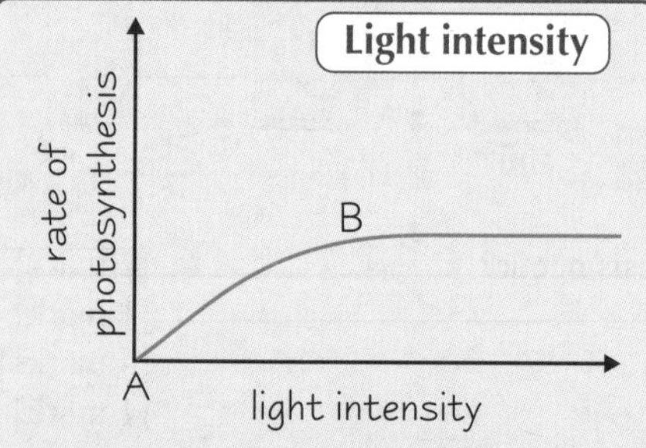

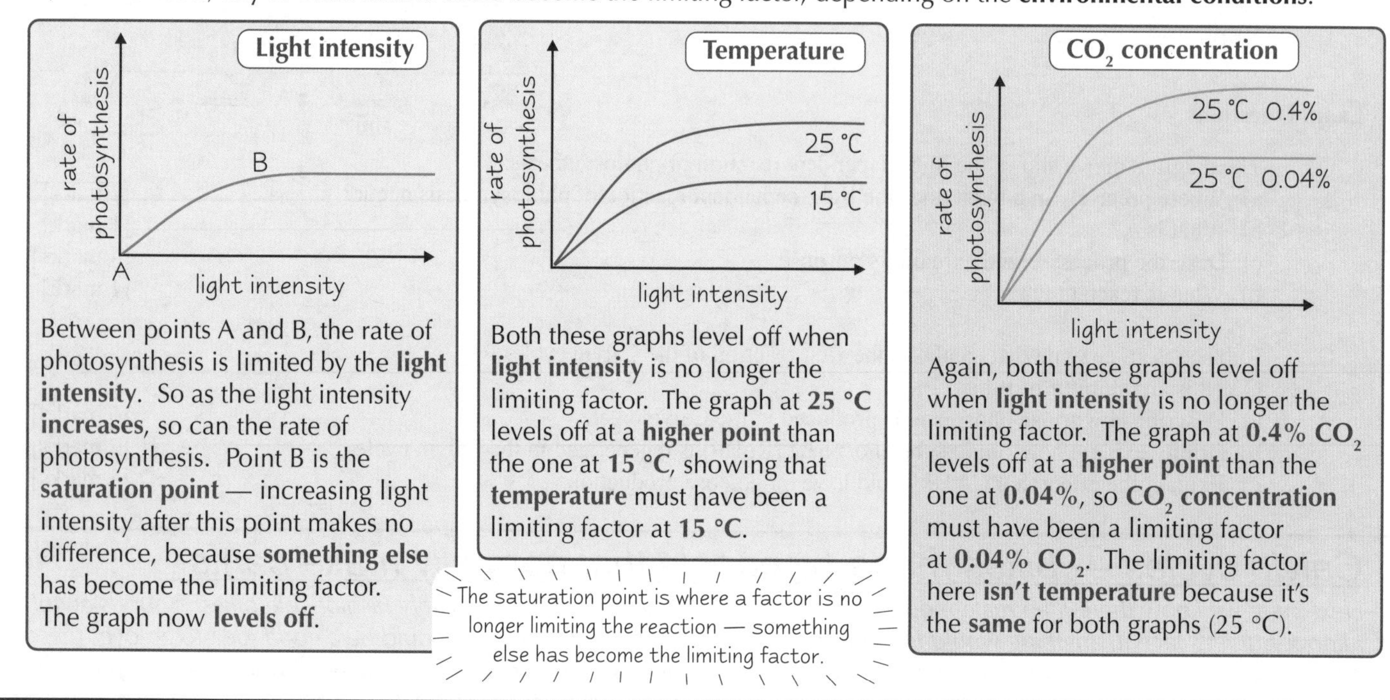

Between points A and B, the rate of photosynthesis is limited by the **light intensity**. So as the light intensity **increases**, so can the rate of photosynthesis. Point B is the **saturation point** — increasing light intensity after this point makes no difference, because **something else** has become the limiting factor. The graph now **levels off**.

Temperature

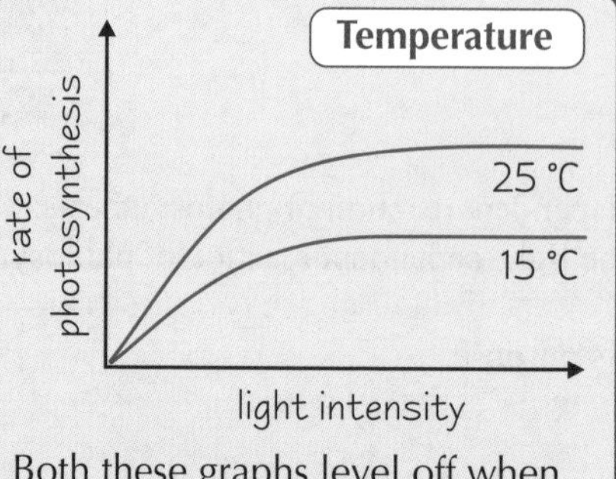

Both these graphs level off when **light intensity** is no longer the limiting factor. The graph at **25 °C** levels off at a **higher point** than the one at **15 °C**, showing that **temperature** must have been a limiting factor at **15 °C**.

The saturation point is where a factor is no longer limiting the reaction — something else has become the limiting factor.

CO_2 concentration

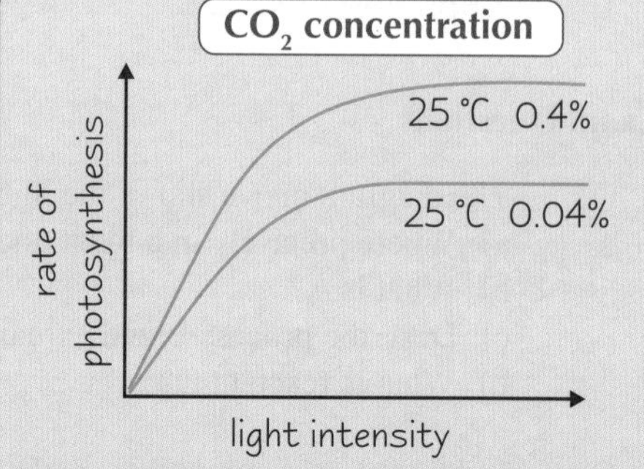

Again, both these graphs level off when **light intensity** is no longer the limiting factor. The graph at **0.4% CO_2** levels off at a **higher point** than the one at **0.04%**, so **CO_2 concentration** must have been a limiting factor at **0.04% CO_2**. The limiting factor here **isn't temperature** because it's the **same** for both graphs (25 °C).

Limiting Factors in Photosynthesis

This page is for OCR Unit 4 only. If you're doing AQA, you can skip straight to the questions.

Light, Temperature and CO_2 Affect the Levels of GP, RuBP and TP

Light intensity, **temperature** and **CO_2 concentration** all **affect** the **rate** of **photosynthesis**, which means they affect the **levels** of **GP**, **RuBP** and **TP** in the **Calvin cycle**.

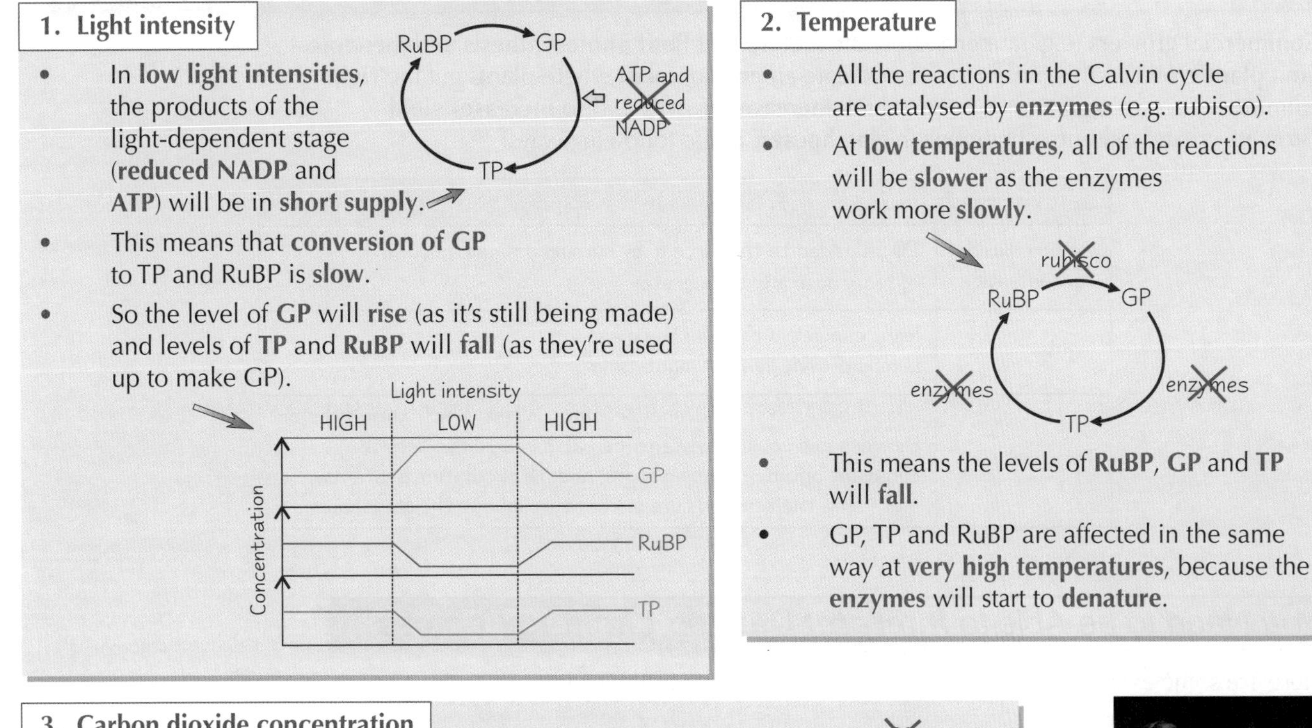

1. Light intensity

- In **low light intensities**, the products of the light-dependent stage (**reduced NADP** and **ATP**) will be in **short supply**.
- This means that **conversion of GP** to TP and RuBP is **slow**.
- So the level of **GP** will **rise** (as it's still being made) and levels of **TP** and **RuBP** will **fall** (as they're used up to make GP).

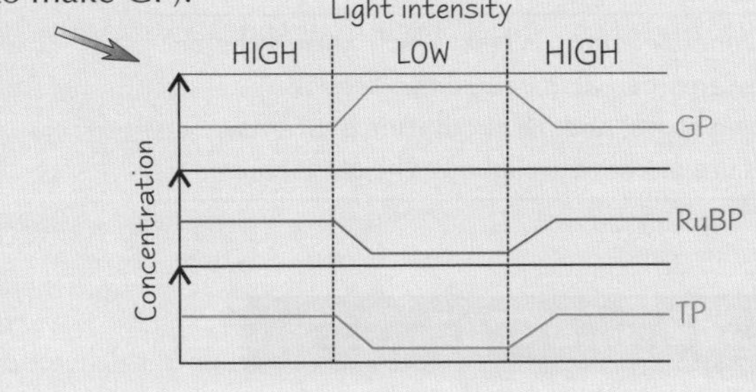

2. Temperature

- All the reactions in the Calvin cycle are catalysed by **enzymes** (e.g. rubisco).
- At **low temperatures**, all of the reactions will be **slower** as the enzymes work more **slowly**.

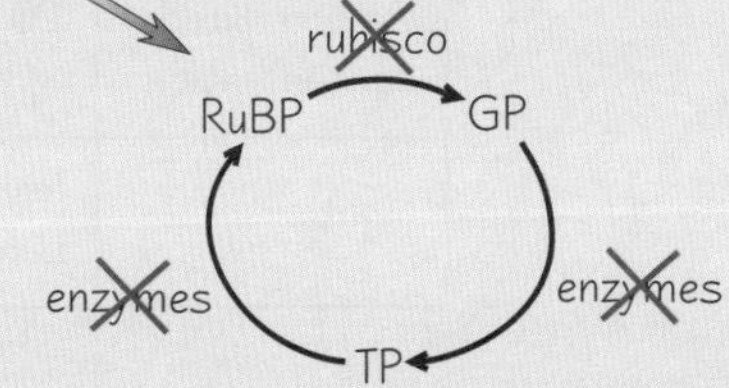

- This means the levels of **RuBP**, **GP** and **TP** will **fall**.
- GP, TP and RuBP are affected in the same way at **very high temperatures**, because the **enzymes** will start to **denature**.

3. Carbon dioxide concentration

- At **low CO_2 concentrations**, **conversion of RuBP** to GP is also **slow** (as there's less CO_2 to combine with RuBP to make GP).
- So the level of **RuBP** will **rise** (as it's still being made) and levels of **GP** and **TP** will **fall** (as they're used up to make RuBP).

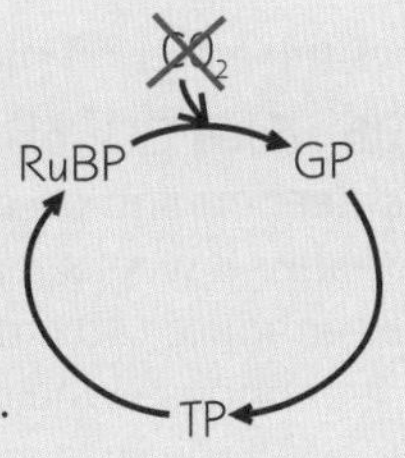

Derek knew that a low light intensity would increase the level of romance.

Practice Questions

Q1 Name two factors that can limit photosynthesis.

Q2 What is the limiting factor for photosynthesis at night?

Q3 How does a low CO_2 concentration in the air affect the level of TP in a plant?

Exam Questions

Q1 An experiment was carried out to investigate how temperature affects photosynthesis.
The rate of photosynthesis was measured at 10 °C, 25 °C and 45 °C.
At which temperature would the rate of photosynthesis have been greatest? Explain your answer. [4 marks]

Q2 A scientist was investigating the effect of different conditions on the levels of GP, TP and RuBP in a plant.
Predict the results of his experiment under the following conditions. Explain your answers.
a) Low light intensity, optimum temperature and optimum CO_2 concentration. [3 marks]
b) Low temperature, optimum light intensity and optimum CO_2 concentration. [3 marks]

I'm a whizz at the factors that limit revision...

... watching Hollyoaks, making tea, watching EastEnders, walking the dog... not to mention staring into space (one of my favourites). These pages aren't that bad though. You just need to learn how light, CO_2 and temperature affect the rate of photosynthesis. Try shutting the book and writing down what you know — you'll be amazed at what you remember.

Limiting Factors in Photosynthesis

This page is for AQA Unit 4 only. If you're doing OCR you can go straight to the next page.

Well, I hope you didn't think we'd finished covering limiting factors.... ohhhhhh no, I could write a whole book on them. But just for you I've kept my exciting facts about them to one page. And look, more graphs.

Growers Use Information About Limiting Factors to Increase Plant Growth

Commercial growers (e.g. farmers) know the **factors** that **limit photosynthesis** and therefore limit **plant growth**. This means they can create an **environment** where plants get the **right amount** of everything that they need, which **increases growth** and so **increases yield**. Growers create optimum conditions in **glasshouses**, in the following ways:

Limiting Factor	Management in Glasshouse
Carbon dioxide concentration	CO_2 is added to the air, e.g. by burning a small amount of propane in a CO_2 generator.
Light	Light can get in through the glass. Lamps provide light at night-time.
Temperature	Glasshouses trap heat energy from sunlight, which warms the air. Heaters and cooling systems can also be used to keep a constant optimum temperature, and air circulation systems make sure the temperature is even throughout the glasshouse.

You Need to be Able to Interpret Data on Limiting Factors

Here are some **examples** of the kind of **data** you might get in the exam:

The graph on the **right** shows the effect on plant growth of **adding carbon dioxide** to a greenhouse.

1) In the greenhouse **with added CO_2** plant **growth** was **faster** (the line is steeper) and on average the plants were **larger** after 8 weeks than they were in the control greenhouse (30 cm compared to only 15 cm in the greenhouse where no CO_2 was added).
2) This is because the plants use CO_2 to produce **glucose** by photosynthesis. The more CO_2 they have, the more glucose they can produce, meaning they can **respire more** and so have **more ATP** for **DNA replication**, **cell division** and **protein synthesis**.

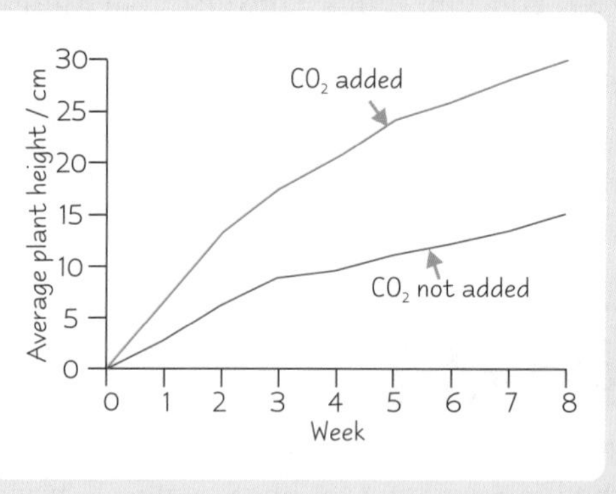

The graph **below** shows the effect of **light intensity** on plant growth, and the effect of two **different types** of **heater**.

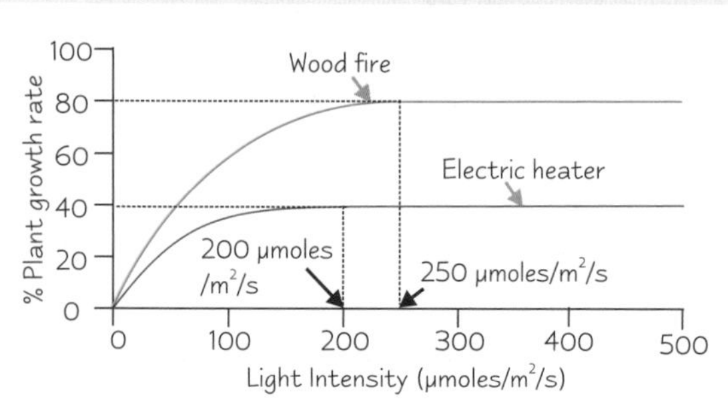

1) At the start of the graph, the **greater** the **light intensity** the **greater** the **plant growth.**
2) At **200 μmoles/m²/s** of light the **bottom** graph flattens out, showing that **CO_2 concentration** or **temperature** is **limiting growth** in these plants.
3) At **250 μmoles/m²/s** of light the **top** graph flattens out. The difference between the two graphs could be because the **wood fire increases** the **temperature more** than the electric heater or because it's **increasing** the **concentration** of **CO_2** in the air (an electric heater **doesn't** release CO_2).

Limiting Factors in Photosynthesis

This page is for OCR Unit 4. If you're doing AQA, you can skip straight to the questions.

Limiting Factors can be Investigated using Pondweed

> Remember photosynthesis produces glucose and oxygen (see page 4).

1) **Canadian pondweed** (***Elodea***) can be used to measure the effect of light intensity, temperature and CO_2 concentration on the **rate of photosynthesis**.
2) The rate at which **oxygen** is **produced** by the pondweed can be easily **measured** and this **corresponds** to the rate of photosynthesis.
3) For example, the **apparatus** below is used to **measure** the **effect** of **light intensity** on photosynthesis.

- A **test tube** containing the **pondweed** and **water** is connected to a **capillary tube** full of water.
- The tube of water is connected to a **syringe**.
- A **source of white light** is placed at a **specific distance** from the pondweed.
- The pondweed is left to photosynthesise for a **set** amount of **time**. As it photosynthesises, the **oxygen released** will **collect** in the **capillary tube**.
- At the end of the experiment, the syringe is used to **draw** the gas **bubble** in the tube **up** alongside a **ruler** and the **length** of the gas bubble (volume of O_2) is **measured**.
- Any **variables** that could affect the results should be **controlled**, e.g. temperature, the time the weed is left to photosynthesise.
- The experiment is **repeated** and the **average** length of gas bubble is calculated, to make the results **more reliable**.
- The whole experiment is then **repeated** with the **light source** placed at **different distances** from the pondweed.

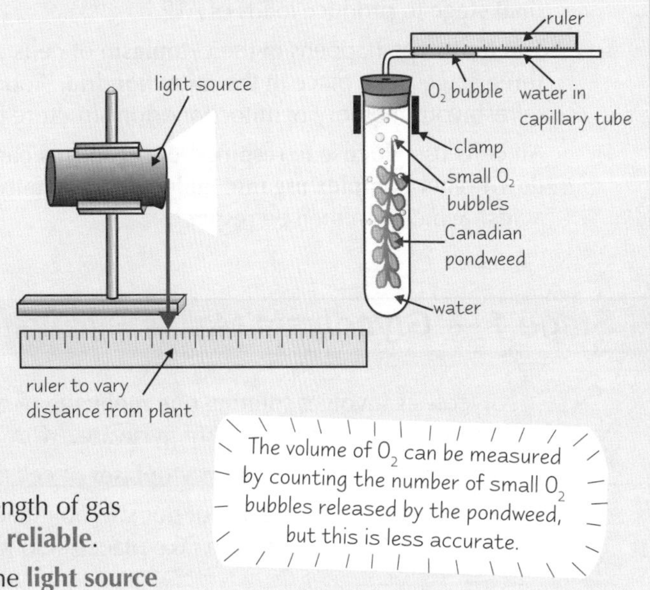

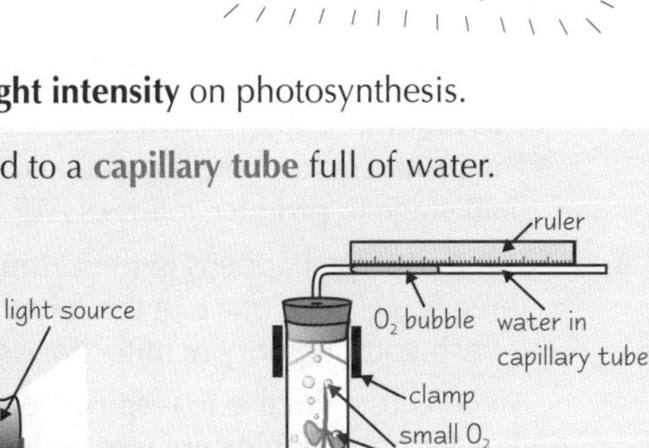

> The volume of O_2 can be measured by counting the number of small O_2 bubbles released by the pondweed, but this is less accurate.

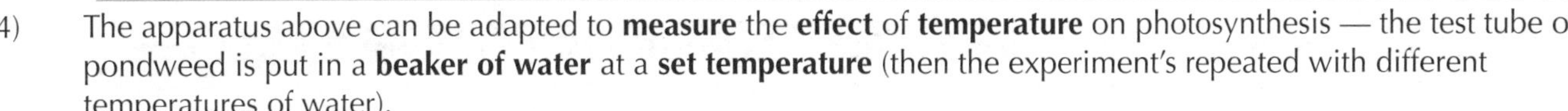

4) The apparatus above can be adapted to **measure** the **effect** of **temperature** on photosynthesis — the test tube of pondweed is put in a **beaker of water** at a **set temperature** (then the experiment's repeated with different temperatures of water).

Practice Questions

Q1 How do commercial growers create an optimum level of CO_2 in a glasshouse?

Q2 In an experiment on the rate of photosynthesis, how can light intensity be varied?

Q3 In the experiment above, give two variables that must be controlled.

Crop	Yield in glasshouse / kg	Yield grown outdoors / kg
Tomato	1000	200
Lettuce	750	230
Potato	850	680
Wheat	780	550

Exam Questions

Q1 The table above shows the yields of various crops when they are grown in glasshouses and when grown outdoors.
a) Yields are usually higher overall in glasshouses.
Describe four ways in which conditions can be controlled in glasshouses to increase yields. [4 marks]
b) Glasshouses are not always financially viable for all crops.
Which crop above benefits the least from being grown in glasshouses? Explain your answer. [2 marks]

Q2 Briefly describe the apparatus and method you would use to investigate how temperature affects photosynthesis in Canadian pondweed. [6 marks]

Aah, Canadian pondweed — a biology student's best friend...

Well... sometimes — usually you end up staring endlessly at it while it produces lots of tiny bubbles. Thrilling. Anyway, an interpreting data question could well come up in the exam — it could be any kind of data, but don't panic if it's not like the graphs on the left-hand page — as long as you understand limiting factors you'll be able to interpret it.

Aerobic Respiration

These pages are for AQA Unit 4, OCR Unit 4 and Edexcel Unit 5.

From the last gazillion pages you know that plants make their own glucose. Unfortunately, that means now you need to learn how plant and animal cells release energy from glucose. It's not the easiest thing in the world to understand, but it'll make sense once you've gone through it a couple of times.

There are **Four Stages** in **Aerobic Respiration**

1) The four stages in aerobic respiration are **glycolysis**, the **link reaction**, the **Krebs cycle** and **oxidative phosphorylation**.
2) The **first three** stages are a **series of reactions**. The **products** from these reactions are **used** in the **final stage** to produce loads of ATP.
3) The **first** stage happens in the **cytoplasm** of cells and the **other three** stages take place in the **mitochondria**. You might want to refresh your memory of mitochondrion structure before you start.
4) All cells use **glucose** to **respire**, but organisms can also **break down** other **complex organic molecules** (e.g. fatty acids, amino acids), which can then be respired.

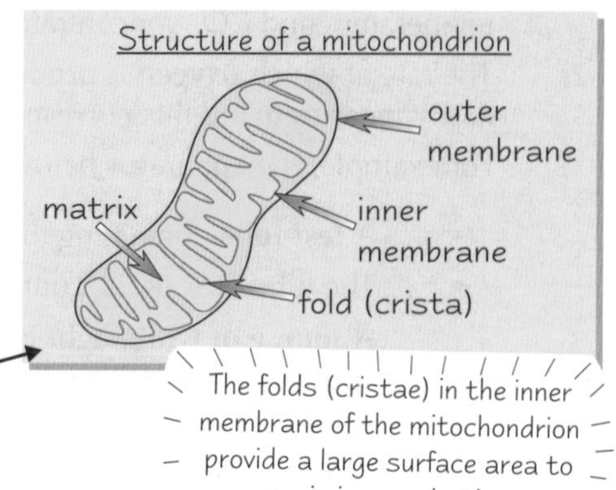

The folds (cristae) in the inner membrane of the mitochondrion provide a large surface area to maximise respiration.

Stage 1 — **Glycolysis** Makes **Pyruvate** from **Glucose**

1) Glycolysis involves splitting **one molecule** of glucose (with 6 carbons — 6C) into **two** smaller molecules of **pyruvate** (3C).
2) The process happens in the **cytoplasm** of cells.
3) Glycolysis is the **first stage** of both aerobic and anaerobic respiration and **doesn't need oxygen** to take place — so it's an **anaerobic** process.

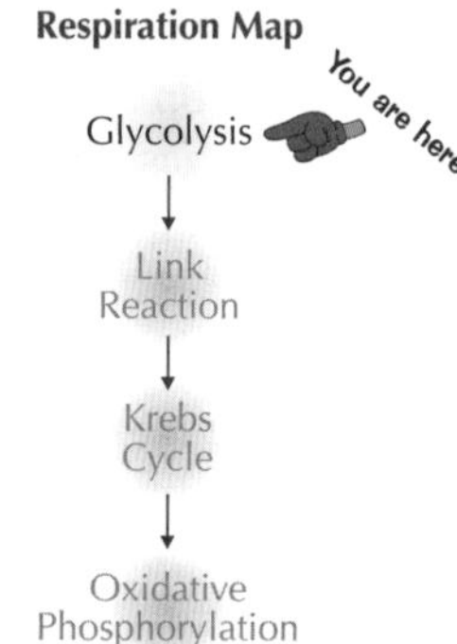

There are **Two Stages** in **Glycolysis** — **Phosphorylation** and **Oxidation**

First, **ATP** is **used** to **phosphorylate glucose** to triose phosphate. Then **triose phosphate** is **oxidised**, **releasing ATP**. Overall there's a **net gain** of **2 ATP**.

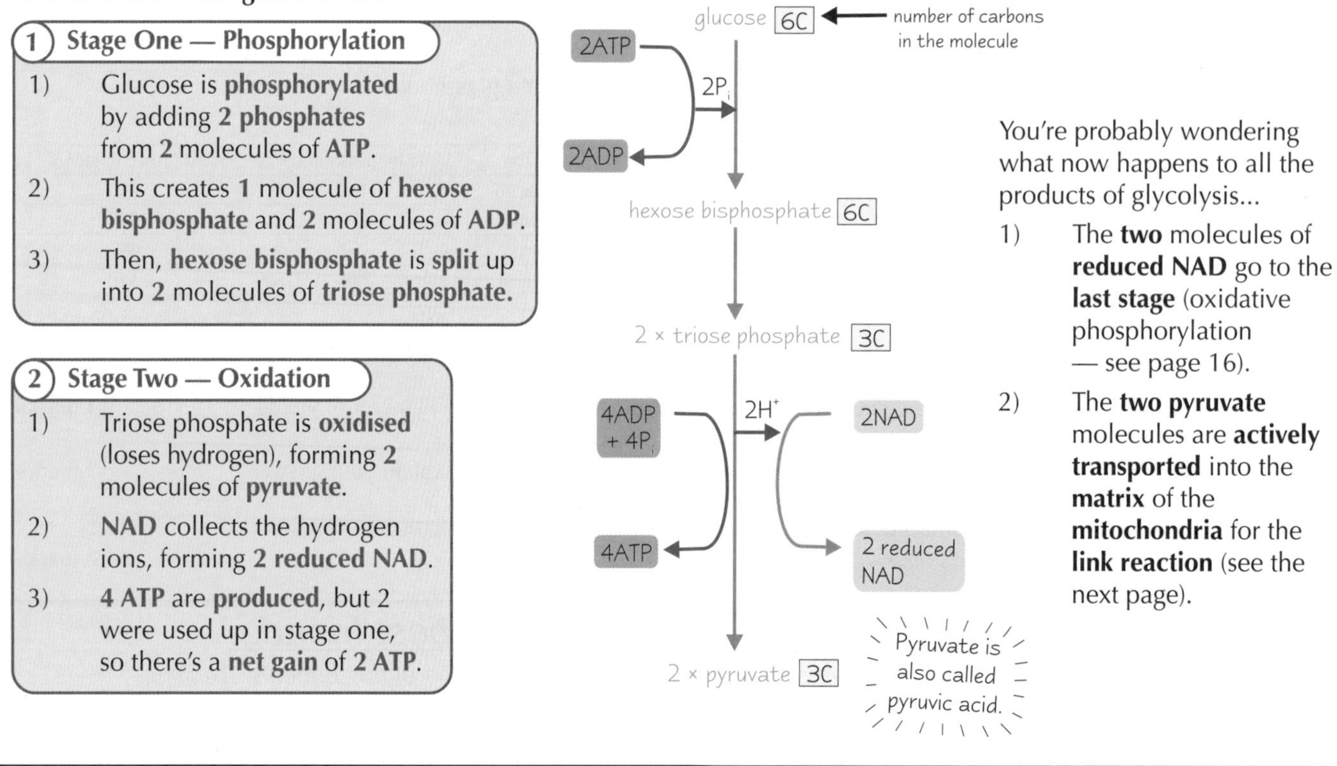

1) Stage One — Phosphorylation

1) Glucose is **phosphorylated** by adding **2 phosphates** from **2** molecules of **ATP**.
2) This creates **1** molecule of **hexose bisphosphate** and 2 molecules of **ADP**.
3) Then, **hexose bisphosphate** is **split** up into **2** molecules of **triose phosphate**.

2) Stage Two — Oxidation

1) Triose phosphate is **oxidised** (loses hydrogen), forming **2** molecules of **pyruvate**.
2) **NAD** collects the hydrogen ions, forming **2 reduced NAD**.
3) **4 ATP** are **produced**, but 2 were used up in stage one, so there's a **net gain** of **2 ATP**.

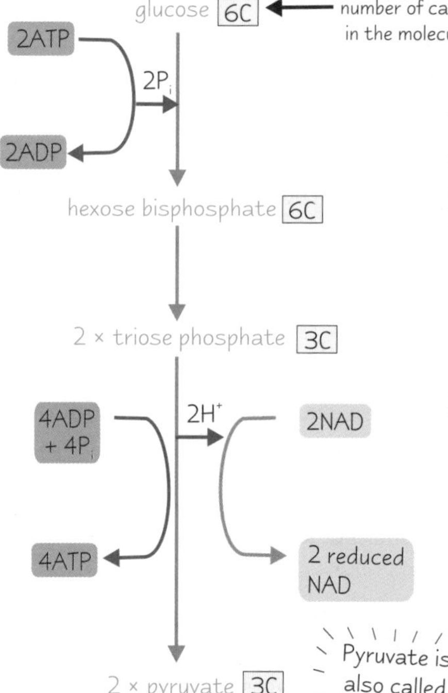

Pyruvate is also called pyruvic acid.

You're probably wondering what now happens to all the products of glycolysis...

1) The **two** molecules of **reduced NAD** go to the **last stage** (oxidative phosphorylation — see page 16).
2) The **two pyruvate** molecules are **actively transported** into the **matrix** of the **mitochondria** for the **link reaction** (see the next page).

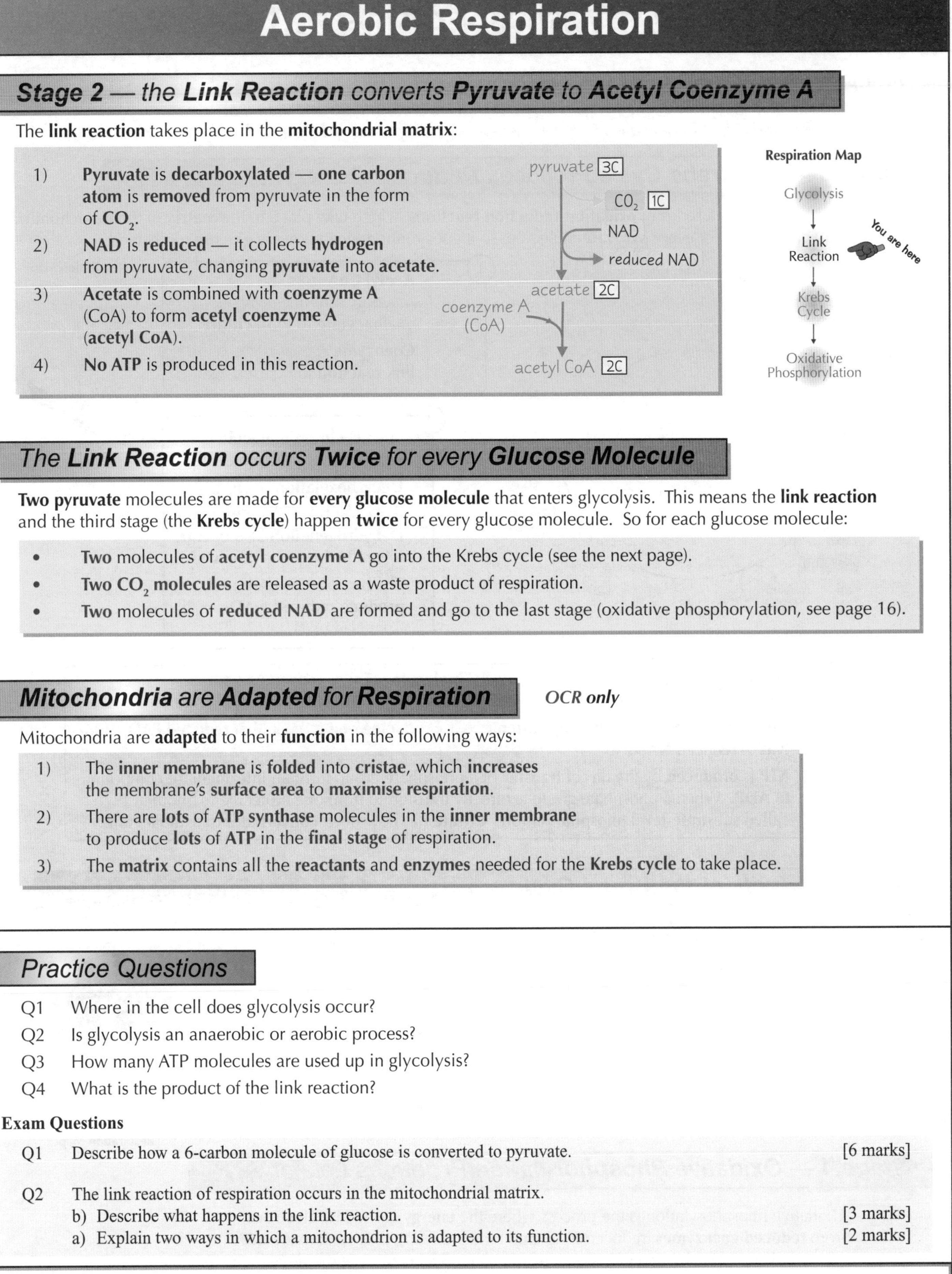

Aerobic Respiration

Stage 2 — the Link Reaction converts Pyruvate to Acetyl Coenzyme A

The **link reaction** takes place in the **mitochondrial matrix**:

1) **Pyruvate** is **decarboxylated — one carbon atom** is **removed** from pyruvate in the form of CO_2.
2) **NAD** is **reduced** — it collects **hydrogen** from pyruvate, changing **pyruvate** into **acetate**.
3) **Acetate** is combined with **coenzyme A** (CoA) to form **acetyl coenzyme A** (**acetyl CoA**).
4) **No ATP** is produced in this reaction.

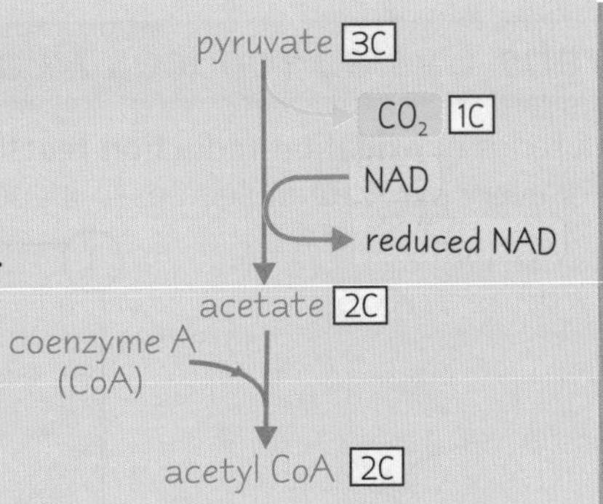

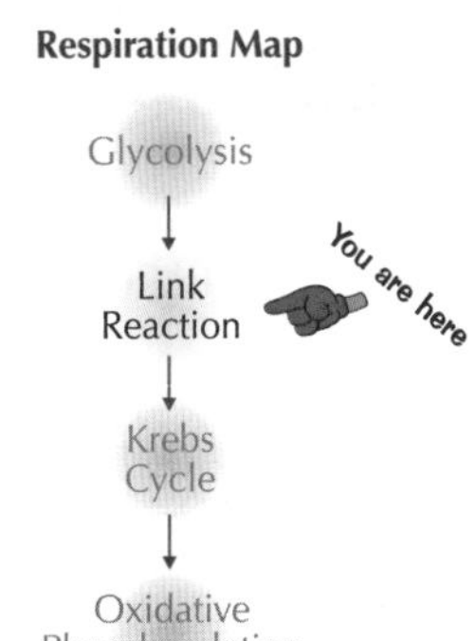

The Link Reaction occurs Twice for every Glucose Molecule

Two pyruvate molecules are made for **every glucose molecule** that enters glycolysis. This means the **link reaction** and the third stage (the **Krebs cycle**) happen **twice** for every glucose molecule. So for each glucose molecule:

- Two molecules of **acetyl coenzyme A** go into the Krebs cycle (see the next page).
- Two CO_2 **molecules** are released as a waste product of respiration.
- Two molecules of **reduced NAD** are formed and go to the last stage (oxidative phosphorylation, see page 16).

Mitochondria are Adapted for Respiration

OCR only

Mitochondria are **adapted** to their **function** in the following ways:

1) The **inner membrane** is **folded** into **cristae**, which **increases** the membrane's **surface area** to **maximise respiration**.
2) There are **lots** of **ATP synthase** molecules in the **inner membrane** to produce **lots** of **ATP** in the **final stage** of respiration.
3) The **matrix** contains all the **reactants** and **enzymes** needed for the **Krebs cycle** to take place.

Practice Questions

Q1 Where in the cell does glycolysis occur?

Q2 Is glycolysis an anaerobic or aerobic process?

Q3 How many ATP molecules are used up in glycolysis?

Q4 What is the product of the link reaction?

Exam Questions

Q1 Describe how a 6-carbon molecule of glucose is converted to pyruvate. [6 marks]

Q2 The link reaction of respiration occurs in the mitochondrial matrix.
b) Describe what happens in the link reaction. [3 marks]
a) Explain two ways in which a mitochondrion is adapted to its function. [2 marks]

No ATP was harmed during this reaction...

Ahhhh... too many reactions. I'm sure your head hurts now, 'cause mine certainly does. Just think of revision as like doing exercise — it can be a pain while you're doing it (and maybe afterwards too), but it's worth it for the well-toned brain you'll have. Just keep going over and over it, until you get the first two stages of respiration straight in your head. Then relax.

Aerobic Respiration

These pages are for AQA Unit 4, OCR Unit 4 and Edexcel Unit 5.

As you've seen, glycolysis produces a net gain of two ATP. Pah, we can do better than that. The Krebs cycle and oxidative phosphorylation are where it all happens — ATP galore.

Stage 3 — the Krebs Cycle Produces Reduced Coenzymes and ATP

The Krebs cycle involves a series of **oxidation-reduction reactions**, which take place in the **matrix** of the **mitochondria**. The cycle happens **once** for **every pyruvate** molecule, so it goes round **twice** for **every glucose** molecule.

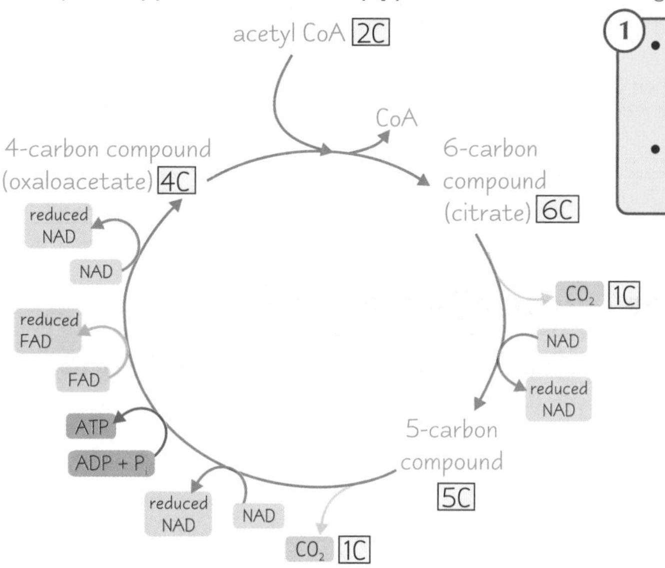

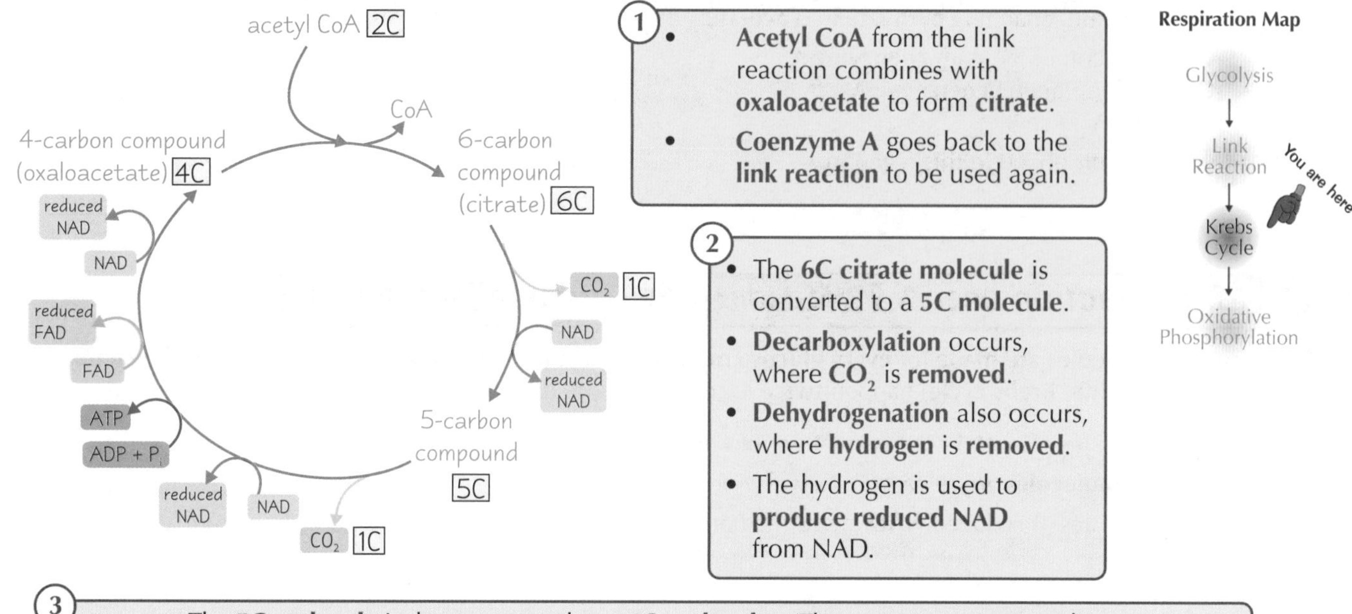

1
- **Acetyl CoA** from the link reaction combines with **oxaloacetate** to form **citrate**.
- **Coenzyme A** goes back to the **link reaction** to be used again.

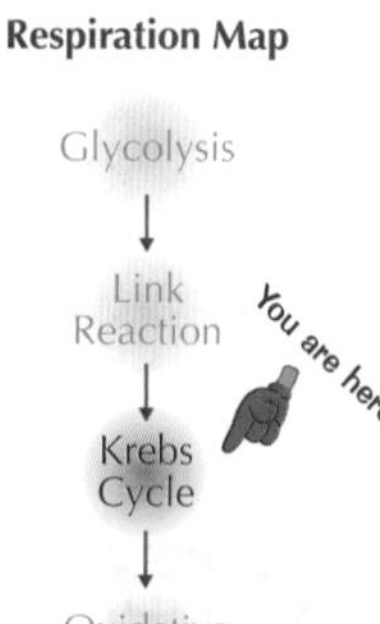

2
- The **6C citrate molecule** is converted to a **5C molecule**.
- **Decarboxylation** occurs, where **CO_2** is **removed**.
- **Dehydrogenation** also occurs, where **hydrogen** is **removed**.
- The hydrogen is used to **produce reduced NAD** from NAD.

3
- The **5C molecule** is then converted to a **4C molecule**. (There are some intermediate compounds formed during this conversion, but you don't need to know about them.)
- **Decarboxylation** and **dehydrogenation** occur, producing **one** molecule of **reduced FAD** and **two** of **reduced NAD**.
- **ATP** is **produced** by the **direct transfer** of a **phosphate** group from an **intermediate** compound **to ADP**. When a phosphate group is directly transferred from one molecule to another it's called **substrate-level phosphorylation**. **Citrate** has now been **converted** into **oxaloacetate**.

Some Products of the Krebs Cycle are Used in Oxidative Phosphorylation

Some products are **reused**, some are **released** and others are used for the **next stage** of respiration:

Product from one Krebs cycle	Where it goes
1 coenzyme A	Reused in the next link reaction
Oxaloacetate	Regenerated for use in the next Krebs cycle
2 CO_2	Released as a waste product
1 ATP	Used for energy
3 reduced NAD	To oxidative phosphorylation
1 reduced FAD	To oxidative phosphorylation

Talking about oxidative phosphorylation was always a big hit with the ladies...

Stage 4 — Oxidative Phosphorylation Produces Lots of ATP

1) Oxidative phosphorylation is the process where the **energy** carried by **electrons**, from **reduced coenzymes** (reduced NAD and reduced FAD), is used to **make ATP**. (The whole point of the previous stages is to make reduced NAD and reduced FAD for the final stage.)
2) Oxidative phosphorylation involves two processes — the **electron transport chain** and **chemiosmosis** (see the next page).

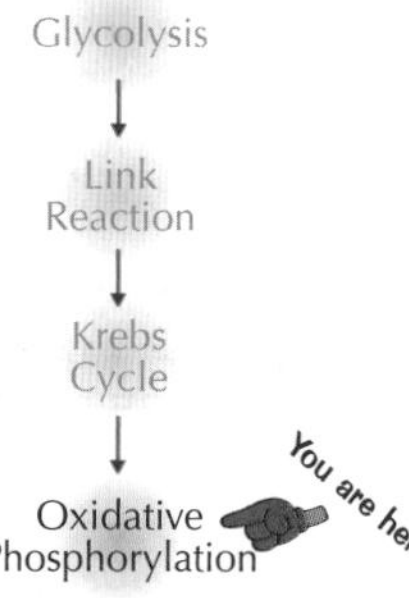

Aerobic Respiration

Protons are Pumped Across the Inner Mitochondrial Membrane

So now on to how **oxidative phosphorylation** actually **works**:

1) **Hydrogen atoms** are released from **reduced NAD** and **reduced FAD** as they're oxidised to NAD and FAD. The H atoms **split** into **protons (H^+)** and **electrons (e^-)**.
2) The **electrons** move along the **electron transport chain** (made up of three **electron carriers**), **losing energy** at each carrier.
3) This energy is used by the electron carriers to **pump protons** from the **mitochondrial matrix into** the **intermembrane space** (the space **between** the inner and outer **mitochondrial membranes**).
4) The **concentration** of **protons** is now **higher** in the **intermembrane space** than in the mitochondrial matrix — this forms an **electrochemical gradient** (a **concentration gradient** of **ions**).
5) Protons **move down** the **electrochemical gradient**, back into the mitochondrial matrix, via **ATP synthase**. This **movement** drives the synthesis of **ATP** from **ADP** and **inorganic phosphate** (P_i).
6) The movement of H^+ ions across a membrane, which generates ATP, is called **chemiosmosis**.
7) In the mitochondrial matrix, at the end of the transport chain, the **protons**, **electrons** and **O_2** (from the blood) combine to form **water**. Oxygen is said to be the final **electron acceptor**.

The regenerated coenzymes are reused in the Krebs cycle.

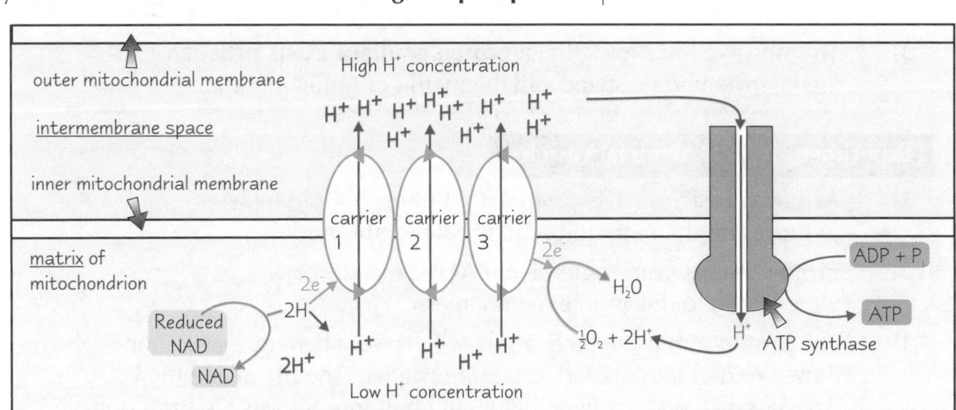

32 ATP Can be Made from One Glucose Molecule

As you know, **oxidative phosphorylation makes ATP** using energy from the reduced coenzymes — **2.5 ATP** are made from each **reduced NAD** and **1.5 ATP** are made from each **reduced FAD**. The table on the right shows **how much** ATP a cell can make from **one molecule** of **glucose** in **aerobic respiration**. (Remember, one molecule of glucose produces 2 pyruvate, so the link reaction and Krebs cycle happen twice.)

Stage of respiration	Molecules produced	Number of ATP molecules
Glycolysis	2 ATP	2
Glycolysis	2 reduced NAD	2 × 2.5 = 5
Link Reaction (×2)	2 reduced NAD	2 × 2.5 = 5
Krebs cycle (×2)	2 ATP	2
Krebs cycle (×2)	6 reduced NAD	6 × 2.5 = 15
Krebs cycle (×2)	2 reduced FAD	2 × 1.5 = 3
		Total ATP = 32

The number of ATP produced per reduced NAD or reduced FAD was thought to be 3 and 2, but new research has shown that the figures are nearer 2.5 and 1.5.

Practice Questions

Q1 Where in the cell does the Krebs cycle occur?

Q2 How many times does decarboxylation happen during one turn of the Krebs cycle?

Q3 What do the electrons lose as they move along the electron transport chain in oxidative phosphorylation?

Exam Question

Q1 Carbon monoxide inhibits the final electron carrier in the electron transport chain.

a) Explain how this affects ATP production via the electron transport chain. [2 marks]

b) Explain how this affects ATP production via the Krebs cycle. [2 marks]

The electron transport chain isn't just a FAD with the examiners...

Oh my gosh, I didn't think it could get any worse... You may be wondering how to learn these pages of crazy chemistry, but basically you have to put in the time and go over and over it. Don't worry though, it WILL pay off, and before you know it you'll be set for the exam. And once you know this section you'll be able to do anything, e.g. world domination...

Respiration Experiments

This page is for OCR Unit 4 only.
If you're doing Edexcel you can go straight to the next page and if you're doing AQA you can skip to page 20.

Congratulations — you've done all the main reactiony bits of respiration, so now it's time for some exciting experiments. When I say 'exciting', I'm using the word loosely. But I've got to say something positive to keep the morale up.

Scientific Experiments Provide Evidence for Chemiosmosis

Before the 1960s, scientists **didn't** understand the **connection** between the **electron transport chain** and **ATP synthesis** in respiration. One idea was that **energy lost** from **electrons moving** down the **electron transport chain** creates a **proton gradient** (a concentration gradient of H^+ ions), which is then used to **synthesise ATP** — this is called the **chemiosmotic theory**. Nowadays, there's quite a lot of **experimental evidence** supporting this theory:

Experiment One — Low pH

1) The **pH** of the **intermembrane space** in mitochondria was found to be **lower** than the pH of the **matrix**.
2) A **lower pH** means the intermembrane space is **more acidic** — it has a **higher concentration of H^+ ions.**
3) This observation shows that a **proton gradient exists** between the intermembrane space and the matrix of mitochondria.

The chemiosmotic theory is the most widely accepted theory for linking the electron transport chain to ATP synthesis.

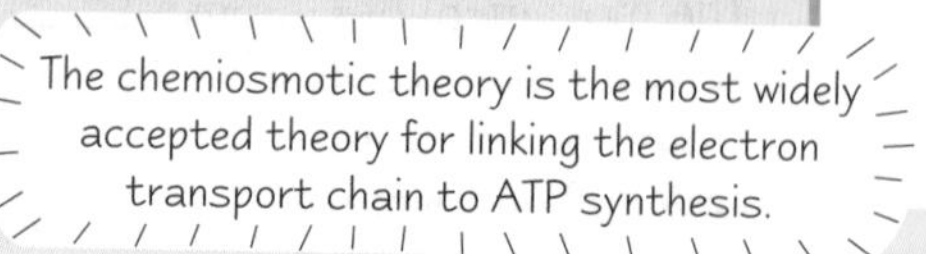

Experiment Two — Artificial Vesicles

1) **Artificial vesicles** were created from **phospholipid bilayers** to **represent** the **inner mitochondrial membrane.**
2) **Proton pumps** from bacteria and **ATP synthase** were added to the vesicle membranes.
3) The **proton pumps** are **activated** by **light**, so when light was shone onto these vesicles they started to **pump protons.** The **pH inside** the vesicles **decreased** — protons were being **pumped into** the vesicle from outside.
4) When **ADP** and **P_i** were **added** to the solution **outside** the vesicles, **ATP** was **synthesised.**
5) This artificial system shows that a **proton gradient** can be **used** to **synthesise ATP** (but doesn't show that this happens in mitochondria).

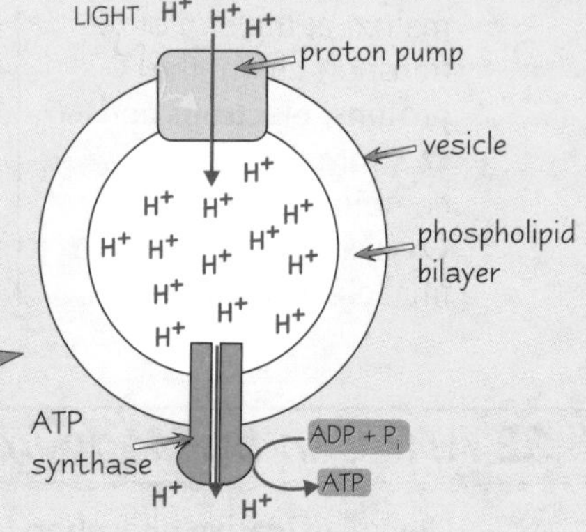

Experiment Three — Mitochondria

1) Mitochondria were put into a **slightly alkaline solution (pH8).**
2) They were left until the **whole** of each mitochondrion (matrix and intermembrane space) **became pH8.**
3) When these mitochondria were given **ADP** and **P_i** **no ATP** was produced.
4) Then the mitochondria were **transferred** to a **more acidic solution** of **pH4** (i.e. one with a **higher concentration** of **protons**).
5) The **outer membrane** of the mitochondrion is **permeable** to **protons** — the protons **moved into** the **intermembrane space**, creating a **proton gradient** across the **inner mitochondrial membrane.**
6) In the presence of **ADP** and **P_i**, **ATP was produced.**
7) This experiment shows that a **proton gradient** can be **used** by mitochondria to **make ATP.**

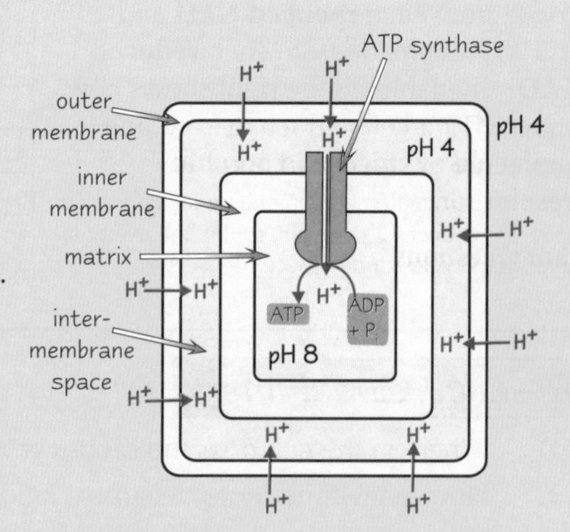

Experiment Four — Uncouplers

1) **Uncouplers** are substances that **destroy** the **proton gradient** across the **inner mitochondrial membrane.**
2) An **uncoupler** was added to mitochondria, along with **reduced NAD**, and **ADP** and **P_i**.
3) **No ATP** was made.
4) This experiment shows that a **proton gradient** is required to **synthesise ATP** in **mitochondria.**

Respiration Experiments

This page is for OCR Unit 4 and Edexcel Unit 5.

The Rate of Respiration can be Measured using a Respirometer

1) The volume of **oxygen taken up** or the volume of **carbon dioxide produced indicates** the **rate** of **respiration**.
2) A **respirometer** measures the rate of **oxygen** being **taken up** — the **more** oxygen taken up, the **faster** the rate of respiration.
3) Here's how you can use a **respirometer** to **measure** the volume of **oxygen taken up** by some **woodlice**:

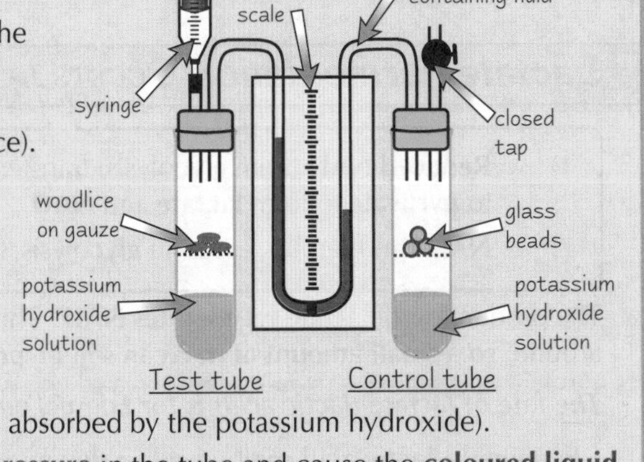

- The apparatus is set up as shown on the right.
- **Each tube** contains **potassium hydroxide** solution (or soda lime), which **absorbs carbon dioxide**.
- The **control tube** is set up in exactly the **same way** as the test tube, but **without** the **woodlice**, to make sure the **results** are **only** due to the woodlice **respiring** (e.g. it contains beads that have the same mass as the woodlice).
- The **syringe** is used to set the **fluid** in the **manometer** to a **known level**.
- The apparatus is **left** for a **set** period of **time** (e.g. 20 minutes).
- During that time there'll be a **decrease** in the **volume** of the **air** in the test tube, due to **oxygen consumption** by the **woodlice** (all the CO_2 produced is absorbed by the potassium hydroxide).
- The decrease in the volume of the air will **reduce the pressure** in the tube and cause the **coloured liquid** in the manometer to **move towards** the test tube.
- The **distance moved** by the **liquid** in a **given time** is **measured**. This value can then be used to **calculate** the **volume of oxygen** taken in by the woodlice **per minute**.
- Any **variables** that could **affect** the results are **controlled**, e.g. temperature, volume of potassium hydroxide solution in each test tube.

4) To produce more **reliable** results the experiment is **repeated** and a **mean volume** of O_2 is calculated.

Practice Questions

Q1 What is the chemiosmotic theory?

Q2 What does a respirometer measure?

Exam Questions

Q1 In the first stage of an experiment, mitochondria were put in a solution at pH 9.1 and left until each mitochondrion had a pH of 9.1 throughout its compartments. In the presence of ADP and P_i, no ATP was produced. During the second stage of the experiment, the same mitochondria were placed in a solution at pH 3.7. In the presence of ADP and P_i, ATP was produced.

a) Why was no ATP produced during the first part of the experiment? [1 mark]

b) What would the pH of the intermembrane space have been during the second stage of the experiment? [1 mark]

c) Do these results support the chemiosmotic theory? Explain your answer. [1 mark]

Q2 A respirometer is set up as shown in the diagram on this page.

a) Explain the purpose of the control tube. [1 mark]

b) Explain what would happen if there was no potassium hydroxide in the tubes. [2 marks]

c) What other substance could be measured to find out the rate of respiration? [1 mark]

My results are dodgy — I'm sure the woodlice are holding their breath...

Okay, that wasn't very funny, but this page doesn't really give me any inspiration. You probably feel the same way. It's just one of those pages that you have to plough through. You could try drawing a few pretty diagrams to get the experiments in your head. And after you've got it sorted do something exciting, like trying to stick your toe in your ear...

Aerobic and Anaerobic Respiration

These pages are for AQA Unit 4, OCR Unit 4 and Edexcel Unit 5.

We're on the home stretch now ladies and gents — these are the last two pages in the section.

There are **Two Types** of **Anaerobic Respiration**

1) **Anaerobic** respiration **doesn't use oxygen**.
2) It **doesn't** involve the **link reaction**, the **Krebs cycle** or **oxidative phosphorylation**.
3) There are **two types** of anaerobic respiration — **alcoholic fermentation** and **lactate fermentation**.
4) These two processes are **similar**, because they both take place in the **cytoplasm**, they both produce **two ATP** per molecule of glucose and they both **start** with **glycolysis** (which produces **pyruvate**).
5) They **differ** in **which organisms** they occur in and what happens to the **pyruvate** (see below).

Lactate Fermentation Occurs in **Mammals** and Produces **Lactate**

1) **Reduced NAD** (from glycolysis) transfers **hydrogen** to **pyruvate** to form **lactate** and **NAD**.
2) **NAD** can then be reused in **glycolysis**.

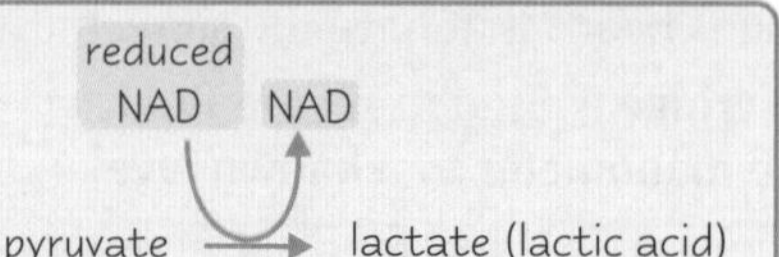

Some bacteria carry out lactate fermentation.

The production of lactate **regenerates NAD**. This means **glycolysis** can **continue** even when there **isn't** much oxygen around, so a **small amount of ATP** can still be **produced** to keep some biological process going... clever.

The fate of lactate (lactic acid) is for Edexcel only.

In **animals, lactic acid** can be **broken down** in two ways:
1) **Cells** can **convert** the lactic acid back to **pyruvate** (which then re-enters aerobic respiration at the **Krebs cycle**).
2) **Liver cells** can **convert** the lactic acid back to **glucose** (which can then be **respired** or **stored**).

Alcoholic Fermentation Occurs in **Yeast Cells** and Produces **Ethanol**

OCR and AQA only

1) CO_2 **is removed** from **pyruvate** to form **ethanal**.
2) **Reduced NAD** (from glycolysis) transfers **hydrogen** to **ethanal** to form **ethanol** and **NAD**.
3) **NAD** can then be reused in **glycolysis**.

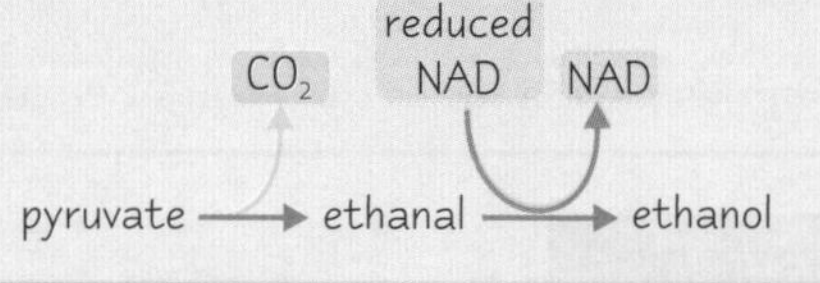

Alcoholic fermentation also occurs in plants.

The production of ethanol also **regenerates NAD** so **glycolysis** can **continue** when there isn't much oxygen around.

Aerobic Respiration **Doesn't Release** as **Much Energy** as **Possible...**

OCR only

In theory, **aerobic respiration** can make **32 ATP** per **glucose molecule** (see page 17).
But in reality the **actual yield** is **lower** because:

1) Some of the **reduced NAD** formed during the **first three stages** of aerobic respiration is used in **other reduction reactions** in the cell instead of in **oxidative phosphorylation**.
2) **Some ATP** is **used up** by **actively transporting** substances **into** the **mitochondria** during respiration, e.g. **pyruvate** (formed at the end of glycolysis), **ADP** and **phosphate** (both needed for making ATP).
3) The **inner mitochondrial membrane** is **leaky** — some **protons** may **leak** into the **matrix** without passing through **ATP synthase** and **without making ATP**.

...but it Still **Releases More Energy** than **Anaerobic** Respiration

OCR only

1) The **ATP yield** from **anaerobic** respiration is **always lower** than from **aerobic** respiration.
2) This is because **anaerobic** respiration **only** includes **one energy-releasing stage** (**glycolysis**), which only produces **2 ATP** per glucose molecule.
3) The energy-releasing reactions of the **Krebs cycle** and **oxidative phosphorylation** need **oxygen**, so they **can't** occur during anaerobic respiration.

Aerobic and Anaerobic Respiration

This page is for OCR Unit 4 only. If you're doing AQA or Edexcel, you can skip straight to the questions.

Cells Can Respire Different Substrates

Proteins and lipids enter respiration at the Krebs cycle.

1) Cells **respire glucose**, but they also respire **other carbohydrates**, **lipids** and **proteins**.
2) Any **biological molecule** that can be **broken down** in **respiration** to **release energy** is called a **respiratory substrate**.
3) When an organism respires a specific **respiratory substrate**, the **respiratory quotient** (RQ) can be **worked out**.
4) The **respiratory quotient** is the volume of **carbon dioxide** produced when that **substrate** is **respired**, **divided** by the volume of **oxygen consumed**, in a set period of **time**.

$$RQ = \frac{\text{Volume of } CO_2 \text{ released}}{\text{Volume of } O_2 \text{ consumed}}$$

5) For example, you can work out the **RQ** for cells that **only respire glucose**:
   - The basic equation for aerobic respiration using glucose is: $C_6H_{12}O_6 + 6O_2 \rightarrow 6CO_2 + 6H_2O$ + energy
   - The RQ of glucose = molecules of **CO_2 released** ÷ molecules of **O_2 consumed**
     = 6 ÷ 6 = 1.
6) Respiratory quotients have been worked out for the respiration of **other respiratory substrates**.
   Lipids and **proteins** have an RQ value **lower than one** because **more oxygen** is needed to oxidise fats and lipids than to oxidise carbohydrates.

Respiratory Substrate	RQ
Lipids (triglycerides)	0.7
Proteins or amino acids	0.9
Carbohydrates	1

*The **Respiratory Quotient** tells you what **Substrate** is being **Respired***

1) The **respiratory quotient** for an organism is **useful** because it tells you **what kind** of **respiratory substrate** an organism is respiring and what **type** of **respiration** it's using (aerobic or anaerobic).
2) For example, under **normal conditions** the **usual RQ** for humans is between **0.7** and **1.0**. An RQ in this range shows that some **fats** (**lipids**) are being used for respiration, as well as **carbohydrates** like glucose. Protein **isn't** normally used by the body for respiration unless there's **nothing else**.
3) **High RQs** (greater than 1) mean that an organism is **short** of **oxygen**, and is having to respire **anaerobically** as well as aerobically.
4) **Plants** sometimes have a **low RQ**. This is because the **CO_2 released** in respiration is **used** for **photosynthesis** (so it's not measured).

Practice Questions

Q1 What molecule is made when CO_2 is removed from pyruvate during alcoholic fermentation?

Q2 Does anaerobic respiration release more or less energy per glucose molecule than aerobic respiration?

Q3 What is a respiratory substrate?

Exam Questions

Q1 A culture of mammalian cells was incubated with glucose, pyruvate and antimycin C.
Antimycin C inhibits an electron carrier in the electron transport chain of aerobic respiration.
Explain why these cells can still produce lactate. [1 mark]

Q2 This equation shows the aerobic respiration of a fat called triolein: $C_{57}H_{104}O_6 + 80O_2 \rightarrow 52H_2O + 57CO_2$
Calculate the respiratory quotient for this reaction. Show your working. [2 marks]

My little sis has an RQ of 157 — she's really clever...

I know, I'm really pushing the boundary between humour and non-humour here. But, at least we've come to the end of the section — and what a section it was. You might think it's unfair finishing it off with nasty calculations, but if you understand how to work out the RQ you'll be one step closer to being sorted for the exam.

Niches and Adaptations

These pages are for AQA Unit 4, OCR Unit 5 and Edexcel Unit 4.

All this ecology-type stuff is pretty wordy, so here are a nice few definitions to get you started. This way, you'll know what I'm banging on about throughout the rest of the section, and that always helps I think.

You Need to Learn Some Definitions to get you Started

Habitat	—	The **place** where an organism **lives**, e.g. a rocky shore or a field.
Population	—	**All** the organisms of **one species** in a **habitat**.
Community	—	Populations of **different species** in a habitat make up a **community**.
Ecosystem	—	**All** the **organisms** living in a **particular area** and all the **non-living** (abiotic) conditions, e.g. a freshwater ecosystem such as a lake. Ecosystems are **dynamic systems** — they're **changing** all the time.
Abiotic conditions	—	The **non-living** features of the ecosystem, e.g. **temperature** and **availability of water**.
Biotic conditions	—	The **living** features of the ecosystem, e.g. the presence of **predators** or **food**.
Abundance	—	The **number of individuals of one species** in a **particular area**.
Distribution	—	**Where** a species is within a **particular area**.

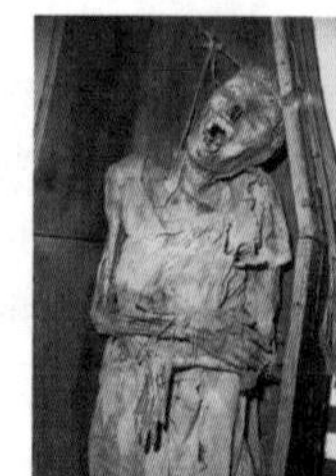

Being a member of the undead made it hard for Mumra to know whether he was a living or a non-living feature of the ecosystem.

If you're doing OCR you can skip straight to the questions — the rest is for AQA and Edexcel.

Every Species Occupies a Different Niche

Don't get confused between habitat (where a species lives) and niche (what it does in its habitat).

1) A **niche** is the **role** of a species within its habitat. It includes:
   - Its **biotic** interactions — e.g. the organisms it **eats**, and those it's **eaten by**.
   - Its **abiotic** interactions — e.g. the **oxygen** an organism breathes in, and the **carbon dioxide** it breathes out.
2) Every species has its own **unique niche** — a niche can only be occupied by **one species**.
3) It may **look** like **two species** are filling the **same niche** (e.g. they're both eaten by the same species), but there'll be **slight differences** (e.g. variations in what they eat). For example:

Common pipistrelle bat

This bat lives throughout Britain on **farmland**, **open woodland**, **hedgerows** and **urban areas**. It feeds by **flying** and catching **insects** using **echolocation** (**high-pitched sounds**) at a **frequency** of around **45 kHz**.

Soprano pipistrelle bat

This bat lives in Britain in **woodland** areas, close to **lakes** or **rivers**. It feeds by **flying** and catching **insects** using **echolocation**, at a **frequency** of **55 kHz**.

It may **look like** both species are filling the **same niche** (e.g. they both eat insects), but there are **slight differences** (e.g. they use **different frequencies** for their echolocation).

4) The **abundance** of different species can be **explained** by the niche concept — two species occupying **similar** niches will **compete** (e.g. for a **food source**), so **fewer individuals** of **both** species will be able to survive in the area. For example, common and soprano pipistrelle bats feed on the **same insects**. This means the **amount of food** available to both species is **reduced**, so there will be **fewer individuals** of **both** species in the same area.
5) The **distribution** of different species can also be **explained** by the niche concept — organisms can only **exist** in habitats where all the **conditions** that make up their **role exist**. For example, the soprano pipistrelle bat couldn't exist in a **desert** because there are **different insects** and **no woodland**.

Niches and Adaptations

If you're doing Edexcel you can skip straight to the questions — the rest is just for AQA.

Organisms are Adapted to Biotic and Abiotic Conditions

1) **Adaptations** are **features** that members of a species have which **increase** their chance of **survival** and **reproduction**, e.g. **giraffes** have **long necks** to help them reach vegetation that's high up. This increases their chances of survival when food is **scarce**.
2) Adaptations can be **physiological** (processes **inside** their body), **behavioural** (the way an organism **acts**) or **anatomical** (**structural features** of their body).
3) Organisms with better adaptations are **more likely** to **survive**, **reproduce** and **pass on** the alleles for their adaptations, so the adaptations become **more common** in the population. This is called **natural selection**.
4) Every species is adapted to **use** an **ecosystem** in a way that **no other** species can. For example, only giant anteaters can **break into** ant nests and **reach** the ants. They have **claws** to rip open the nest, and a **long, sticky tongue** which can move **rapidly** in and out of its mouth to **pick up** the ants.
5) Organisms are **adapted** to both the **abiotic conditions** (e.g. how much **water** is available) and the **biotic conditions** (e.g. what **predators** there are) in their ecosystem.

Here are a few ways that **different organisms** are **adapted** to the **abiotic** or the **biotic** conditions in their ecosystems:

Adaptations to abiotic conditions

- **Otters** have **webbed paws** — this means they can both **walk** on land and **swim** effectively. This increases their chance of survival because they can **live** and **hunt** both on land and in water.
- **Whales** have a **thick layer** of **blubber** (fat) — this helps to keep them **warm** in the **coldest seas**. This increases their chance of survival because they can **live** in places where food is plentiful.
- **Brown bears hibernate** — they **lower their metabolism** (all the chemical reactions taking place in their body) over **winter**. This increases their chance of survival because they can **conserve energy** during the **coldest** months.

Adaptations to biotic conditions

- **Sea otters** use **rocks** to **smash open** shellfish and clams. This increases their chance of survival because it gives them **access** to **another source** of food.
- **Scorpions dance** before **mating** — this makes sure they **attract** a **mate** of the **same species**. This increases their chance of reproduction by making **successful mating** more likely.
- Some **bacteria** produce **antibiotics** — these **kill other species** of bacteria in the **same area**. This increases their chance of survival because there's **less competition** for **resources**.

Take your partner 1, 2, 3, swing them round a sycamore tree.

Practice Questions

Q1 Define the term ecosystem.

Q2 Give the term for the living features of an ecosystem.

Q3 What are adaptations?

Q4 Give one example of an adaptation.

Exam Question

Q1 Two species of lizard (X and Y) live in the same area. Both feed on the same insects and are eaten by the same predator species. Species X feeds mainly during the morning and species Y feeds mainly during the afternoon.

a) Explain the term 'niche'. [2 marks]

b) 'Lizards X and Y occupy the same niche.' Explain why this statement is incorrect. [3 marks]

Unique quiche niche — say it ten times really fast...

As I said, pretty wordy — there's not even a diagram in sight, but I'll tell you what, you'll be missing it when you get onto the really sciencey stuff later. You just need to learn and relearn all the key words here, then when they ask you to interpret some bat-related babble in the exam, you'll know exactly what they're flapping on about. Niche... I mean, nice work.

Investigating Populations and Abiotic Conditions

***These pages are for** AQA Unit 4, OCR Unit 5 **and** Edexcel Unit 4.*

Examiners aren't happy unless you're freezing to death in the rain in a field somewhere in the middle of nowhere. Still, it's better than being stuck in the classroom being bored to death learning about fieldwork techniques...

You need to be able to **Investigate Populations** of **Organisms**

Investigating **populations** of organisms involves looking at the **abundance** and **distribution** of **species** in a particular **area**.

1) **Abundance** — the **number of individuals** of **one species** in a **particular area**.
   The abundance of **mobile organisms** and **plants** can be estimated by simply counting the **number** of individuals in samples taken. There are other measures of abundance that can be used too:
   - **Frequency** — the **number of samples** a species is **recorded in**, e.g. 70% of samples.
   - **Percentage cover** (for plants only) — **how much** of the area you're investigating is **covered** by a species.
2) **Distribution** — this is **where** a particular species is within the **area you're investigating**.

You need to take a **Random Sample** from the **Area You're Investigating**

Most of the time it would be too **time-consuming** to measure the **number of individuals** and the **distribution** of every species in the **entire area** you're investigating, so instead you take **samples**:

1) **Choose** an **area** to **sample** — a **small** area **within** the area being investigated.
2) Samples should be **random** to **avoid bias**, e.g. if you were investigating a field you could pick random sample sites by dividing the field into a **grid** and using a **random number generator** to select **coordinates.**
3) Use an **appropriate technique** to take a sample of the population (see pages 24-28).
4) **Repeat** the process, taking as many samples as possible. This gives a more **reliable** estimate for the **whole area.**
5) The **number of individuals** for the **whole area** can then be **estimated** by taking an **average** of the data collected in each sample and **multiplying** it by the size of the whole area. The **percentage cover** for the whole area can be estimated by taking the average of all the samples.

Finally! 26 542 981 poppies. What do you mean I didn't need to count them all?

Frame Quadrats can be used to **Investigate Plant Populations**

1) A **frame quadrat** is a **square** frame divided into a **grid** of 100 **smaller squares** by strings attached across the frame.
2) They're **placed on the ground** at **random points** within the area you're investigating. This can be done by selecting **random coordinates** (see above).
3) The **number of individuals** (or the **species frequency**) of each species is recorded in **each quadrat**.
4) The **percentage cover** of a **plant species** can also be measured by counting how much of the quadrat is **covered** by the plant species — you count a square if it's **more than half-covered**. Percentage cover is a **quick** way to investigate populations and you **don't** have to **count** all the **individual** plants.
5) Frame quadrats are useful for **quickly** investigating areas with species that **fit within a small quadrat** — most frame quadrats are **1 m by 1 m**.
6) Areas with **larger plants** and **trees** need **very large** quadrats. Large quadrats **aren't** always in a frame — they can be marked out with a **tape measure**.

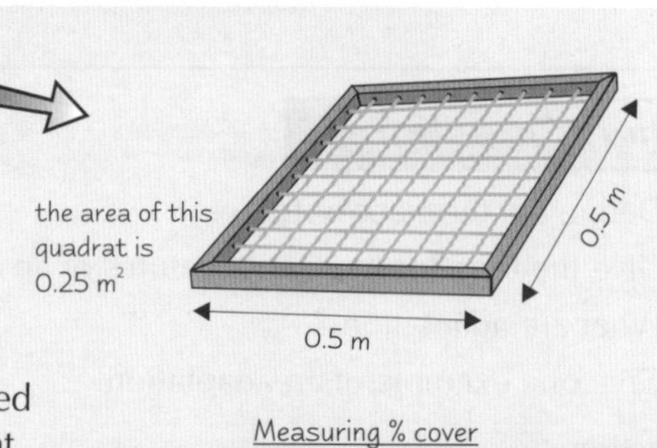

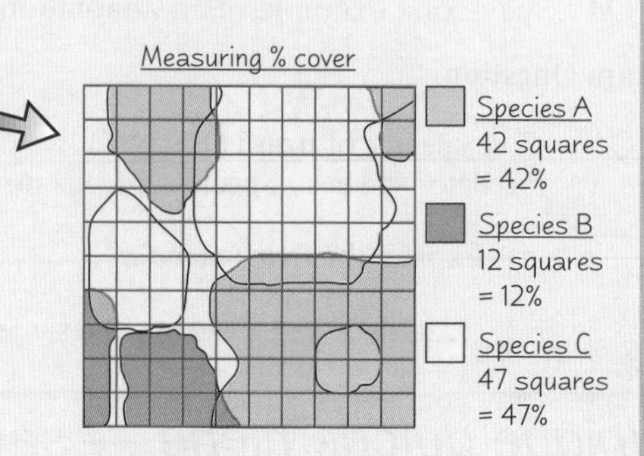

Investigating Populations and Abiotic Conditions

Point Quadrats can also be used to *Investigate Plant Populations*

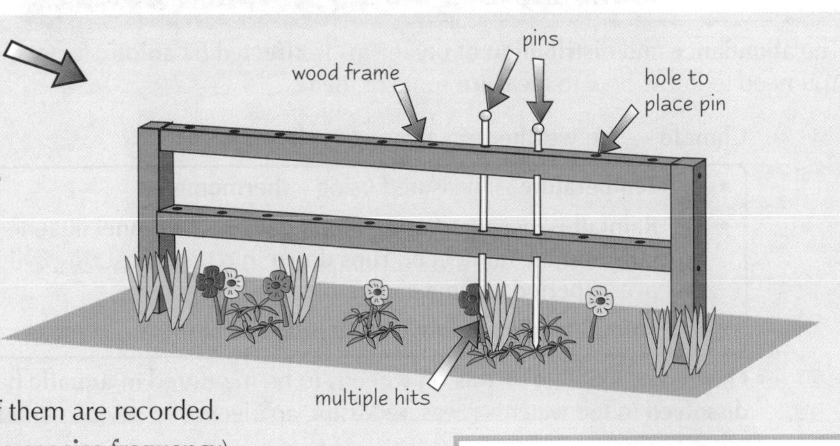

1) A **point quadrat** is a **horizontal bar** on **two legs** with a series of holes at set intervals along its length.
2) Point quadrats are **placed on the ground** at **random points** within the area you're investigating.
3) **Pins** are dropped through the holes in the frame and **every plant** that each pin **touches** is **recorded**. If a pin touches several **overlapping** plants, **all** of them are recorded.
4) The **number of individuals** (or the **species frequency**) of each species is recorded in **each quadrat**.

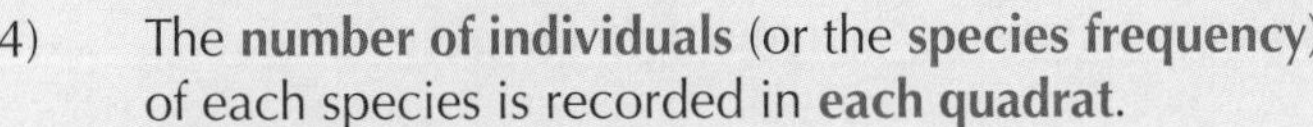

5) The **percentage cover** of a species can also be measured by calculating the **number of times** a pin has touched a species as a **percentage** of the **total number** of pins dropped.
6) Point quadrats are especially useful in areas where there's lots of **dense vegetation** close to the ground.

It's a horizontal bar with two legs alright, but where do we put the pins?

Transects are used to *Investigate* the *Distribution* of *Plant Populations*

You can use **lines** called **transects** to help find out how plants are **distributed across** an area, e.g. how species **change** from a hedge towards the middle of a field.
You need to know about **three** types of transect:

1) **Line transects** — a **tape measure** is placed **along** the transect and the species that **touch** the tape measure are **recorded**.

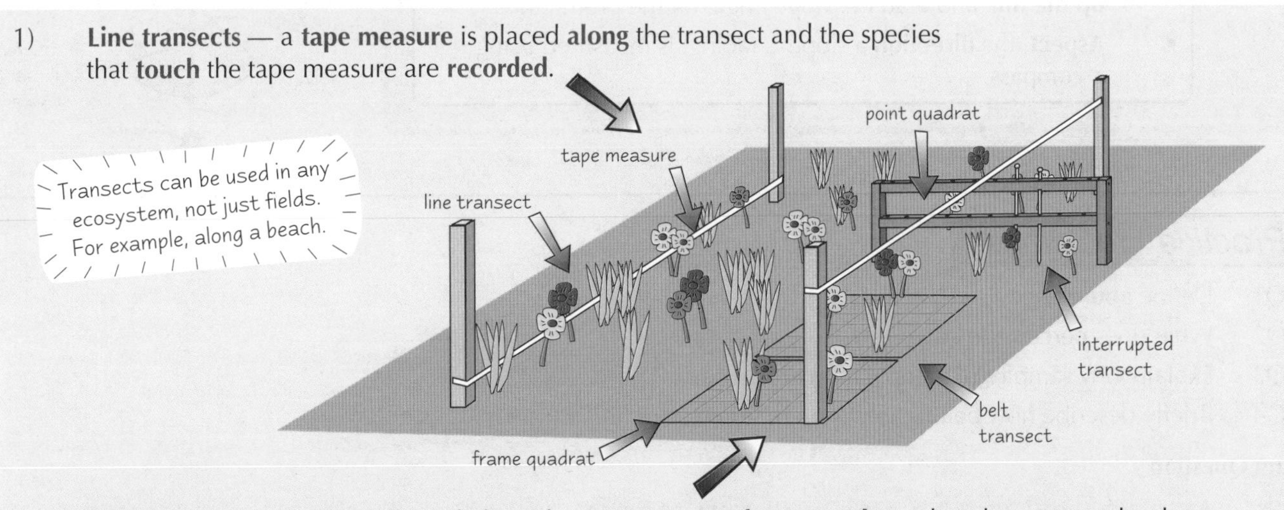

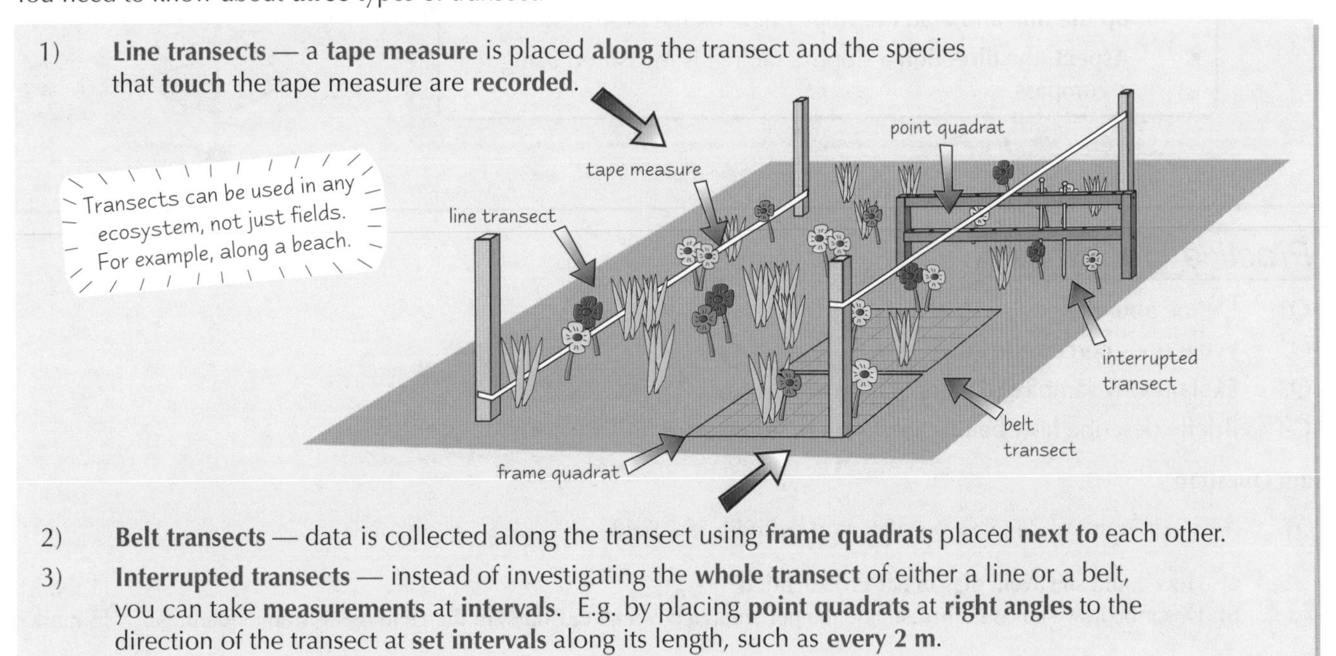

2) **Belt transects** — data is collected along the transect using **frame quadrats** placed **next to** each other.
3) **Interrupted transects** — instead of investigating the **whole transect** of either a line or a belt, you can take **measurements** at **intervals**. E.g. by placing **point quadrats** at **right angles** to the direction of the transect at **set intervals** along its length, such as **every 2 m**.

Investigating Populations and Abiotic Conditions

If you're doing AQA or OCR you can skip straight to the questions — this bit's just for Edexcel.

You can also Measure Different Abiotic Factors

The **abundance** and **distribution** of organisms is **affected** by **abiotic** factors.
You need to know how to **measure** some of them:

1) **Climate** — the **weather** of a region:
   - **Temperature** is measured using a **thermometer**.
   - **Rainfall** is measured using a **rain gauge** — a **funnel** attached to a **measuring cylinder**. The rain **falls into** the funnel and **runs down** into the measuring cylinder. The **volume** of water collected over a **period of time** can be measured.
   - **Humidity** (the amount of **water vapour** in the air) is measured using an electronic **hygrometer**.
2) **Oxygen availability** — this only needs to be measured in **aquatic habitats**. The **amount** of oxygen **dissolved** in the water is measured using an electronic device called an **oxygen sensor**.
3) **Solar input** (light intensity) is measured using an electronic device called a **light sensor**.
4) **Edaphic factors** (**soil** conditions):
   - **pH** is measured using **indicator liquid** — a **sample** of the soil is **mixed** with an indicator liquid that **changes colour** depending on the pH. The colour is matched against a **chart** to determine the pH of the soil. Electronic **pH monitors** can also be used.
   - **Moisture content** — the **mass** of a sample of soil is measured **before** and **after** being **dried out** in an **oven** at **80-100 °C** (until it reaches a **constant mass**). The difference in mass as a **percentage** of the **original** mass of the soil is then calculated. This shows the water content of the soil sample.
5) **Topography** — the **shape** and **features** of the Earth's surface:
   - **Relief** (how the **height** of the land changes across a surface) can be measured by taking **height readings** using a **GPS** (global positioning system) device at **different points** across the surface. You can also use **maps** with **contour lines** (lines that join points that are the same height).
   - **Slope angle** (how **steep** a slope is) is measured using a **clinometer**. A simple clinometer is just a piece of **string** with a **weight** on the end attached to the centre of a **protractor**. You **point** the flat edge of the protractor **up the hill**, and read the slope angle off the protractor.
   - **Aspect** (the **direction** a slope is facing) is measured using a **compass**.

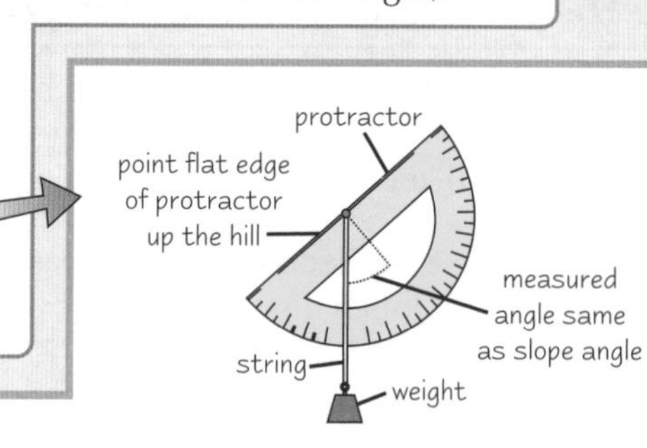

Practice Questions

Q1 Define abundance.

Q2 What does percentage cover show?

Q3 Explain why samples of a population are taken.

Q4 Briefly describe how belt transects are different from line transects.

Exam Question

Q1 A student wants to sample a population of daffodils in a field.

a) How could she avoid bias in her investigation? [1 mark]

b) Describe how she could investigate the percentage cover of daffodils in the field using frame quadrats. [3 marks]

What did the quadrat say to the policeman — I've been framed...

If you want to know what it's really like doing these investigations then read these pages outside in the pouring rain. Doing it while you're tucked up in a nice warm, dry exam hall won't seem so bad after that, take my word for it.

Investigating Populations and Analysing Data

This page is for AQA Unit 4 only. If you're doing OCR or Edexcel take a trip to p. 30, and collect £200 if you pass Go.

More practical fun on this page, but wait for it — there's some data interpretation coming up too. Sadly, dealing with data is pretty important — I'd bet my beloved bike that there'll be at least one data interpretation question in the exam.

Different Methods are Used to Investigate Mobile Organisms

To investigate the **abundance** and **distribution** of mobile organisms like **insects**, you need to **catch them** first. Here are a few of the ways it's done:

1 *Pitfall Traps and Pooters are used to Investigate Ground Insects*

Pitfall traps

1) **Pitfall traps** are **steep-sided containers** that are sunk in a **hole** in the ground. The top is **partially open**.
2) Insects **fall** into the container and **can't get out** again — they're **protected** from **rain** and **some predators** by a **raised lid.**
3) The sample can be affected by **predators small enough** to fall into the pitfall trap though — they may **eat** other insects, **affecting** the **results.**

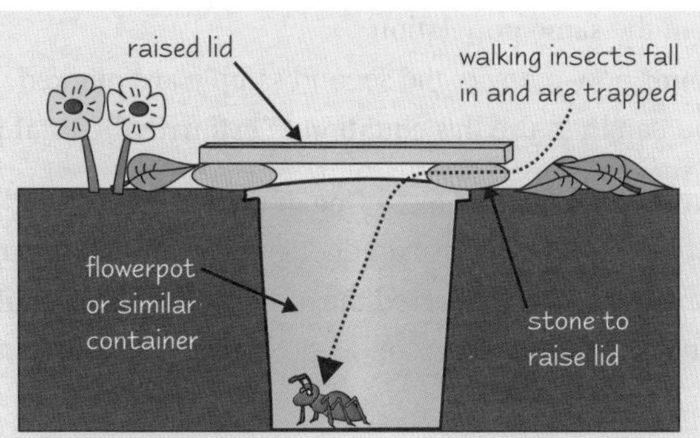

Pooters

1) **Pooters** are **jars** that have **rubber bungs** sealing the top, and **two tubes** stuck through the bung.

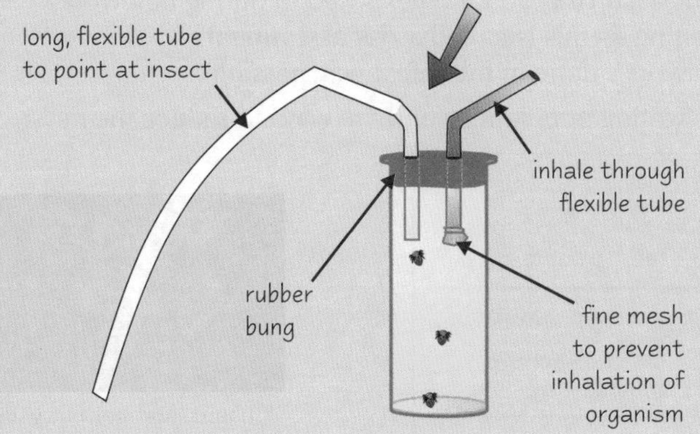

2) The **shorter tube** has **mesh** over the end that's in the jar. The **longer tube** is **open** at both ends.
3) When you **inhale** through the shorter tube, **air is drawn** through the longer tube. If you **place** the end of the **longer tube** over an insect it'll be **sucked** into the jar.
4) It can take a **long time** (or **lots of people**) to get a **large sample** using pooters. Some species may be **missed** if the sample **isn't large enough**.

2 *Beating Trays are used to Investigate Insects Found in Vegetation*

1) A **beating tray** is a **tray** or **sheet** held **under** a plant or tree.
2) The plant or tree is **shaken** and a sample of insects **falls onto** the beating tray.
3) You can take **large samples** using beating trays, giving **good estimates** of the **abundance** of each species.
4) However, the **sample** may **not be random** because most of it will be made up of insects that **fall easily** when the vegetation is **shaken**.

Beating trays are usually white so you can see the insects.

Ken was going to beat Malcolm with his tray in a minute if he kept getting all the tips.

Investigating Populations and Analysing Data

These pages are for AQA Unit 4 only.

Mark-Release-Recapture is Used to Measure Abundance

Mark-release-recapture is a method used to measure the **abundance** of more **mobile** species. Here's how it's done:

1) **Capture** a sample of a species using an **appropriate technique**, e.g. you could use pitfall traps to capture mobile ground insects (see p. 27), and **count** them.
2) **Mark** them in a harmless way, e.g. by putting a spot of **paint** on them, or by **removing** a tuft of **fur**.
3) **Release** them back into their habitat.
4) Wait a week, then take a **second sample** from the **same population**.
5) **Count** how many of the second sample **are marked**.
6) You can then use this **equation** to **estimate** the **total** population size.

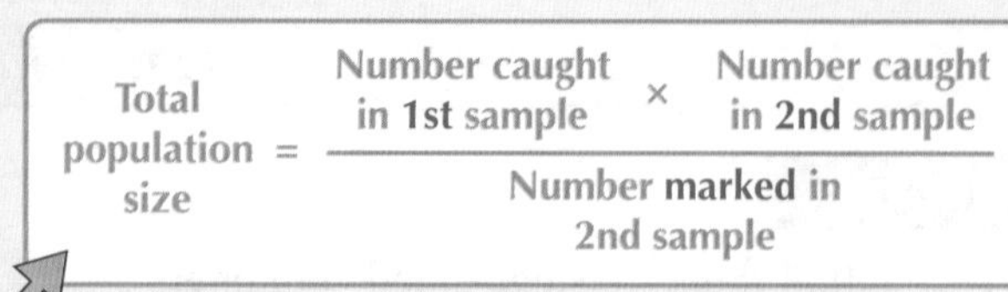

The **accuracy** of this method (how **free of errors** it is) depends on a few **assumptions**:

1) The marked sample has had enough **time** and **opportunity** to **mix** back in with the population.
2) The marking hasn't affected the individuals' **chances of survival**, and is **still visible**.
3) **Changes** in **population size** due to **births**, **deaths** and **migration** are **small** during the period of the study.

You Need to Carry Out a Risk Assessment for all Practical Work

When you're carrying out fieldwork to investigate populations you expose yourself to **risks** — things that could **potentially** cause you **harm**. You need to think about **what risks** you'll be exposed to during fieldwork, so you can **plan** ways to **reduce** the **chance** of them happening — this is called a **risk assessment**. Risk assessments are always carried out to ensure that fieldwork's done in the **safest way possible**.

Here are some **examples** of the fieldwork risks when investigating populations and the **ways** to **reduce** the risks:

Risk	Ways to reduce the risk
Falls and slips	Wear suitable footwear for the terrain, e.g. wellies on wet or boggy ground, and take care on rough terrain. Make sure the study area isn't near any cliffs or on steep ground.
Bad weather	Check the weather forecast beforehand and take precautions, e.g. wear warm or waterproof clothing on cold or wet days. If the weather is too bad, do the fieldwork another day.
Stings and bites	Wear insect repellent or, if you have an allergy, take medication with you.

OK chaps, just get out your lightning-proof suits and let's crack on with the fieldwork.

There are Ethical Issues to Consider When Doing Fieldwork

All fieldwork **affects** the **environment** where it's carried out, e.g. lots of people **walking around** may cause **soil erosion**. Some people don't think it's **right** to **damage** the **environment** when doing fieldwork, so investigations should be planned to have the **smallest impact possible**, e.g. people should restrict **where they walk** to the area being studied.

Some fieldwork **affects** the **organisms** being studied, e.g. **capturing** an organism for study may cause it **stress**. Some people don't think it's **right** to **distress** organisms **at all** when doing fieldwork, so investigations should be planned so that organisms are treated with **great care**, and are **kept** and **handled** as **little** as possible. They should also be **released** as soon as possible after they have been captured.

Investigating Populations and Analysing Data

*You need to **Analyse** and **Interpret Data** on the **Distribution of Organisms***

Here's an **example** of the kind of thing you might get in the **exam**:

A group of students investigated how the **distribution** of plant species changed with **distance** from a path. They used a **belt transect** (see p. 25) and measured **percentage cover** of plant species in each quadrat. The students also carried out a **survey** at the **same** location to record how many people **strayed away** from the path, and **how far** they strayed.

Here's a **table** and a **graph** showing their results.

You might have to:

1) **Describe the data:**
   - The table shows **low** percentage cover of plants **near** the path, e.g. **2 m** from the path it was **12%**, but **higher** percentage cover **away** from the path, e.g. **20 m** from the path it was **100%**.
   - The graph shows **lots** of walkers **near** the path, e.g. **0-4 m** from the path there were **79** walkers, but **fewer** walkers away from the path, e.g. **16-20 m** from the path there **were none**.

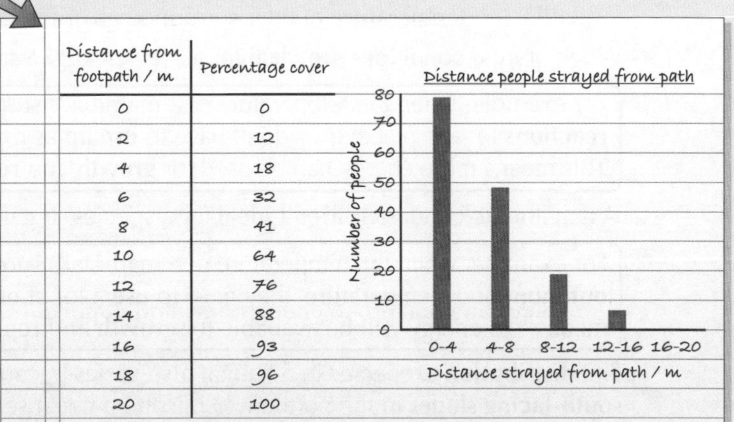

Distance from footpath / m	Percentage cover
0	0
2	12
4	18
6	32
8	41
10	64
12	76
14	88
16	93
18	96
20	100

2) **Draw conclusions:**
   - There's a **positive correlation** between **distance** from the path and **percentage cover** of plants — as distance from the path **increases**, the percentage cover of plants **increases**.
   - There's a **negative correlation** between **distance** from the path and the **number** of people that walk there — as distance from the path **increases**, the number of people that walk there **decreases**.
   - There's a **negative correlation** between the **number of walkers** and the **percentage cover** of plants — the **higher** the number of people that walk over an area, the **lower** the percentage cover of plants.

   You **can't conclude** that the **lower percentage cover** of plants **near** the path is **caused** by the **higher number** of **people** walking there. There could be **other factors** involved that affect the percentage cover of plants, e.g. the path may be covered by **stones** or **gravel**, so plants won't grow **on** or **near** the path regardless of how many people walk on it.

See pages 214-216 for more on interpreting data.

3) **Suggest explanations for your conclusions:**

   As you move **away** from the path the **number** of people that trample the ground **decreases** because people tend to **follow the path**. As you move **away** from the path the **percentage cover** of plants **increases** because plants **grow** and **survive better** where they're trodden on **less**.

Practice Questions

Q1 Give one drawback of using beating trays to investigate insect populations.

Q2 Name one fieldwork risk when investigating populations.

Exam Question

Number of snails caught in first sample	Number of snails caught in second sample	Number of marked snails caught in second sample
52	38	14

Q1 The size of a snail population was investigated using the mark-release-recapture method. The table shows the results.

a) Describe the method that could have been used to collect the data. [5 marks]

b) Calculate the total population size. [2 marks]

Risks associated with this book — laughter, increased intelligence...

Mark-release-recapture isn't too bad — it's exactly what it sounds like. Risk assessments aren't too bad either — they usually just involve a bit of common-sense thinking to work out what might be dangerous. As always, interpreting data can be a pain — just make sure you're clear that even if two things correlate, it doesn't mean that one is caused by the other.

Variation in Population Size

These pages are for AQA Unit 4, OCR Unit 5 and Edexcel Unit 4.

Uh-oh, anyone who loves cute little bunny-wunnys look away now — these pages are about how the population sizes of organisms fluctuate and the reasons why. One of the reasons, I'm sad to say, is because the little rabbits get eaten.

Population Size Varies Because of Abiotic Factors...

Remember — abiotic factors are the non-living features of the ecosystem.

1) **Population size** is the **total number** of organisms of **one species** in a **habitat**.
2) The **population size** of any species **varies** because of **abiotic** factors, e.g. the amount of **light**, **water** or **space** available, the **temperature** of their surroundings or the **chemical composition** of their surroundings.
3) When abiotic conditions are **ideal** for a species, organisms can **grow fast** and **reproduce successfully**.

   For example, when the temperature of a mammal's surroundings is the ideal temperature for **metabolic reactions** to take place, they don't have to **use up** as much energy **maintaining** their **body temperature**. This means more energy can be used for **growth** and **reproduction**, so their population size will **increase**.

4) When abiotic conditions **aren't ideal** for a species, organisms **can't** grow as **fast** or reproduce as **successfully**.

   For example, when the temperature of a mammal's surroundings is significantly **lower** or **higher** than their **optimum** body temperature, they have to **use** a lot of **energy** to maintain the right **body temperature**. This means less energy will be available for **growth** and **reproduction**, so their population size will **decrease**.

5) The **distribution** of species in a habitat also **varies** because of **abiotic** factors, e.g. some **plants** only grow on **south-facing slopes** in the northern hemisphere because that's where **solar input** (light intensity) is **greatest**.

...and Because of Biotic Factors

Biotic factors are the living features of the ecosystem.

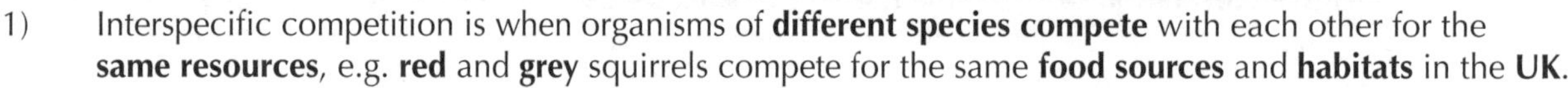

1 Interspecific Competition — Competition Between Different Species

1) Interspecific competition is when organisms of **different species compete** with each other for the **same resources**, e.g. **red** and **grey** squirrels compete for the same **food sources** and **habitats** in the **UK**.
2) Interspecific competition between two species can mean that the **resources available** to **both** populations are **reduced**, e.g. if they share the **same** source of food, there will be **less** available to both of them. This means both populations will be **limited** by a lower amount of food. They'll have less **energy** for **growth** and **reproduction**, so the population sizes will be **lower** for both species. E.g. in areas where both **red** and **grey** squirrels live, both populations are **smaller** than they would be if there was **only one** species there.
3) Interspecific competition can also affect the **distribution** of species. If **two** species are competing but one is **better adapted** to its surroundings than the other, the less well adapted species is likely to be **out-competed** — it **won't** be able to **exist** alongside the better adapted species. E.g. since the introduction of the **grey squirrel** to the UK, the native **red squirrel** has **disappeared** from large areas. The grey squirrel has a better chance of **survival** because it's **larger** and can store **more fat** over winter.

2 Intraspecific Competition — Competition Within a Species

Intraspecific competition is when organisms of the **same species compete** with each other for the **same resources**.

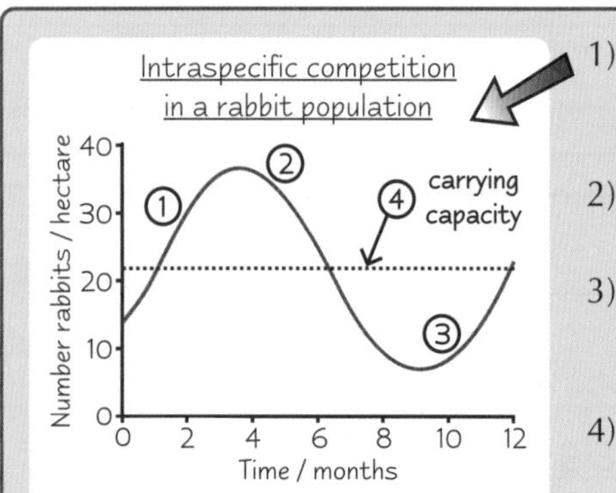

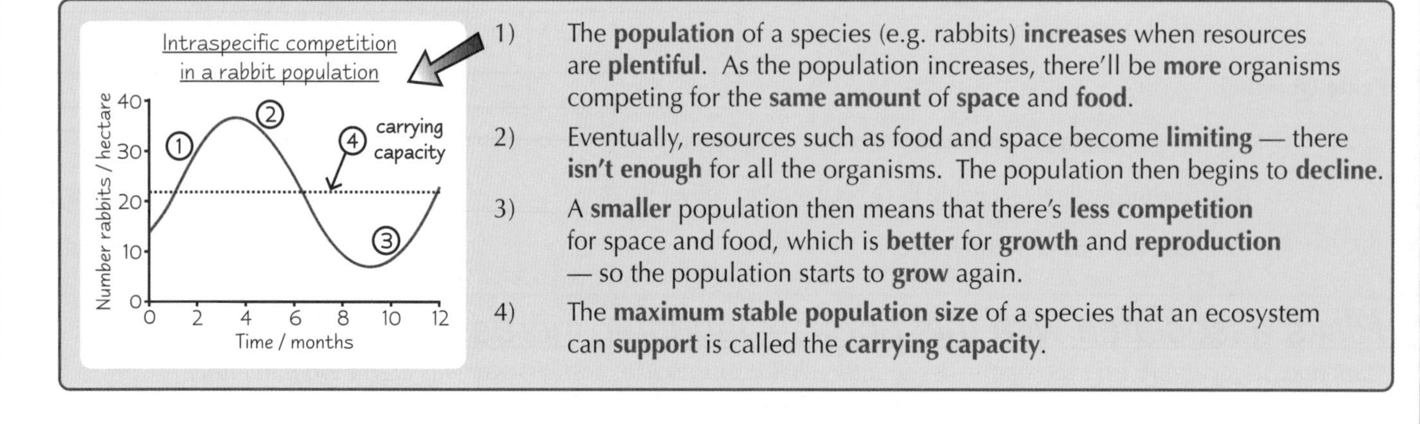

1) The **population** of a species (e.g. rabbits) **increases** when resources are **plentiful**. As the population increases, there'll be **more** organisms competing for the **same amount** of **space** and **food**.
2) Eventually, resources such as food and space become **limiting** — there **isn't enough** for all the organisms. The population then begins to **decline**.
3) A **smaller** population then means that there's **less competition** for space and food, which is **better** for **growth** and **reproduction** — so the population starts to **grow** again.
4) The **maximum stable population size** of a species that an ecosystem can **support** is called the **carrying capacity**.

Variation in Population Size

3 Predation — Predator and Prey Population Sizes are Linked

Predation is where an organism (the predator) kills and eats another organism (the prey), e.g. lions kill and eat (**predate** on) buffalo. The **population sizes** of predators and prey are **interlinked** — as the population of one **changes**, it **causes** the other population to **change**:

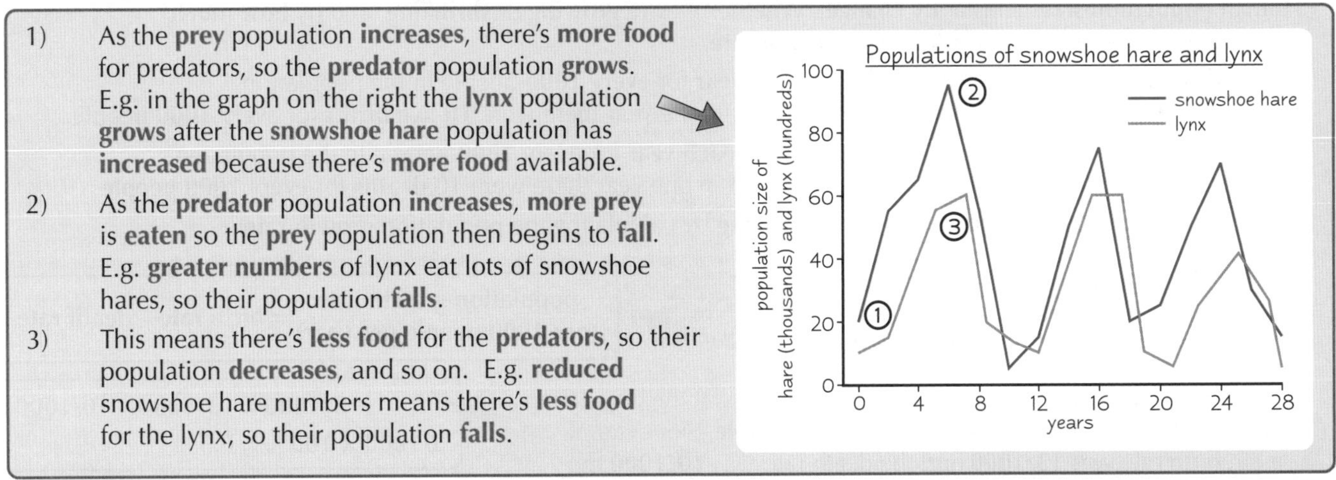

1) As the **prey** population **increases**, there's **more food** for predators, so the **predator** population **grows**. E.g. in the graph on the right the **lynx** population **grows** after the **snowshoe hare** population has **increased** because there's **more food** available.
2) As the **predator** population **increases**, **more prey** is **eaten** so the **prey** population then begins to **fall**. E.g. **greater numbers** of lynx eat lots of snowshoe hares, so their population **falls**.
3) This means there's **less food** for the **predators**, so their population **decreases**, and so on. E.g. **reduced** snowshoe hare numbers means there's **less food** for the lynx, so their population **falls**.

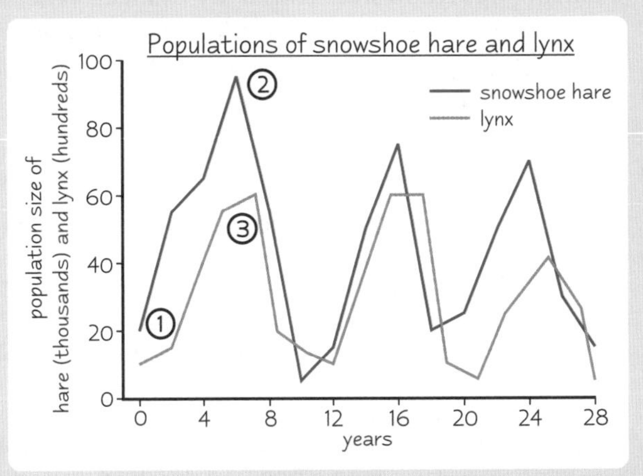

Predator-prey relationships are usually more **complicated** than this though because there are **other factors** involved, like availability of **food** for the **prey**. E.g. it's thought that the population of snowshoe hare initially begins to **decline** because there's **too many** of them for the amount of **food available**. This is then **accelerated** by **predation** from the lynx.

Limiting Factors Stop the Population Size of a Species Increasing

OCR only

1) Limiting factors can be **abiotic**, e.g. the amount of **shelter** in an ecosystem **limits** the population size of a species because there's only enough shelter for a **certain number** of individuals.
2) Limiting factors can also be **biotic**, e.g. **interspecific competition limits** the population size of a species because the amount of **resources** available to a species is **reduced**.

Practice Questions

Q1 What is interspecific competition?

Q2 What will be the effect of interspecific competition on the population size of a species?

Q3 Give one example of interspecific competition.

Q4 Define intraspecific competition.

Q5 What is a limiting factor?

Exam Question

Q1 The graph on the right shows the population size of a predator species and a prey species over a period of 30 years.

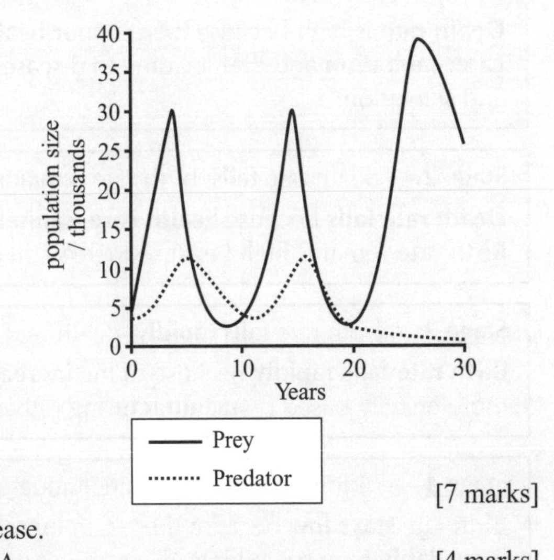

a) Using the graph, describe and explain how the population sizes of the predator and prey species vary over the first 20 years. [7 marks]

b) The numbers of species B declined after year 20 because of a disease. Describe and explain what happened to the population of species A. [4 marks]

Predator-prey relationships — they don't usually last very long...

You'd think they could have come up with names a little more different than inter- and intraspecific competition. I always remember it as int-er means diff-er-ent species. The factors that affect population size are divided up nicely for you here — abiotic factors, competition and predation — just like predators like to nicely divide up their prey into bitesize chunks.

Human Populations

These pages are for AQA Unit 4 only. If you're doing OCR or Edexcel you've finished the section. Have a breather.
These pages are about how human populations change, so they're about joyful births... and not so joyful deaths.

Human Population Growth is Calculated using Birth and Death Rates

Human population sizes constantly **change**. Whether they're **growing** or **shrinking** (and by **how much**) depends on the population's **birth rate** and **death rate**.

1) **Birth rate** — the number of **live births each year** for **every 1000** people in the population, e.g. a birth rate of **10/1000** would mean that in one year there were **10 live births** for every **1000 people**.
2) **Death rate** — the number of people that **die each year** for every **1000** people in the population, e.g. a death rate of **10/1000** would mean that in one year there were **10 deaths** for every **1000 people**.

You can work out **how fast** the population's **changing** by calculating the **population growth rate**:

Population growth rate is how much the **population** size **increases** or **decreases** in a **year**. You can work it out using the **birth** and **death rate**:

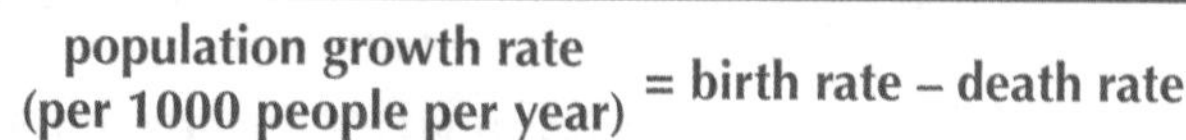

population growth rate
(per 1000 people per year) = birth rate – death rate

This gives you the **overall** (**net**) **number of people** that the population **grows** or **shrinks by** in a **year** for every **1000 people**. For example, if the birth rate was **13/1000** and the death rate was **10/1000** the population would grow by **3 people** for every **1000 people each year** (or **3/1000** people per year). It's normally given as a **percentage**, so a growth rate of **3/1000** people per year would be **0.3%** (3/1000 × 100%).

3/1000 = 13/1000 – 10/1000
3/1000 × 100% = 0.3%

The Demographic Transition Model shows Trends in Human Populations

The **Demographic Transition Model** (**DTM**) is a graph that shows changes in **birth rate**, **death rate** and **total population size** for a **human population** over a **long period** of time. It's divided into **five** stages:

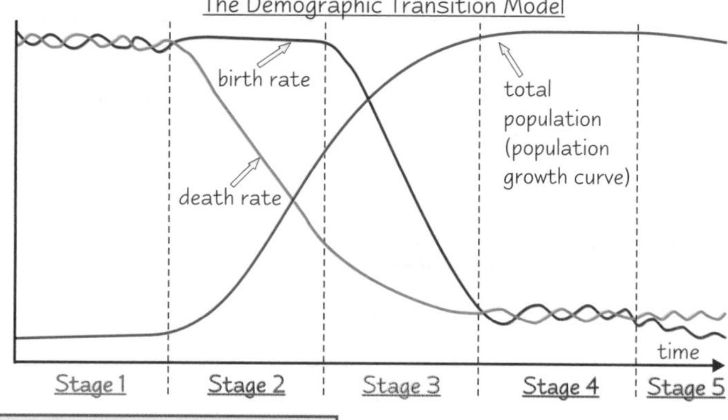

Stage 1 — birth rate and death rate fluctuate at a **high level**. The population stays **low**.
Birth rate is high because there's **no birth control** or **family planning** and **education** is **poor**. Lots of children **die young** (high **infant mortality**), so parents have more children so enough **survive** to **work** on farms, as well as **look after** them in later life.
Death rate is high because there's **poor health care**, **sanitation** and **diet**, leading to disease and starvation.

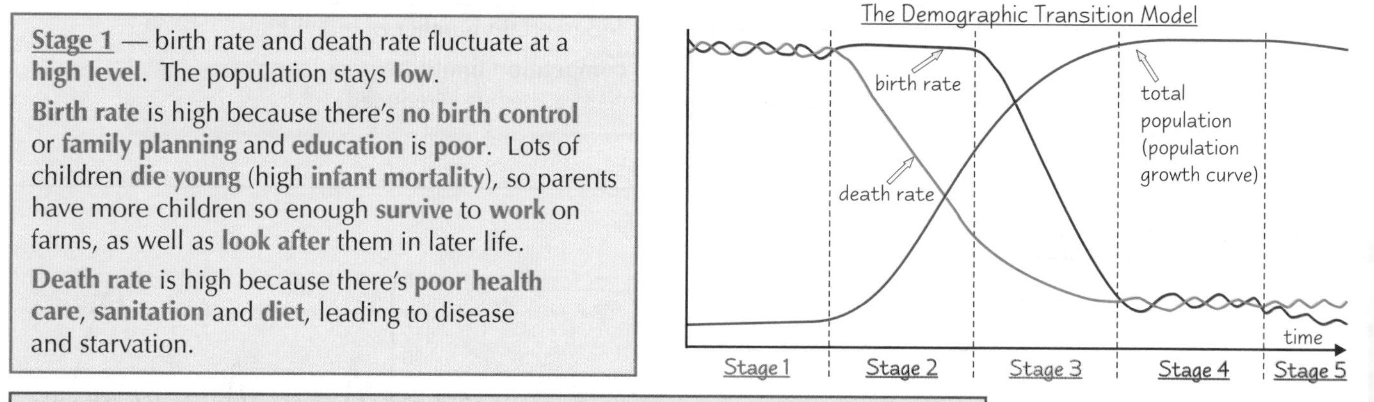

Stage 2 — death rate **falls**, birth rate **remains high**. The population **increases rapidly**.
Death rate falls because **health care**, **sanitation** and **diet improve**.
Birth rate remains **high** because there's still little **birth control** or **family planning**.

'Demographic' means it's to do with human populations.

Stage 3 — birth rate **falls rapidly**, death rate **falls more slowly**. The population **increases** at a **slower rate**.
Birth rate falls rapidly because of the **increased** use of **birth control** and **family planning**. Also, the economy becomes more heavily based on **manufacturing** rather than agriculture, so **fewer** children are needed to work on **farms**.

Stage 4 — birth rate and death rate fluctuate at a **low level**. The population remains **stable** but **high**.
Birth rate **stays low** because there's an **increased demand** for **luxuries** and **material possessions**, so **less money** is **available** to raise children. They're not needed to work to **provide income**, so parents have **fewer children**.

Stage 5 — birth rate begins to **fall**, death rate **remains stable**. The population begins to **decrease**.
Birth rate falls because children are **expensive** to raise and people often have **dependent elderly relatives**.
Death rate remains **steady** despite continued health care advances as **larger generations** of **elderly people die**.

Human Populations

Human Population Data can be Plotted in Different Ways

1) Population Growth Curves show Change in Population Size

Population change can be shown by a **population growth curve** (the **DTM** has one, see previous page). They're made by plotting data for **population size** against **time**.

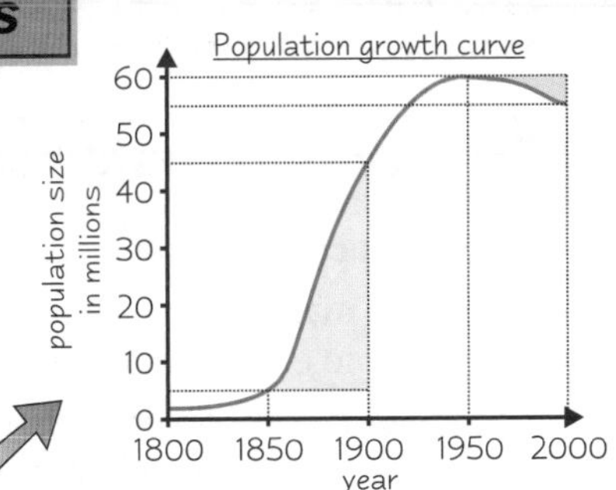

1) **Growth curves** show whether the population was **increasing** or **decreasing** by the direction of the curve (**up** or **down**).
2) The **steepness** of the curve shows **how fast** the population was **changing** (the **steeper** the curve, the **faster** it was changing). You can use the curve to **calculate** the **rate of change**. For example, between **1850** and **1900** this population **increased** from **5** to **45** million. An increase of **40** million in **50** years meant the population **increased** at a rate of **800 000 people per year** (40 000 000 ÷ 50 = 800 000). Between **1950** and **2000**, the population **decreased** by **5** million in **50** years, so it **decreased** at a rate of **100 000 people per year** (5 000 000 ÷ 50 = 100 000).

2) Survival Curves show Survival Rates

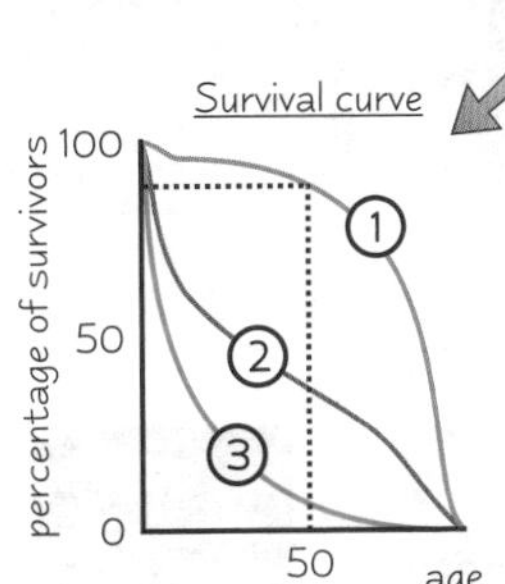

Survival curves show the **percentage** of all the individuals that were **born** in a population that are still **alive** at any **given age**. This gives a **survival rate** for any given age, e.g. population 1 has a survival rate of around **90%** for individuals at the age of **50** — 90% of people survive to be 50.

> **Population 1** — **few** people die at a young age, **lots** of people **survive** to an old age.
> **Population 2** — **many** people die at a young age, but **some survive** to an old age.
> **Population 3** — **most people die** at a young age, very **few survive** to an old age.

Life expectancy is the **age** that a person born into a population is expected to **live to** — it's worked out by calculating the **average age** that people **die**.

3) Age-sex Pyramids show Population Structure

Population structure can be shown using **age-sex pyramids**. These show how many **males** and **females** there are in different **age groups** within a population.

1) This population has **a lot** of **young people** with **very few** surviving to **old age** — there's a **low** life expectancy (**DTM stage 1**).

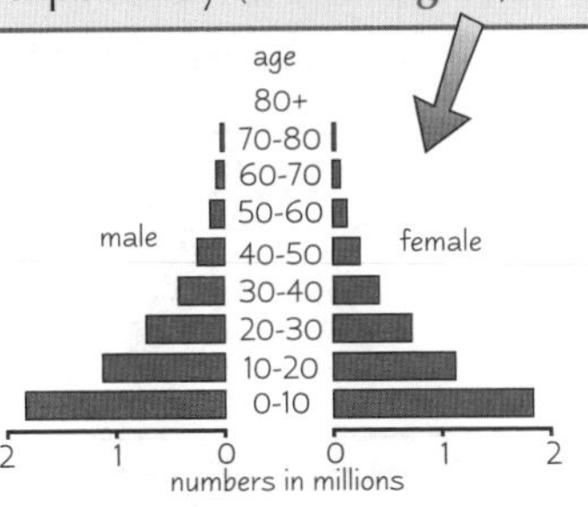

2) This population has **a lot** of **young people** with **more surviving** to **old age** — life expectancy is **higher** (**DTM stage 2**).

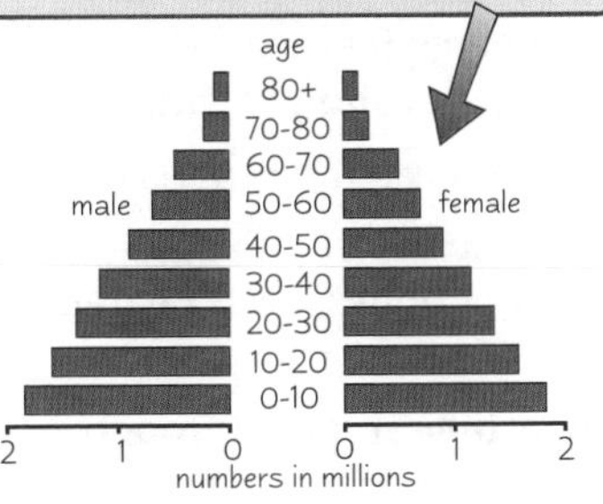

3) This population has **fewer young people** with **a lot of older people** — life expectancy is **high** (**DTM stage 5**).

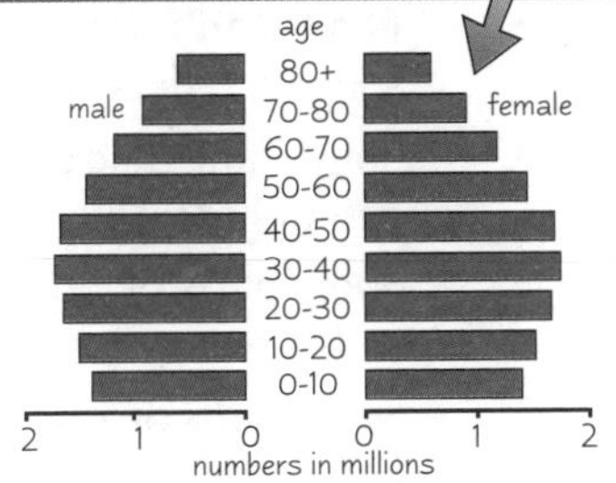

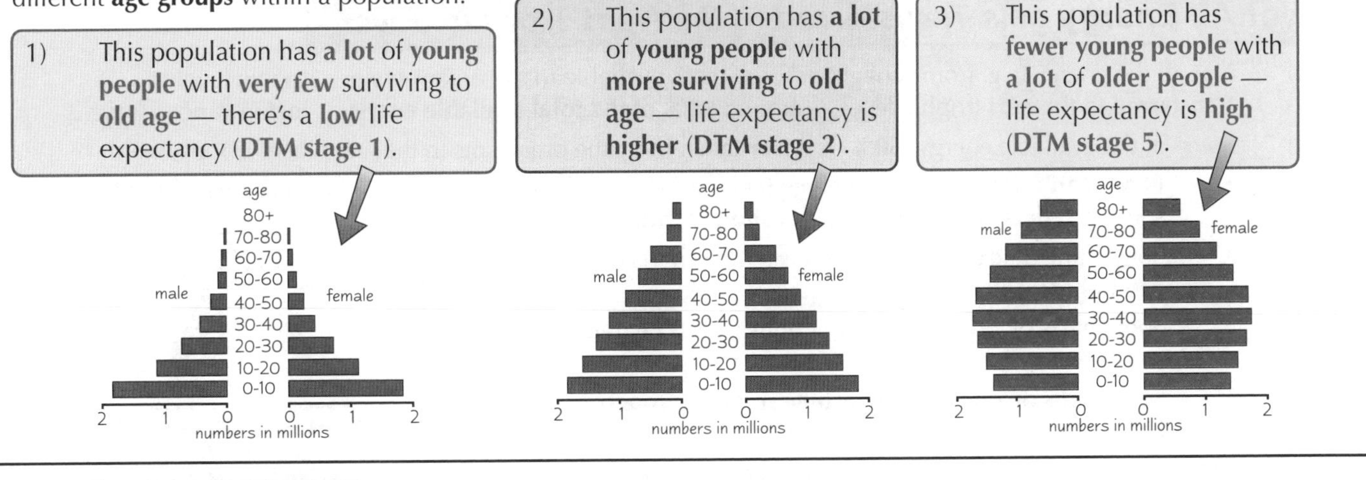

Practice Questions

Q1 How do you calculate population growth rate from birth and death rate?

Q2 What is shown by an age-sex pyramid?

Exam Question

Q1 Describe the differences in population size and structure at stage 1 compared to stage 5 of the DTM. [4 marks]

Population's growth rate — almost 20 cm a year now he's a teenager...

Boy, when it comes to human populations these biologists love their graphs. Even if you feel like your brain's turning to custard, you need to understand what the graphs are showing — you might have to interpret them in the exam.

Energy Transfer and Productivity

These pages are for AQA Unit 4, OCR Unit 5 and Edexcel Unit 4.

Some organisms get their energy from the sun and some get it from other organisms, and it's all very friendly. Yeah right.

Energy is Transferred Through Ecosystems

1) An **ecosystem** includes all the **organisms** living in a particular area and all the **non-living** (abiotic) conditions.
2) The **main route** by which energy **enters** an ecosystem is **photosynthesis** (e.g. by plants, see p. 4). (Some energy enters sea ecosystems when bacteria use chemicals from deep sea vents as an energy source.)
3) During photosynthesis plants **convert sunlight energy** into a form that can be **used** by other organisms — plants are called **producers** (because they produce **organic molecules** using sunlight energy).
4) Energy is **transferred** through the **living organisms** of an ecosystem when organisms **eat** other organisms, e.g. producers are eaten by organisms called **primary consumers**. Primary consumers are then eaten by **secondary consumers** and secondary consumers are eaten by **tertiary consumers**.
5) Each of the stages (e.g. producers, primary consumers) are called **trophic levels**.
6) **Food chains** and **food webs** show how energy is **transferred** through an ecosystem.
7) **Food chains** show **simple lines** of energy transfer.
8) **Food webs** show **lots** of **food chains** in an ecosystem and how they **overlap**.
9) Energy locked up in the things that **can't be eaten** (e.g. bones, faeces) gets recycled back into the ecosystem by microorganisms called **decomposers** — they **break down dead** or **undigested** material.

Consumers are organisms that eat other organisms.

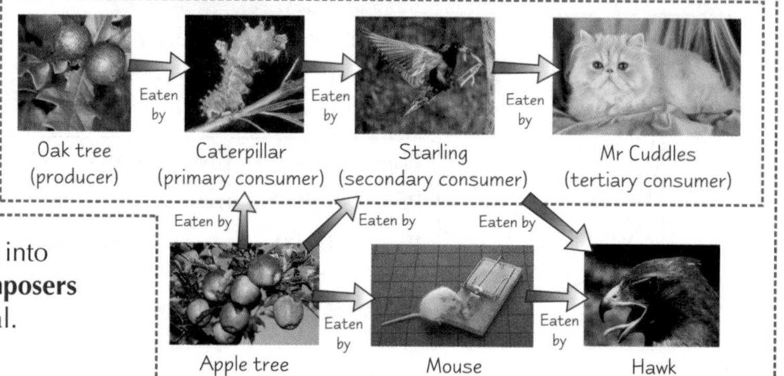

Not All Energy gets Transferred to the Next Trophic Level

1) **Not all** the energy (e.g. from sunlight or food) that's available to the organisms in a trophic level is **transferred** to the **next** trophic level — around **90%** of the **total available energy** is **lost** in various ways.
2) Some of the available energy (**60%**) is **never taken in** by the organisms in the first place. For example:
   - Plants **can't use** all the light energy that reaches their **leaves**, e.g. some is the **wrong wavelength**, some is **reflected**, and some **passes straight through** the leaves.
   - Some sunlight can't be used because it hits parts of the plant that **can't photosynthesise**, e.g. the bark of a tree.
   - Some **parts** of food, e.g. **roots** or **bones**, **aren't eaten** by organisms so the energy isn't taken in.
   - Some parts of food are **indigestible** so **pass through** organisms and come out as **waste**, e.g. **faeces**.
3) The rest of the available energy (**40%**) is **taken in** (**absorbed**) — this is called the **gross productivity**. But not all of this is available to the next trophic level either.
   - **30%** of the **total energy** available (75% of the gross productivity) is **lost to the environment** when organisms use energy produced from **respiration** for **movement** or body **heat**. This is called **respiratory loss**.
   - **10%** of the **total energy** available (25% of the gross productivity) becomes **biomass** (e.g. it's **stored** or used for **growth**) — this is called the **net productivity**.
4) **Net productivity** is the amount of energy that's **available** to the **next trophic level**.

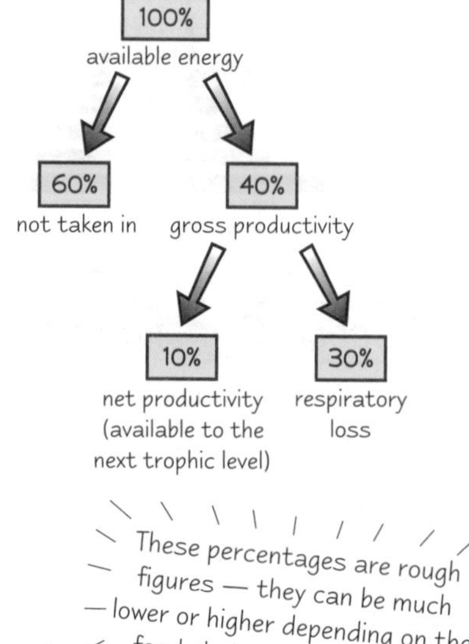

These percentages are rough figures — they can be much lower or higher depending on the food chain and trophic level.

Energy Transfer and Productivity

Productivity and Energy Transfer Between Trophic Levels can be Calculated

You can **calculate** the **net productivity** of a trophic level when you know the **gross productivity** and the **respiratory losses** of the trophic level. Here's the **equation**:

net productivity = gross productivity – respiratory loss

Here's an example of how **net productivity** is **calculated**:

The rabbits in an ecosystem receive **20 000 $kJm^{-2}yr^{-1}$** of energy, but don't take in **12 000 $kJm^{-2}yr^{-1}$** of it, so their gross productivity is **8000 $kJm^{-2}yr^{-1}$** (20 000 – 12 000). They lose **6000 $kJm^{-2}yr^{-1}$** using energy from **respiration**. You can use this to **calculate** the **net productivity** of the rabbits:

net productivity = 8000 – 6000
= 2000 $kJm^{-2}yr^{-1}$

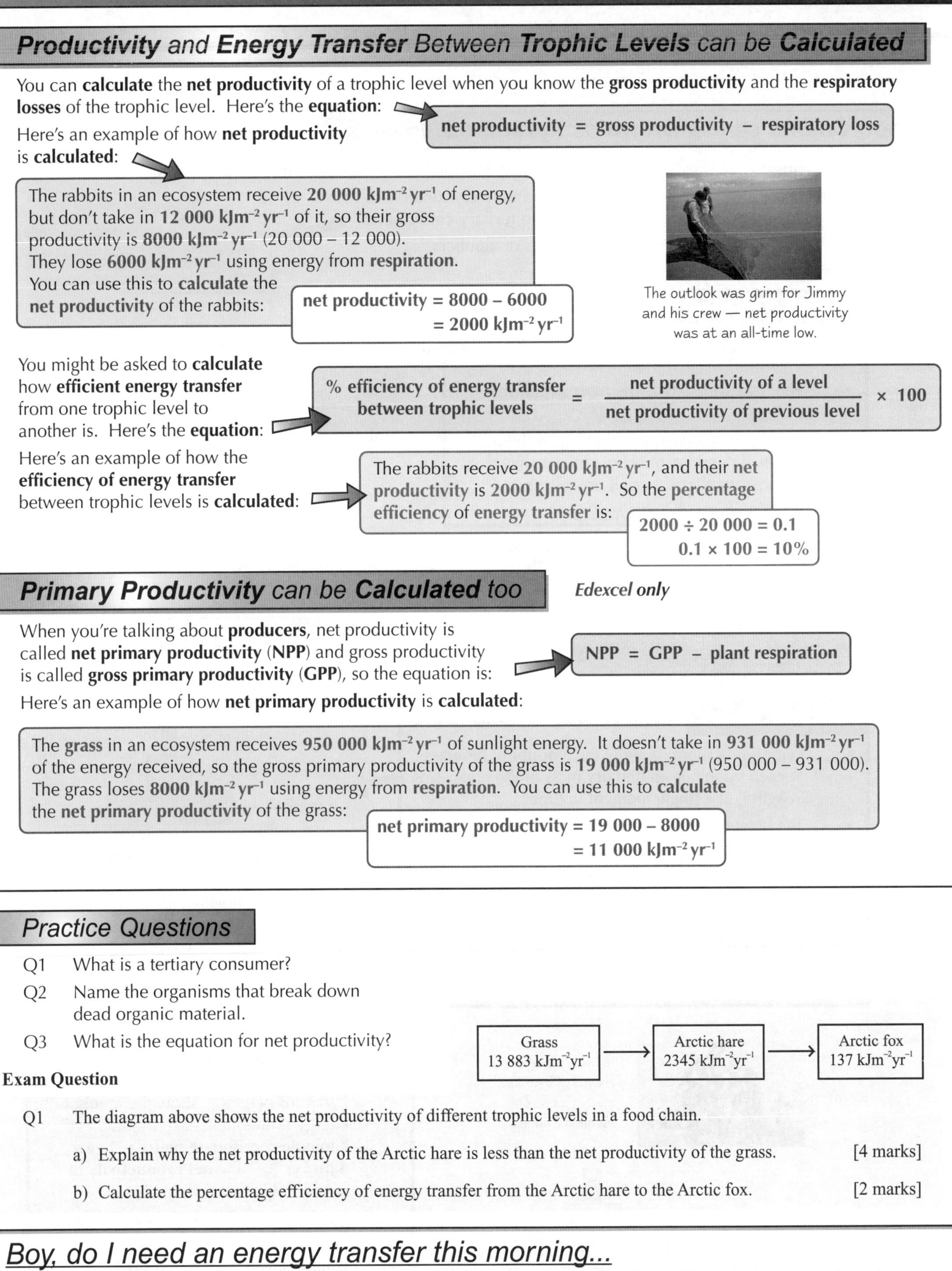

The outlook was grim for Jimmy and his crew — net productivity was at an all-time low.

You might be asked to **calculate** how **efficient energy transfer** from one trophic level to another is. Here's the **equation**:

$$\text{\% efficiency of energy transfer between trophic levels} = \frac{\text{net productivity of a level}}{\text{net productivity of previous level}} \times 100$$

Here's an example of how the **efficiency of energy transfer** between trophic levels is **calculated**:

The rabbits receive **20 000 $kJm^{-2}yr^{-1}$**, and their **net productivity** is **2000 $kJm^{-2}yr^{-1}$**. So the **percentage efficiency** of **energy transfer** is:

2000 ÷ 20 000 = 0.1
0.1 × 100 = 10%

Primary Productivity can be Calculated too

Edexcel only

When you're talking about **producers**, net productivity is called **net primary productivity** (**NPP**) and gross productivity is called **gross primary productivity** (**GPP**), so the equation is:

NPP = GPP – plant respiration

Here's an example of how **net primary productivity** is **calculated**:

The **grass** in an ecosystem receives **950 000 $kJm^{-2}yr^{-1}$** of sunlight energy. It doesn't take in **931 000 $kJm^{-2}yr^{-1}$** of the energy received, so the gross primary productivity of the grass is **19 000 $kJm^{-2}yr^{-1}$** (950 000 – 931 000). The grass loses **8000 $kJm^{-2}yr^{-1}$** using energy from **respiration**. You can use this to **calculate** the **net primary productivity** of the grass:

net primary productivity = 19 000 – 8000
= 11 000 $kJm^{-2}yr^{-1}$

Practice Questions

Q1 What is a tertiary consumer?

Q2 Name the organisms that break down dead organic material.

Q3 What is the equation for net productivity?

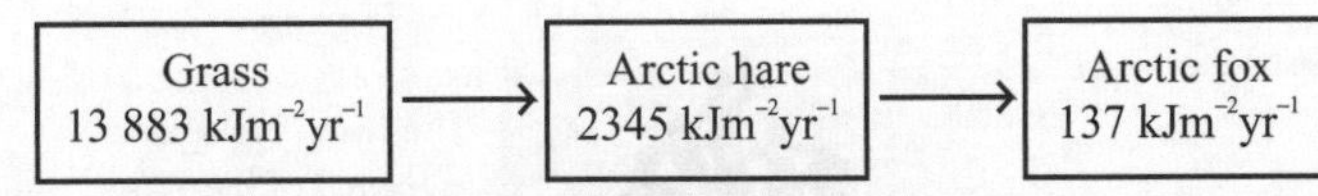

Exam Question

Q1 The diagram above shows the net productivity of different trophic levels in a food chain.

a) Explain why the net productivity of the Arctic hare is less than the net productivity of the grass. [4 marks]

b) Calculate the percentage efficiency of energy transfer from the Arctic hare to the Arctic fox. [2 marks]

Boy, do I need an energy transfer this morning...

It's really important to remember that energy transfer through an ecosystem isn't 100% efficient — most gets lost along the way so the next organisms don't get all the energy. Food chains are a nice simple way of picturing what happens, but you need to remember that real ecosystems are a bit more complicated, so food webs are needed too.

Pyramid Diagrams and Energy Transfer

This page is for AQA Unit 4 only, and page 37 is for OCR Unit 5. If you're doing Edexcel, go and put the kettle on because you've finished this section.

Food chains can be shown using nice colourful pyramid diagrams. Boy, that makes the whole topic sooo much more fun.

Food Chains can be Drawn as Pyramid Diagrams

AQA only

1) **Food chains** can be shown by drawing **pyramids** with each block representing a **trophic level**.
2) **Producers** are always on the **bottom**, then **primary consumers** are above them, followed by **secondary consumers** then **tertiary consumers**.
3) There are **three** types of pyramid — pyramids of **numbers**, **biomass** and **energy**:

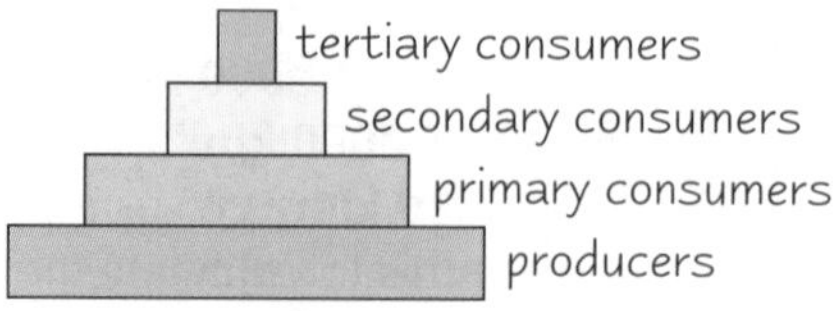

tertiary consumers
secondary consumers
primary consumers
producers

Pyramids of Numbers

- Pyramids of numbers show the **number** of organisms in each trophic level.
- The **bigger** the block, the **more organisms** there are.
- They're not always **pyramid-shaped** though — **small numbers** of **big organisms** (like trees) or **large numbers** of **small organisms** (like parasites) change the shape.

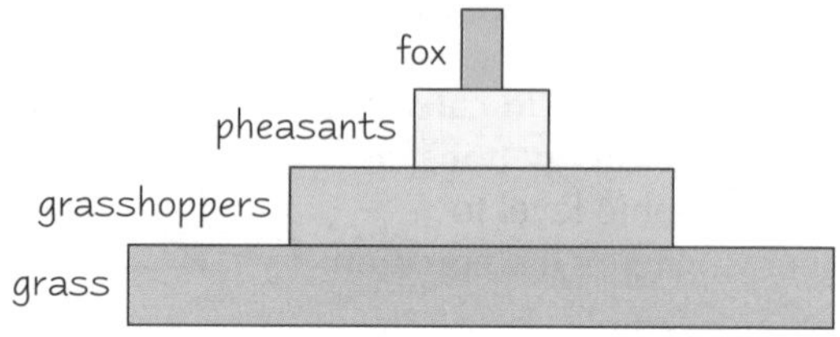

fox
pheasants
grasshoppers
grass

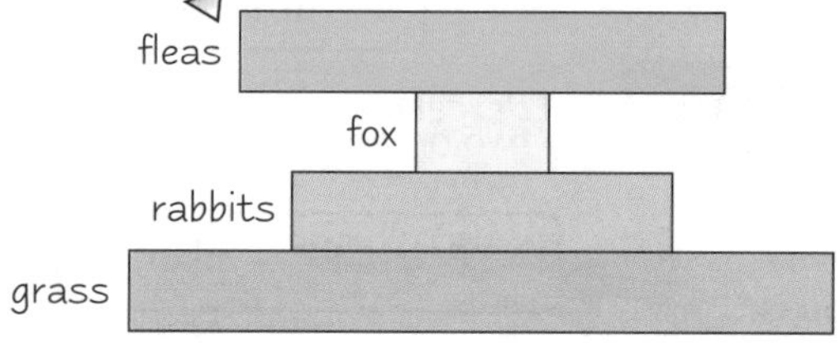

fleas
fox
rabbits
grass

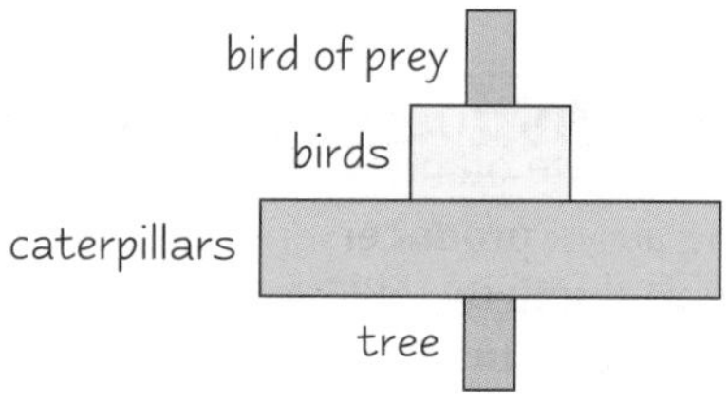

bird of prey
birds
caterpillars
tree

Pyramids of Biomass

- Pyramids of biomass show the **amount** of **biomass** in each trophic level (the **dry mass** of the organisms in **kgm^{-2}**) at a **single moment** in time.
- The **bigger** the block, the **more biomass** there is.
- They **nearly** always come out pyramid-shaped. An exception is when they're based on **plant plankton** (microorganisms that photosynthesise) — the amount of plant plankton is quite **small** at any **given moment**, but because they have a **short life span** and **reproduce very quickly** there's **a lot** around over a **period of time**.

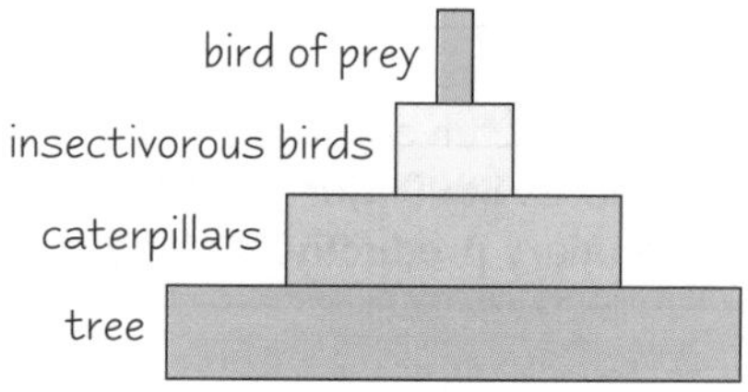

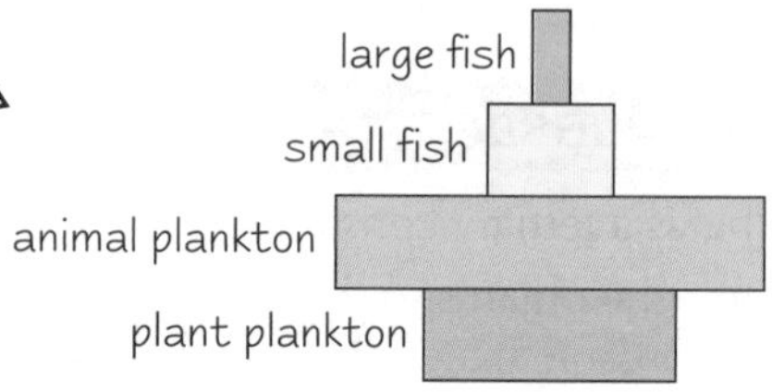

bird of prey
insectivorous birds
caterpillars
tree

large fish
small fish
animal plankton
plant plankton

The flying biologists' pyramid of biomass always came out pyramid-shaped.

Pyramids of Energy

- Pyramids of energy show the **amount** of **energy** available in each trophic level in **kilojoules** per **square metre** per **year** (**kJm^{-2} yr^{-1}**) — the **net productivity** of each trophic level (see p. 34).
- The **bigger** the block, the **more energy** there is.
- Pyramids of energy are **always** pyramid-shaped.

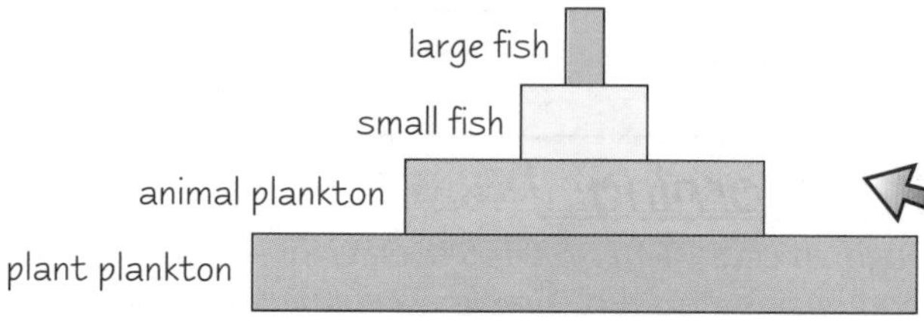

large fish
small fish
animal plankton
plant plankton

Pyramid Diagrams and Energy Transfer

If you're doing AQA you can skip straight to the questions — the rest is for OCR only.

Energy Transfer Between Trophic Levels can be Measured

1) To **measure** the **energy transfer** between two trophic levels you need to **calculate** the **difference** between the amount of **energy** in each level (the net productivity of each level).
2) There are a couple of ways to **calculate** the **amount of energy** in a trophic level:

> 1) You can **directly measure** the amount of **energy** (in **joules**) in the organisms by **burning** them in a **calorimeter**. The amount of **heat given off** tells you **how much** energy is in them.
> 2) You can **indirectly measure** the amount of **energy** in the organisms by measuring their **dry mass** (their **biomass**). Biomass is created **using energy**, so it's an **indicator** of how much energy an organism **contains**.

You can also estimate the amount of energy in a trophic level using the living mass of the organisms.

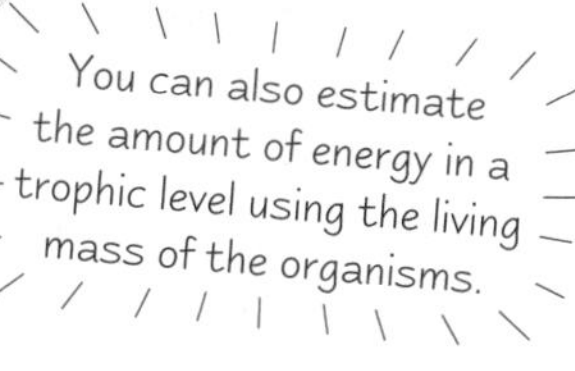

3) First you calculate the amount of energy or biomass in a **sample** of the organisms, e.g. a 1 m^2 area of **wheat** or a single **mouse** that feeds on the wheat.
4) Then you **multiply** the results from the **sample** by the **size** of the **total population** (e.g. a 10 000 m^2 **field** of wheat or the **number** of mice in the population) to give the **total** amount of energy in the organisms at that **trophic level**.
5) The **difference** between the amount of energy in the organisms at the two trophic levels is the amount of energy **transferred**.
6) There are **problems** with this method though. For example, the consumers (mice) might have **taken in energy** from sources **other than** the producer measured (wheat). This means the difference between the two figures calculated **wouldn't** be an **accurate** estimate of the energy transferred between **only those two** organisms. To **accurately** estimate energy transfer between two trophic levels, you'd need to include **all** the individual organisms at each trophic level.

Now, I will perform the energy transfer mime.

Practice Questions

Q1 Which organisms always make up the bottom level of a pyramid diagram?

Q2 What do pyramids of biomass show?

Q3 What do you need to calculate to find the energy transfer between two trophic levels?

Q4 Give one way of indirectly measuring the energy transfer between two trophic levels.

Organism	Number of individuals
Flea	150
Hedgehog	5
Slug	300
Oak tree	1

Exam Question

Q1 Above is a data table for a food chain.

a) Draw a pyramid diagram for this data. What kind of pyramid have you drawn? [2 marks]

b) Draw the rough shape of a pyramid of biomass for this food chain. [1 mark]

c) What do pyramids of energy show? [1 mark]

No animals were harmed during the making of these pages...

Pyramids of numbers, biomass and energy are simple ways of showing food chains and how the different trophic levels are related. You can also measure the energy transfer between trophic levels, but it's a tricky business. Not sure I like the sound of these calorimeters — if I was a rabbit, I'd think twice about volunteering to help collect data on energy transfer.

Farming Practices and Productivity

These pages are for AQA Unit 4 and OCR Unit 5.

Farmers may still wear wellies and say ooh-ar, but farming's all about the very serious business of increasing productivity.

Intensive Farming Systems are More Productive than Natural Ecosystems

1) A **natural ecosystem** is an ecosystem that **hasn't been changed** by **human activity**.
2) The **energy input** of a natural ecosystem is the **amount of sunlight** captured by the producers in the ecosystem.
3) Humans can **manipulate** (change) the **transfer** of energy through ecosystems, e.g. some **farmers** use **intensive farming methods** to **increase** energy transfer.
4) **Intensive farming** involves changing an ecosystem by **controlling** the **biotic** or **abiotic conditions**, e.g. the presence of pests or the amount of nutrients available, to make it **more favourable** for crops or livestock.
5) This means intensively farmed **crops** or **livestock** can have **greater net productivity** (a greater **amount** of **biomass**) than **organisms** in **natural ecosystems**.
6) The **energy input** might be **greater** in an intensively farmed area than in a natural ecosystem, e.g. cattle may be given food that's **higher in energy** than their natural food. Or it might be the **same** as a natural ecosystem, e.g. a field of crops still receives the **same** amount of **sunlight** as a natural field.

Intensive Farming Practices Increase Productivity

Intensive farming methods **increase productivity** in different ways:

1) They can **increase** the **efficiency** of **energy conversion** — more of the energy organisms **have** is used for **growth** and less is used for **other activities**, e.g. recovering from **disease** or **movement**.
2) They can remove **growth-limiting factors** — **more** of the energy **available** can be used for **growth**.
3) They can **increase energy input** — more energy is **added** to the ecosystem so there's **more energy** for **growth**.

Here are **three** of the main intensive farming practices used:

1 Killing Pest Species

Pests are organisms that **reduce** the **productivity** of **crops** by reducing the amount of energy available for **growth**. This means the crops are **less efficient** at **converting energy**. Here are **three** ways that farmers reduce pest numbers:

Using chemical pesticides

- **Herbicides** kill **weeds** that **compete** with agricultural crops for **energy**. Reducing competition means crops receive **more energy**, so they grow **faster** and become **larger**, **increasing** productivity.
- **Fungicides** kill **fungal infections** that **damage** agricultural crops. The crops **use more** energy for **growth** and **less** for fighting infection, so they grow **faster** and become **larger**, **increasing** productivity.
- **Insecticides** kill **insect** pests that **eat** and **damage** crops. Killing insect pests means **less biomass** is **lost** from crops, so they grow to be **larger**, which means productivity is **greater**.

Using biological agents

Biological agents reduce the **numbers of pests**, so crops lose **less energy** and **biomass**, **increasing** productivity.

- **Natural predators** introduced to the ecosystem **eat** the pest species, e.g. ladybirds eat greenfly.
- **Parasites** live in or lay their **eggs** on a **pest insect**. Parasites either **kill** the insect or **reduce** its ability to **function**, e.g. some species of wasps lay their eggs inside caterpillars — the eggs hatch and **kill** the caterpillars.
- **Pathogenic** (disease-causing) **bacteria** and **viruses** are used to kill pests, e.g. the bacterium ***Bacillus thuringiensis*** produces a **toxin** that kills a wide range of **caterpillars**.

Using integrated systems

Integrated systems use **both chemical pesticides** (e.g. insecticides) and **biological agents** (e.g. parasites).

1) The **combined effect** of using both can reduce pest numbers **even more** than either method **alone**, meaning **productivity** is **increased** even more.
2) Integrated systems can **reduce costs** if one method is **particularly expensive** — the expensive method can be used **less** because the two methods are used **together**.
3) Integrated systems can **reduce** the **environmental impact** of things like pesticides, because **less** is used.

Farming Practices and Productivity

2 Using Fertilisers

Fertilisers are chemicals that provide crops with **minerals** needed **for growth**, e.g. **nitrates**. Crops **use up** minerals in the soil as they **grow**, so their growth is **limited** when there **aren't enough** minerals. Adding fertiliser **replaces** the lost minerals, so **more energy** from the ecosystem can be used to grow, **increasing** the **efficiency** of energy conversion.

1) **Natural** fertilisers are **organic** matter — they include **manure** and **sewage sludge** (that's "muck" to you and me).
2) **Artificial** fertilisers are **inorganic** — they contain **pure chemicals** (e.g. ammonium nitrate) as powders or pellets.

3 Rearing Livestock Intensively

Rearing livestock **intensively** involves **controlling** the **conditions** they live in, so **more** of their **energy** is used for **growth** and **less** is used for **other activities** — the **efficiency** of energy conversion is increased so **more biomass** is produced and productivity is **increased**.

Here are a couple of **examples**:

1) Animals may be kept in **warm, indoor** pens where their **movement** is **restricted**. **Less energy** is **wasted** keeping **warm** and **moving around**.
2) Animals may be given **feed** that's **higher in energy** than their natural food. This **increases** the **energy input**, so **more energy** is available for **growth**.

The benefits are that **more food** can be produced in a **shorter** space of time, often at **lower cost**. However, enhancing productivity by intensive rearing raises **ethical issues**. For example, some people think the **conditions** intensively reared animals are kept in cause the animals **pain**, **distress** or restricts their **natural behaviour**, so it **shouldn't be done**.

Intensive Farming Raises *Environmental* and *Economic Issues*

AQA only

Environmental issues

1) Chemical pesticides and biological agents can **directly** affect (**damage** or **kill**) other **non-pest species**.
2) Chemical pesticides may **indirectly** affect other **non-pest species**, e.g. eating lots of **primary consumers** that each contain a **small amount** of **chemical pesticide** can be enough to **poison** a **secondary consumer**.
3) Natural predators introduced to an ecosystem may **become** a **pest species** themselves.
4) Fertiliser can be washed into **rivers** and **ponds**, **killing fish** and **plant life** (**eutrophication**, see p. 40).
5) Using fertilisers changes the **balance** of **nutrients** in the soil — **too much** of a particular nutrient can cause crops and other plants to **die**.

Economic issues

1) It may not be **profitable** for some farmers to use chemical pesticides — their **cost** may be **greater** than the **extra money** made from **increased productivity**.
2) Biological agents may be less **cost-effective** than chemical pesticides, i.e. they may increase productivity **less** in the **short term** for the **same amount** of money invested.
3) Farmers need to apply exactly the **right amount** of fertiliser. **Too much** and money is **wasted** as excess fertiliser is **washed away**. **Too little** and productivity **won't** be increased, so **less money** can be made from **selling** the crop.

Practice Questions

Q1 What does intensive farming involve?

Q2 What are fertilisers?

Q3 Give one example of how farmers increase the productivity of animals.

Exam Question

Q1 Organic farmers don't use artificial chemicals on their land. Describe and explain how an organic farmer might increase productivity by reducing pest numbers on their farm. [5 marks]

Farming practices — baa-aa-aa-rmy...

Crikey, so farming's not just about getting up early to feed the chooks then — farmers are manipulating the transfer of energy to produce as much food as they can. They use all kinds of intensive methods — they even have mysterious agents working for them, agents who get rid of the pests and don't ask any questions (I'm talking about biological agents by the way).

The Nitrogen Cycle and Eutrophication

These pages are for AQA Unit 4 and OCR Unit 5.
There's a bit of cycling to do here. Don't worry though — there aren't any actual hills to climb...

The Nitrogen Cycle shows how Nitrogen is Passed on and Recycled

Plants and animals **need** nitrogen to make **proteins** and **nucleic acids** (DNA and RNA). The atmosphere's made up of about 78% nitrogen, but plants and animals **can't use it** in that form — they need **bacteria** to **convert** it into **nitrogen compounds** first. The **nitrogen cycle** shows how nitrogen is **converted** into a useable form and then **passed** on between different **living** organisms and the **non-living** environment.

The nitrogen cycle includes **food chains** (nitrogen is passed on when organisms are eaten), and four different processes that involve bacteria — **nitrogen fixation**, **ammonification**, **nitrification** and **denitrification**:

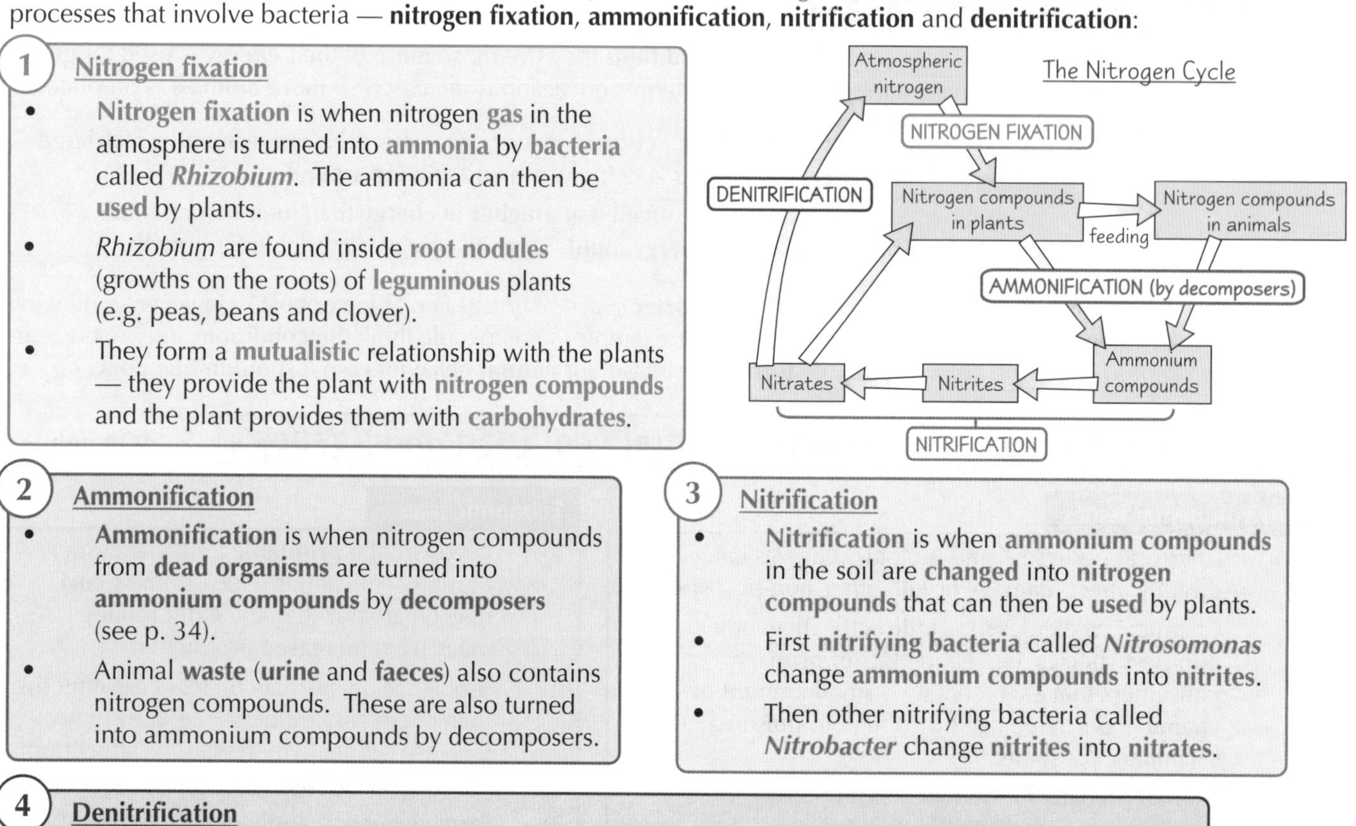

1 Nitrogen fixation

- **Nitrogen fixation** is when nitrogen **gas** in the atmosphere is turned into **ammonia** by **bacteria** called ***Rhizobium***. The ammonia can then be **used** by plants.
- *Rhizobium* are found inside **root nodules** (growths on the roots) of **leguminous** plants (e.g. peas, beans and clover).
- They form a **mutualistic** relationship with the plants — they provide the plant with **nitrogen compounds** and the plant provides them with **carbohydrates**.

The Nitrogen Cycle

Atmospheric nitrogen
NITROGEN FIXATION
Nitrogen compounds in plants
feeding
Nitrogen compounds in animals
DENITRIFICATION
AMMONIFICATION (by decomposers)
Ammonium compounds
Nitrites
Nitrates
NITRIFICATION

2 Ammonification

- **Ammonification** is when nitrogen compounds from **dead organisms** are turned into **ammonium compounds** by **decomposers** (see p. 34).
- Animal **waste** (**urine** and **faeces**) also contains nitrogen compounds. These are also turned into ammonium compounds by decomposers.

3 Nitrification

- **Nitrification** is when **ammonium compounds** in the soil are **changed** into **nitrogen compounds** that can then be **used** by plants.
- First **nitrifying bacteria** called ***Nitrosomonas*** change **ammonium compounds** into **nitrites**.
- Then other nitrifying bacteria called ***Nitrobacter*** change **nitrites** into **nitrates**.

4 Denitrification

- **Denitrification** is when nitrates in the soil are **converted** into **nitrogen gas** by **denitrifying bacteria** — they use nitrates in the soil to carry out **respiration** and produce nitrogen gas.
- This happens under **anaerobic conditions** (where there's **no** oxygen), e.g. in **waterlogged** soils.

Parts of the nitrogen cycle can also be carried out **artificially** and on an **industrial** scale. The **Haber process** produces **ammonia** from **atmospheric nitrogen** — it's used to make things like **fertilisers**.

Nitrogen Fertilisers can Leach into Water and Cause Eutrophication

AQA only

Leaching is when **water-soluble** compounds in the soil are **washed away**, e.g. by rain or irrigation systems. They're often **washed** into nearby **ponds** and **rivers**. If **nitrogen fertiliser** is leached into waterways (e.g. when **too much** is applied to a field) it can cause **eutrophication**:

1) **Nitrates leached** from fertilised fields stimulate the **growth** of **algae** in ponds and rivers.
2) Large amounts of algae **block light** from reaching the plants below.
3) Eventually the **plants die** because they're **unable** to photosynthesise enough.
4) **Bacteria** feed on the dead plant matter.
5) The **increased** numbers of **bacteria reduce** the **oxygen** concentration in the water by carrying out **aerobic respiration**.
6) **Fish** and other aquatic organisms **die** because there **isn't enough dissolved oxygen**.

Hey, who turned out the lights?

The Nitrogen Cycle and Eutrophication

If you're doing OCR you can skip straight to the questions — the rest is for AQA only.

*You Need to be able to **Analyse, Interpret** and **Evaluate Data** on **Eutrophication***

You've got to know how to **analyse data**, so here's an example of the kind of thing you might get in your exam:

A study was conducted to investigate the effect, on a nearby **river**, of adding **fertiliser** to **farmland**.

The **oxygen** and **algal** content of a river that runs past a field where **nitrate fertiliser** had been applied, was measured **at the field** and up to a distance of **180 m** away. A similar **control river** next to an **unfertilised** field was also studied. The results are shown in the graphs on the right.

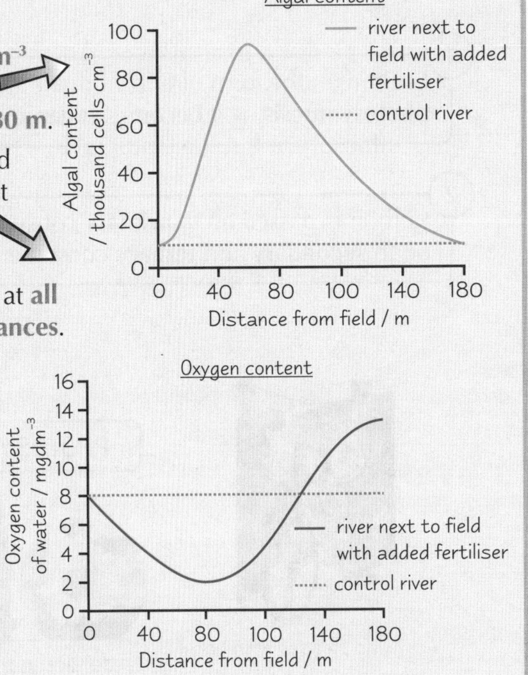

1) **Describe the data:**

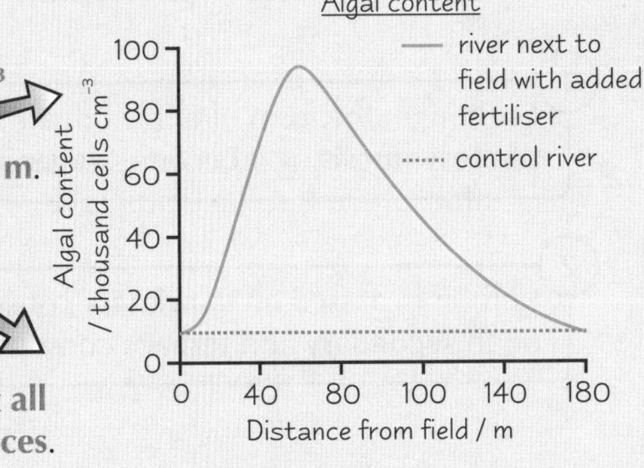

The **algal content** of the water **increases** sharply from **10 000 cells cm^{-3}** at the field to **95 000 cells cm^{-3}** at a distance of **60 m** from the field. Algal content then **decreases** beyond **60 m** to **10 000 cells cm^{-3}** at **180 m.**

The **oxygen content** of the water **decreases** from **8 mgdm^{-3}** at the field to **2 mgdm^{-3}** at a distance of **80 m** from the field. The oxygen content then **increases** beyond **80 m** up to **13 mgdm^{-3}** at 180 m, where it begins to **level off.**

The control river showed a **steady algal content** of **10 000 cells cm^{-3}** at **all distances**, as well as a **steady oxygen content** of **8 mgdm^{-3}** at **all distances.**

2) **Draw a conclusion:**

There's a **negative correlation** between the algal content and the oxygen content of the water — as the algal content **increases**, the oxygen content **decreases**, and vice versa.

3) **Evaluate the methodology:**

A control river was used which helps to control the effect of **some variables**, e.g. water temperature. But it doesn't remove the effect of **all** variables, e.g. **different organisms** may live in the control river, which could **affect** the algal or oxygen content.

The experiment only looked at **two rivers**, which means the **sample size** was **small**. Studying other rivers may have produced **different results** and a **different conclusion. More experiments** and results would be needed to make the data **more reliable.**

4) **Suggest an explanation for your conclusion:**

The results **suggest leaching** of the fertiliser and **eutrophication** have occurred. **Nitrate fertilisers** from the field could have **leached** into the river and caused the algal content of the river to **increase** by **stimulating** algal growth. The increased algal content could have **prevented light** from reaching plants **below**, causing them to die and be decomposed by **bacteria**. The bacteria **use up** the oxygen in the river when carrying out **aerobic respiration**, resulting in **decreased** dissolved oxygen levels.

Practice Questions

Q1 What is nitrification?

Q2 What is leaching?

Q3 Briefly describe eutrophication.

Exam Question

Q1 The diagram on the right shows the nitrogen cycle.

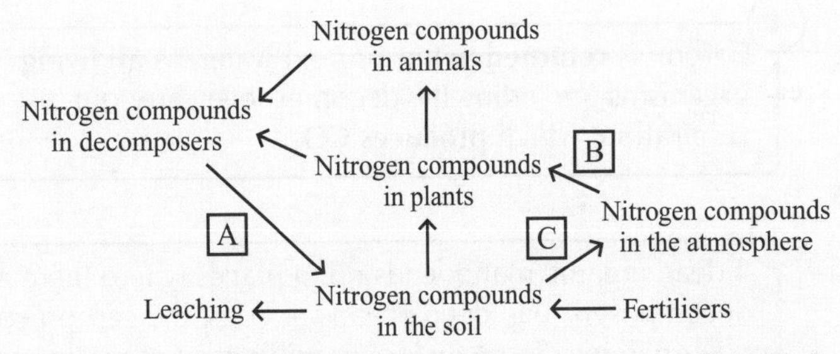

a) Name the processes labelled A and C in the diagram. [2 marks]

b) Name and describe process B in detail. [3 marks]

Nitrogen fixation — cheaper than a shoe fixation...

The nitrogen cycle's not as bad as it seems — divide up the four processes of nitrogen fixation, ammonification, nitrification and denitrification and learn them separately. Then before you know it, you'll have learnt the whole cycle. Easy peesy.

The Carbon Cycle

These pages are for AQA Unit 4 only.

Carbon molecules are found in plants, animals, your petrol tank and on your burnt toast. They get cycled round and the concentration of them in the atmosphere fluctuates up and down. They also do the hokey-cokey and they turn around. That's what they're all about.

*The **Carbon Cycle** shows how **Carbon** is **Passed On** and **Recycled***

All organisms need carbon to make **essential compounds**, e.g. plants use CO_2 in photosynthesis to make glucose. The **carbon cycle** is how carbon **moves** through **living organisms** and the **non-living environment**. It involves four processes — **photosynthesis**, **respiration**, **decomposition** and **combustion**:

1. **Carbon** (in the form of CO_2 from **air** and **water**) is **absorbed** by plants when they carry out **photosynthesis** — it becomes **organic compounds** (compounds that contain carbon) in **plant tissues**.

2. Carbon is **passed on** to **primary consumers** when they **eat** the plants. It's passed on to **secondary** and **tertiary consumers** when they eat other consumers.

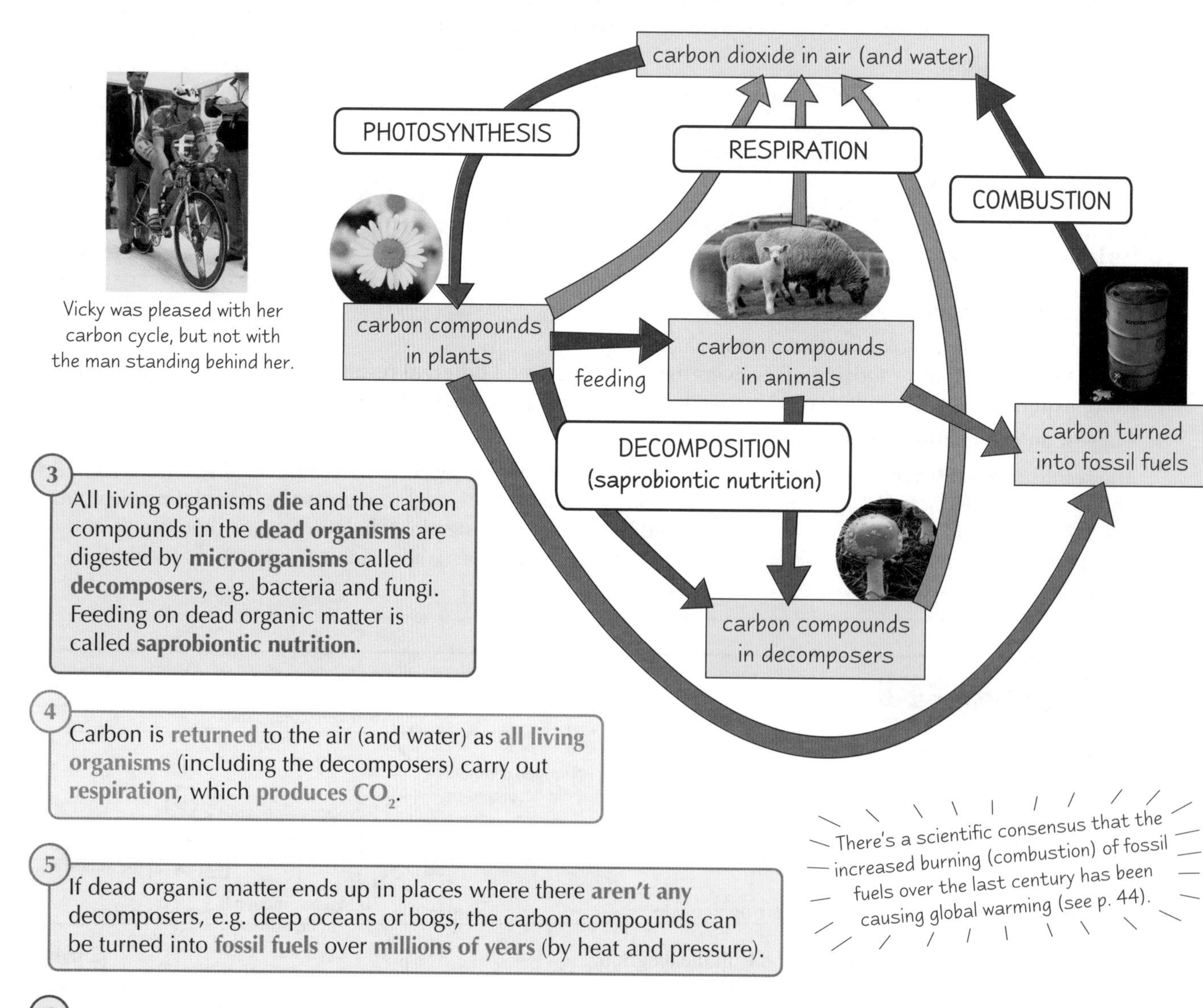

Vicky was pleased with her carbon cycle, but not with the man standing behind her.

3. All living organisms **die** and the carbon compounds in the **dead organisms** are digested by **microorganisms** called **decomposers**, e.g. bacteria and fungi. Feeding on dead organic matter is called **saprobiontic nutrition**.

4. Carbon is **returned** to the air (and water) as **all living organisms** (including the decomposers) carry out **respiration**, which **produces CO_2**.

5. If dead organic matter ends up in places where there **aren't any** decomposers, e.g. deep oceans or bogs, the carbon compounds can be turned into **fossil fuels** over **millions of years** (by heat and pressure).

There's a scientific consensus that the increased burning (combustion) of fossil fuels over the last century has been causing global warming (see p. 44).

6. The carbon in fossil fuels is **released** when they're **burnt** — this is called **combustion**.

The Carbon Cycle

Respiration and *Photosynthesis* Cause Fluctuations in CO_2 Concentration

Respiration (which is carried out by **all** organisms) **adds** CO_2 to the atmosphere. **Photosynthesis removes** CO_2 from the atmosphere. The **amount** of respiration and photosynthesis going on **varies** on a **daily** and a **yearly** basis, so the amount of **atmospheric CO_2 changes**.

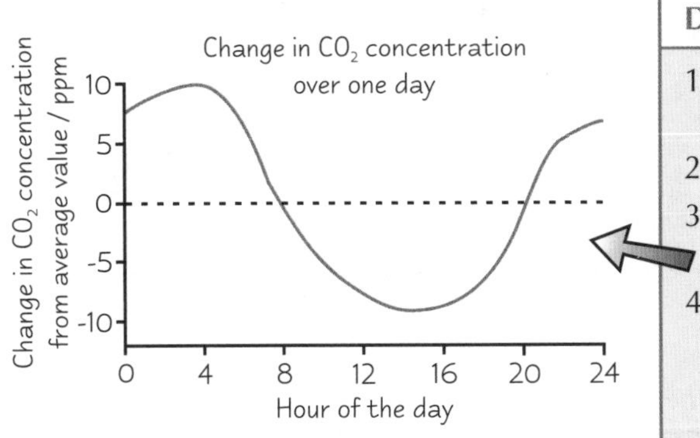

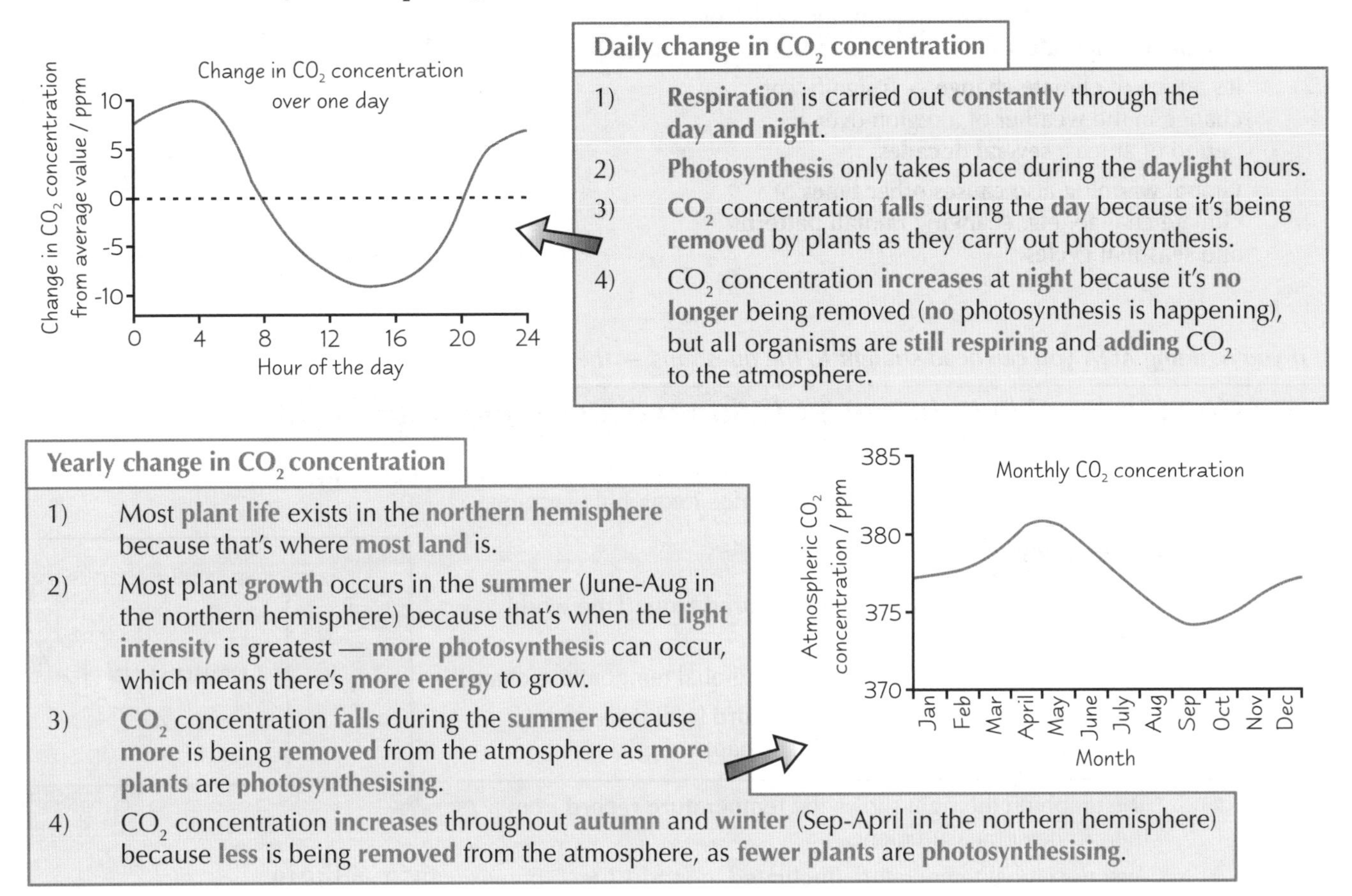

Daily change in CO_2 concentration

1) **Respiration** is carried out **constantly** through the **day and night**.
2) **Photosynthesis** only takes place during the **daylight** hours.
3) CO_2 concentration **falls** during the **day** because it's being **removed** by plants as they carry out photosynthesis.
4) CO_2 concentration **increases** at **night** because it's **no longer** being removed (**no** photosynthesis is happening), but all organisms are **still respiring** and **adding** CO_2 to the atmosphere.

Yearly change in CO_2 concentration

1) Most **plant life** exists in the **northern hemisphere** because that's where **most land** is.
2) Most plant **growth** occurs in the **summer** (June-Aug in the northern hemisphere) because that's when the **light intensity** is greatest — **more photosynthesis** can occur, which means there's **more energy** to grow.
3) CO_2 concentration **falls** during the **summer** because **more** is being **removed** from the atmosphere as **more plants** are **photosynthesising**.
4) CO_2 concentration **increases** throughout **autumn** and **winter** (Sep-April in the northern hemisphere) because **less** is being **removed** from the atmosphere, as **fewer plants** are **photosynthesising**.

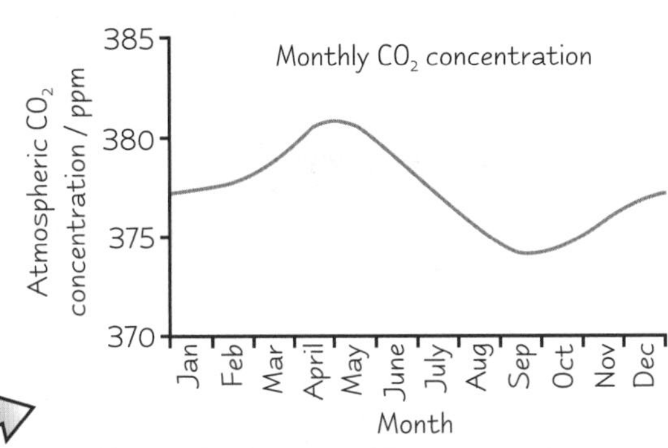

Practice Questions

Q1 How does respiration vary throughout the day?

Q2 Why does the atmospheric CO_2 concentration decrease during daylight hours?

Q3 Why does the atmospheric CO_2 concentration increase throughout autumn and winter in the northern hemisphere?

Exam Question

Q1 The diagram on the right shows the carbon cycle.

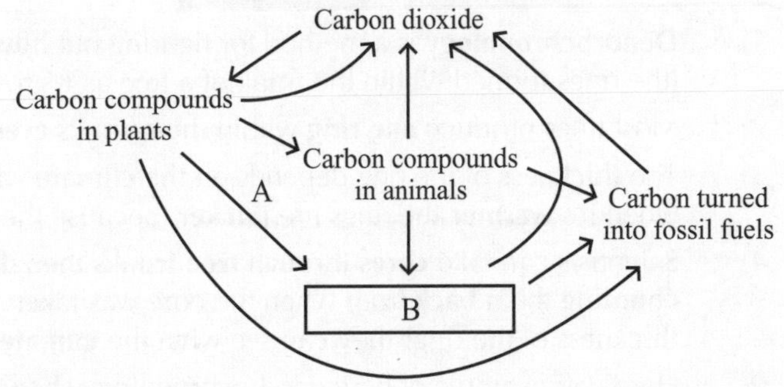

a) Name A and B in the diagram. [2 marks]

b) Use the diagram to help you describe how carbon is cycled through living organisms and the non-living environment. [6 marks]

Daily pattern of my concentration — low during lessons and revision...

The carbon cycle might look a bit messy, but it isn't as complicated as it looks. Just like the nitrogen cycle, it helps to break it down into four processes — photosynthesis, respiration, combustion and decomposition — and learn them separately. Then there's just the slight complication of the fluctuating level of CO_2 in the atmosphere to worry about.

Introduction to Global Warming

These pages are for AQA Unit 4 and Edexcel Unit 4.

A2 level student, meet global warming — take a few pages to get to know each other...

Global Warming is the Recent Rise in Global Temperature

1) **Global warming** is the term used for the **rapid increase** in **global temperature** over the **last century**.
2) It's a type of **climate change** — a significant change in the **weather** of a region over a period of at least **several decades**.
3) Global warming also **causes other types** of climate change, e.g. changing **rainfall patterns** and **seasonal cycles**.

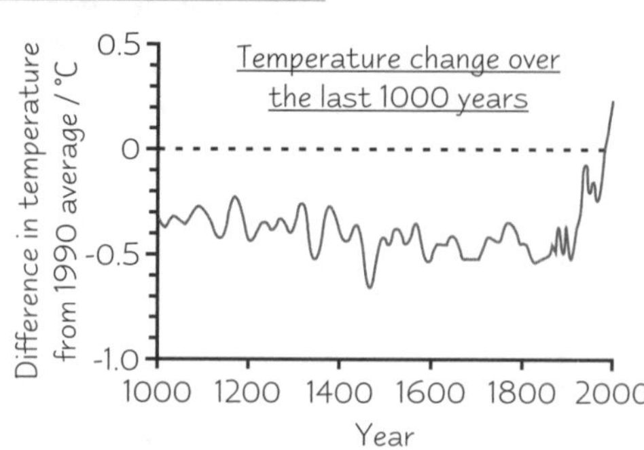

If you're doing AQA you can head straight to the questions — the rest is for Edexcel.

You Need to be able to Interpret Evidence for Global Warming

There are **different types** of **evidence** that can be used to show that global warming **is happening**. You need to be able to interpret three types of evidence for global warming:

1 Temperature Records

1) Since the 1850s **temperature** has been **measured** around the world using **thermometers**.
2) This gives a **reliable** but **short-term** record of global temperature change.
3) Here's an example of how a **temperature record** from thermometer measurements **shows** that global warming **is happening**:

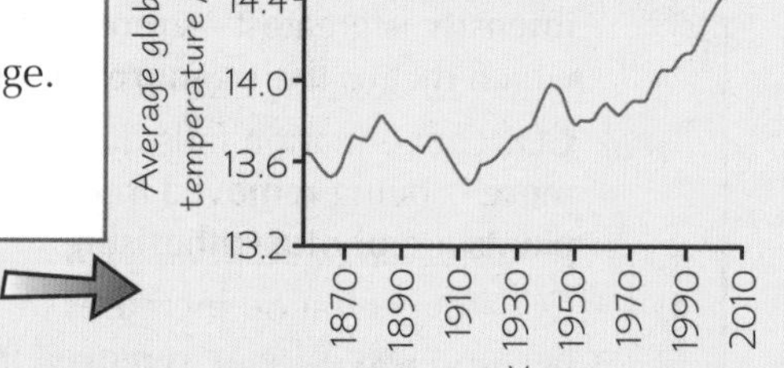

> 1) The graph on the right shows the **temperature record** from thermometer measurements.
> 2) Average global temperature **fluctuated** around **13.6 °C** between **1850** and **1910**.
> 3) It has **steadily increased** (with a couple of fluctuations) from **13.6 °C** in **1910** to around **14.4 °C** today.
> 4) The **general trend** of **increasing** global temperature over the last century (since 1910) is **evidence** for **global warming**.

2 Dendrochronology (Tree rings)

1) **Dendrochronology** is a method for figuring out **how old** a **tree** is using **tree rings** (the rings formed within the trunk of a tree as it grows).
2) Most trees produce **one ring** within their trunks **every year**.
3) The **thickness** of the ring depends on the **climate** when the ring was formed — when it's **warmer** the rings are **thicker** (because the conditions for growth are better).
4) Scientists can take **cores** through **tree trunks** then **date** each ring by **counting** them **back** from when the core was taken. By looking at the **thickness** of the rings they can see what the **climate** was like **each year**.
5) Here's an example of how dendrochronology **shows** that global warming **is happening**:

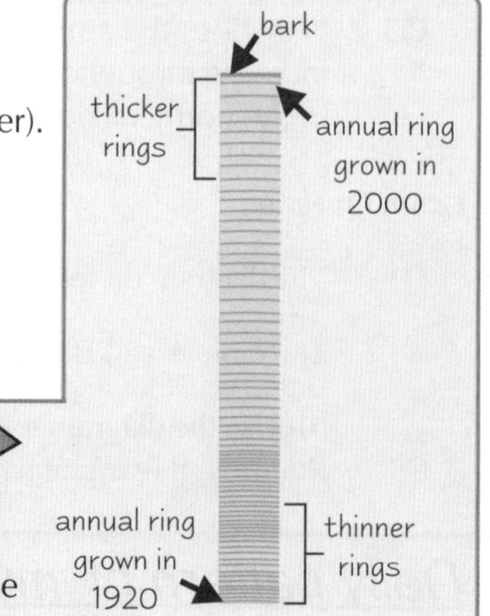

> 1) The diagram on the right shows a **core** taken from a **tree** in 2000.
> 2) The **most recent** rings are the **thickest** and the rings get **steadily thinner** the further in the **past** they were formed.
> 3) The **trend** of increasingly thicker rings from **1920** to **2000** suggests that the climate where the tree lived had become **warmer** over the **last century**.

Introduction to Global Warming

3) Pollen in Peat Bogs

Pollen in peat bogs can be used to show how **temperature** has **changed** over **thousands** of years. Here's how it works:

1) **Pollen** is often **preserved** in **peat bogs** (acidic wetland areas).
2) Peat bogs accumulate in **layers** so the **age** of the preserved **pollen increases** with **depth**.
3) Scientists can take **cores** from peat bogs and extract **pollen grains** from the different aged layers. They then **identify** the **plant species** the pollen came from.
4) Only **fully grown** (mature) plant species **produce pollen**, so the samples only show the species that were **successful** at that time.
5) Scientists know the **climates** that different plant species live in **now**. When they find preserved pollen from **similar plants**, it indicates that the **climate** was **similar** when that pollen was **produced**.
6) Because plant species **vary** with **climate** the preserved pollen will **vary** as climate **changes** over time.
7) So a gradual **increase** in **pollen** from a plant species that's **more successful** in **warmer climates** would show a **rise** in **temperature** (a decrease in pollen from a plant that needs cold conditions would show the same thing).
8) Here's an example of how pollen in peat bogs can provide **evidence** for global warming events in the **past**:

Depth of sample (metres)	Approximate age of sample (years)	Number of pollen grains in sample	
		Oak	Fir
0.5	3100	253	28
1.0	4200	194	121
1.5	5700	138	167
2.0	7100	51	231

1) The table shows data on **samples** of **pollen** taken from a **core** of a **peat bog**.
2) Between **7100** and **3100 years ago** the number of **oak tree** pollen grains **increased** from **51** grains to **253** grains.
3) This suggests that the **climate** in the area had become **better** for **oak trees** — **more** oak trees **reached maturity** and **produced pollen.**
4) Between **7100** and **3100** years ago the number of **fir tree** pollen grains in the sample **decreased** from **231** grains to **28** grains.
5) This suggests that the **climate** in the area had become **worse** for **fir trees** — **fewer** fir trees **reached maturity** and **produced pollen.**
6) Today, **oak trees** are mainly found in **temperate** (mild) regions, and **fir trees** are mainly found in **cooler** regions.
7) This suggests that the **temperature** around the peat bog **increased** over this time period — a **warming event** had occurred.

Practice Questions

Q1 Define global warming.

Q2 Explain why only pollen from successful plant species is preserved in peat bogs.

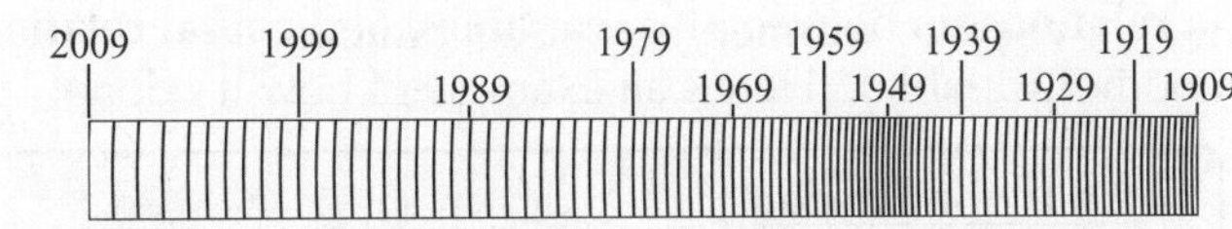

Exam Questions

Q1 The diagram on the right shows a core taken from a pine tree in 2009.
Describe what the core is showing, and explain how this provides evidence for global warming. [6 marks]

Q2 How can the pollen of present-day species be used to show what the climate was like in the past? [2 marks]

I'm actually a dendrochronologist by trade — oh, you've fallen asleep...

A lot of people get global warming confused with climate change — make sure you know that global warming's just the rapid increase in global temperature over the last century. I bet you'd already thought of using a thermometer to show that global warming is happening — using tree rings and pollen isn't as obvious, but they're important sources of evidence.

Causes of Global Warming

***These pages are for** AQA Unit 4 **and** Edexcel Unit 4.*

Now you know what it is, it's time to find out what causes it...

Global Warming is Caused by Human Activity

1) The **scientific consensus** is that the recent increase in global temperature (global warming) is **caused** by **human activity**.
2) Human activity has caused global warming by **enhancing** the **greenhouse effect** — the effect of greenhouse gases absorbing outgoing **energy**, so that less is **lost** to space.
3) The greenhouse effect is **essential** to keep the planet warm, but **too much** greenhouse gas in the atmosphere means the planet **warms up**.
4) **Two** of the main greenhouse gases are **CO_2** and **methane**:

Carbon dioxide (CO_2)

- **Atmospheric CO_2** concentration has **increased rapidly** since the **mid-19th century** from **280 ppm** (parts per million) to nearly **380 ppm**. The concentration had been **stable** for the previous **10 000 years**.
- CO_2 concentration is **increasing** as more **fossil fuels** like coal, oil, natural gas and petrol are **burnt**, e.g. in power stations or in cars. Burning fossil fuels **releases CO_2**.
- CO_2 concentration is also **increased** by the **destruction** of **natural sinks** (things that keep CO_2 **out** of the atmosphere by storing **carbon**). E.g. trees are a big CO_2 sink — they store the carbon as **organic compounds**. CO_2 is **released** when trees are **burnt**, or when **decomposers break down** the organic compounds and **respire** them.

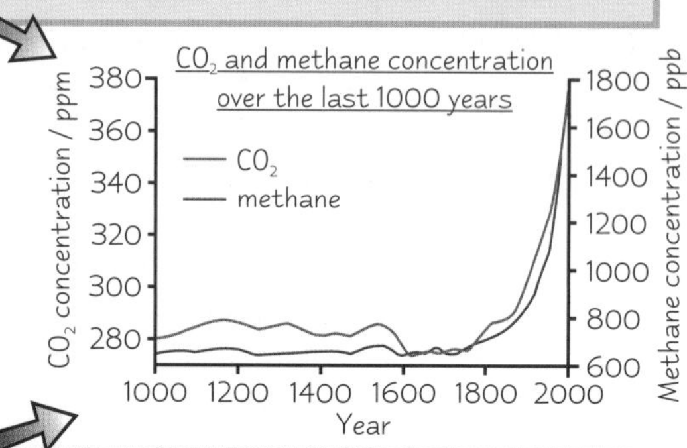

Methane (CH_4)

- **Atmospheric methane** concentration has **increased rapidly** since the **mid-19th century** from **700 ppb** (parts per billion) to **1700 ppb** in **2000**. The level had been **stable** for the previous **850 years**.
- Methane concentration is **increasing** because **more** methane is being **released** into the atmosphere, e.g. because **more fossil fuels** are being **extracted**, there's more **decaying waste** and there are **more cattle** which give off methane as a **waste gas**.
- Methane can also be released from **natural stores**, e.g. **frozen ground** (permafrost). As temperatures **increase** it's thought these stores will **thaw** and release **large amounts** of methane into the atmosphere.

An increase in **human activities** like **burning fossil fuels** (for industry and in cars), **farming** and **deforestation** has **increased** atmospheric concentrations of CO_2 and methane. This has **enhanced** the greenhouse effect and **caused** a rise in average global temperature — **global warming**.

***If you're doing** AQA **it's time to do the questions — the rest is for** Edexcel.*

You Need to be able to Interpret Evidence for the Causes of Global Warming

You need to be able to **interpret data** on atmospheric CO_2 concentration and temperature, and recognise **correlations** (a **relationship** between two variables) and **causal relationships** (where a change in one variable **causes** a change in another variable). Here's an example of how it's done:

Temperature and atmospheric CO_2 concentration

1) **Describe the data:**

   The **temperature fluctuated** between **1958** and **2008**, but the general trend was a **steady increase** from around **13.9 °C** to around **14.4 °C**. The **atmospheric CO_2 concentration** also showed a trend of **increasing** from around **315 ppm** in **1958** to around **385 ppm** in **2008**.

2) **Draw a conclusion:**

   There's a **positive correlation** between the temperature and CO_2 concentration. The increasing **CO_2 concentration** could be **linked** to the increasing **temperature**. However, you **can't conclude** from this data that it's a **causal relationship** — **other factors** may have been involved, e.g. changing solar activity. **Other studies** would need to be carried out to **investigate** the effects of other factors.

Causes of Global Warming

Some People **Disagree** About Whether **Humans** are **Causing Global Warming**

1) It's agreed that **global warming** is **happening** — there **has** been a rapid rise in global temperature over the past century.
2) It's also agreed that **human activity** is **increasing** the **atmospheric CO_2 concentration**.
3) The **scientific consensus** is that the **increase** in atmospheric CO_2 concentration **is causing** the **increase** in global temperature (i.e. humans are causing global warming).
4) But a **handful** of scientists have drawn a **different conclusion** from the **data** on atmospheric CO_2 concentration and temperature — they think that the **increase** in atmospheric CO_2 concentration **isn't** the **main cause** of the **increase** in global temperature.
5) The conclusions scientists reach can be affected by **how good** the **data** is that they're basing their conclusions on (i.e. how **reliable** it is), **how much evidence** there is for a certain theory, and also sometimes by **bias**.
6) Biased conclusions **aren't objective** — they've been **influenced** by an **opinion**, instead of being **purely** based on **scientific evidence**.
7) For example, the conclusions research scientists reach may be **biased towards** the goals of the **organisation funding** their work:
   - A scientist working for an **oil company** may be more likely to say humans **aren't** causing global warming — this would help to **keep oil sales high**.
   - A scientist working for a **renewable energy company** may be more likely to say humans **are** causing global warming — this would **increase sales** of energy produced from renewable sources, e.g. from wind turbines.

Trust me, humans definitely aren't causing global warming. Now, let's talk about where to build my new oil refinery.

Practice Questions

Q1 What is the greenhouse effect?

Q2 State one human activity that increases atmospheric carbon dioxide concentration.

Q3 State two human activities that increase atmospheric methane concentration.

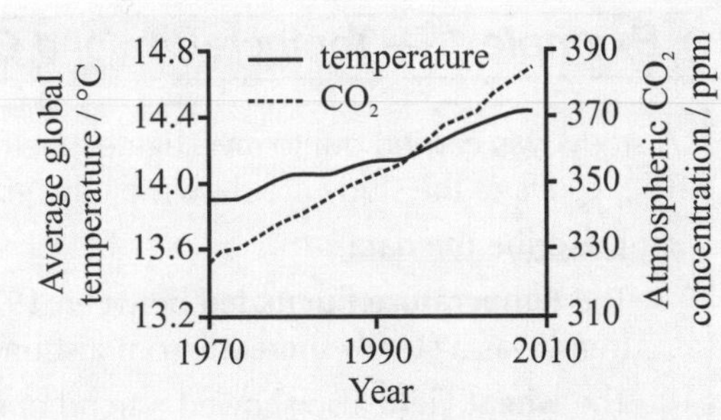

Exam Question

Q1 The graph shows the average global temperature and atmospheric CO_2 concentration from 1970 to 2008.

a) Describe the changes that the graph is showing. *[4 marks]*

b) Draw a conclusion about the relationship between atmospheric CO_2 concentration and temperature shown on the graph. *[2 marks]*

Earth not hot enough for you — spice it up with a dash of CO_2...

Another fine mess that humanity's gotten itself into — too much of things like driving, leaving TVs on standby, making big piles of rubbish and cows farting has caused global warming. I suspect you've come to the conclusion that revising global warming isn't the most fun in the world. However, I suspect that your conclusion is biased — it's so much fun...

Effects of Global Warming

These pages are for AQA Unit 4 only. If you're doing Edexcel jump to page 50.
Global warming might mean you can wear a bikini in Scotland, but it's bad news for some organisms...

Global Warming Could Affect all Organisms

Increasing **CO_2 concentration** is causing **global warming**, which is leading to other climate changes, e.g. different **rainfall patterns** and changes to **seasonal weather patterns**. All organisms could be **affected** by this, but **different organisms** could be affected in **different ways**:

Crop yield

The **increasing CO_2 concentration** that's **causing** global warming could **also** be **causing** an **increase** in **crop yields** (the **amount** of crops produced from an area). CO_2 concentration is a **limiting factor** for photosynthesis (see p. 10), so increasing global CO_2 concentration could mean crops grow **faster**, **increasing** crop yields.

Insect pests

1) Climate change may affect the **life cycle** of some insect species. For example, it's thought that increasing global temperature (**global warming**) means some insects go through their **larval stage** quicker and emerge as **adults earlier**, e.g. some butterflies may spend **10** fewer days as larvae for every **1 °C** rise in temperature.
2) Climate change may also affect the **numbers** of some insect species:
   - Some species are becoming **more** abundant, e.g. **warmer** and **wetter** summers in some places have led to an **increase** in the number of **mosquitoes**.
   - Other species may become **less** abundant, e.g. some **tropical** insect species can only thrive in **specific temperature ranges**, so if it gets **too hot** fewer insects may be able to **reproduce successfully**.

Wild animals and plants

1) Climate change could affect the **distribution** of many wild **animal** and **plant** species:
   - Some species may become **more** widely distributed, e.g. species that need **warmer temperatures** may spread **further** as the conditions they **thrive** in exist over a **wider** area.
   - Other species may become **less** widely distributed, e.g. species that need **cooler temperatures** may have **smaller** ranges as the conditions they **thrive** in exist over a **smaller** area.
2) Climate change could also affect the **number** of wild animals and plants:
   - Some species are becoming **more** abundant, e.g. **boarfish** are increasing in number in parts of the Atlantic Ocean where sea temperature is **rising**.
   - Other species are becoming **less** abundant, e.g. **polar bears** need frozen sea ice to hunt and **global warming** is causing more sea ice to **melt**. It's thought that the number of polar bears is **decreasing** because there isn't enough sea ice for them to hunt on.

You Need to be able to Analyse Data on the Effects of Global Warming

Analysing data's pretty important when looking at the **effects** of global warming. Here are a few examples:

1 Example 1 — Temperature and Crop Yield

A study was carried out to investigate whether **rising growing season temperature** is affecting **crop yields**. The results of the study are shown on the graph. You might be asked to:

1) **Describe the data:**

The **temperature fluctuated** between **1970** and **2000**, but the general trend was a **steady increase** from just under **17 °C** to just under **18 °C**.

The **wheat yield** also showed a trend of **increasing** from around **1.6 tons** per hectare in **1970** to around **2.7 tons** per hectare in **2000**.

2) **Draw a conclusion:**

The graph shows a **positive correlation** between **temperature** and **wheat yield**. The increasing growing season temperature could be **linked** to the increasing wheat yields.

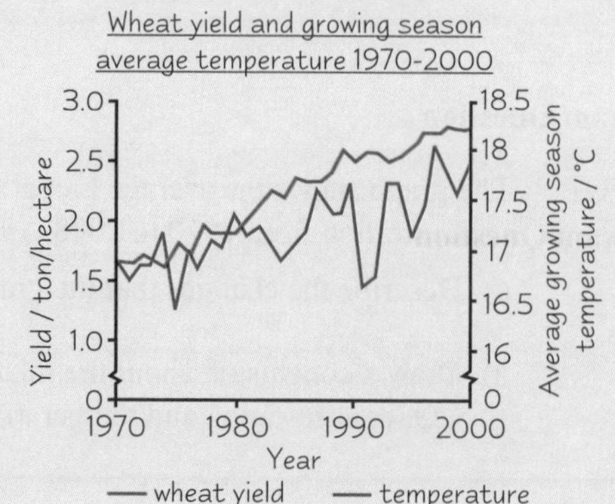

Even though the graph shows a **correlation**, you can't conclude that the increase in temperature **caused** the increase in wheat yield — there could have been **other factors** involved. This study actually found that the rising growing season temperature had a **negative effect** on wheat yields, but **improvements in technology** during the **same period** meant that crop yields **increased overall**.

Effects of Global Warming

2 Example 2 — Temperature and Insect Numbers

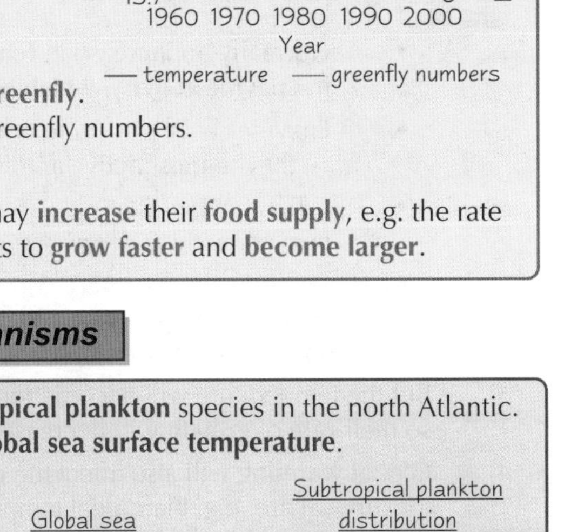

A study counted the **number** of **greenfly** in an area from 1960 to 2000. A separate study collected data on **global temperature** at the same time. The results are shown on the graph. You might be asked to:

Temperature and number of greenfly caught
Average global temperature / °C
14.5
14.3
14.1
13.9
13.7
Number of greenfly caught
600
500
400
300
200
100
0
1960 1970 1980 1990 2000
Year
— temperature — greenfly numbers

1) **Describe the data:**

   The **temperature fluctuated** between **1960** and **2000**, but the general trend was a **steady increase** from just over **13.8 °C** to just over **14.4 °C**.

   The **number of greenfly** also **fluctuated** with a generally **increasing** trend from around **110** in **1960** to just around **480** in **2000**.

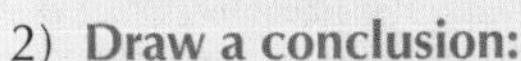

2) **Draw a conclusion:**

   There's a **positive correlation** between **temperature** and **numbers of greenfly**. The increasing global temperature could be **linked to** the increasing greenfly numbers.

3) **Suggest an explanation for your conclusion:**

   Greenfly numbers **could** be increasing because higher temperatures may **increase** their **food supply**, e.g. the rate of **photosynthesis** may **increase** at higher temperatures, allowing plants to **grow faster** and **become larger**.

3 Example 3 — Temperature and the Distribution of Organisms

A study was carried out to investigate the changing **distribution** of **subtropical plankton** species in the north Atlantic. The results are shown below, along with data that's been collected on **global sea surface temperature**. You might be asked to:

1) **Describe the data:**

   Sea surface temperature fluctuated around the average between **1950** and **1978**, then there was a **steady increase** between 1978 and 2000, up to just over **0.3 °C** greater than the average.

   Subtropical plankton species were found in the sea **south of the UK** in 1958-1981. By 2000-2002 their distribution had moved **further north** along the west coast of the UK and Ireland to the **Arctic Ocean**.

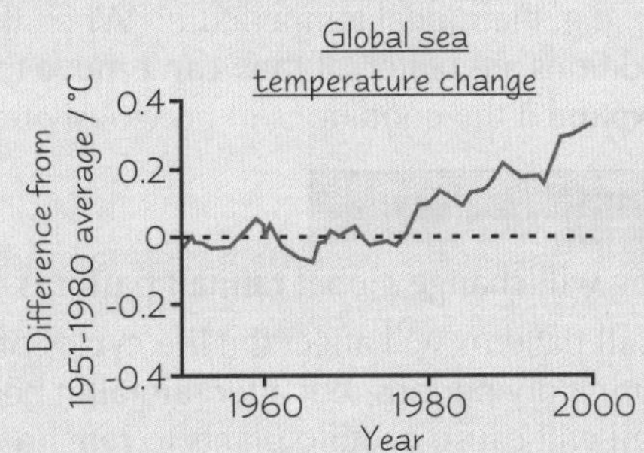

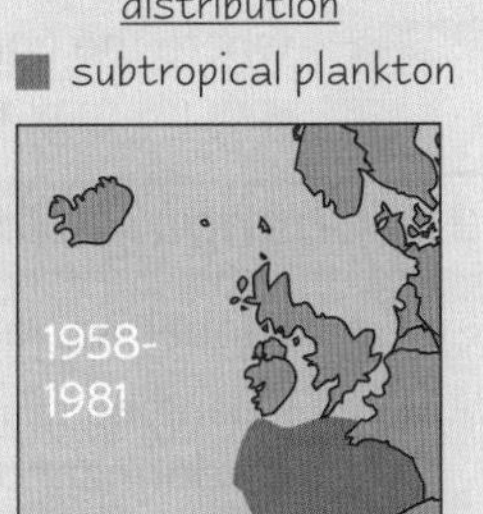

2) **Draw a conclusion:**

   There's a link between **rising global sea surface temperature** and the **northward** change in **distribution** of subtropical plankton.

   The data shows a **link**, but you can't say that the increase in temperature **caused** the change in distribution — there could have been **other factors** involved, e.g. **overfishing** could have removed plankton **predator species**.

Practice Questions

Q1 Give one way that climate change is affecting populations of insect pests.

Q2 Give one way that climate change is affecting wild animal species.

Exam Question

Q1 The graph on the right shows CO_2 concentration and corn yield.

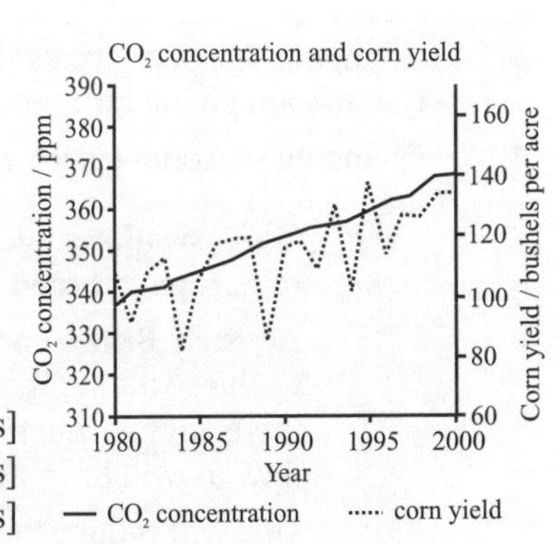

a) Describe what the graph is showing. [4 marks]

b) Draw a conclusion. [1 marks]

c) Use your knowledge to suggest an explanation for your conclusion. [2 marks]

Global warming effects — not as much fun as special effects...

Boy, that wasn't fun, but the business of analysing data is an important one if you want to profit in your exam. There could be lots of questions on data, so have a good read through these examples, and never mix up correlation and cause.

Effects of Global Warming

These pages are for Edexcel Unit 4 only. If you're doing AQA, you've finished global warming — put your feet up.

The world's getting hotter, we know that much — but global warming has a couple of other tricks up its sleeve...

Global Warming Has **Different Effects**

Global warming will **directly affect plants** and **animals**. It will also change **global rainfall patterns** and the **timing of seasonal cycles**, which will also affect plants and animals:

1 Rising Temperature

1) An **increase** in **temperature** will affect the **metabolism** of **all** organisms:

- Normally an **increase** in **temperature** causes an **increase** in **enzyme activity**, which **speeds up** metabolic reactions.

- Enzymes have a specific **optimum temperature** — they're **most active** at this temperature.

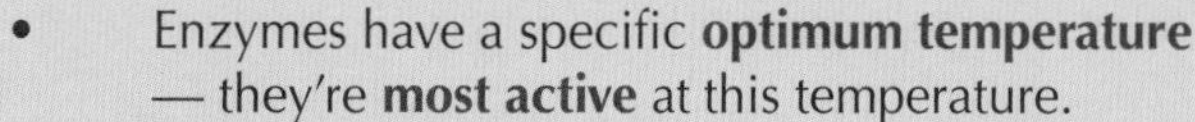

- When temperature increases **above** the optimum temperature enzyme activity **decreases**, which **slows down** metabolic reactions.

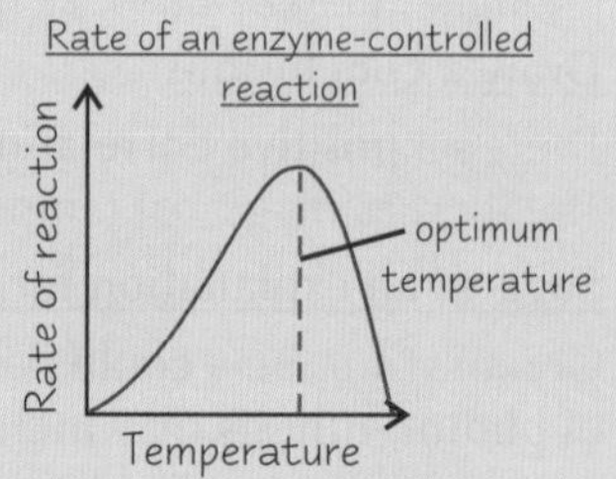

2) So an **increase** in **temperature** will mean the metabolic reactions in some organisms will **speed up**, so their **rate** of **growth** will **increase**. This also means they'll **progress** through their **life cycle faster**.

3) But the temperature will become **too high** for some organisms. Their metabolic reactions will **slow down**, so their **rate** of **growth** will **decrease**. This also means they'll **progress** through their **life cycle slower**.

4) Global warming will also affect the **distribution** of some species — all species exist where their **ideal conditions** for survival are, e.g. their ideal temperature. When these conditions **change**, they'll have to **move** to a **new area** where the conditions are better. If they **can't move** they may **die out** in that area. Also, the **range** of some species may **expand** if the conditions in previously uninhabitable areas change.

2 Changing Rainfall Patterns

1) Global warming will **change** global **rainfall patterns** — some areas will get **more rain**, others will get **less rain**.

2) Changing rainfall patterns will affect the **life cycles** of some organisms, e.g. ocotillo is a desert plant — it's dormant during dry periods, but after rainfall it becomes active and grows new leaves. Reduced rainfall will cause ocotillo plants to remain dormant for longer periods.

3) Changing rainfall patterns will also affect the **distribution** of some species, e.g. deserts could increase in area because of decreases in rainfall — species that aren't adapted to live in deserts will have to move to new areas or they'll die out.

3 Seasonal Cycles

1) Global warming is thought to be changing the **timing of the seasons**, e.g. when winter changes to spring.

2) Organisms are **adapted** to the timing of the seasons and the **changes** that happen, e.g. changes in temperature, rainfall and the availability of food.

3) Changing seasonal cycles will affect the **life cycles** of some organisms, e.g. some red squirrels in Canada are giving birth nearly three weeks earlier than usual because of an earlier availability of food.

4) Changing seasonal cycles will also affect the **distribution** of some species, for example:

> 1) Some **swallows** live in **South Africa** over the **winter** and fly to different parts of **Europe** to **breed** at the start of **spring** (when more food is available).
>
> 2) An **early British spring** will produce **flowers** and **insects** earlier than usual, so the swallows that migrate to Britain at the normal time will **arrive** when there **isn't** as much **food available** (there'll be **fewer insects** because the flowers will have **disappeared** earlier).
>
> 3) This will **reduce** the number of swallows that are born in Britain, and could **eventually** mean that the population of **swallows** that migrate to Britain will **die out**. The **distribution** of swallows in Europe will have **changed**.

I told you we should've come back earlier.

Effects of Global Warming

You Need to Know *How to Investigate* the Effect of *Temperature* on *Organisms*

Global warming will affect the **development** of **plants** and **animals**. You need to know how to **investigate** the effect of temperature on **seedling growth rate** and **brine shrimp hatch rate**:

1) Seedling Growth Rate

1) **Plant** some seedlings in **soil trays** and **measure** the **height** of each seedling.
2) Put the trays in **incubators** at **different temperatures**.
3) Make sure **all other variables** (e.g. the water content of the soil, light intensity and CO_2 concentration) are the **same** for **each tray**.
4) After a period of incubation record the **change in height** of each seedling. The **average growth rate** in each tray can be calculated in the following way:

$$\frac{\textbf{average change in seedling height in each tray}}{\textbf{incubation period}}$$

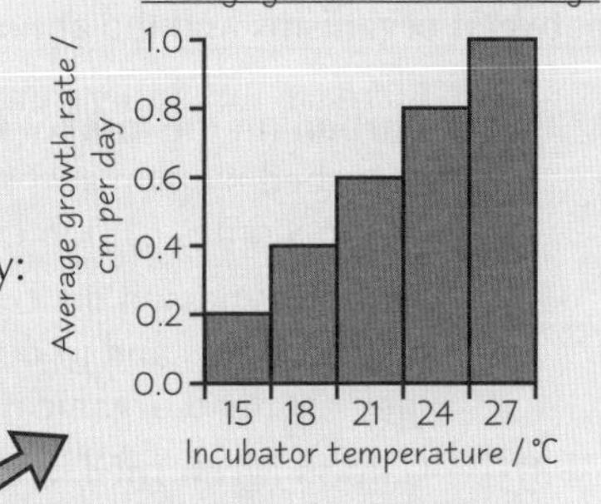

5) For example, the **graph on the right** shows that as **temperature increases**, seedling **growth rate increases** — from **0.2 cm per day** at **15 °C** to **1.0 cm per day** at **27 °C**. You can **conclude** that **higher temperatures** cause **faster growth rates** (it's a **causal relationship**) because **all** other variables were **controlled**.

2) Brine Shrimp Hatch Rate

Brine shrimp are also known as Sea-Monkeys®.

1) Put an **equal number** of brine shrimp eggs in **water baths** set at **different temperatures**.
2) Make sure **all other variables** (e.g. the volume of water, the salinity of the water and O_2 concentration) are the **same** for **each water bath**.
3) The **number** of **hatched brine shrimp** in each water bath are recorded every five hours. The **hatch rate** in each water bath can be calculated in the following way:

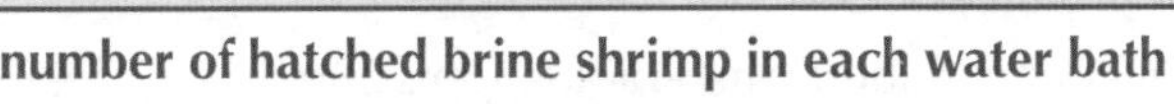

$$\frac{\textbf{number of hatched brine shrimp in each water bath}}{\textbf{number of hours}}$$

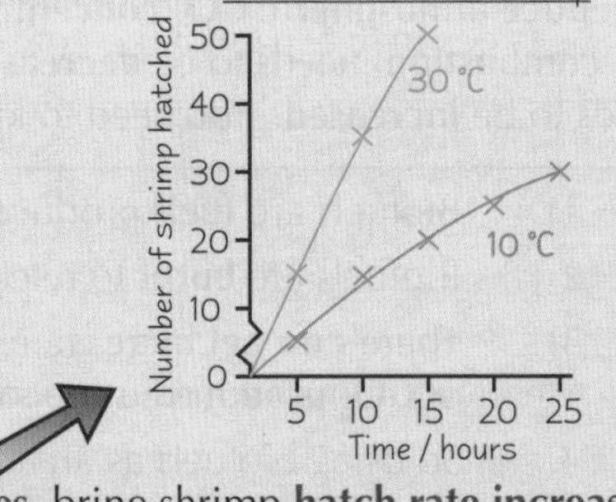

4) For example, the **graph on the right** shows that as **temperature increases**, brine shrimp **hatch rate increases**, e.g. at **30 °C** the initial hatch rate is **3 per hour** and at **10 °C** it's **1 per hour**. You can **conclude** that **higher temperatures** cause **faster hatch rates** (it's a **causal relationship**) because **all** other variables were **controlled**.

Practice Questions

Q1 Give one example of how changing rainfall patterns could affect the distribution of a plant or an animal.

Q2 Give one example of how the timing of the seasonal cycles could affect the life cycle of a plant or an animal.

Exam Questions

Q1 A potato tuber moth completes its life cycle faster at 21 °C than at 16 °C.
Explain why this is the case. [4 marks]

Q2 Describe how a student could investigate the effect of global warming on seedling growth rate. [5 marks]

I know what you're thinking — why do I need to know about brine shrimp...

Higher temperatures make a weekend away at an English coastal town sound more appealing, but I doubt many plants or animals will appreciate it too much. An earlier spring sounds good too — Easter eggs in February, anyone? That's if the Easter Bunny hasn't died of starvation. Thanks a lot global warming... oops, it's got all morbid and serious.

Reducing Global Warming

These pages are for Edexcel Unit 4 only.

Since global warming will have some pretty dire consequences, some humans are having a pop at reducing it.

There are Different Ways to **Reduce Atmospheric CO_2 Concentration**

Increasing atmospheric CO_2 concentration is one of the **causes** of global warming (see p. 46). Scientists need to know how **carbon compounds** are **recycled** between **organisms** and the **atmosphere** so they can come up with ways to **reduce atmospheric CO_2 concentration**. The **movement** of carbon **between organisms** and the **atmosphere** is called the **carbon cycle**:

1) **Carbon** (in the form of **CO_2** from the **atmosphere**) is **absorbed** by plants when they carry out **photosynthesis** — it becomes carbon compounds in **plant tissues**.
2) Carbon is **passed on** to **animals** when they **eat** the plants and to **decomposers** when they eat **dead organic matter**.
3) Carbon is **returned** to the atmosphere as **all living organisms** carry out **respiration**, which **produces CO_2**.
4) If dead organic matter ends up in places where there **aren't any decomposers**, e.g. deep oceans or bogs, the carbon compounds can be turned into **fossil fuels** over **millions of years** (by heat and pressure).
5) The carbon in fossil fuels is **released** as **CO_2** when they're **burnt** — this is called **combustion**.

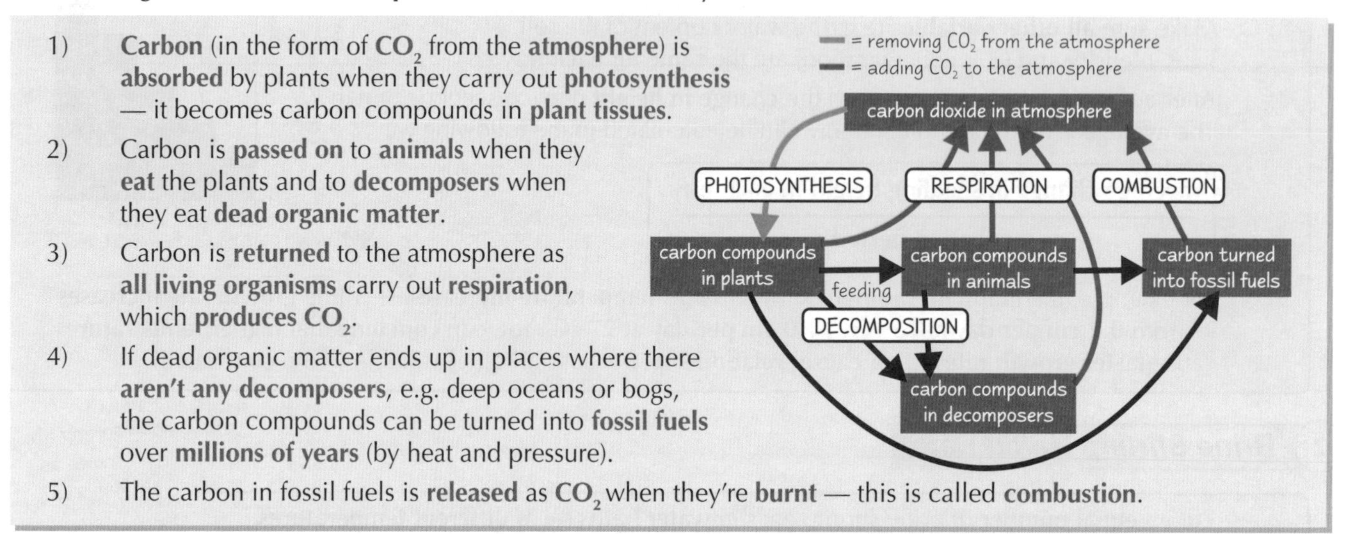

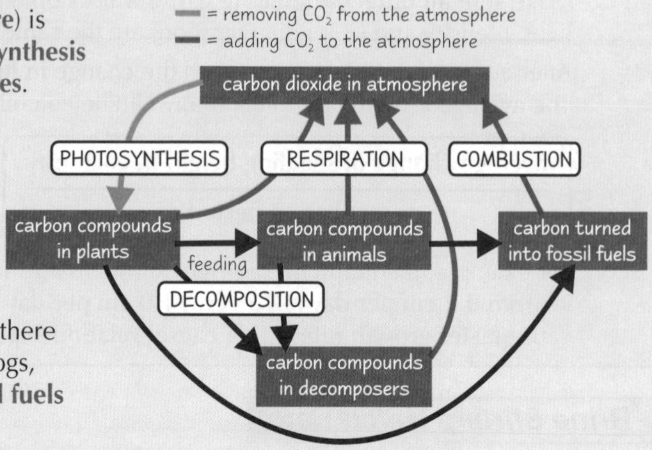

= removing CO_2 from the atmosphere
= adding CO_2 to the atmosphere
carbon dioxide in atmosphere
PHOTOSYNTHESIS
RESPIRATION
COMBUSTION
carbon compounds in plants
feeding
carbon compounds in animals
carbon turned into fossil fuels
DECOMPOSITION
carbon compounds in decomposers

To **reduce atmospheric CO_2 concentration** either the **amount** of CO_2 **going into** the atmosphere (due to **respiration** and **combustion**) needs to be **decreased** or the **amount** of CO_2 being **taken out** of the atmosphere (by **photosynthesis**) needs to be **increased**. You need to know about two methods of reducing atmospheric CO_2 concentration:

Biofuels

1) Biofuels are **fuels** produced from **biomass** — material that **is** or **was recently living**.
2) Biofuels are **burnt** to release energy, which **produces CO_2**.
3) There's **no net increase** in atmospheric CO_2 concentration when biofuels are burnt — the amount of CO_2 **produced** is the **same** as the amount of CO_2 **taken in** when the material was **growing**.
4) So using biofuels as an **alternative** to fossil fuels **stops** the **increase** in atmospheric CO_2 concentration caused by burning fossil fuels.

Reforestation

1) Reforestation is the planting of **new trees** in **existing forests** that have been **depleted**.
2) **More trees** means **more CO_2** is **removed** from the atmosphere by **photosynthesis**.
3) CO_2 is **converted** into carbon compounds and **stored** as plant tissues in the trees. This means more carbon is **kept out** of the atmosphere, so there's **less CO_2** contributing to global warming.

People **Disagree** *About* **How to Reduce Global Warming**

Scientists have come up with lots of **strategies** that all **reduce global warming**. There's **debate** about which strategies are the **right** ones to use because **different people** have **different viewpoints**. Here are a few examples of why different people might support or oppose **increasing** the use of **biofuels** and **wind turbines** for energy production:

Increase the use of biofuels

- Some **farmers** might **support** this strategy — some governments **fund** the **farming** of **crops** for biofuels.
- **Drivers** might **support** this strategy — the **price** of **biofuels** is usually **lower** than **oil-based fuels**.
- **Consumers** might **oppose** this strategy — using **farmland** to grow **crops** for biofuels could cause **food shortages**.
- **Conservationists** might **oppose** this strategy — **forests** have been **cleared** to grow **crops** for biofuels.

Reducing Global Warming

Increase the use of wind turbines

- Companies that make **wind turbines** would **support** this strategy — their **sales** would **increase**.
- **Environmentalists** might **support** this strategy — wind turbines produce electricity **without increasing** atmospheric CO_2 concentration.
- **Local communities** might **oppose** this strategy — some people think wind turbines **ruin** the **landscape**.
- **Bird conservationists** might **oppose** this strategy — many **birds** are **killed** by **flying into** wind turbines.

Data about Global Warming can be Extrapolated to make Predictions

Data that's **already** been collected on atmospheric CO_2 concentration can be **extrapolated** — used to make **predictions** about how it will **change** in the **future**. These predictions can then be used to **model** the amount of **global warming** that might happen in the **future**. For example:

1) An **international** group of scientists called the Intergovernmental Panel on Climate Change (**IPCC**) has **extrapolated data** on atmospheric CO_2 concentration to produce a number of **emissions scenarios** — **predictions** of how human CO_2 emissions will **change** up until 2100.
2) Scenarios include:
   - Emissions continuing to **grow** as they are now ('**business as usual**').
   - Emissions **increasing** by a lot (scenario 1, maximum emissions).
   - Emissions being **controlled** by **management strategies**.
   - Emissions **not increasing** much more (scenario 5, minimum emissions).
3) They can put all these different scenarios into **global climate models** (computer models of how the climate works), to see **how much** global **temperature** will **rise** with each scenario.

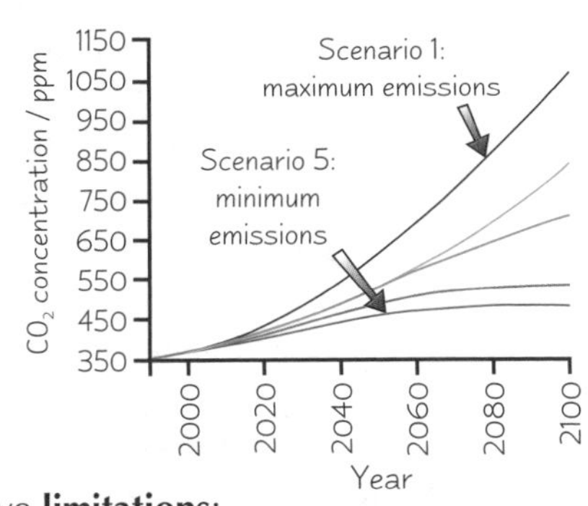

Models of future global warming based on extrapolated CO_2 concentration data have **limitations**:

1) We don't actually know how CO_2 emissions will **change**, i.e. which emissions scenario is most **accurate**.
2) We don't know exactly how much each emissions scenario will **cause** the global **temperature** to **rise by**.
3) The change in atmospheric CO_2 concentration due to **natural causes** (without human influence) **isn't known**.
4) We don't know what attempts there will be to **manage** atmospheric CO_2 concentration, or how **successful** they'll be.

In this scenario, an athlete is floored by emissions.

Practice Questions

Q1 What is reforestation?

Q2 Suggest one group of people who might oppose increasing the use of biofuels to reduce global warming.

Q3 Suggest one group of people who might support increasing the use of wind turbines to reduce global warming.

Exam Questions

Q1 Explain what biofuels are and describe how they help to reduce atmospheric CO_2 concentration. [4 marks]

Q2 Describe the limitations to models of global warming based on extrapolated CO_2 concentration data. [4 marks]

A massive, damp flannel — that's how I'd reduce global warming...

There's always talk of the next big idea to reduce global warming, but I reckon it's just hot air. Ho ho ho, sorry about that. We'll all probably have melted by the time everyone agrees on the best way to reduce global warming, but you can start doing your bit now — stop using your private jet to attend lessons and dedicate your life to planting trees.

Succession

These pages are for AQA Unit 4, OCR Unit 5 and Edexcel Unit 4.
Repeat after me: successful succession involves several simple successive seral stages.

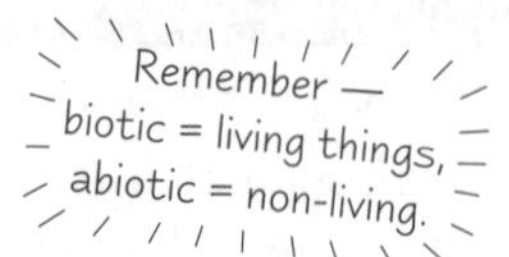

Succession is the **Process of Ecosystem Change**

Succession is the process by which an **ecosystem changes** over **time**. The **biotic conditions** (e.g. **plant** and **animal communities**) change as the **abiotic conditions** change (e.g. **water** availability). There are **two** types of succession:

1) **Primary succession** — this happens on land that's been **newly formed** or **exposed**, e.g. where a **volcano** has erupted to form a **new rock surface**, or where **sea level** has **dropped** exposing a new area of land. There's **no soil** or **organic material** to start with, e.g. just bare rock.
2) **Secondary succession** — this happens on land that's been **cleared** of all the **plants**, but where the **soil remains**, e.g. after a **forest fire** or where a forest has been **cut down by humans**.

Succession Occurs in **Stages** called **Seral Stages**

1) **Primary succession** starts when species **colonise** a new land surface. **Seeds** and **spores** are blown in by the **wind** and begin to **grow**. The **first species** to colonise the area are called **pioneer species** — this is the **first seral stage**.
   - The **abiotic conditions** are **hostile** (**harsh**), e.g. there's no soil to **retain water**. Only pioneer species **grow** because they're **specialised** to cope with the harsh conditions, e.g. **marram grass** can grow on sand dunes near the sea because it has **deep roots** to get water and can **tolerate** the salty environment.
   - The pioneer species **change** the **abiotic conditions** — they **die** and **microorganisms decompose** the dead **organic material** (**humus**). This forms a **basic soil**.
   - This makes conditions **less hostile**, e.g. the basic soil helps to **retain water**, which means **new organisms** can move in and grow. These then die and are decomposed, adding **more** organic material, making the soil **deeper** and **richer in minerals**. This means **larger plants** like **shrubs** can start to grow in the deeper soil, which retains **even more** water.
2) **Secondary succession** happens in the **same way**, but because there's already a **soil layer** succession starts at a **later seral stage** — the pioneer species in secondary succession are **larger plants**, e.g. shrubs.
3) At each stage, **different** plants and animals that are **better adapted** for the improved conditions move in, **out-compete** the plants and animals that are already there, and become the **dominant species** in the ecosystem.
4) As succession goes on, the ecosystem becomes **more complex**. New species move in **alongside** existing species, which means the **species diversity** (the number of **different species** and the **abundance** of each species) **increases**.
5) The amount of **biomass** also **increases** because plants at later stages are **larger** and **more dense**, e.g. **woody trees**.
6) The **final seral stage** is called the **climax community** — the ecosystem is supporting the **largest** and **most complex** community of plants and animals it can. It **won't change** much more — it's in a **steady state**.

This example shows primary succession on bare rock, but succession also happens on sand dunes, salt marshes and even on lakes.

Example of primary succession — bare rock to woodland

1) **Pioneer species colonise** the rocks. E.g. **lichens** grow on and **break down** rocks, **releasing minerals**.
2) The lichens **die** and are **decomposed** helping to form a **thin soil**, which thickens as more **organic material** is formed. This means other species such as **mosses** can **grow**.
3) **Larger plants** that need **more water** can move in as the soil **deepens**, e.g. **grasses** and **small flowering plants**. The soil **continues to deepen** as the larger plants die and are decomposed.
4) **Shrubs**, **ferns** and **small trees** begin to grow, **out-competing** the grasses and smaller plants to become the **dominant** species. **Diversity increases**.
5) Finally, the soil is **deep** and **rich** enough in **nutrients** to support **large trees**. These become the dominant species, and the **climax community** is formed.

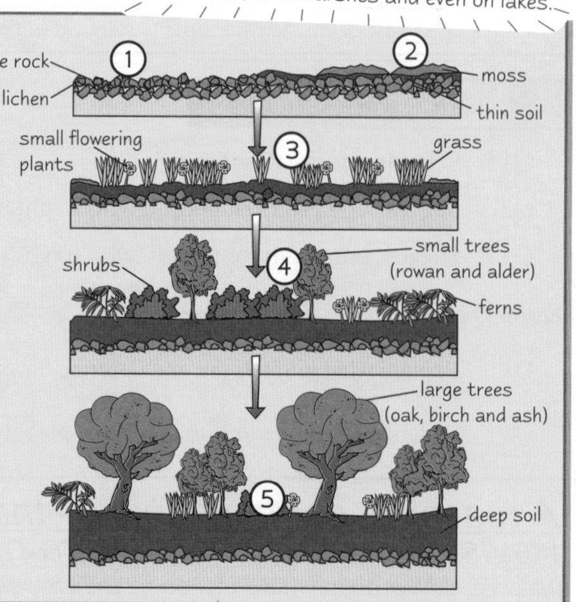

Succession

Different Ecosystems have Different Climax Communities

Which species make up the climax community depends on what the **climate's** like in an ecosystem. The climax community for a **particular** climate is called its **climatic climax**. For example:

In a **temperate climate** there's **plenty** of **available water**, **mild temperatures** and not much **change** between the seasons. The climatic climax will contain **large trees** because they **can grow** in these conditions once **deep soils** have developed. In a **polar climate** there's **not much available water**, temperatures are **low** and there are **massive changes** between the seasons. Large trees **won't ever** be able to grow in these conditions, so the climatic climax contains only **herbs** or **shrubs**, but it's still the **climax community**.

Succession can be Prevented or Deflected

Human activities can **prevent succession**, stopping the normal climax community from **developing**. When succession is stopped **artificially** like this, the climax community is called a **plagioclimax**. **Deflected succession** is when succession is prevented by human activity, but the plagioclimax that develops is one that's **different** to any of the **natural seral stages** of the ecosystem — the path of succession has been **deflected** from its natural course. For example:

A **regularly mown** grassy field **won't develop** woody plants, even if the climate of the ecosystem could support them. The **growing points** of the woody plants are **cut off** by the lawnmower, so larger plants **can't establish** themselves — only the grasses can **survive** being mowed, so the **climax community** is a **grassy field**. A grassy field isn't a **natural seral stage** — there should also be things like small flowering plants, so succession has been **deflected**.

Conservation Often Involves Managing Succession

AQA only

Conservation (the **protection** and **management** of ecosystems) sometimes involves **preventing succession** in order to **preserve** an ecosystem in its **current** seral stage. For example, there are large areas of **moorland** in **Scotland** that provide **habitats** for many species of plants and animals. If the moorland was left to **natural processes**, succession would lead to a **climax community** of **spruce forest**. This would mean the **loss** of the moorland habitat and could lead to the loss of some of the plants and animals that **currently** live there. Preventing succession keeps the moorland ecosystem **intact**. There are a couple of ways to **manage succession** to **conserve** the moorland ecosystem:

1) **Animals** are allowed to **graze** on the land. As described above, the **growing points** of the woody plants are **cut off** by the grazing animals, so larger plants **can't establish** themselves and the vegetation is kept **low**.
2) **Managed fires** are lit. After the fires, **secondary succession** will occur on the moorland — the species that grow back **first** (**pioneer species**) are the species that are being **conserved**, e.g. heather. Larger species will take **longer** to grow back and will be **removed again** the next time the moor's burnt.

Practice Questions

Q1 What is the difference between primary and secondary succession?

Q2 What is the name given to species that are the first to colonise an area during succession?

Q3 What is meant by a climax community?

Exam Question

Q1 A farmer has a field where he plants crops every year. When the crops are fully grown he removes them all and then ploughs the field (churns up all the plants and soil so the field is left as bare soil). The farmer has decided not to plant crops or plough the field for several years.

a) Describe, in terms of succession, what will happen in the field over time. [6 marks]

b) Explain why succession doesn't usually take place in the farmer's field. [2 marks]

Revision succession — bare brain to a woodland of knowledge...

When answering questions on succession, examiners are pretty keen on you using the right terminology — that means saying "pioneer species" instead of "the first plants to grow there". If you can manage that, then you can manage succession.

Conservation

These pages are for AQA Unit 4. If you're doing OCR go to p. 60, and if you're doing Edexcel go to the next section.

Conservation is important for us and the environment — won't somebody think of the polar bears...

Conserving Species and Habitats is Important for Many Reasons

Conservation is the **protection** and **management** of **species** and **habitats** (**ecosystems**).
It's **important** for **many reasons**:

1) **Species** are **resources** for lots of things that **humans need**, e.g. **rainforests** contain species that provide things like **drugs, clothes** and **food**. If the species and their habitats **aren't** conserved, the resources that we use now will be **lost**. Resources that **may be useful** in the **future** could also be **lost**.
2) Some people think we should conserve species simply because it's the **right thing to do**, e.g. most people think organisms have a **right to exist**, so they shouldn't become extinct as a result of **human activity**.
3) Many species and habitats bring **joy** to lots of people because they're **attractive** to **look at**. The species and habitats may be **lost** if they **aren't** conserved, so **future generations** won't be able to enjoy them.
4) Conserving species and habitats can help to prevent **climate change**. E.g. when trees are **burnt**, CO_2 is **released** into the atmosphere, which contributes to global warming. If they're conserved, this **doesn't happen**.
5) Conserving species and habitats helps to **prevent** the **disruption** of **food chains**. Disruption of food chains could mean the **loss** of **resources**. E.g. some species of **bear feed** on **salmon**, which feed on **herring** — if the number of herring **decreases** it can affect **both** the salmon and the bear populations.

Not everyone agrees with every conservation measure though — there's often **conflict** when conservation **affects people's livelihoods**, e.g. conservation of the Siberian tiger in Russia affects people who make money from killing the tigers and selling their fur (there's conflict between the conservationists and the hunters).

There are Many Different Ways to Conserve Species and Habitats

Different species and habitats need to be conserved in **different ways**.
Here are a few examples of **some** of the different **conservation methods** that can be used.

1 Plants can be Conserved using Seedbanks

The seedbank — 0% APR on branch transfers.

1) A **seedbank** is a **store** of lots of **seeds** from lots of **different plant species**.
2) They help to conserve species by storing the seeds of **endangered** plants.
3) They also help to conserve **different varieties** of each species by storing a **range** of seeds from plants with **different characteristics**, e.g. seeds from tall sunflowers and seeds from short sunflowers.
4) If the plants become **extinct** in the wild the stored seeds can be used to **grow new plants**.
5) Seedbanks are a **good way** of conserving plant species — **large numbers** of species can be conserved because seeds don't need **much space**. Seeds can also be **stored anywhere** and for a **long time**, as long as it's **cool** and **dry**.
6) But there are **disadvantages** — the seeds have to be regularly tested to see if they're still **viable** (whether they can grow into a plant), which can be **expensive** and **time-consuming**.

2 Fish species can be Conserved using Fishing Quotas

1) **Fishing quotas** are **limits** to the **amount** of certain fish species that fishermen are **allowed** to **catch**.
2) **Scientists** study different species and decide **how big** their populations need to be for them to **maintain** their numbers. Then they decide **how many** it's **safe** for fishermen to take without reducing the population **too much**.
3) **International agreements** are made (e.g. the Common Fisheries Policy in the EU) that state the **amount** of fish **each country** can take, and **where** they're allowed to take them from.
4) Fishing quotas help to **conserve** fish species by **reducing** the numbers that are **caught** and **killed**, so the populations aren't **reduced** too much and the species aren't at risk from becoming **extinct**.
5) There are **problems** with fishing quotas though — many fishermen **don't agree** with the scientists who say that the fish numbers are **low**. Some also think introducing quotas will cause **job losses**.

Conservation

3) Animals can be Conserved using Captive Breeding Programmes

1) Captive breeding programmes involve breeding animals in **controlled environments**.
2) Species that are **endangered**, or already **extinct in the wild**, can be **bred** in captivity to help **increase their numbers**, e.g. pandas are bred in captivity because their numbers are **critically low** in the wild.
3) There are some **problems** with captive breeding programmes though, e.g. animals can have **problems breeding** outside their **natural habitat**, which can be hard to **recreate** in a zoo. For example, pandas don't reproduce as **successfully** in captivity as they do in the wild.
4) Animals bred in captivity can be **reintroduced to the wild**. This **increases** their **numbers** in the wild, which can help to conserve their **numbers** or bring them **back** from the **brink of extinction**.
5) Reintroducing animals into the wild can cause **problems** though, e.g. reintroduced animals could bring **new diseases** to habitats, **harming** other species **living there**.

No way, I'm not breeding with him. He's ugly and his breath smells of bamboo.

4) Any organism can be Conserved by Relocation

1) **Relocating** a species means **moving** a population of a species to a **new location** because they're directly under **threat**, e.g. from poaching, or the **habitat** they're living in is under threat, e.g. from rising sea levels.
2) The species is moved to an area where it's **not at risk** (e.g. a protected national park, see below), but with a **similar environment** to where it's come from, so the species is still able to **survive**.
3) It's often used for species that only exist in **one place** (if that population **dies out**, the species will be **extinct**).
4) It helps to **conserve** species because they're relocated to a place where they're **more likely** to **survive**, so their numbers may **increase**.
5) Relocating species can cause **problems** though, e.g. native species in the new area may be **out-competed** by the species that's moved in and become **endangered** themselves.

5) Habitats can be Conserved using Protected Areas

1) **Protected areas** such as **national parks** and **nature reserves** protect habitats (and so protect the **species** in them) by **restricting urban development**, **industrial development** and **farming**.
2) Habitats in **protected areas** can be **managed** to conserve them, e.g. by **coppicing** — **cutting** down trees in a way that lets them **grow back**, so they don't need to be **replanted**. This helps to conserve the woodland, but allows some wood to be **harvested**.
3) There are **problems** with using protected areas to conserve habitats though, e.g. national parks are also used as **tourist destinations** (many are **funded** by **revenue** from the tourists that visit). This means there's conflict between the need to **conserve** the habitats and the need to allow people to **visit** and **use** them.

Practice Questions

Q1 Suggest why conservation of species and habitats is important for humans.

Q2 What is a seedbank?

Exam Question

Q1 A conservationist has argued that the deforestation of tropical rainforests will have terrible consequences for human beings and the environment.

a) Use your knowledge to outline the reasons for the conservation of tropical rainforests. [6 marks]

b) Name two suitable methods that could be used to conserve species from a tropical rainforest. [2 marks]

Captive breeding — you will procreate, or else...

There's lots of debate about conservation — what should be conserved and what's the best way to do it. That means there's a chance to get top marks in your exam — examiners love it when you talk about both sides of something. For example, even if you don't agree with captive breeding you need to say that it's got both positive and negative points.

Conservation Evidence and Data

These pages are for AQA Unit 4 only. *And now my pretties, it's time for some data. Mwa ha ha...*

You May Have to **Evaluate Evidence** *and* **Data** *About* **Conservation Issues**

You need to be able to **evaluate** any **evidence** or **data** about **conservation** projects and research that the examiners throw at you — so here's an example I made earlier:

In recent years, **native British bluebells** have become **less common** in woodland areas. It's thought that this is due to the presence of **non-native Spanish bluebells**, which compete with the native species for a **similar niche.** An experiment was carried out to see if **removing** the invasive Spanish species would help to **conserve** the native species. Each year for 15 years the **percentage cover** of native species was estimated in a **50 m by 50 m** area of **woodland** using random sampling and 250, **1 m² quadrats.** After five years, **all** the Spanish bluebells were **removed.** A **similar sized control woodland** in which the Spanish bluebells remained **untouched** was also studied. The results are shown on the right. You might be asked to:

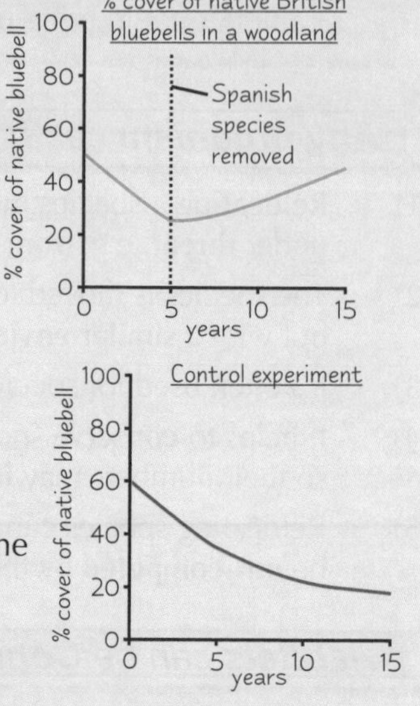

1) Describe the data:
   - For the first **five years**, the **percentage cover** of **native bluebells fell** from **50%** to around **25%**. After the Spanish species was **removed**, it **increased** from around **25%** to around **45%** in **ten years.**
   - The **control experiment** shows a fairly **steady drop** in native bluebell percentage cover from **60%** to **20%** over the 15 years.

% cover of native British bluebells in a woodland
% cover of native bluebell: 0, 20, 40, 60, 80, 100
years: 0, 5, 10, 15
Spanish species removed

Control experiment
% cover of native bluebell: 0, 20, 40, 60, 80, 100
years: 0, 5, 10, 15

2) Draw conclusions:

   The removal of Spanish bluebells **resulted** in an **increase** in the percentage cover of **native bluebells** over a **ten year period.** This suggests that the **recent decrease** in native British bluebells is due to **competition** with the Spanish bluebells.

3) Evaluate the method:
   - The effects of some **other variables** (e.g. **changing weather**) were **removed** by the **control experiment**, where the percentage cover of native bluebells continued to fall throughout the 15 year study. This makes the data **more reliable.**
   - The **study area** and **sample size** were quite **large**, giving **more accurate** data.
   - **Random sampling** removed bias — the data's **more likely** to be an **accurate estimate** of the **whole area.**

There's more about interpreting data on pages 214-216.

You Need to be Able to **Consider Conflicting Evidence**

1) The **evidence** from **one study** alone **wouldn't usually be enough** to conclude that there's a **link** between decreasing percentage cover of native bluebells, and the presence of Spanish bluebells.
2) **Similar studies** would be carried out to **investigate** the link. If these studies came to the **same conclusion**, the conclusion would become **increasingly accepted.**
3) Sometimes studies come up with **conflicting evidence** though — evidence that leads to a **different conclusion** than other studies. For example:

Another study was carried out to **investigate** the effect on native bluebells of **removing** Spanish bluebells. It was **similar** to the study above except a **20 m by 20 m** area was sampled using a random sample of **20 quadrats**, and **no control** woodland was used. You might be asked to:

1) Describe the data:

   In the first five years, the **percentage cover** of **native bluebells fell** from **50%** to around **25%**. After the Spanish species was **removed**, it **kept decreasing** to around **15%** after the **full 15** years.

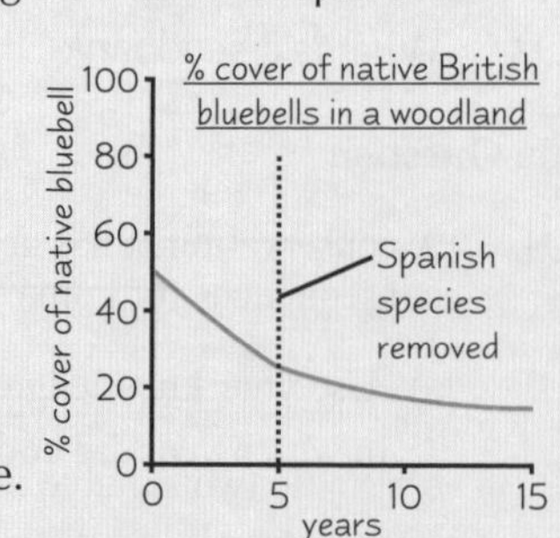

2) Draw conclusions:

   The **removal** of the Spanish bluebells had **no effect** on the **decreasing** percentage cover of native bluebells — which **conflicts** with the study above.

3) Evaluate the method:
   - There **wasn't** a **control** woodland, so the **continuing decrease** in native bluebell cover after the removal of the Spanish bluebells could be due to **another factor**, e.g. cold weather in years 5-10.
   - The **study area** and **sample size** were quite small, giving a **less accurate** total percentage cover.

Conservation Evidence and Data

Conservation Relies on Science to Make Informed Decisions

1) Scientists carry out **research** to provide **information** about conservation issues.
2) This information can then be used to make **informed decisions** about **which** species and habitats **need** to be conserved, and the **best way** to conserve them.
3) For example, the study at the **top of the previous page** showed that **native bluebell** coverage increased after the removal of **Spanish bluebells**, which **suggests** that the decrease in native bluebell coverage is due to **competition** with the Spanish species. It provides evidence that there's a **conservation issue** (native bluebells are decreasing) and a way to **solve it** (remove the Spanish species).

Many conservation **decisions** have been made using the results of **scientific research** — take a look at these examples:

Scientific results	Decision
Between 1970 and 1989 the number of African elephants dropped from around 3 million to around 50 000 because they were being hunted for their ivory tusks.	In 1989, the Convention on International Trade in Endangered Species banned ivory trade to end the demand for elephant tusks, so that fewer elephants would be killed for their tusks.
The commonly used pesticide DDT was found to have contributed to the loss of half the peregrine falcon population in the UK in the 1950s and 1960s. DDT built up in the food chain and caused the falcon eggs to have thin shells. This meant the eggs were crushed and the chicks weren't hatched.	The use of DDT as a pesticide was banned in the UK in 1984 to try to conserve and increase peregrine falcon numbers.
The numbers of some species of sea turtle have dropped so low that they're now endangered. Many eggs are removed from the beaches by poachers before the turtles hatch and reach the sea.	Conservation agencies have set up hatching programmes where eggs are taken away from beaches and looked after until they hatch. The young turtles are then released into the sea.
A reduction in the size of hedgerows in farmers' fields was found to cause a decrease in biodiversity in the British countryside.	The government provides subsidies to encourage farmers to plant hedgerows and leave margins of ground unharvested around fields. This increases the size of hedgerows and conserves biodiversity.
Whale numbers were found to have dropped massively due to whale hunting.	Commercial whaling was banned in 1986 by the International Whaling Commission in order to conserve whale numbers.

Practice Questions

Q1 What is conflicting evidence?

Q2 Give one example of scientific evidence that has informed decision-making about conservation issues.

Q3 Give one example of a decision that has been made as a result of scientific evidence about conservation issues.

Exam Question

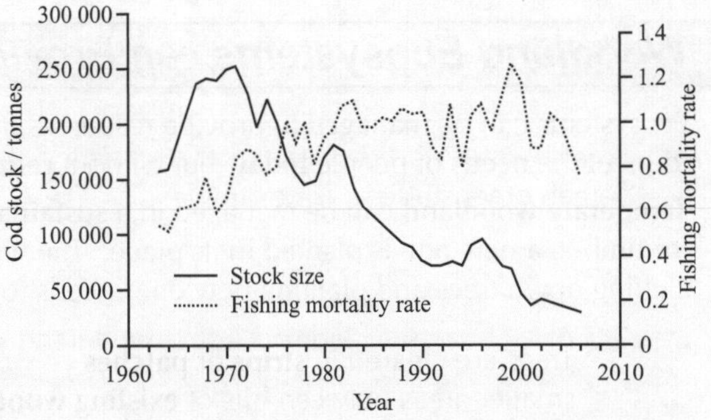

Q1 The graph shows the stock of spawning cod in the North Sea and the rate of mortality caused by fishing since 1960.

a) Describe the results shown by the graph. [4 marks]

b) Suggest a conclusion that could be drawn from the graph. [2 marks]

c) Scientists have stated that 150 000 tonnes is the minimum stock needed to preserve a cod population. In which year did cod stocks first fall below this level? [1 mark]

d) How might this data be used to make informed decisions about the conservation of cod stocks? [1 mark]

I'm considering conflict after these pages, I tell you...

Ah hah ha, aaaaah ha ha ha... oh, I think I need to stop my evil laugh now. I quite enjoyed that. Evaluating evidence and data's an important nut to crack — you might have to do it in your exam for conservation or for another topic altogether.

Conservation of Ecosystems

These pages are for OCR Unit 5 only.

It's important that ecosytems are conserved so the resources we use from them to make lots of nice things don't run out.

We Need to *Conserve Ecosystems*

1) **Conservation** is the **protection** and **management** of **ecosystems** so that the **natural resources** in them can be **used** without them **running out**. E.g. using rainforests for timber without any species becoming **extinct** and without any habitats being **destroyed**. This means the natural resources will still be available for **future generations**.
2) It's a **dynamic process** — conservation methods need to be **adapted** to the **constant changes** (caused **naturally** and by **humans**) that occur within ecosystems.
3) Conservation involves the **management** of ecosystems — controlling how **resources** are **used** and **replaced**.
4) Conservation can also involve **reclamation** — **restoring ecosystems** that have been **damaged** or **destroyed** so they can be **used again**, e.g. restoring **forests** that have been **cut down** so they can be used again.
5) Conservation is **important** for many reasons:

Economic: **Ecosystems** provide **resources** for lots of things that **humans need**, e.g. **rainforests** contain species that provide things like **drugs, clothes** and **food**. These resources are **economically important** because they're **traded** on a **local** and **global** scale. If the ecosystems **aren't** conserved, the resources that we use now will be **lost**, so there will be **less trade** in the future.

Social: Many ecosystems bring **joy** to lots of people because they're **attractive** to **look at** and people **use** them for **activities**, e.g. birdwatching and walking. The species and habitats in the ecosystems may be **lost** if they **aren't** conserved, so **future generations** won't be able to use and enjoy them.

Ethical:
1) Some people think we should conserve ecosystems simply because it's the **right thing to do**, e.g. most people think organisms have a **right to exist**, so they shouldn't become extinct as a result of **human activity**.
2) Some people think we have a **moral responsibility** to conserve ecosystems for **future generations**, so they can enjoy and use them.

Cast your mind back to AS biology — the reasons for conservation are similar to the reasons for conserving biodiversity.

6) **Preservation** is different from conservation — it's the **protection** of ecosystems so they're kept **exactly as they are**. Nothing is **removed** from a preserved ecosystem and they're only **used** for activities that **don't damage** them. For example, **Antarctica** is a preserved ecosystem because it's protected from **exploitation** by humans — it's only used for **limited tourism** and **scientific research**, not **mining** or other **industrial** activities.

Woodland Ecosystems can *Provide Resources* in a *Sustainable Way*

Ecosystems can be **managed** to provide resources in a way that's **sustainable** — this means enough resources are taken to meet the **needs** of people **today**, but without **reducing the ability** of people in the **future** to meet their own needs.

Temperate woodland can be managed in a **sustainable way** — for every tree that's **cut down** for timber, a **new one** is planted in its place. The woodland should never become **depleted**. Cutting down trees and planting new ones needs to be done **carefully** to be **successful**:

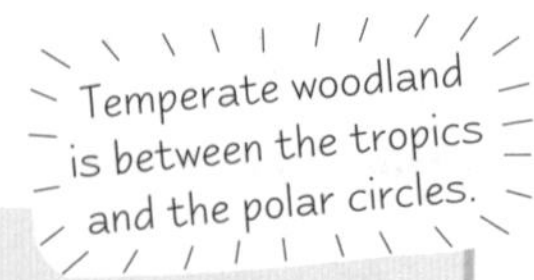

Temperate woodland is between the tropics and the polar circles.

1) Trees are cleared in **strips** or **patches** — woodland grows back **more quickly** in smaller areas between bits of **existing woodland** than it does in larger, **open areas**.
2) The cleared strips or patches aren't **too large** or **exposed** — lots of **soil erosion** can occur on large areas of **bare ground**. If the soil is eroded, newly planted trees **won't** be able to **grow**.
3) Timber is sometimes harvested by **coppicing** — **cutting** down trees in a way that lets them **grow back**. This means new trees don't need to be planted.
4) Only **native species** are planted — they grow most **successfully** because they're **adapted** to the climate.
5) Planted trees are attached to **posts** to provide **support**, and are grown in **plastic tubes** to stop them being **eaten** by grazing animals — this makes it **more likely** the trees will **survive** to become mature adults.
6) Trees **aren't** planted too **close together** — this means the trees aren't **competing** with each other for **space** or **resources**, so they're more likely to **survive**.

Conservation of Ecosystems

Human Activities Affect Ecosystems like the Galapagos Islands

Humans often need to **conserve** or **preserve** ecosystems because our **activities** have **badly affected** them, e.g. large areas of the **Amazon rainforest** have been **cleared** without being **replaced**, **destroying** the ecosystem.

Human activities have had a negative effect on the **Galapagos Islands**, a small group of islands in the **Pacific Ocean** about 1000 km off the coast of South America. Many species of animals and plants have evolved there that **can't** be found **anywhere else**, e.g. the **Galapagos giant tortoise** and the **Galapagos sea lion**. Here are some examples of how the **animal** and **plant populations** there have been affected by human activity:

1) **Explorers** and **sailors** that visited the Galapagos Islands in the **19th century** directly affected the populations of some animals by **eating them**. For example, a type of **giant tortoise** found on **Floreana Island** was hunted to **extinction** for food.
2) **Non-native animals introduced** to the islands **eat** some native species. This has caused a decrease in the populations of native species. For example, non-native **dogs**, **cats** and **black rats** eat young **giant tortoises** and **Galapagos land iguanas**. **Pigs** also destroy the nests of the iguanas and **eat their eggs**. **Goats** have eaten much of the **plant life** on some of the islands.
3) **Non-native plants** have also been introduced to the islands. These **compete** with native plant species, causing a decrease in their populations. For example, **quinine trees** are **taller** than some native plants — they **block out light** to the native plants, which then **struggle** to **survive**.
4) **Fishing** has caused a decrease in the populations of some of the **sea life** around the Galapagos Islands. For example, the populations of **sea cucumbers** and **hammerhead sharks** have been reduced because of **overfishing**. **Galapagos green turtle** numbers have also been reduced by overfishing and they're also killed **accidentally** when they're caught in **fishing nets**. They're now an **endangered species**.
5) A recent increase in **tourism** (from **41 000** tourists in **1991** to around **160 000** in **2008**) has led to an increase in **development** on the islands. For example, the **airport** on Baltra island has been redeveloped to receive more tourists. This causes **damage** to the ecosystems as **more land** is **cleared** and **pollution** is **increased**.
6) The **population** on the islands has also **increased** due to the increased **opportunities** from tourism. This could lead to further **development** and so more **damage** to the ecosystems.

Darwin (the sea lion) worried he was about to be affected by human activity.

Practice Questions

Q1 Why does conservation need to be dynamic?

Q2 What is meant by reclamation?

Q3 How is preservation different from conservation?

Q4 What does managing an ecosystem in a sustainable way mean?

Q5 Give one way that temperate woodlands are managed to make sure newly planted trees grow.

Exam Questions

Q1 Explain why conservation is important for economic, social and ethical reasons. [3 marks]

Q2 Explain how the following human activities have affected specific native animal or plant populations on the Galapagos Islands.

a) Introduction of non-native animal species. [2 marks]

b) Introduction of non-native plant species. [2 marks]

c) Fishing. [2 marks]

If I can sustain this revision it'll be a miracle...

Never mind ecosystems, I'm more interested in preserving my sanity after all this hard work. I know it doesn't seem all that sciencey, but you can still study biology without a lab coat, some Petri dishes and a rack of test tubes. Sustainability's a funny one to get your head around, but luckily you just need to know about how it applies to temperate woodlands.

Inheritance

This section is for AQA Unit 4 and OCR Unit 5 only.

If you've ever wondered what causes colour blindness, how gender is controlled or how genetic diseases are passed on, then this is the section for you. If you've never wondered this and don't really care — tough. You still need to know it.

You Need to Know These Genetic Terms

'Codes for' means 'contains the instructions for'.

TERM	DESCRIPTION
Gene	A sequence of bases on a DNA molecule that codes for a protein (polypeptide), which results in a characteristic, e.g. the gene for eye colour.
Allele	A different version of a gene. Most plants and animals, including humans, have two alleles of each gene, one from each parent. The order of bases in each allele is slightly different — they code for different versions of the same characteristic. They're represented using letters, e.g. the allele for brown eyes (B) and the allele for blue eyes (b).
Genotype	The genetic constitution of an organism — the alleles an organism has, e.g. BB, Bb or bb for eye colour.
Phenotype	The expression of the genetic constitution and its interaction with the environment — an organism's characteristics, e.g. brown eyes.
Dominant	An allele whose characteristic appears in the phenotype even when there's only one copy. Dominant alleles are shown by a capital letter. E.g. the allele for brown eyes (B) is dominant — if a person's genotype is Bb or BB, they'll have brown eyes.
Recessive	An allele whose characteristic only appears in the phenotype if two copies are present. Recessive alleles are shown by a lower case letter. E.g. the allele for blue eyes (b) is recessive — if a person's genotype is bb, they'll have blue eyes.
Codominant	Alleles that are both expressed in the phenotype — neither one is recessive, e.g. the alleles for haemoglobin (see page 63).
Locus	The fixed position of a gene on a chromosome. Alleles of a gene are found at the same locus on each chromosome in a pair.
Homozygote	An organism that carries two copies of the same allele, e.g. BB or bb.
Heterozygote	An organism that carries two different alleles, e.g. Bb.
Linkage	When alleles located on the same chromosome are inherited together (see page 71).

Genetic Diagrams Show the Possible Genotypes of Offspring

Individuals have **two alleles** for **each gene**. **Gametes** (sex cells) contain only **one allele** for each gene. When gametes from two parents fuse together, the alleles they contain form the **genotype** of the **offspring** produced. **Genetic diagrams** can be used to **predict** the **genotypes** and **phenotypes** of the offspring produced if two parents are **crossed** (bred). You need to know how to use genetic diagrams to predict the results of various crosses, including **monohybrid crosses**.

Monohybrid inheritance is the inheritance of a **single characteristic** (gene) controlled by **different alleles**. **Monohybrid crosses** show the **likelihood** of alleles (and so different versions of the characteristic) being **inherited** by offspring of particular parents. The genetic diagram below shows how **wing length** is inherited in fruit flies:

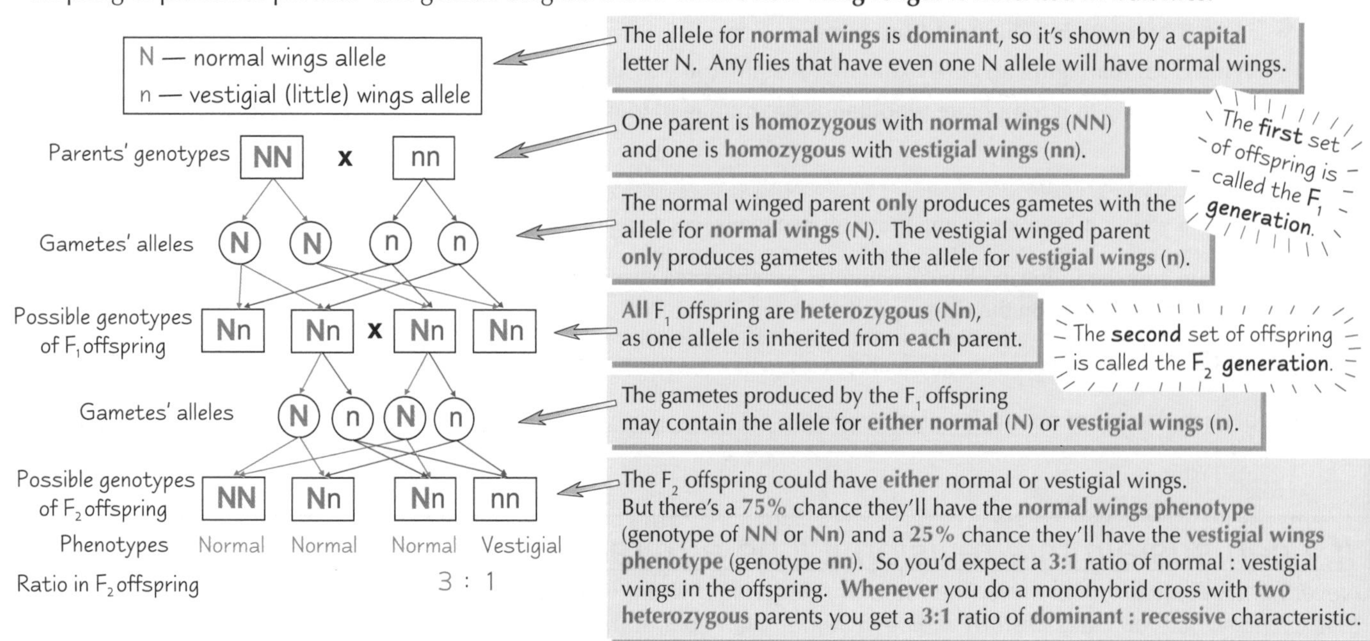

Inheritance

A **Punnett square** is just another way of showing a **genetic diagram** — they're also used to predict the **genotypes** and **phenotypes** of offspring. The Punnett squares below show the same crosses from the previous page:

1) First work out the alleles the **gametes** would have.

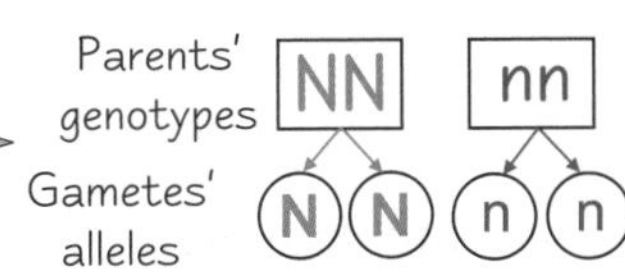

2) Next **cross** the **parents' gametes** to show the possible genotypes of the F_1 **generation** — **all heterozygous**, Nn.

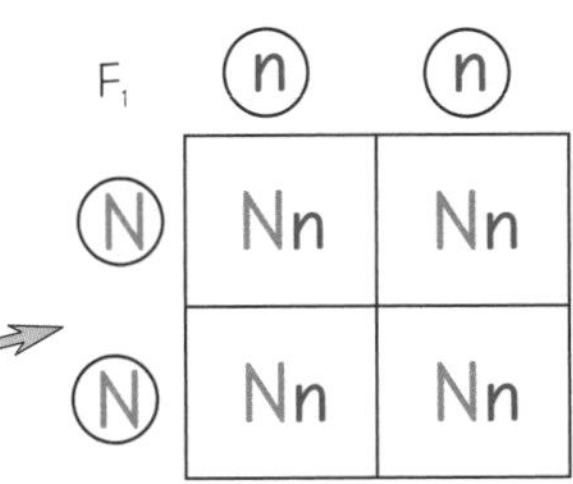

3) Then **cross** the **gametes' alleles** of the F_1 **generation** to show the possible **genotypes** of the F_2 **generation.** The Punnett square shows a **75%** chance that offspring will have **normal wings** and a **25%** chance that they'll have **vestigial wings**, i.e. a **3:1 ratio.**

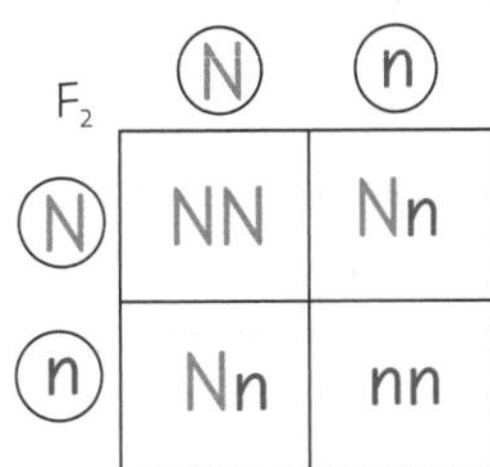

- 1 in 4 chance of offspring having the genotype NN (normal wings)
- 2 in 4 chance of offspring having the genotype Nn (normal wings)
- 1 in 4 chance of offspring having the genotype nn (vestigial wings)
- So, phenotype ratio normal : vestigial = 3 : 1

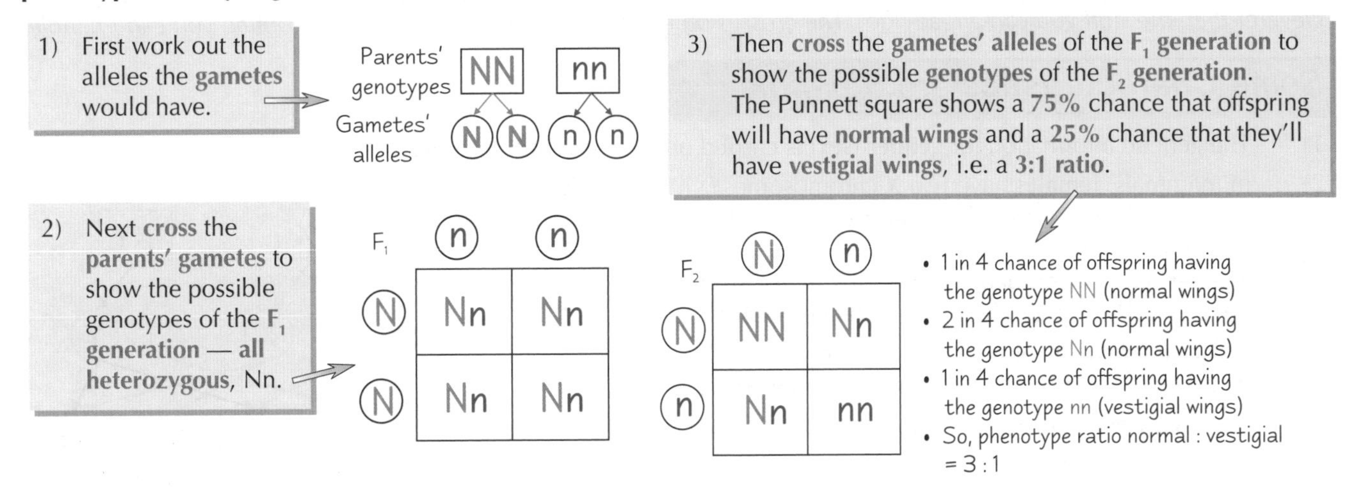

Some Genes Have Codominant Alleles

You need to be able to work out genetic diagrams for codominant alleles too.

Occasionally, alleles show **codominance** — **both alleles** are expressed in the **phenotype**, **neither one** is recessive. One example in humans is the allele for **sickle-cell anaemia:**

1) People who are **homozygous** for **normal haemoglobin** (**H^NH^N**) don't have the disease.
2) People who are **homozygous** for **sickle haemoglobin** (**H^SH^S**) have **sickle-cell anaemia** — all their **blood cells** are **sickle-shaped** (crescent-shaped).
3) People who are **heterozygous** (**H^NH^S**) have an **in-between** phenotype, called the **sickle-cell trait** — they have **some** normal haemoglobin and some sickle haemoglobin. The two alleles are **codominant** because they're **both** expressed in the **phenotype**.
4) The **genetic diagram** on the right shows the possible offspring from **crossing** two parents with **sickle-cell trait** (**heterozygous**).

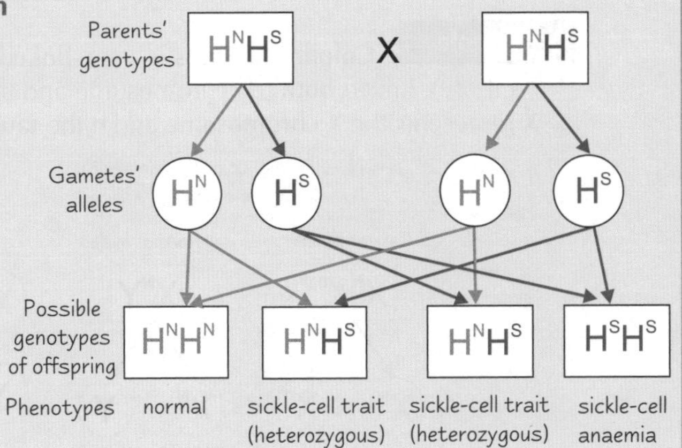

Practice Questions

Q1 What is meant by the term genotype?

Q2 What is meant by the term phenotype?

Q3 What is meant by the term codominance?

Exam Question

Q1 In pea plants, seed texture (round or wrinkled) is passed from parent to offspring by monohybrid inheritance. The allele for round seeds is represented by R and the allele for wrinkled seeds is represented by r.

a) Draw a genetic diagram to show the possible genotypes of F_1 offspring produced by crossing a homozygous round seed pea plant with a homozygous wrinkled seed pea plant. [3 marks]

b) What ratio of round to wrinkled seeds would you expect to see in the F_2 generation? [3 marks]

If there's a dominant revision allele I'm definitely homozygous recessive...

OK, so there are a lot of fancy words on these pages and yes, you do need to know them all. Sorry about that. But don't despair — once you've learnt what the words mean and know how genetic diagrams work it'll all just fall into place.

Inheritance

These pages are for AQA Unit 4 and OCR Unit 5.

Now you know how these genetic diagram thingies work, you can use them to work out all kinds of clever stuff — even cleverer than the stuff you can already do. The crosses on these pages are a bit trickier, but nothing you can't handle.

Some **Characteristics** are **Sex-linked**

1) The genetic information for **gender** (**sex**) is carried on two **sex chromosomes**.
2) In mammals, **females** have **two X** chromosomes (XX) and **males** have **one X** chromosome and **one Y** chromosome (XY). The genetic diagram on the right shows how gender is **inherited**. The probability of having **male offspring** is **50%** and the probability of having **female offspring** is **50%**.

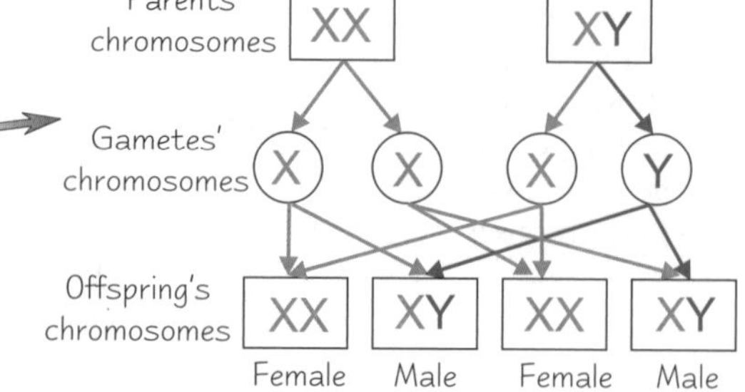

3) A **characteristic** is said to be **sex-linked** when the allele that codes for it is located on a **sex chromosome**.
4) The **Y chromosome** is **smaller** than the X chromosome and carries **fewer genes**. So most genes on the sex chromosomes are **only carried** on the X chromosome (called **X-linked** genes).
5) As **males** only have **one X chromosome** they often only have **one allele** for sex-linked genes. So because they **only** have one copy they **express** the **characteristic** of this allele even if it's **recessive**. This makes males **more likely** than females to show **recessive phenotypes** for genes that are sex-linked.
6) Genetic disorders caused by **faulty alleles** located on sex chromosomes include **colour blindness** and **haemophilia**. The faulty alleles for both of these disorders are carried on the X chromosome and so are called **X-linked disorders**. **Y-linked disorders** do exist but are **less common**.

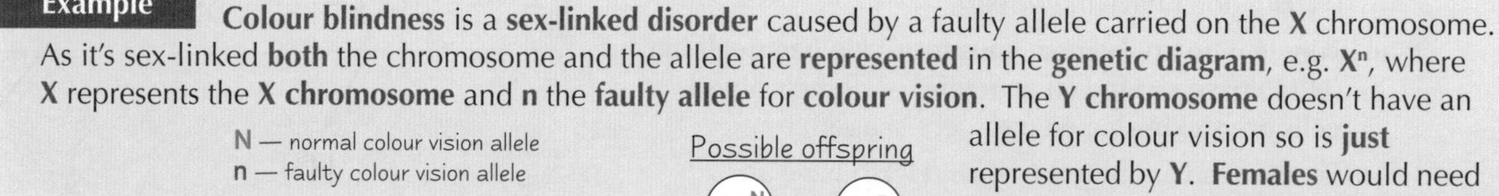

Example **Colour blindness** is a **sex-linked disorder** caused by a faulty allele carried on the **X** chromosome. As it's sex-linked **both** the chromosome and the allele are **represented** in the **genetic diagram**, e.g. X^n, where **X** represents the **X chromosome** and **n** the **faulty allele** for **colour vision**. The **Y chromosome** doesn't have an allele for colour vision so is **just** represented by **Y**. **Females** would need **two copies** of the **recessive allele** to be colour blind, while **males** only need **one copy**. This means colour blindness is **much rarer** in **women** than **men**.

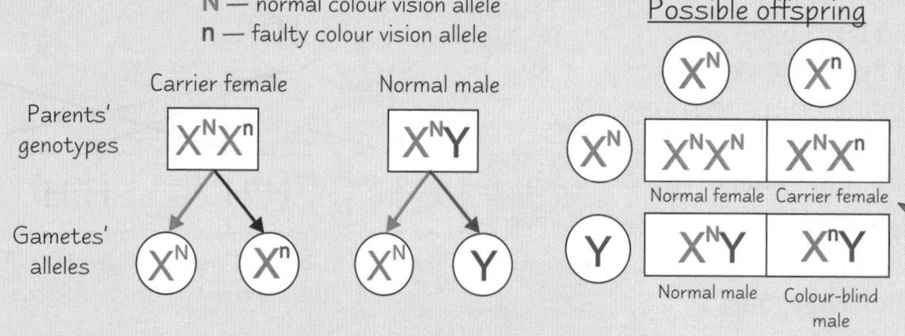

A carrier is a person carrying an allele which is not expressed in the phenotype but that can be passed on to offspring.

Some **Genes** Have **Multiple Alleles**

AQA only. If you're doing OCR you can go to the questions.

Inheritance is **more complicated** when there are **more than two** alleles of the same gene (**multiple alleles**).

Example In the **ABO blood group system** in humans there are **three alleles** for blood type:

I^O is the allele for blood group **O**. I^A is the allele for blood group **A**. I^B is the allele for blood group **B**.

Allele I^O is **recessive**. Alleles I^A and I^B are **codominant** — people with genotype I^AI^B will have blood group **AB**.

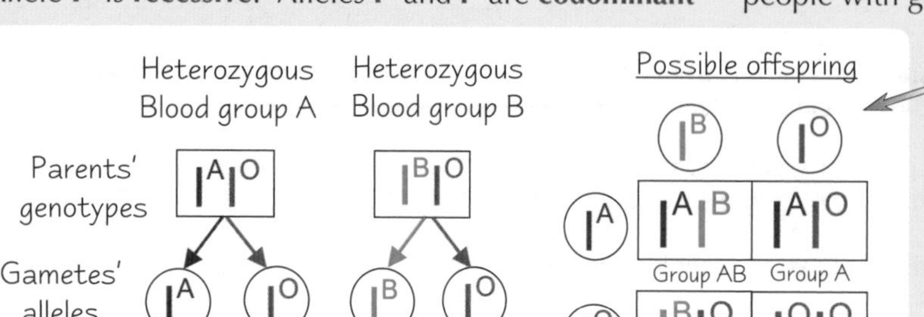

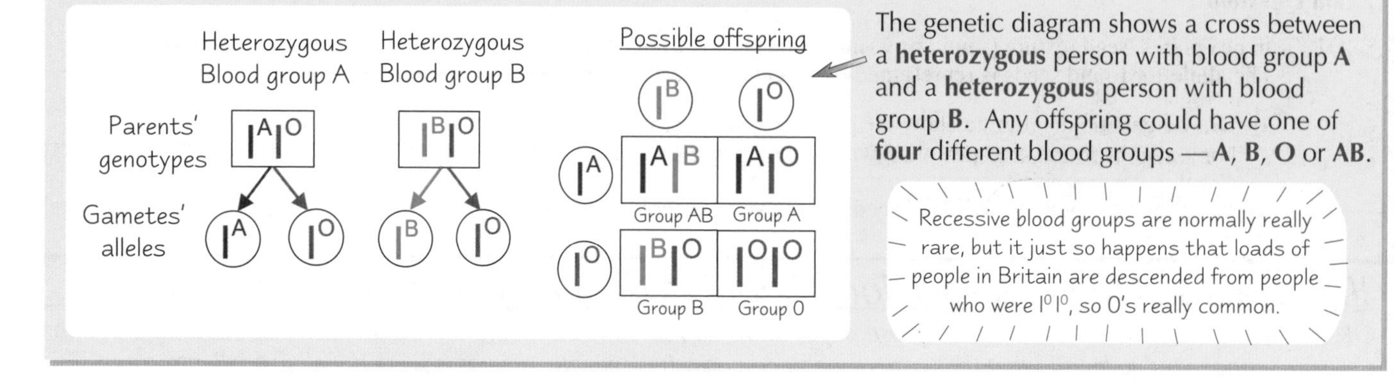

The genetic diagram shows a cross between a **heterozygous** person with blood group **A** and a **heterozygous** person with blood group **B**. Any offspring could have one of **four** different blood groups — **A, B, O** or **AB**.

Recessive blood groups are normally really rare, but it just so happens that loads of people in Britain are descended from people who were I^OI^O, so O's really common.

Inheritance

Genetic Pedigree Diagrams Show How Traits Run in Families

AQA only

Genetic pedigree diagrams show an **inherited trait** (characteristic) in a group of **related individuals**. You might have to **interpret** genetic pedigree diagrams to work out the **genotypes** or **potential phenotypes** of individuals:

Example

Cystic fibrosis (CF) is an inherited disorder that's caused by a faulty **recessive** allele (f) — it codes for a **faulty chloride ion channel**. A person will only have the disorder if they're **homozygous** for the allele (ff) — they must inherit one recessive allele **from each parent**. If a person is **heterozygous** (Ff), they **won't** have CF but they'll be a **carrier**.

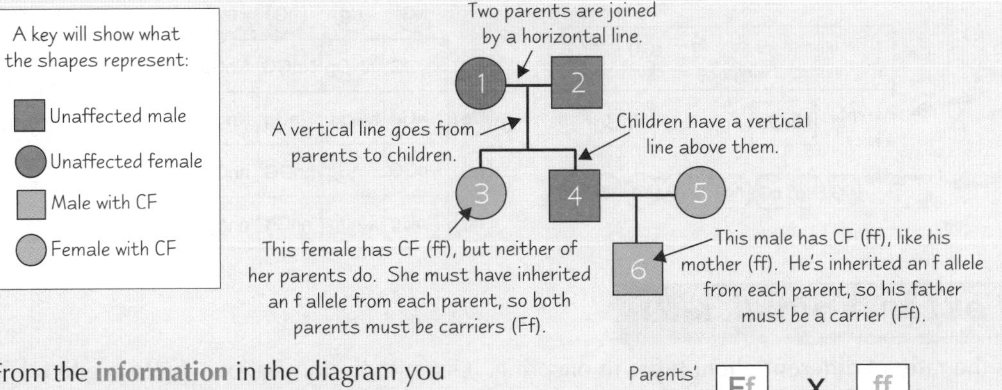

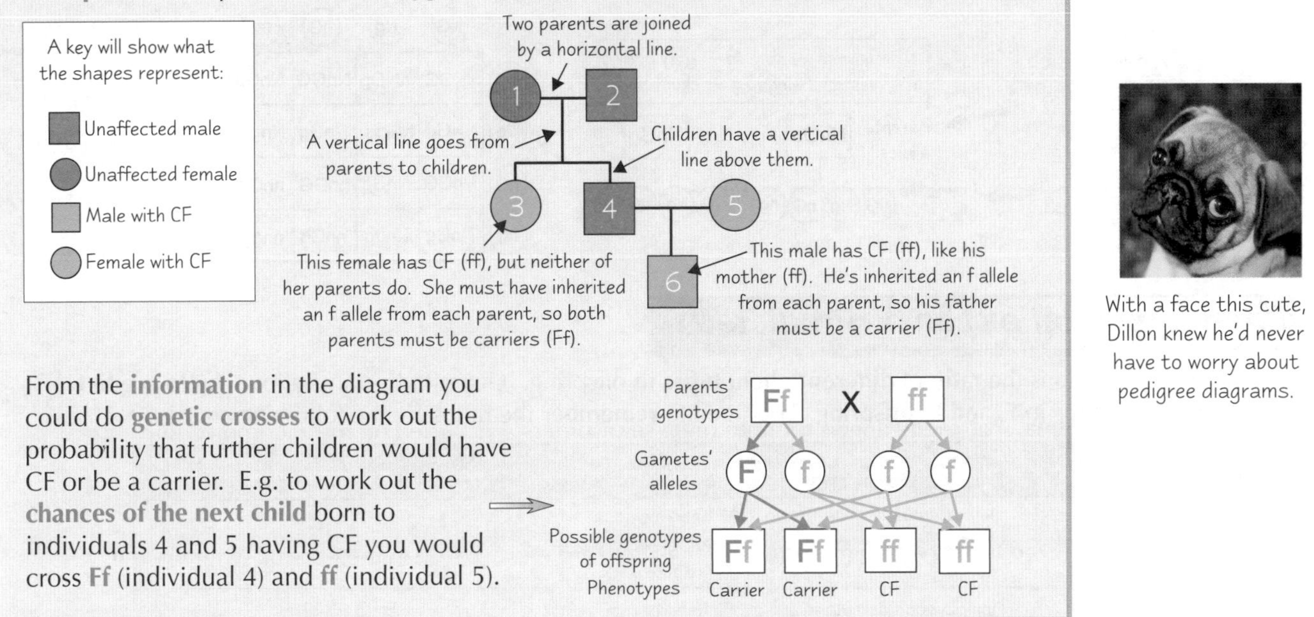

From the **information** in the diagram you could do **genetic crosses** to work out the probability that further children would have CF or be a carrier. E.g. to work out the **chances of the next child** born to individuals 4 and 5 having CF you would cross **Ff** (individual 4) and **ff** (individual 5).

Parents' genotypes: Ff X ff

Gametes' alleles: F f f f

Possible genotypes of offspring: Ff Ff ff ff

Phenotypes: Carrier Carrier CF CF

With a face this cute, Dillon knew he'd never have to worry about pedigree diagrams.

Practice Questions

Q1 What is a sex-linked gene?

Q2 What is a carrier?

Q3 What do pedigree diagrams show?

1 2 3 4 5 6 7

Unaffected male
Unaffected female
Male with ADA deficiency
Female with ADA deficiency

Exam Questions

Q1 Haemophilia A is a sex-linked genetic disorder caused by a recessive allele carried on the X chromosome. Explain why haemophilia A is more common in males than females. [3 marks]

Q2 Using a genetic diagram, show the probability of a heterozygous person with blood group A and a homozygous person with blood group B having a child with blood group B. [4 marks]

Q3 ADA deficiency is an inherited metabolic disorder caused by a recessive allele (a). Use the genetic pedigree diagram above to answer the following questions:

a) Give the possible genotype(s) of individual 2. [1 mark]

b) What is the genotype of individual 6? Explain your answer. [2 marks]

c) What is the probability that the next child born to individuals 5 and 6 will have ADA deficiency? Show your working. [4 marks]

Sex-linkage — it's all starting to sound a little bit kinky...

Congratulations — you've made it to the end of two tricky pages — go and get a cup of tea and a biscuit to recuperate. It's difficult stuff, but if you work through the diagrams in a logical way you'll get there in the end.

Phenotypic Ratios and Epistasis

The rest of this section is for OCR Unit 5 only.

Right, this stuff is fairly hard, so if you don't get it first time don't panic.
Make sure you're happy with the genetic diagrams on the previous pages before you get stuck into these two.

Genetic Diagrams can Show how More Than One Characteristic is Inherited

You can use genetic diagrams to work out the chances of offspring inheriting certain **combinations** of characteristics. For example, you can look at how **two different genes** are inherited — **dihybrid inheritance**. The diagram below is a dihybrid cross showing how wing size **and** colour are inherited in **fruit flies**.

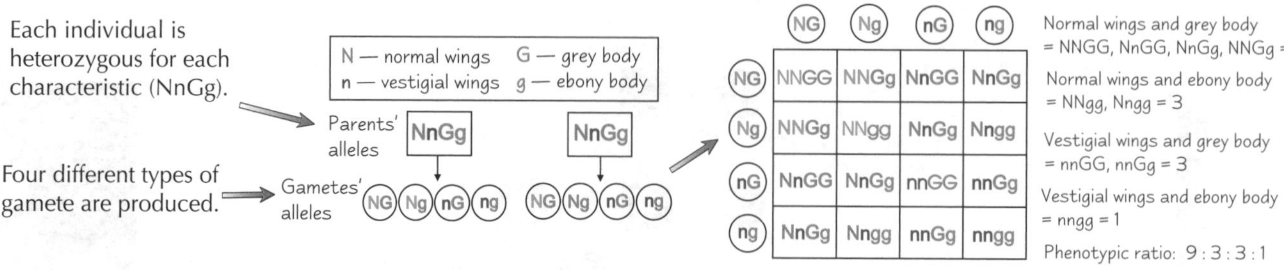

Phenotypic Ratios can be Predicted

The **phenotypic ratio** is the **ratio** of **different phenotypes** in offspring. Genetic diagrams allow you to **predict** the phenotypic ratios in **F_1 and F_2 offspring**. You need to **remember** the ratios for the following crosses:

Type of cross	Parents	Phenotypic ratio in F_1	Phenotypic ratio in F_2
Monohybrid	Homozygous dominant × homozygous recessive (e.g. NN × nn)	All heterozygous offspring (e.g. Nn)	3 : 1 dominant : recessive
Dihybrid	Homozygous dominant × homozygous recessive (e.g. NNGG × nngg)	All heterozygous offspring (e.g. NnGg)	9 : 3 : 3 : 1 dominant both : dominant 1st recessive 2nd : recessive 1st dominant 2nd : recessive both
Codominant	Homozygous for one allele × homozygous for the other allele (e.g. H^NH^N x H^SH^S)	All heterozygous offspring (e.g. H^NH^S)	1 : 2 : 1 homozygous for one allele : heterozygous : homozygous for the other allele

Sometimes you **won't** get the **expected** (predicted) phenotypic ratio — it'll be quite different. This can be because of **epistasis** (coming up next) or **linkage** (see page 71).

An Epistatic Gene Masks the Expression of Another Gene

1) **Many different genes** can control the **same** characteristic — they **interact** to form the phenotype.
2) This can be because the **allele** of one gene **masks** (blocks) **the expression** of the alleles of other genes — this is called **epistasis**.

Example 1 In humans a **widow's peak** (see picture) is controlled by one gene and **baldness** by others. If you have the **alleles** that code for baldness, it **doesn't matter** whether you have the allele for a widow's peak or not, as you have **no hair**. The baldness genes are **epistatic** to the widow's peak gene, as the baldness genes **mask** the expression of the widow's peak gene.

I'm still dashing, even with my widow's peak.

Example 2 **Flower pigment** in a plant is controlled by two genes. **Gene 1** codes for a **yellow pigment** (Y is the dominant yellow allele) and **gene 2** codes for an enzyme that **turns** the yellow pigment **orange** (R is the dominant orange allele). If you **don't have** the **Y** allele it **won't matter** if you have the R allele or not as the flower **will be colourless**. Gene 1 is **epistatic** to gene 2 as it can **mask** the expression of gene 2.

Colourless molecule → (gene 1 (YY or Yy)) → Yellow pigment → (gene 2 (RR or Rr)) → Orange pigment

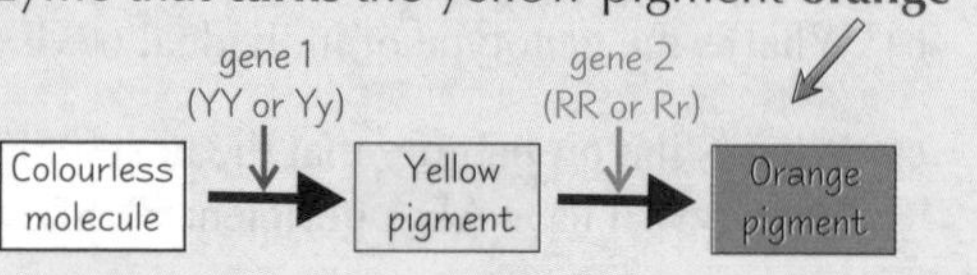

3) **Crosses** involving epistatic genes **don't result** in the **expected phenotypic ratios** given above, e.g. if you cross **two heterozygous orange** flowered plants (YyRr) from the above example you wouldn't get the expected **9 : 3 : 3 : 1** phenotypic ratio for a **normal dihybrid cross** (see next page).

Phenotypic Ratios and Epistasis

*You can **Predict** the **Phenotypic Ratios** for Some **Epistatic Genes***

Just as you can **predict** the phenotypic ratios for **normal dihybrid crosses** (see previous page), you can predict the phenotypic ratios for dihybrid crosses involving some **epistatic genes** too:

A dihybrid cross involving a recessive epistatic allele — 9 : 3 : 4

Having **two copies** of the **recessive** epistatic allele **masks (blocks)** the expression of the **other gene**. If you cross a **homozygous recessive** parent with a **homozygous dominant** parent you will get a **9 : 3 : 4** phenotypic ratio of **dominant both : dominant epistatic recessive other : recessive epistatic** in the **F_2 generation.**

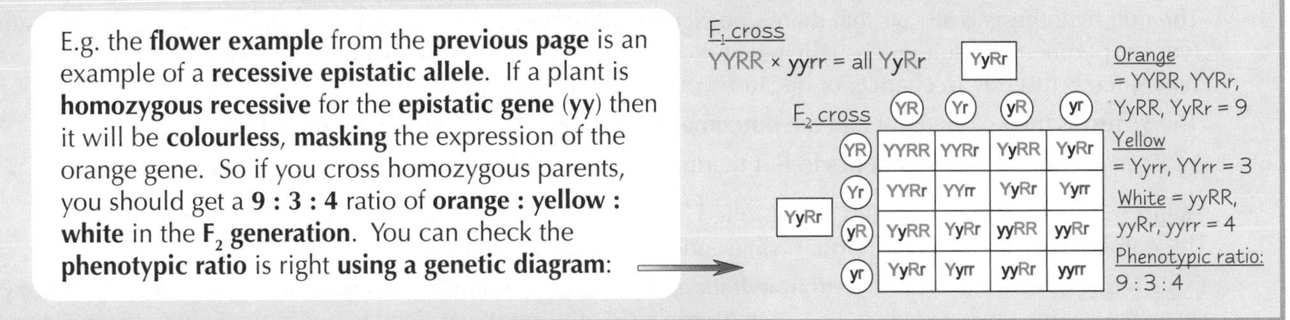

E.g. the **flower example** from the **previous page** is an example of a **recessive epistatic allele**. If a plant is **homozygous recessive** for the **epistatic gene** (**yy**) then it will be **colourless**, **masking** the expression of the orange gene. So if you cross homozygous parents, you should get a **9 : 3 : 4** ratio of **orange : yellow : white** in the **F_2 generation**. You can check the **phenotypic ratio** is right **using a genetic diagram**:

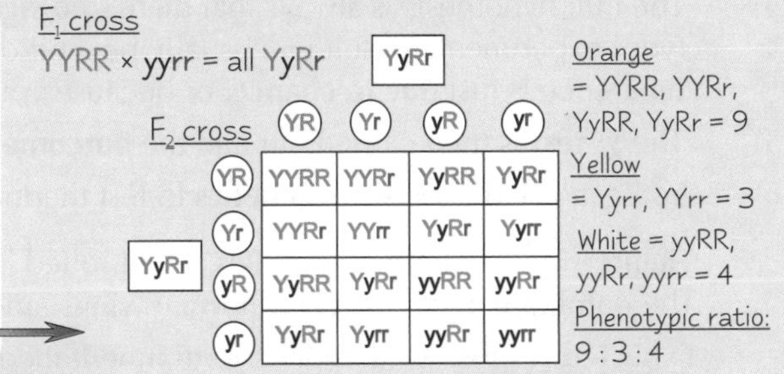

	YR	Yr	yR	yr
YR	YYRR	YYRr	YyRR	YyRr
Yr	YYRr	YYrr	YyRr	Yyrr
yR	YyRR	YyRr	yyRR	yyRr
yr	YyRr	Yyrr	yyRr	yyrr

A dihybrid cross involving a dominant epistatic allele — 12 : 3 : 1

Having **at least one** copy of the **dominant epistatic** allele **masks (blocks)** the expression of the other gene. Crossing a **homozygous recessive** parent with a **homozygous dominant** parent will produce a **12 : 3 : 1** phenotypic ratio of **dominant epistatic : recessive epistatic dominant other : recessive both** in the F_2 generation.

E.g. **squash colour** is controlled by two genes — the **colour epistatic gene** (**W/w**) and the **yellow gene** (**Y/y**). The **no-colour, white** allele (**W**) is **dominant** over the **coloured** allele (**w**), so **WW** or **Ww** will be **white** and **ww** will be **coloured**. The yellow gene has the **dominant yellow** allele (**Y**) and the **recessive green** allele (**y**). So if the plant has **at least one W**, then the squash **will be white**, **masking** the expression of the yellow gene. So if you cross **wwyy** with **WWYY**, you'll get a **12 : 3 : 1** ratio of **white : yellow : green** in the F_2 generation. Here's a **genetic diagram** to prove it:

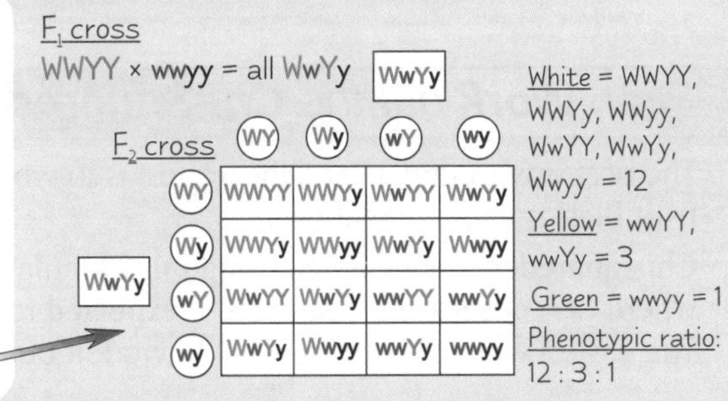

	WY	Wy	wY	wy
WY	WWYY	WWYy	WwYY	WwYy
Wy	WWYy	WWyy	WwYy	Wwyy
wY	WwYY	WwYy	wwYY	wwYy
wy	WwYy	Wwyy	wwYy	wwyy

Practice Questions

Q1 What phenotypic ratio would be produced in the F_1 generation and the F_2 generation by the cross aabb × AABB (assuming no epistasis)?

Q2 Describe epistasis.

Exam Questions

Q1 Colour (R red, r pink) and lines (G green, g white) are controlled by two genes in the Snozcumber plant. Draw a genetic diagram of the cross: homozygous for red and white lines × homozygous for pink and green lines. [3 marks]

Q2 Coat colour in mice is controlled by two genes. Gene 1 controls whether fur is coloured (C) or albino (c). Gene 2 controls whether the colour is grey (G) or black (g). Gene 1 is epistatic over gene 2. Describe and explain the phenotypic ratio produced in the F_2 generation from a CCGG × ccgg cross. [4 marks]

Q3 Hair type in Dillybopper beetles is controlled by two genes: hair (H bald, h hair) and type (S straight, s curly). The F_2 offspring of a cross are shown in the table. Explain the phenotypic ratio shown by the cross. [3 marks]

Homozygous curly hair (hhss) crossed with a homozygous bald (HHSS)

Phenotypes of the F_2 offspring produced		
Bald	Straight hair	Curly hair
36	9	3

Biology students — 9 : 1 phenotypic ratio normal : geek...

I don't know about you but I think I need a lie-down after these pages. Epistasis is a bit of a tricky topic, but you just need to understand what it is and learn the phenotypic ratios for the different types of epistasis — dominant and recessive.

The Chi-Squared Test

These pages are for OCR Unit 5 only.

Just when you thought it was safe to turn the page... I stick in some maths. Surprise!

The *Chi-Squared Test* Can Be Used to *Check* the *Results* of *Genetic Crosses*

1) The **chi-squared (χ^2) test** is a **statistical test** that's used to see if the **results** of an experiment **support** a **theory**.
2) First, the theory is used to **predict** a **result** — this is called the **expected result**.
Then, the experiment is carried out and the **actual result** is recorded — this is called the **observed result**.
3) To see if the results support the theory you have to make a **hypothesis** called the **null hypothesis**.
4) The null hypothesis is always that there's **no significant difference** between the observed and expected results (your experimental result will usually be a bit different from what you expect, but you need to know if the difference is just **due to chance**, or because your **theory is wrong**).
5) The **χ^2 test** is then carried out and the **outcome** either **supports** or **rejects** the **null hypothesis**.
6) You can use the χ^2 test in **genetics** to test theories about the **inheritance** of **characteristics**. For example:

Theory: **Wing length** in fruit flies is controlled by a **single gene** with **two alleles** (**monohybrid inheritance**). The **dominant** allele (N) gives **normal** wings, and the **recessive** allele (n) gives **vestigial** wings.

Expected results: With monohybrid inheritance, if you cross a **homozygous dominant** parent with a **homozygous recessive** parent, you'd expect a **3 : 1 phenotypic ratio** of **normal : vestigial** wings in the F_2 generation (see p. 62).

Observed results: The **experiment** (of crossing a homozygous dominant parent with a homozygous recessive parent) is **carried out** on fruit flies and the **number of offspring** with normal and vestigial wings is **counted**.

Null hypothesis: There's **no significant difference** between the observed and expected results.

(If the χ^2 test shows the observed and expected results are **not significantly different** the null hypothesis is **accepted** — the data supports the **theory** that wing length is controlled by **monohybrid inheritance**.)

First, *Work* out the *Chi-Squared Value...*

The best way to understand the χ^2 test is to work through an example — here's one for testing the **wing length** of **fruit flies** as explained above.

Chi-squared χ^2 is calculated using this formula: $\chi^2 = \sum \frac{(O-E)^2}{E}$

where **O** = **observed** result and **E** = **expected** result.

The easiest way to calculate χ^2 is to work it out in **stages** using a table:

You don't need to learn the formula for chi-squared — it'll be given to you in the exam.

1. First, the **number of offspring** (out of a total of 160) **expected** for each phenotype is worked out. E for normal wings: 160 (total) ÷ 4 (ratio total) × 3 (predicted ratio for normal wings) = 120. E for vestigial wings: 160 ÷ 4 × 1 = 40.

Phenotype	Ratio	Expected Result (E)	Observed Result (O)
Normal wings	3	120	
Vestigial wings	1	40	

2. Then the **actual number** of offspring **observed** with each phenotype (out of the 160 offspring) is **recorded**, e.g. 111 with normal wings.

Phenotype	Ratio	Expected Result (E)	Observed Result (O)
Normal wings	3	120	111
Vestigial wings	1	40	49

3. The results are used to work out χ^2, taking it **one step at a time**:

   a. First calculate **O – E** (**subtract** the **expected result** from the **observed result**) for each phenotype. E.g. for normal wings: 111 – 120 = –9.

   b. Then the resulting numbers are **squared**, e.g. $9^2 = 81$

   c. These figures are divided by the **expected results**, e.g. 81 ÷ 120 = 0.675.

   d. Finally, the numbers are **added** together to get χ^2, e.g. 0.675 + 2.025 = **2.7**.

Phenotype	Ratio	Expected Result (E)	Observed Result (O)	O – E	$(O-E)^2$	$\frac{(O-E)^2}{E}$
Normal wings	3	120	111	–9	81	0.675
Vestigial wings	1	40	49	9	81	2.025
					$\sum\frac{(O-E)^2}{E} =$	2.7

Remember, you need to work it out for each phenotype first, then add all the numbers together.

The Chi-Squared Test

...Then Compare it to the Critical Value

1) To find out if there **is** no significant difference between your observed and expected results you need to **compare** your **χ^2 value** to a **critical value**.
2) The critical value is the value of χ^2 that corresponds to a 0.05 (**5%**) level of **probability** that the **difference** between the observed and expected results is **due to chance**.
3) If your χ^2 value is **smaller** than the critical value then there **is no significant difference** between the observed and expected results — the **null hypothesis** is **accepted**. E.g. for the example on the previous page the χ^2 value is **2.7**, which is **smaller** than the critical value of **3.84** — there's **no significant difference** between the observed and expected results. This means the **theory** that wing length in fruit flies is controlled by **monohybrid inheritance** is **supported**.
4) If your χ^2 value is **larger** than the critical value then there **is a significant difference** between the observed and expected results (something **other than chance** is causing the difference) — the **null hypothesis** is **rejected**.
5) In the exam you might be **given** the **critical value** or asked to **work it out** from a **table**:

Using a χ^2 table:

If you're not given the critical value, you may have to find it yourself from a **χ^2 table** — this shows a range of **probabilities** that correspond to different **critical values** for different **degrees of freedom** (explained below). Biologists normally use a **probability** level of **0.05** (5%), so you only need to look in that column.

- First, the **degrees of freedom** for the experiment are worked out — this is the **number of classes** (number of phenotypes) **minus one**. E.g. 2 – 1 = 1.
- Next, the **critical value** corresponding to a **probability** of **0.05** at **one degree of freedom** is found in the table — here it's **3.84**.
- Then just **compare** your χ^2 value of **2.7** to this critical value, as explained above.

degrees of freedom	no. of classes	Critical values					
1	2	0.46	1.64	2.71	3.84	6.64	10.83
2	3	1.39	3.22	4.61	5.99	9.21	13.82
3	4	2.37	4.64	6.25	7.82	11.34	16.27
4	5	3.36	5.99	7.78	9.49	13.28	18.47
probability that result is due to chance only		0.50 (50%)	0.20 (20%)	0.10 (10%)	0.05 (5%)	0.01 (1%)	0.001 (0.1%)

Practice Questions

Q1 What is a χ^2 test used for?

Q2 What can the results of the χ^2 test tell you?

Q3 How do you tell if the difference between your observed and expected results is due to chance?

Exam Question

Q1 A scientist is investigating petal colour in a flower. It's thought to be controlled by two separate genes (dihybrid inheritance), the colour gene — B = blue, b = purple, and the spots gene — W = white, w = yellow. A cross involving a homozygous dominant parent and a homozygous recessive parent should give a 9 : 3 : 3 : 1 ratio in the F_2 generation. The scientist observes the number of offspring showing each of four phenotypes in 240 F_2 offspring. Her results are shown in the table.
Her null hypothesis is that there is no significant difference between the observed and expected ratios.

a) Complete the table to calculate χ^2 for this experiment. [4 marks]

b) The critical value for this experiment is 7.82. Explain whether the χ^2 value supports or rejects the null hypothesis. [2 marks]

Phenotype	Ratio	Expected Result (E)	Observed Result (O)	O – E	O – E^2	$\frac{(O-E^2)}{E}$
Blue with white spots	9	135	131			
Purple with white spots	3	45	52			
Blue with yellow spots	3	45	48			
Purple with yellow spots	1	15	9			
					$\sum\frac{(O-E)^2}{E}$ =	

The expected result of revising these pages — boredom...

...the observed result — boredom (except for the maths geeks among you). Don't worry if you're not brilliant at maths though, you don't have to be to do the chi-squared test — just make sure you know the steps above off by heart. You could even practise going through the example on these pages without looking at the book... go on, you know you want to.

Meiosis

These pages are for OCR Unit 5 only.

If you thought mitosis was exciting at AS, you'll have a blast with meiosis...

DNA from One Generation is Passed to the Next by Gametes

1) **Gametes** are the **sperm** cells in males and **egg** cells in females. They join together at **fertilisation** to form a **zygote**, which divides and develops into a **new organism**.
2) Normal **body cells** have the **diploid number** (**2n**) of chromosomes — meaning each cell contains **two** of each chromosome, one from the mum and one from the dad.
3) **Gametes** have the **haploid** (**n**) number — there's only one copy of each chromosome.
4) At **fertilisation**, a **haploid sperm** fuses with a **haploid egg**, making a cell with the **normal diploid number** of chromosomes (2n).

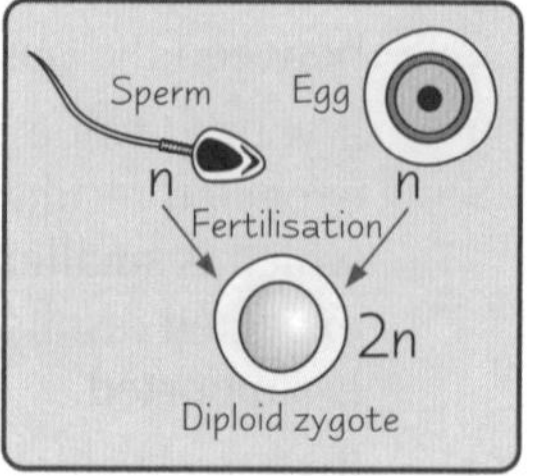

Gametes are Formed by Meiosis

Meiosis is a type of **cell division** that happens in the reproductive organs to **produce gametes**. Cells that divide by meiosis are **diploid** to start with, but the cells that are formed from meiosis are **haploid** — the chromosome number halves. Cells formed by meiosis are all **genetically different** because each new cell ends up with a **different combination** of chromosomes.

Gametes Divide Twice in Meiosis

Before meiosis, **interphase** happens — the cell's DNA unravels and **replicates** so there are **two** copies of each chromosome in each cell. Each copy of the chromosome is called a **chromatid** and a pair are called **sister chromatids** — they're joined in the middle by a **centromere**. After interphase, the cells enter meiosis where they **divide twice** — the first division is called **meiosis I** and the second is called **meiosis II**. There are **four similar stages** to each division called **prophase**, **metaphase**, **anaphase** and **telophase**:

Meiosis I

Prophase I
The **chromosomes condense**, getting shorter and fatter. **Homologous chromosomes pair up** — number 1 with number 1, 2 with 2, 3 with 3 etc. **Crossing-over** occurs (see next page). Tiny bundles of protein called **centrioles** start moving to opposite ends of the cell, forming a network of protein fibres across it called the **spindle**. The **nuclear envelope** (the membrane around the nucleus) **breaks down**.

Metaphase I
The **homologous pairs line up** across the **centre** of the cell and **attach** to the **spindle fibres** by their **centromeres**.

Anaphase I
The **spindles contract**, pulling the **pairs apart** (**one** chromosome goes to **each end** of the cell).

Telophase I
A **nuclear envelope** forms around each group of chromosomes and the **cytoplasm divides** so there are now **two haploid daughter cells**.

A pair of homologous chromosomes is also called a bivalent.

Meiosis II

The two daughter cells undergo **prophase II, metaphase II, anaphase II** and **telophase II** — which are pretty much the same as the ones in **meiosis I** except with **half** the number of chromosomes. In **anaphase II**, the **sister chromatids are separated** — each **new** daughter cell inherits **one chromatid** from **each chromosome. Four haploid** daughter cells are produced.

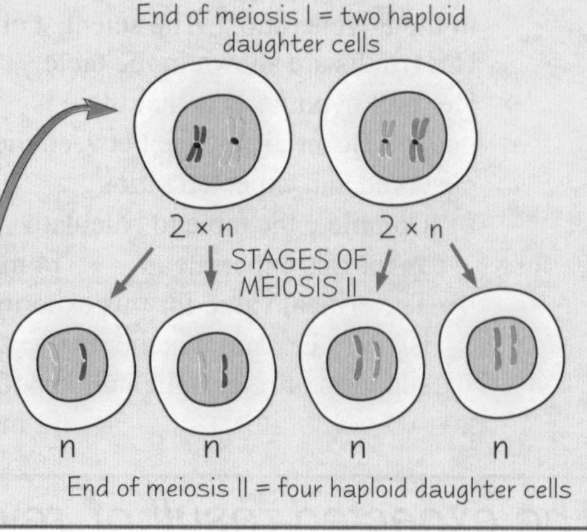

We've only shown 4 chromosomes — humans really have 46 (23 pairs).

Meiosis

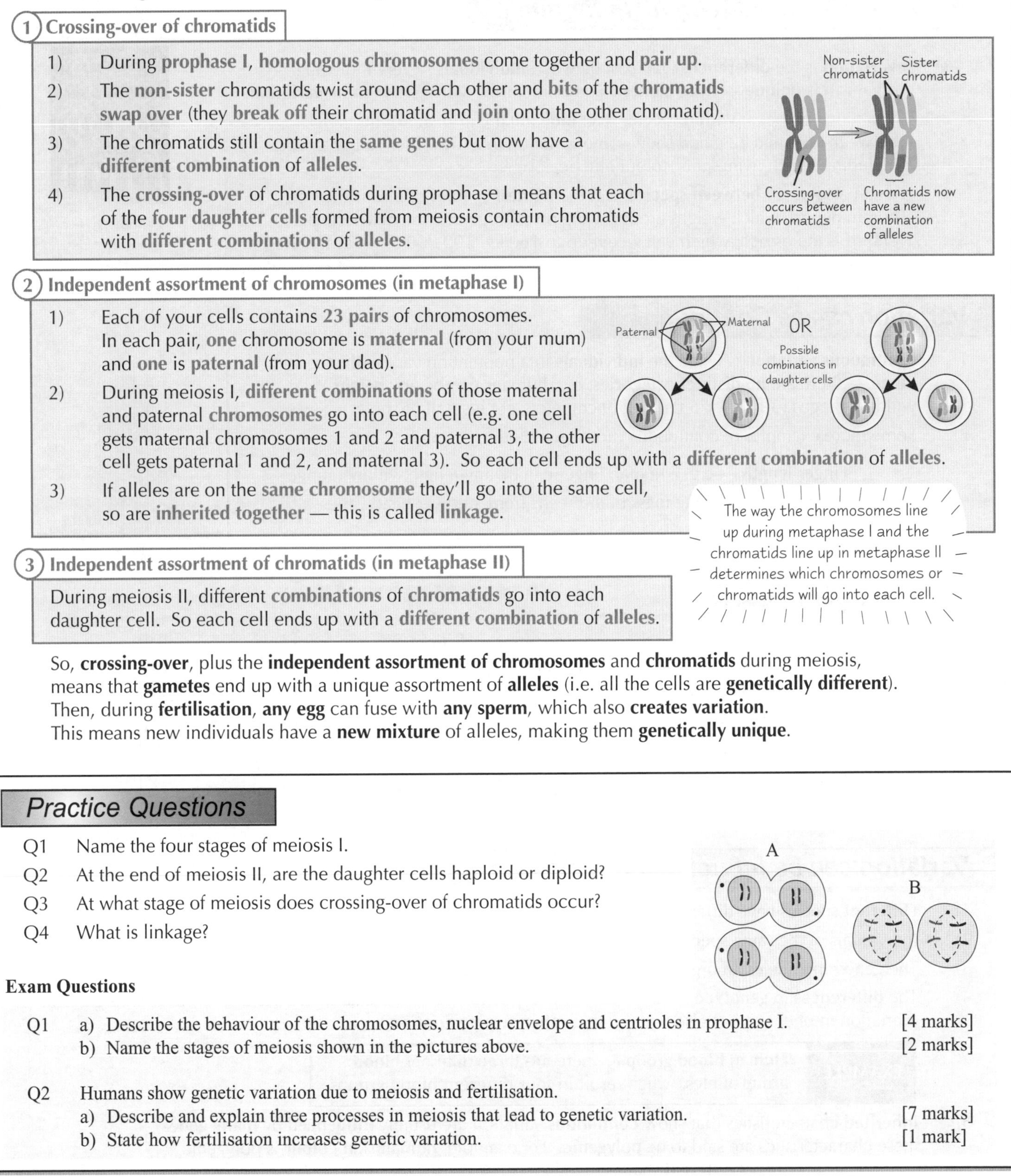

Meiosis Produces Cells that are Genetically Different

Genetic variation is the **differences** that exist between **individuals' genetic material**. The reason meiosis is important is that it **creates** genetic variation — it makes gametes that are all genetically different. It does this in three ways:

1) Crossing-over of chromatids

1) During **prophase I**, **homologous chromosomes** come together and **pair up**.
2) The **non-sister** chromatids twist around each other and **bits** of the **chromatids swap over** (they **break off** their chromatid and **join** onto the other chromatid).
3) The chromatids still contain the **same genes** but now have a **different combination** of **alleles**.
4) The **crossing-over** of chromatids during prophase I means that each of the **four daughter cells** formed from meiosis contain chromatids with **different combinations** of **alleles**.

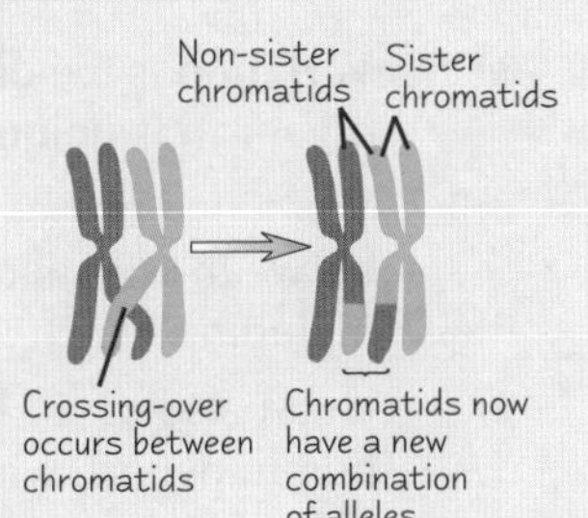

2) Independent assortment of chromosomes (in metaphase I)

1) Each of your cells contains **23 pairs** of chromosomes. In each pair, **one** chromosome is **maternal** (from your mum) and **one** is **paternal** (from your dad).
2) During meiosis I, **different combinations** of those maternal and paternal **chromosomes** go into each cell (e.g. one cell gets maternal chromosomes 1 and 2 and paternal 3, the other cell gets paternal 1 and 2, and maternal 3). So each cell ends up with a **different combination** of **alleles**.
3) If alleles are on the **same chromosome** they'll go into the same cell, so are **inherited together** — this is called **linkage**.

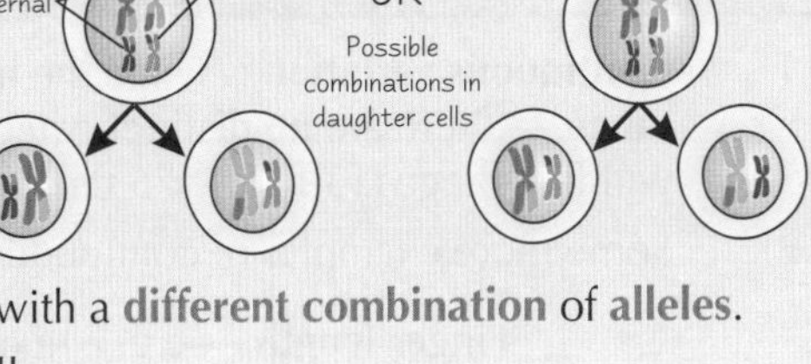

The way the chromosomes line up during metaphase I and the chromatids line up in metaphase II determines which chromosomes or chromatids will go into each cell.

3) Independent assortment of chromatids (in metaphase II)

During meiosis II, different **combinations** of **chromatids** go into each daughter cell. So each cell ends up with a **different combination** of **alleles**.

So, **crossing-over**, plus the **independent assortment of chromosomes** and **chromatids** during meiosis, means that **gametes** end up with a unique assortment of **alleles** (i.e. all the cells are **genetically different**). Then, during **fertilisation**, **any egg** can fuse with **any sperm**, which also **creates variation**. This means new individuals have a **new mixture** of alleles, making them **genetically unique**.

Practice Questions

Q1 Name the four stages of meiosis I.

Q2 At the end of meiosis II, are the daughter cells haploid or diploid?

Q3 At what stage of meiosis does crossing-over of chromatids occur?

Q4 What is linkage?

Exam Questions

Q1 a) Describe the behaviour of the chromosomes, nuclear envelope and centrioles in prophase I. [4 marks]

b) Name the stages of meiosis shown in the pictures above. [2 marks]

Q2 Humans show genetic variation due to meiosis and fertilisation.

a) Describe and explain three processes in meiosis that lead to genetic variation. [7 marks]

b) State how fertilisation increases genetic variation. [1 mark]

Physics — that's what I call crossing-over to the dark side...

You're probably sat there thinking about the good old days of AS, where meiosis didn't seem that hard... But, as your teachers will say, this is sooooo much more interesting. And I'm afraid that even if you don't agree with that, you still have to get your head around this lot. Go over it again and again until you start dreaming about chromosomes...

Variation

These pages are for OCR Unit 5 only. If you're doing AQA or Edexcel go straight to page 74.

Some people are tall, others are short. Some people wear glasses, others don't. Some people like peanut butter sandwiches, others... well, you get the picture. Basically variety is the spice of life and here's why we're all different.

Variation Exists Between **All Individuals**

1) **Variation** is the **differences** that exist between **individuals**. Every individual organism is **unique** — even **clones** (such as identical twins) show **some variation**.
2) Variation can occur **within species**, e.g. **individual** European robins weigh **between** 16 g and 22 g and show some variation in many other characteristics including length, wingspan, colour and beak size.
3) It can also occur **between species**, e.g. the **lightest** species of bird is the bee hummingbird, which weighs around 1.6 g on average and the **heaviest** species of bird is the ostrich, which can weigh up to 160 kg (100 000 times as much).

Variation — a concept lost on the army.

Variation can be **Continuous**...

1) **Continuous variation** is when the **individuals** in a population vary **within a range** — there are **no distinct categories**, e.g. **humans** can be **any height** within a range (139 cm, 175 cm, 185.9 cm, etc.), not just tall or short.
2) Some more examples of continuous variation include:
   - **Finger length** — e.g. a human finger can be any length within a range.
   - **Plant mass** — e.g. the mass of the seeds from a flower head varies within a range.

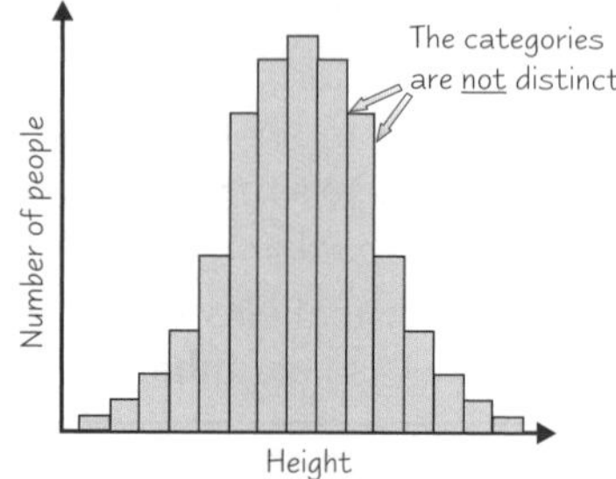

...or **Discontinuous**

1) **Discontinuous variation** is when there are two or more **distinct categories** — each individual falls into **only one** of these categories, there are **no intermediates**.
2) Here are some examples of discontinuous variation:
   - **Sex** — e.g. animals can be either male or female.
   - **Blood group** — e.g. humans can be group A, B, AB or O.

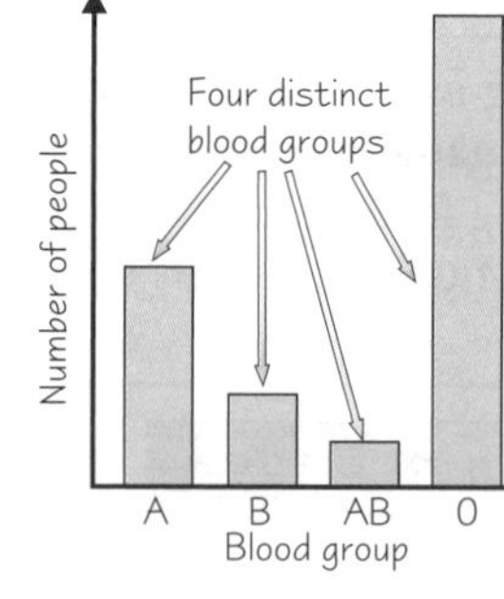

Variation can be **Influenced** by Your **Genes**...

1) **Different species** have **different genes**.
2) Individuals of the **same species** have the **same genes**, but **different versions** of them (called **alleles**).
3) The genes and alleles an organism has make up its **genotype**.
4) The **differences** in **genotype** result in **variation** in **phenotype** — the **characteristics** displayed by an organism. (Variation in phenotype is also referred to as **phenotypic variation**.)

EXAMPLE	**Human blood group** — there are **three** different **blood group alleles**, which result in **four different blood groups**.

5) **Inherited** characteristics that show **continuous** variation are usually **influenced** by **many genes** — these characteristics are said to be **polygenic**. For example, **human skin colour** is polygenic — it comes in **loads** of **different shades** of colour.
6) **Inherited** characteristics that show **discontinuous** variation are usually influenced by only **one gene** (or a **small number** of genes), e.g. **violet flower colour** (either coloured or white) is controlled by only one gene. Characteristics controlled by **only one gene** are said to be **monogenic**.

Variation

...the Environment...

Variation can also be caused by **differences in the environment**, e.g. climate, food, lifestyle. Characteristics controlled by environmental factors can **change** over an organism's life.

EXAMPLES

1) **Accent** — this is determined by **environmental factors only**, including **where you live** now, where you **grew up** and the accents of **people around you**.
2) **Pierced ears** — this is also **only** determined by **environmental factors**, e.g. **fashion, peer pressure**.

...or Both

Genetic factors determine genotype and the characteristics an organism's **born with**, but **environmental factors** can **influence** how some characteristics **develop**. Most phenotypic variation is caused by the **combination** of **genotype** and **environmental factors**. Phenotypic variation influenced by both usually shows **continuous variation**.

EXAMPLES

1) **Height of pea plants** — pea plants come in **tall** and **dwarf** forms (**discontinuous** variation), which is determined by **genotype**. However, the **exact height** of the tall and dwarf plants **varies** (**continuous** variation) because of **environmental factors** (e.g. **light intensity** and **water availability** affect how tall a plant grows).

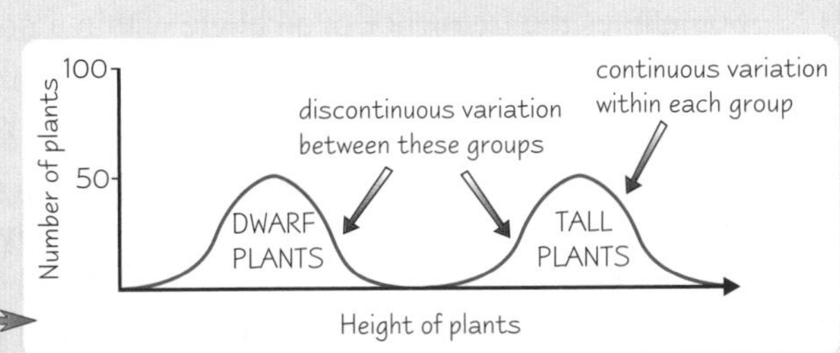

2) **Human body mass** — this is **partly genetic** (large parents often have large children), but it's also **strongly influenced** by **environmental factors**, like **diet** and **exercise**. Body mass **varies** within a **range**, so it's **continuous** variation.

Practice Questions

Q1 What is variation?

Q2 Describe what is meant by discontinuous variation and give one example.

Q3 Briefly describe what is meant by variation caused by genotype.

Puppy	Mass / kg	Colour
1	10.04	yellow
2	10.23	chocolate
3	15.65	black
4	18.99	black
5	9.45	black

Puppy	Mass / kg	Colour
6	10.39	yellow
7	10.55	chocolate
8	15.87	chocolate
9	16.99	black
10	10.47	yellow

Puppy	Mass / kg	Colour
11	9.25	black
12	11.06	black
13	12.45	yellow
14	14.99	yellow
15	10.93	chocolate

Exam Questions

Q1 The mass and coat colour of 15 Labrador puppies is shown in the table.

a) What type of variation (continuous or discontinuous) is shown by the coat colour of the puppies? [1 mark]

b) Calculate the range of puppy mass. [1 mark]

c) Which of the characteristics described in the table is most likely to be influenced by both genotype and the environment? Explain your answer. [2 marks]

Q2 Give an example of a human characteristic influenced by both genotype and the environment. Explain your answer. [2 marks]

Revision boredom shows discontinuous variation — always bored with it...

Hopefully you remember a lot of the info on these pages from AS, but I'm afraid you still need to know it off by heart for your A2 exam. Test yourself on examples of continuous and discontinuous variation — you never know when a sneaky question could pop up on them. Then, rest your brain so it's well and truly ready for a bit of natural selection...

Natural Selection and Genetic Drift

***This page is for** AQA Unit 4, OCR Unit 5 **and** Edexcel Unit 4.*

Variation between individuals of a species means that some organisms are better adapted to their environment than others — so they're more likely to survive and reproduce. Which leads us nicely on to the topic of natural selection...

Members of a Population Share a Gene Pool

1) A **species** is defined as a group of **similar organisms** that can **reproduce** to give **fertile offspring**.
2) A **population** is a group of organisms of the **same species** living in a **particular area**.
3) Species can exist as **one** or **more populations**, e.g. there are populations of the American black bear (*Ursus americanus*) in parts of America and in parts of Canada.
4) The **gene pool** is the complete range of **alleles** present in a **population**.
5) **New alleles** are usually generated by **mutations** in **genes**.
6) How **often** an **allele occurs** in a population is called the **allele frequency**. It's usually given as a **percentage** of the total population, e.g. 35%, or a **number**, e.g. 0.35.

Yogi wanted everyone to know what population he was in.

Allele Frequency is Affected by Differential Reproductive Success

1) Sometimes the **frequency** of an **allele** within a population **changes** — this is called **evolution**.
2) This can happen when the allele codes for a characteristic that **affects** the **chances** of an organism **surviving**.
3) Not all individuals are as likely to **reproduce** as each other because there's **variation** in the alleles they have.
4) This means that there's **differential reproductive success** in a population — individuals that have an allele that **increases** their **chance of survival** are **more likely** to **survive**, **reproduce** and **pass on** their genes (including the **beneficial** allele), than individuals with different alleles.
5) This means that a **greater proportion** of the next generation **inherit** the **beneficial allele**.
6) They, in turn, are **more likely** to **survive**, **reproduce** and **pass on** their genes.
7) So the **frequency** of the beneficial allele **increases** from generation to generation.
8) This process is called **natural selection**.

Different Types of Natural Selection Lead to Different Frequency Patterns

Stabilising selection and **directional selection** are **types** of **natural selection** that affect **allele frequency** in different ways:

Stabilising selection is where individuals with alleles for characteristics towards the **middle** of the range are more likely to **survive** and **reproduce**. It occurs when the environment **isn't changing**, and it **reduces the range** of possible **phenotypes**.

EXAMPLE In any **mammal population** there's a **range** of **fur length**. In a **stable climate**, having fur at the **extremes** of this range **reduces** the **chances** of **surviving** as it's harder to maintain the **right body temperature**. Animals with alleles for **average fur length** are the **most** likely to **survive**, **reproduce** and **pass on** their alleles. So these alleles **increase** in **frequency**. The **proportion** of the **population** with **average fur length increases** and the **range** of fur lengths **decreases**.

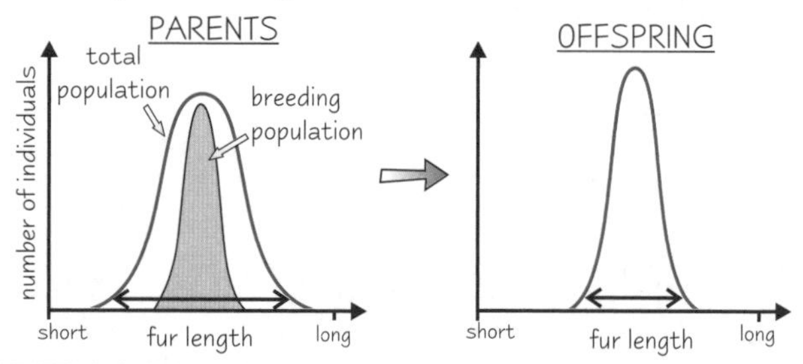

Directional selection is where individuals with alleles for characteristics of an **extreme type** are more likely to **survive** and **reproduce**. This could be in response to an **environmental change**.

EXAMPLE **Cheetahs** are the **fastest** animals on land. It's likely that this characteristic was developed through **directional selection**, as individuals that have **alleles** for **speed** are **more likely** to **catch prey** than slower individuals. So they're **more likely** to **survive**, **reproduce** and **pass on** their alleles. Over time the **frequency** of alleles for **high speed increases** and the population becomes **faster**.

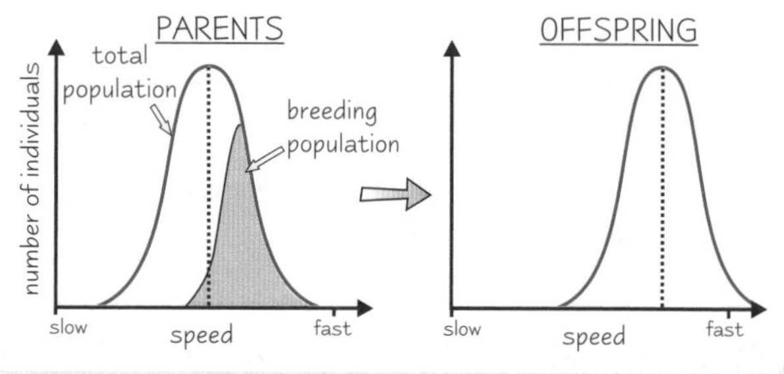

Natural Selection and Genetic Drift

This page is for OCR Unit 5 only. Those of you doing AQA and Edexcel can go straight to the questions.

Evolution Also Occurs via **Genetic Drift**

1) **Natural selection** is just **one** process by which **evolution** occurs.
2) Evolution **also** occurs due to **genetic drift** — instead of **environmental factors** affecting which individuals **survive**, **breed** and pass on their alleles, **chance** dictates **which alleles** are **passed on**. Here's how it works:

- Individuals within a population show **variation** in their **genotypes** (e.g. A and B).
- By **chance**, the **allele** for **one genotype** (B) is **passed on** to the offspring **more often** than others.
- So the number of individuals with the allele **increases**.
- If by chance the same allele is passed on more often again and again, it can lead to **evolution** as the allele becomes **more common** in the population.

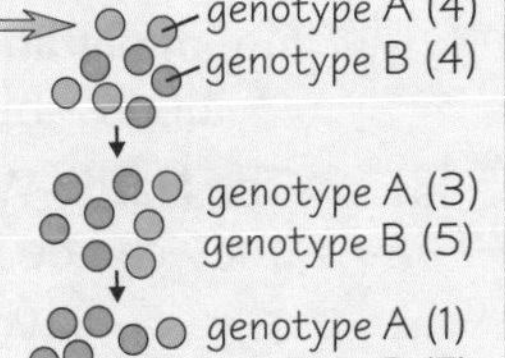

genotype A (4)
genotype B (4)
genotype A (3)
genotype B (5)
genotype A (1)
genotype B (7)

3) Natural selection and genetic drift work **alongside each other** to drive evolution, but one process can drive evolution **more** than the other depending on the **population size**.
4) **Evolution by genetic drift** usually has a **greater effect** in **smaller populations** where **chance** has a **greater influence**. In larger populations any chance factors tend to **even out** across the whole population.
5) The evolution of **human blood groups** is a good example of **genetic drift**:

Different **Native American tribes** show different **blood group frequencies**. For example, **Blackfoot Indians** are mainly **group A**, but **Navajos** are mainly **group O**. Blood group doesn't affect **survival** or **reproduction**, so the differences **aren't** due to evolution by natural selection. In the past, human populations were much **smaller** and were often found in **isolated groups**. The blood group differences were due to evolution by genetic drift — by **chance** the allele for **blood group O** was **passed on more often** in the **Navajo tribe**, so over time this **allele** and blood group became **more common**.

6) Evolution by genetic drift also has a greater effect if there's a **genetic bottleneck** — e.g. when a large population **suddenly becomes smaller** because of a **natural disaster**. For example:

The **mice** in a **large population** are either **black or grey**. The coat colour **doesn't** affect their **survival** or **reproduction**. A **large flood** hits the population and the **only survivors** are **grey** mice and **one black** mouse. **Grey** becomes the **most common colour** due to **genetic drift**.

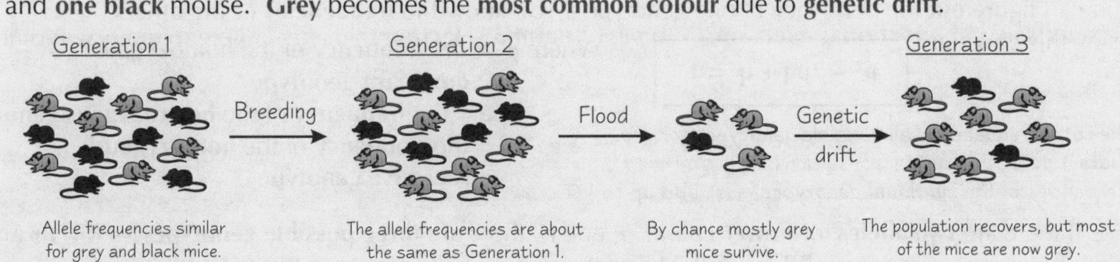

Generation 1
Generation 2
Generation 3
Breeding
Flood
Genetic drift
Allele frequencies similar for grey and black mice.
The allele frequencies are about the same as Generation 1.
By chance mostly grey mice survive.
The population recovers, but most of the mice are now grey.

Practice Questions

Q1 What term describes a change in the allele frequency within a population?

Q2 What is directional selection?

Q3 What is genetic drift?

Exam Question

Q1 During the 1800s, air pollution in Manchester rose as lots of coal-powered factories were built. Before the 1800s, paler coloured peppered moths were more common in the area. Towards the end of the 1800s, darker coloured peppered moths had become more common. Suggest what type of natural selection gave rise to the increase in darker peppered moths. Explain your answer. [2 marks]

Differential reproductive success — not PC, but it sorts the hot from the not...

The trickiest thing here is tying all the information together in your head. Basically, natural selection (and genetic drift for those of you doing OCR) drives evolution. Then there are two types of natural selection — stabilising and directional.

The Hardy-Weinberg Principle

This page is for AQA Unit 4 and OCR Unit 5. Those of you doing Edexcel can go straight to page 78.
Now you know what allele frequency is you need to be able to calculate it. So switch your maths brain on now...

The **Hardy-Weinberg Principle** Predicts That **Allele Frequencies Won't Change**

1) The **Hardy-Weinberg principle** predicts that the **frequencies** of **alleles** in a population **won't change** from **one generation** to the **next**.
2) But this prediction is **only true** under **certain conditions** — it has to be a **large population** where there's **no immigration**, **emigration**, **mutations** or **natural selection**. There also needs to be **random mating** — all possible genotypes can breed with all others.
3) The **Hardy-Weinberg equations** (see below) are based on this principle. They can be used to **estimate the frequency** of particular **alleles** and **genotypes** within populations.
4) The Hardy-Weinberg equations can also be used to test whether or not the Hardy-Weinberg principle **applies** to **particular alleles** in **particular populations**, i.e. to test whether **selection** or any **other factors** are **influencing** allele frequencies — if frequencies **do change** between generations in a large population then there's a pressure of some kind (see next page).

The **Hardy-Weinberg Equations** Can be Used to **Predict Allele Frequency...**

1) You can **figure out** the frequency of one allele if you **know the frequency of the other**, using this equation:

$$p + q = 1$$

Where: **p** = the **frequency** of the **dominant** allele
q = the **frequency** of the **recessive** allele

The **total frequency** of **all possible alleles** for a characteristic in a certain population is **1.0**. So the frequencies of the **individual alleles** (the dominant one and the recessive one) must **add up to 1.0**.

2) E.g. a species of plant has either **red** or **white** flowers. Allele **R** (red) is **dominant** and allele **r** (white) is **recessive**. If the frequency of **R** is **0.4**, then the frequency of **r** is 1 – 0.4 = **0.6.**

Make sure you **learn** both equations.

...**Genotype Frequency**...

1) You can **figure out** the frequency of one genotype if you **know the frequencies of the others**, using this equation:

$$p^2 + 2pq + q^2 = 1$$

Where p^2 = the **frequency** of the **homozygous dominant genotype**
2pq = the **frequency** of the **heterozygous genotype**
q^2 = the **frequency** of the **homozygous recessive genotype**

The **total frequency** of **all possible genotypes** for one characteristic in a certain population is **1.0**. So the frequencies of the **individual genotypes** must **add up to 1.0**.

2) E.g. if there are **two alleles** for **flower colour** (R and r), there are **three possible genotypes** — **RR, Rr** and **rr.** If the frequency of genotype **RR** (p^2) is **0.34** and the frequency of genotype **Rr** (**2pq**) is **0.27**, the frequency of genotype **rr** (q^2) must be 1 – 0.34 – 0.27 = **0.39.**

...and the **Percentage** of a **Population** that has a **Certain Genotype**

EXAMPLE

The **frequency** of **cystic fibrosis** (genotype ff) in the UK is currently approximately **1 birth in 2000**. From this information you can estimate the **proportion** of people in the UK that are cystic fibrosis **carriers** (Ff). To do this you need to find the **frequency** of **heterozygous genotype Ff**, i.e. **2pq**, using **both** equations:

First calculate q:
- Frequency of cystic fibrosis (homozygous recessive, ff) is 1 in 2000
- ff = q^2 = 1 ÷ 2000 = 0.0005
- So, q = $\sqrt{0.0005}$ = 0.022

Next calculate p:
- using p + q = 1, p = 1 – q
- p = 1 – 0.022
- p = 0.978

Then calculate 2pq:
- 2pq = 2 × 0.978 × 0.022
- 2pq = 0.043

The **frequency** of **genotype Ff** is **0.043**, so the **percentage** of the UK population that are **carriers** is **4.3%**.

The Hardy-Weinberg Principle

This page is for AQA Unit 4 only. Those of you doing OCR can go straight to the questions.

They can Also **Show if External Factors** are **Affecting Allele Frequency**

Example

If the **frequency** of **cystic fibrosis** is measured **50 years later** it might be found to be **1 birth in 3000**. From this information you can estimate the **frequency** of the **recessive allele** (f) in the population, i.e. **q**.

To calculate q:

- Frequency of cystic fibrosis (homozygous recessive, ff) is 1 in 3000
- ff = $q^2 = 1 \div 3000 = 0.00033$
- So, $q = \sqrt{0.00033} = 0.018$

The frequency of the recessive allele is now **0.018**, compared to **0.022** currently (see previous page). As the frequency of the allele has **changed** between generations the **Hardy-Weinberg principle doesn't apply** so there must have been some **factors** affecting **allele frequency**, e.g. **immigration, migration, mutations** or **natural selection.**

You Need to be able to **Interpret Data** Relating to the **Effect of Selection**

1) For example, there are **two common forms** of the peppered moth in the UK — one **dark coloured** and one **pale**.
2) The allele for **dark colouring** is **dominant** (**M**) over the allele for **pale colouring** (**m**).
3) The table shows how **allele frequency** in a population of **peppered moths** changed between 1852 and 1860 as the number of **coal-powered factories** in the area increased.
4) The frequency of the **m** allele **decreases** from **0.75** to **0.39** as the number of factories increases. So the frequency of the **M allele** must **increase** from **0.25** to **0.61** (remember: p + q = 1, see previous page).
5) As the allele frequencies are **changing**, it's likely there's selective pressure **for** dark colouring. This could be because of **pollution** — more factories means more pollution, which **darkens** buildings etc. The **dark coloured moths** would be better **camouflaged**, making them **more likely** to **survive**, **reproduce** and pass on **M**.

Year	Number of coal-powered factories	Frequency of m allele
1852	1	0.75
1854	3	0.70
1856	5	0.63
1858	7	0.47
1860	10	0.39

Practice Questions

Q1 What does the Hardy-Weinberg principle predict about allele frequency?

Q2 What conditions are needed for the Hardy-Weinberg principle to apply?

Q3 Which term represents the frequency of the dominant allele in the Hardy-Weinberg equations?

Q4 Which term represents the frequency of the heterozygous genotype in the Hardy-Weinberg equations?

Exam Questions

Q1 A species of dog has either a black or brown coat. Allele B (black) is dominant and allele b (brown) is recessive. If the frequency of the b allele is 0.23, what is the frequency of the B allele? [1 mark]

Q2 Cleft chins are controlled by a single gene with two alleles. The allele coding for a cleft chin (C) is dominant over the allele coding for a non-cleft chin (c). In a particular population the frequency of the homozygous dominant genotype for cleft chin is 0.14.

a) What is the frequency of the recessive allele in the population? [3 marks]

b) What is the frequency of the homozygous recessive genotype in the population? [1 mark]

c) What percentage of the population have a cleft chin? [2 marks]

This stuff's surely not that bad — Hardly worth Weining about...

Not many of you will be thrilled with the maths content of these pages, but don't worry 'cause you just need to know the equations off by heart and what the terms mean in the equations. Then in the exam you'll be able to put the numbers in the correct places in the equation and, hey presto, you'll have your answer. Oh, and don't forget to take a calculator...

Speciation

This page is for AQA Unit 4, OCR Unit 5 and Edexcel Unit 4.

Evolution leads to the development of lots of different species. Unfortunately for some species, the biologists had run out of good names, e.g. Colon rectum *(a type of beetle) and* Aha ha *(an Australian wasp). Oh dear.*

Speciation** is the Development of a **New Species

1) A **species** is defined as a group of **similar organisms** that can **reproduce** to give **fertile offspring**.
2) **Speciation** is the development of a **new species**.
3) It occurs when **populations** of the **same species** become **reproductively isolated** — changes in allele frequencies cause changes in phenotype that mean they can **no longer breed** together to produce **fertile offspring**.

Geographical Isolation** and **Natural Selection** Lead to **Speciation

1) Geographical isolation happens when a **physical barrier divides** a population of a species — **floods**, **volcanic eruptions** and **earthquakes** can all cause barriers that isolate some individuals from the main population.
2) **Conditions** on either side of the barrier will be slightly **different**. For example, there might be a **different climate** on each side.

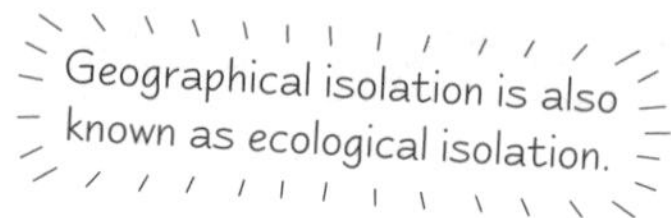

3) Because the environment is **different** on each side, different **characteristics** (phenotypes) will become more common due to natural selection:
   - Because different **characteristics** will be **advantageous** on each side, the **allele frequencies** will change in each population, e.g. if one allele is more advantageous on one side of the barrier, the frequency of that allele on that side will **increase**.
   - **Mutations** will take place **independently** in each population, also changing the **allele frequencies**.
   - The changes in allele frequencies will lead to changes in **phenotype frequencies**, e.g. the advantageous characteristics (**phenotypes**) will become more common on that side.
4) Eventually, individuals from different populations will have changed so much that they won't be able to breed with one another to produce **fertile** offspring — they'll have become **reproductively isolated**.
5) The two groups will have become separate **species**.

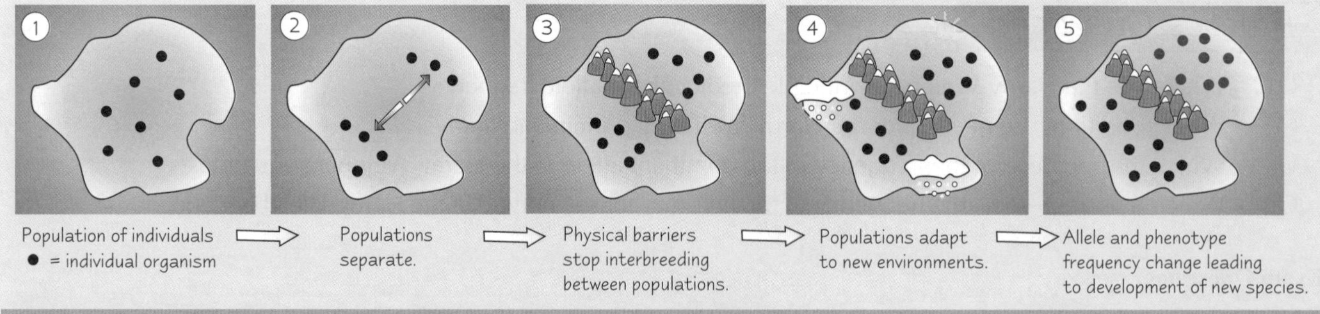

***Reproductive Isolation** Occurs in Many Ways*

Reproductive isolation occurs because the **changes** in the alleles and phenotypes of the two populations **prevent** them from **successfully breeding together**. These changes include:

1) **Seasonal changes** — individuals from the same population develop different **flowering** or **mating** seasons, or become **sexually active** at **different times** of the year.
2) **Mechanical changes** — changes in **genitalia** prevent successful mating.
3) **Behavioural changes** — a group of individuals develop **courtship rituals** that **aren't attractive** to the main population.

Janice's courtship ritual was still successful in attracting mates.

A population **doesn't** have to become **geographically isolated** to become **reproductively isolated**. Random mutations could occur **within a population**, resulting in the changes mentioned above, **preventing** members of that population breeding with other members of the species.

Speciation

This page is for OCR Unit 5. Those of you doing AQA or Edexcel can go straight to the questions.

There are Different Ways to Classify Species

1) The traditional definition of a species is a group of **similar organisms** that can **reproduce** to give **fertile offspring**. This way of defining a species is called the **biological species concept**.
2) Scientists can have problems when using this definition, e.g. problems deciding **which species** an organism belongs to or if it's a new, **distinct species**.
3) This is because you can't always see their **reproductive behaviour** — you can't always tell if different organisms can reproduce to give **fertile offspring**. For example:
    1) They might be **extinct**, so you **can't** study their reproductive behaviour.
    2) They might **reproduce asexually** — they never **reproduce together** even if they belong to the same species, e.g. bacteria.
    3) There might be **practical** and **ethical issues** involved — you can't see if some organisms reproduce successfully in the wild (due to geography) and you can't study them in a lab (because it's unethical, e.g. humans and chimps are classed as separate species but has anyone ever tried mating them...).
4) Because of these problems, scientists sometimes use the **phylogenetic species concept** to classify organisms.
5) Phylogenetics is the **study** of the **evolutionary history** of groups of organisms (you might remember it from AS).
6) All organisms have **evolved** from shared common ancestors (**relatives**). The **more closely related** two species are, the **more recently** their last common ancestor will be.
7) Phylogenetics tells us **what's related** to what and how **closely related** they are.
8) Scientists can use phylogenetics to decide **which species** an organism belongs to or if it's a **new species** — if it's **closely related** to members of another species then it's probably the **same species**, but if it's **quite different** to any known species it's probably a **new species**.
9) There are also **problems** with classifying organisms using this concept, e.g. there's no cut-off to say how different two organisms have to be to be different species. For example, **chimpanzees** and **humans** are **different species** but about **94%** of our DNA is exactly the **same**.

The phylogenetic concept is also called the cladistic or evolutionary species concept.

Practice Questions

Q1 What is speciation?

Q2 What two concepts can be used to classify a species?

Exam Question

Q1 The diagram shows an experiment conducted with fruit flies. One population was split in two and each population was fed a different food. After many generations the two populations were placed together and it was observed that they were unable to breed together.

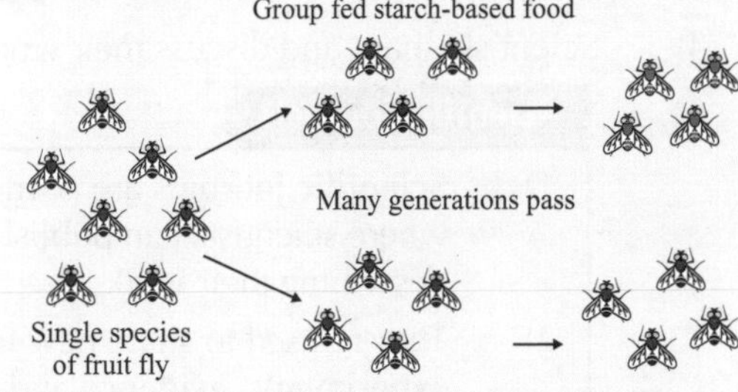

a) What evidence shows that speciation occurred? [1 mark]

b) Explain why the experiment resulted in speciation. [3 marks]

c) Suggest two possible reasons why members of the two populations were not able to breed together. [2 marks]

d) During the experiment, populations of fruit flies were artificially isolated. Suggest one way that populations of organisms could become isolated naturally. [1 mark]

Chess club members — self-enforced reproductive isolation...

These gags get better and better... Anyway, it's a bit of a toughie getting your head round the different mechanisms that can produce a new species. It doesn't help that reproductive isolation can happen on its own OR as a result of geographical isolation. Also, when reproductive isolation is caused by a seasonal change it's sometimes called seasonal isolation.

Evolution Evidence and Artificial Selection

This page is for Edexcel Unit 4 only. Those of you doing OCR, go straight to the next page.
If you're doing AQA you're done for this section (yay).

In science, you can't just do some research and then wave it about like it's pure fact. Oh no. A load of other scientists stick their noses in first — and then if you're lucky, your evidence will be accepted. Take evolution for example...

There's **Plenty of Evidence** to **Support Evolution**

The theory of evolution has been around for quite a long time now and there's plenty of **evidence** to support it. Fairly **new** evidence includes some from **molecular biology** — the study of **molecules** such as **DNA** and **proteins**:

DNA evidence — The theory of evolution suggests that all organisms have **evolved** from shared **common ancestors**. Closely related species **diverged** (evolved to become different species) **more recently**. Evolution is caused by **gradual changes** in the **base sequence** of organisms' **DNA**. So, organisms that diverged away from each other more recently should have **more similar DNA**, as **less time** has passed for changes in the DNA sequence to occur. This is exactly what scientists have found.

Example — Humans, chimps and mice all evolved from a common ancestor. Humans and mice diverged a **long time ago**, but humans and chimps diverged **quite recently**. The **DNA base sequence** of humans and chimps is 94% the same, but human and mouse DNA is only 85% the same.

Proteomics — this is the study of **proteins**, e.g. the study of the **size**, **shape** and **amino acid sequence** of **proteins**. The **sequence** of **amino acids** in a protein is **coded for** by the **DNA sequence** in a gene. **Related** organisms have **similar DNA sequences** and so **similar amino acid sequences** in their proteins. So organisms that diverged away from each other **more recently** should have **more similar proteins**, as **less time** has passed for changes to occur. This is what scientists have found.

The **Scientific Community Validates Evidence** About **Evolution**

1) The job of a scientist is to collect **data** and use it to **test theories** and **ideas** — the data either **supports** the theory (it's **evidence for it**) or it doesn't (it's **evidence to disprove it**). E.g. DNA and proteomic data has been collected that provides evidence for the theory of evolution.
2) The **scientific community** is all the scientists around the world, e.g. researchers, technicians and professors.
3) Scientists within the scientific community **accept** the theory of **evolution** because they've **shared** and **discussed** the evidence for evolution to make sure it's **valid** and **reliable**.
4) Scientists share and discuss their work in **three main ways**:

Scientific journals

1) **Scientific journals** are **academic magazines** where scientists can publish **articles** describing their work.
2) They're used to share new **ideas**, **theories**, **experiments**, **evidence** and **conclusions**.
3) Scientific journals allow other scientists to repeat experiments and see if they get the **same results** using the **same methods**.
4) If the results are **replicated** over and over again, the scientific community can be pretty confident that the evidence collected is **reliable**.

Peer review

1) **Before** scientists can get their work **published** in a journal it has to undergo something called the **peer review process**.
2) This is when **other scientists** who work in that area (**peers**) read and **review** the work.
3) The peer reviewer has to **check** that the work is **valid** and that it **supports** the **conclusions**.
4) Peer review is used by the scientific community to try and make sure that any scientific evidence that's published is **valid** and that experiments are carried out to the **highest possible standards**.

Conferences

Scientific conferences are **meetings** that scientists attend so they can **discuss** each other's work. Scientists with important or interesting results might be invited to present their work in the form of a **lecture** or **poster presentation**. Other scientists can then **ask questions** and **discuss** their work with them **face to face**. Conferences are valuable because they're an **easy way** for the latest theories and evidence to be **shared** and **discussed**.

Evolution Evidence and Artificial Selection

This page is for OCR Unit 5 only. Those of you doing Edexcel can go straight to the questions.

Artificial Selection Involves *Breeding* Individuals with *Desirable Traits*

Artificial selection is when **humans select individuals** in a population to **breed together** to get **desirable traits**. There are two examples you need to **learn**:

Artificial selection is also called selective breeding, which you might remember from AS.

Modern Dairy Cattle

Modern **dairy cows** produce **many litres of milk** a day as a result of **artificial selection**:

1) Farmers **select** a **female** with a **very high milk yield** and a **male** whose **mother** had a very high milk yield and **breed** these two **together**.
2) Then they **select** the **offspring** with the **highest milk yields** and **breed** them **together**.
3) This is continued over **several generations** until a **very high milk-yielding cow** is produced.

Bread Wheat

Bread wheat (*Triticum aestivum*) is the plant from which **flour** is produced for **bread-making**. It produces a **high yield** of wheat because of **artificial selection** by **humans**:

1) Wheat plants with a **high wheat yield** (e.g. large ears) are **bred together**.
2) The **offspring** with the **highest yields** are then **bred together**.
3) This is continued over **several generations** to produce a plant that has a **very high yield**.

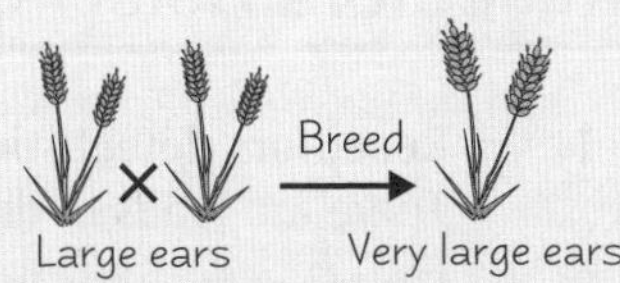

You Need to be Able to Compare Natural Selection *and* Artificial Selection

You need to be able to describe the **similarities** and **differences** between **natural** and **artificial selection**:

<u>Similarities:</u>

- Both change the **allele frequencies** in the next generation — the **alleles** that **code** for the **beneficial/desirable characteristic** will become **more common** in the next generation.
- Both may make use of **random mutations** when they occur — if a random mutation produces an **allele** that gives a **beneficial/desirable phenotype**, it will be **selected for** in the next generation.

<u>Differences:</u>

- In natural selection, the organisms that reproduce are **selected by the environment** but in artificial selection this is **carried out by humans**.
- Artificial selection aims for a **predetermined result**, e.g. a farmer aims for a higher yield of milk, but in natural selection the **result** is **unpredictable**.
- Natural selection makes the species **better adapted** to the **environment**, but artificial selection makes the species **more useful** for **humans**.

Practice Questions

Q1 What is proteomics?

Q2 Describe two similarities between natural and artificial selection.

Exam Questions

Q1 Scientific journals publish evidence for theories, such as the theory of evolution. Explain how the scientific community checks that evidence published in scientific journals is valid and reliable. [4 marks]

Q2 Modern beef cattle (raised for meat production) produce a very high meat yield. Explain how artificial selection by farmers could have led to this. [3 marks]

Peer review — checking out who's got the latest mobile phone...

Right-o, for both of these pages you need to keep testing yourself on the information until you're so sick of it you want to throw the book out of the window (takes about 3 goes). Then you need to retrieve your book for the next section...

Nervous and Hormonal Communication

***This section is for** AQA Unit 5, OCR Units 4 & 5 **and** Edexcel Unit 5.*

Right, it's time to get your brain cells fired up and get your teeth stuck into a mammoth — a mammoth section, that is...

Responding to their Environment Helps Organisms Survive

1) **Animals increase** their **chances** of **survival** by **responding** to **changes** in their **external environment**, e.g. by **avoiding harmful environments** such as places that are too hot or too cold.
2) They also **respond** to **changes** in their **internal environment** to make sure that the **conditions** are always **optimal** for their **metabolism** (all the chemical reactions that go on inside them).
3) **Plants** also **increase** their **chances** of **survival** by **responding** to **changes** in their **environment** (see p. 106).
4) Any **change** in the internal or external **environment** is called a **stimulus.**

Receptors Detect Stimuli and Effectors Produce a Response

1) **Receptors detect stimuli** — they can be **cells** or **proteins** on **cell surface membranes.** There are **loads** of **different types** of receptors that detect **different stimuli.**
2) **Effectors** are cells that bring about a **response** to a **stimulus**, to produce an **effect.** Effectors include **muscle cells** and cells found in **glands**, e.g. the **pancreas.**
3) Receptors **communicate** with effectors via the **nervous system** (see below) or the **hormonal system** (see p. 84), or sometimes using **both.**

Cells Communicate with Each Other by Cell Signalling

OCR Unit 4 only

1) Cells need to **communicate** with each other to **control processes** inside the body and to **respond** to **stimuli.**
2) **Cells communicate** by a process called **cell signalling** (the cells **signal** to each other).
3) **Nervous** and **hormonal communication** are both **examples** of cell signalling.

The Nervous System Sends Information as Electrical Impulses

1) The **nervous system** is made up of a **complex network** of cells called **neurones.** There are **three main types** of neurone:
   - **Sensory neurones** transmit electrical impulses from **receptors** to the **central nervous system** (**CNS**) — the **brain** and **spinal cord.**
   - **Motor neurones** transmit electrical impulses from the **CNS** to **effectors.**
   - **Relay neurones** transmit electrical impulses **between** sensory neurones and motor neurones.

Electrical impulses are also called nerve impulses.

2) A stimulus is detected by **receptor cells** and an **electrical impulse** is sent along a **sensory neurone.**
3) When an **electrical impulse** reaches the end of a neurone chemicals called **neurotransmitters** take the information across to the **next neurone**, which then sends an **electrical impulse** (see p. 92).
4) The **CNS processes** the information, **decides what to do** about it and sends impulses along **motor neurones** to an **effector.**

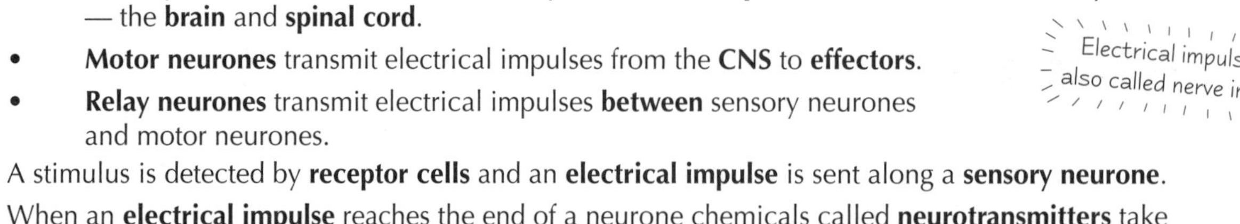

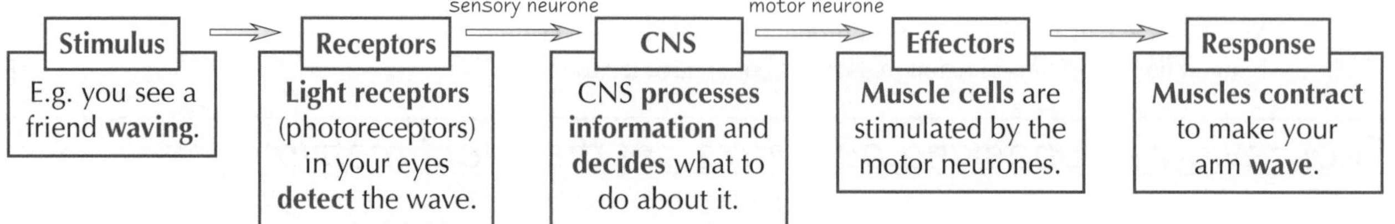

Nervous and Hormonal Communication

*Nervous System Communication is **Localised**, **Short-lived** and **Rapid***

1) When an **electrical impulse** reaches the end of a neurone, **neurotransmitters** are **secreted directly** onto **cells** (e.g. muscle cells) — so the nervous response is **localised**.
2) **Neurotransmitters** are **quickly removed** once they've done their job, so the response is **short-lived**.
3) Electrical impulses are **really fast**, so the response is **rapid** — this allows animals to **react quickly** to stimuli.

*The **Nervous System** is **Split** into **Two Main Systems***

AQA and OCR Unit 5 only

The **central nervous system** (**CNS**) — made up of the **brain** and the **spinal cord**.

The **peripheral nervous system** — made up of the neurones that connect the CNS to the **rest** of the **body**. It also has two different systems:

The **somatic nervous system** controls **conscious** activities, e.g. running and playing video games.

The **autonomic nervous system** controls **unconscious** activities, e.g. digestion. It's got two divisions that have **opposite effects** on the body:

The **sympathetic** nervous system gets the body **ready for action**. It's the '**flight or fight**' system.

The **parasympathetic** nervous system **calms** the body down. It's the '**rest and digest**' system.

Harold thought it was about time his sympathetic nervous system took over.

*Reflexes are **Rapid**, **Automatic Responses** to **Stimuli***

AQA only

1) A **reflex** is where the body **responds** to a stimulus **without** making a **conscious decision** to respond.
2) Because you don't have to **spend time deciding** how to respond, information travels **really fast** from **receptors** to **effectors**.
3) So simple reflexes help organisms to **avoid damage** to the body because they're **rapid**.
4) The **pathway** of neurones linking receptors to effectors in a reflex is called a **reflex arc**. You need to **learn** a **simple reflex arc** involving three neurones — a **sensory**, a **relay** and a **motor** neurone.

E.g. the **hand-withdrawal response** to **heat**

- **Thermoreceptors** in the skin **detect** the **heat** stimulus.
- The **sensory neurone** carries impulses to the **relay neurone**.
- The **relay neurone** connects to the **motor neurone**.
- The **motor neurone** sends **impulses** to the **effector** (your biceps muscle).
- Your **muscle contracts** to **stop** your hand being **damaged**.

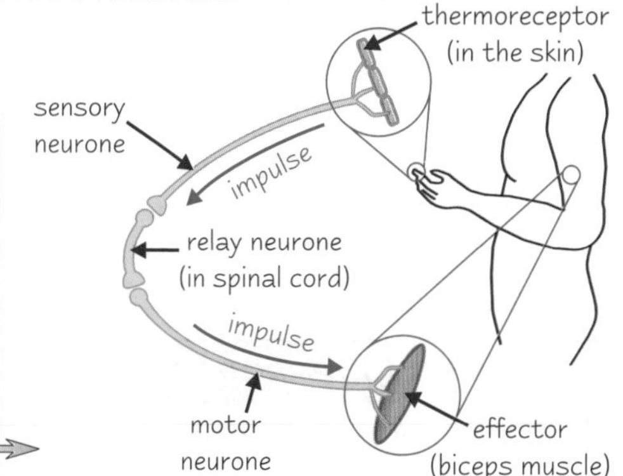

*The **Nervous System** Controls **Pupil Dilation** and **Constriction***

Edexcel only

1) Your **pupils dilate** (get bigger) so you can **see better** when there's **not much light**, and they **constrict** (shrink) to **protect** your eyes from **bright light**.
2) These responses are **controlled** by the **nervous system**.
3) When the light is **too bright** it's **detected** by **photoreceptors** in the eye and impulses are sent along the **optic nerve** (a bundle of neurones) to the brain. The brain **unconsciously processes** the information and sends impulses along **motor neurones** to the **circular muscles** in the **iris**. The circular muscles **contract** to **constrict** the pupils.
4) When there's **not enough light** it's **detected** by **photoreceptors** in the eye and impulses are sent along the **optic nerve** to the brain. The brain **unconsciously processes** the information and sends impulses along **motor neurones** to the **radial muscles** in the **iris**. The radial muscles **contract** to **dilate** the pupils.

Because the brain unconsciously processes the information, the responses are reflexes.

Nervous and Hormonal Communication

These pages are for AQA Unit 5, OCR Unit 4 and Edexcel Unit 5 — if you're doing OCR Unit 5, go to the questions.

The *Hormonal System* Sends Information as *Chemical Signals*

1) The **hormonal system** is made up of **glands** and **hormones**:

- A **gland** is a group of cells that are specialised to **secrete** a useful substance, such as a **hormone**. E.g. the **pancreas** secretes **insulin**.
- **Hormones** are '**chemical messengers**'. Many hormones are **proteins** or **peptides**, e.g. **insulin**. Some hormones are **steroids**, e.g. **progesterone**.

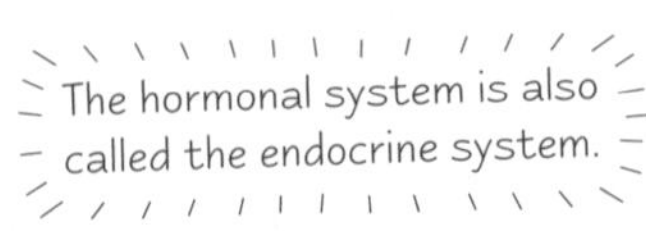

2) **Hormones** are **secreted** when a **gland** is **stimulated**:

- Glands can be **stimulated** by a **change** in **concentration** of a specific **substance** (sometimes **another hormone**).
- They can also be **stimulated** by **electrical impulses**.

3) Hormones **diffuse directly into** the **blood**, then they're **taken** around the body by the **circulatory system**.
4) They **diffuse out** of the blood **all over** the **body** but each hormone will only **bind** to **specific receptors** for that hormone, found on the membranes of some cells (called **target cells**).
5) The hormones trigger a **response** in the **target cells** (the **effectors**).

Tissue that contains target cells is called target tissue.

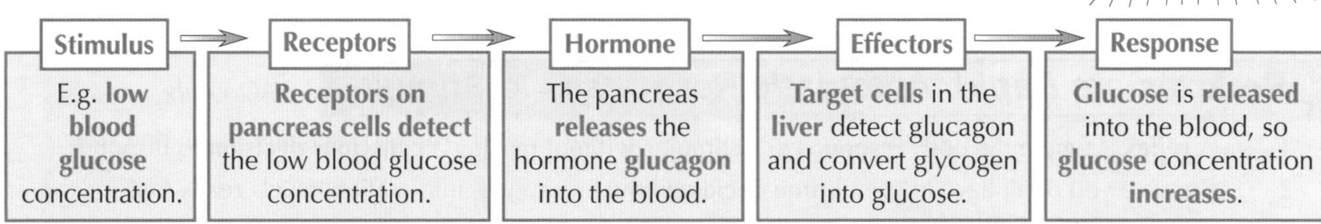

Hormonal System Communication is *Slower, Long-lasting* and *Widespread*

1) Hormones **aren't** released directly onto their target cells — they must **travel** in the **blood** to get there. This means that chemical communication (by hormones) is **slower** than electrical communication (by nerves).
2) They **aren't broken down as quickly** as neurotransmitters, so the **effects** of hormones can **last** for much **longer**.
3) Hormones are transported **all over** the **body**, so the response may be **widespread** if the target cells are widespread.

Practice Questions

These questions cover pages 82-84.

Q1 Why do organisms respond to changes in their environment?

Q2 Give two types of effector.

Q3 Which part of the nervous system controls unconscious activities?

Q4 What is a gland?

Exam Questions

Q1 Bright light causes circular iris muscles in an animal's eyes to contract, which constricts the pupils and protects the eyes. Describe and explain the roles of receptors and effectors in this response. [5 marks]

Q2 Describe three ways in which nervous communication is different from hormonal communication. [3 marks]

Vacancy — talented gag writer required for boring biology topics...

Actually, this stuff is really quite fascinating once you realise just how much your body can do without you even knowing. Just sit back and relax, let your nerves and hormones do the work... Ah, apart from the whole revision thing — your body can't do that without you knowing, unfortunately. Get your head around these pages before you tackle the rest of the section.

Receptors

This page is for AQA Unit 5, OCR Unit 4 and Edexcel Unit 5. If you're doing OCR Unit 5, go to p. 92.

So now you know why organisms respond it's time for the (slightly less interesting but equally important) details... first up — receptors.

Receptors are Specific to One Kind of Stimulus

1) Receptors are **specific** — they only **detect one particular stimulus**, e.g. light, pressure or glucose concentration.
2) There are **many different types** of receptor that each detect a **different type of stimulus**.
3) Some receptors are **cells**, e.g. photoreceptors are receptor cells that connect to the nervous system. Some receptors are **proteins** on **cell surface membranes**, e.g. glucose receptors are proteins found in the cell membranes of some pancreatic cells.
4) Here's a bit more about how receptor cells that communicate information via the **nervous system** work:

- When a nervous system receptor is in its **resting state** (not being stimulated), there's a **difference in voltage** (**charge**) between the **inside** and the **outside** of the cell — this is generated by ion pumps and ion channels (see p. 88). The **difference in voltage** across the membrane is called the **potential difference**.
- The **potential difference** when a cell is at **rest** is called its **resting potential**. When a stimulus is detected, the cell membrane is **excited** and becomes **more permeable**, allowing **more ions** to move **in** and **out** of the cell — **altering** the **potential difference**. The **change** in **potential difference** due to a stimulus is called the **generator potential**.
- A **bigger stimulus** excites the membrane more, causing a **bigger movement** of ions and a **bigger change** in potential difference — so a **bigger generator potential** is produced.
- If the **generator potential** is **big enough** it'll trigger an **action potential** — an electrical impulse along a neurone (see p. 89). An action potential is only triggered if the generator potential reaches a certain level called the **threshold** level. Action potentials are all one size, so the **strength** of the **stimulus** is measured by the **frequency** of **action potentials**.
- If the stimulus is **too weak** the generator potential **won't reach** the **threshold**, so there's **no action potential**.

Sensory Receptors Convert Stimulus Energy into Electrical Impulses

OCR only

1) **Different stimuli** have **different forms** of **energy**, e.g. light energy or chemical energy.
2) But your **nervous system** only sends information in the form of **electrical impulses**.
3) **Sensory receptors convert** the energy of a **stimulus** into **electrical energy**.
4) So, sensory receptors act as **transducers** — something that **converts** one form of energy into another.

Pacinian Corpuscles are Pressure Receptors in Your Skin

AQA only

1) **Pacinian corpuscles** are **mechanoreceptors** — they detect **mechanical stimuli**, e.g. **pressure** and **vibrations**. They're found in your **skin**.
2) Pacinian corpuscles contain the end of a **sensory neurone**, imaginatively called a **sensory nerve ending**.
3) The sensory nerve ending is **wrapped** in loads of layers of connective tissue called **lamellae**.
4) When a Pacinian corpuscle is **stimulated**, e.g. by a tap on the arm, the lamellae are **deformed** and **press** on the **sensory nerve ending**.
5) This causes **deformation** of **stretch-mediated sodium channels** in the sensory neurone's cell membrane. The sodium ion channels **open** and **sodium ions diffuse into** the cell, creating a **generator potential**.
6) If the **generator potential** reaches the **threshold**, it triggers an **action potential**.

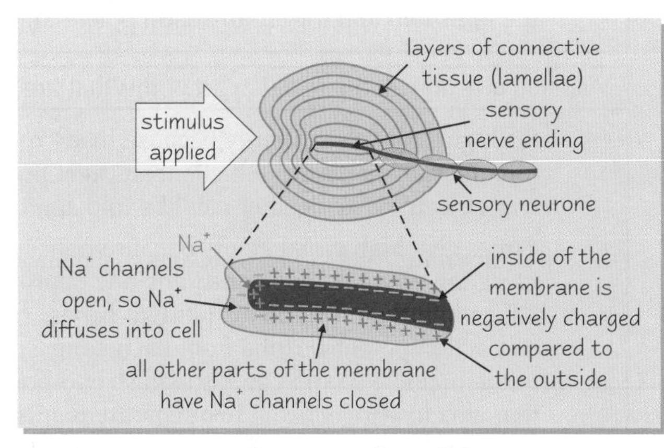

Receptors

This page is for** AQA Unit 5 **and** Edexcel Unit 5. **If you're doing** OCR, **you can skip straight to the questions.

Photoreceptors are Light Receptors in Your Eye

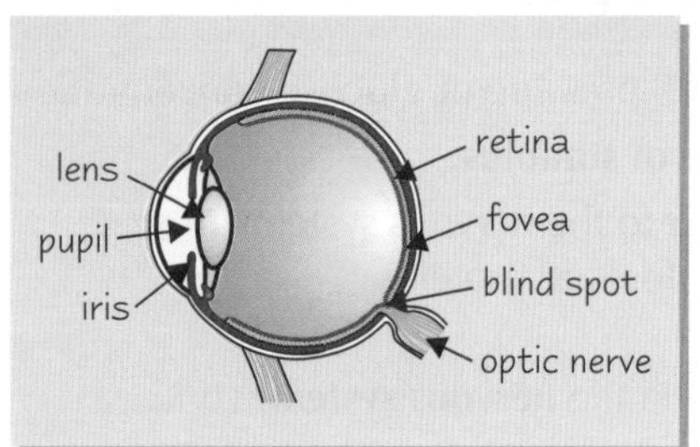

1) **Light** enters the eye through the **pupil**. The **amount** of light that enters is **controlled** by the muscles of the **iris**.
2) Light rays are **focused** by the **lens** onto the **retina**, which lines the inside of the eye. The retina contains **photoreceptor cells** — these **detect light**.
3) The **fovea** is an area of the retina where there are **lots** of **photoreceptors**.
4) **Nerve impulses** from the photoreceptor cells are carried from the **retina** to the **brain** by the **optic nerve**, which is a bundle of **neurones**. Where the optic nerve leaves the eye is called the **blind spot** — there **aren't** any **photoreceptor cells**, so it's **not sensitive** to **light**.

Photoreceptors Convert Light into an Electrical Impulse

1) **Light** enters the eye, hits the **photoreceptors** and is **absorbed** by **light-sensitive pigments**.
2) Light bleaches the pigments, causing a **chemical change**.
3) This triggers a **nerve impulse** along a **bipolar neurone**.
4) Bipolar neurones connect **photoreceptors** to the **optic nerve**, which takes impulses to the **brain**.

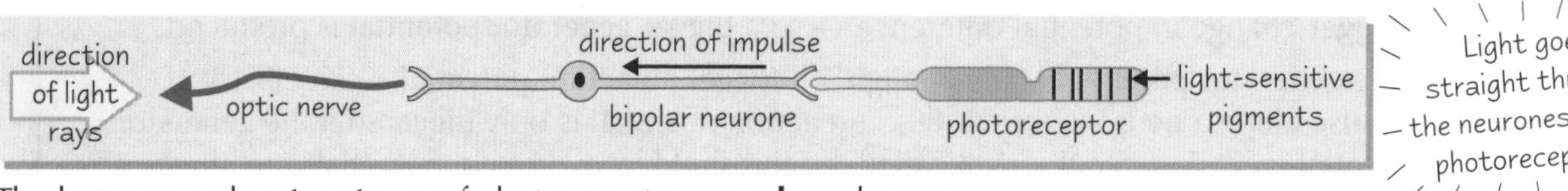

Light goes straight through the neurones to the photoreceptors.

5) The human eye has **two types** of photoreceptor — **rods** and **cones**.
6) Rods are mainly found in the **peripheral** parts of the **retina**, and cones are found **packed together** in the **fovea**.
7) Rods only give information in **black and white** (monochromatic vision), but cones give information in **colour** (trichromatic vision). There are three types of cones — **red-sensitive**, **green-sensitive** and **blue-sensitive**. They're stimulated in **different proportions** so you see different colours.

Rods are More Sensitive, but Cones let you See More Detail

AQA only

Sensitivity

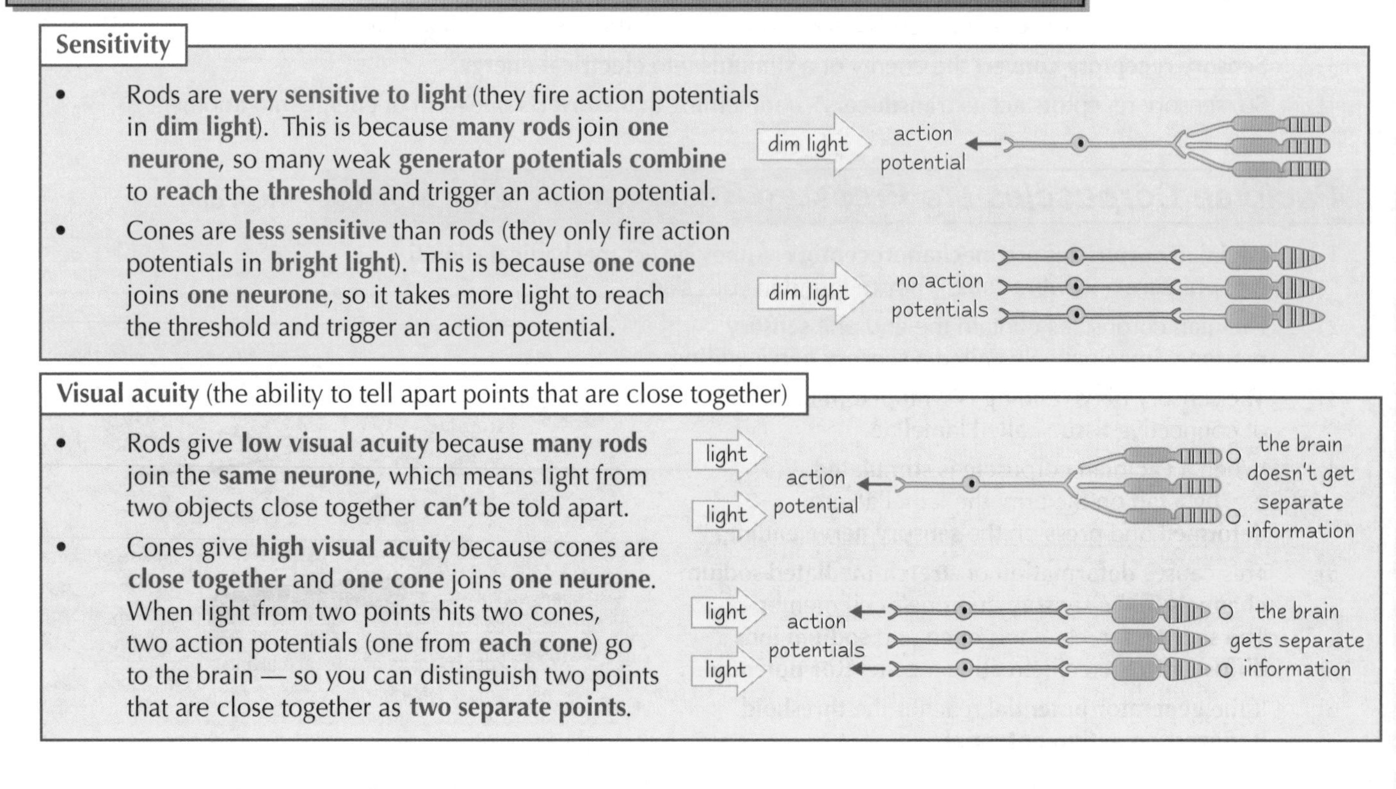

- Rods are **very sensitive to light** (they fire action potentials in **dim light**). This is because **many rods** join **one neurone**, so many weak **generator potentials combine** to **reach** the **threshold** and trigger an action potential.
- Cones are **less sensitive** than rods (they only fire action potentials in **bright light**). This is because **one cone** joins **one neurone**, so it takes more light to reach the threshold and trigger an action potential.

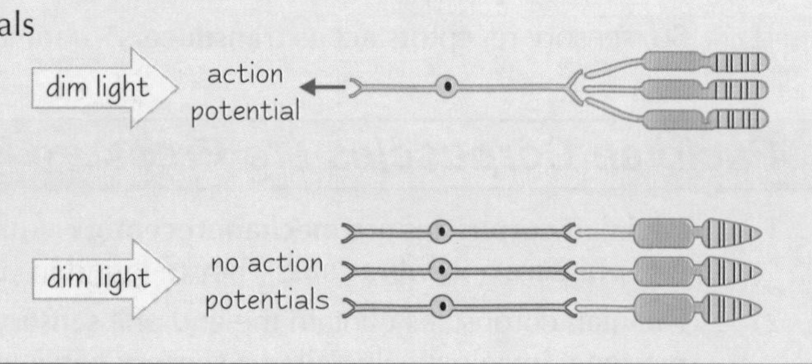

Visual acuity (the ability to tell apart points that are close together)

- Rods give **low visual acuity** because **many rods** join the **same neurone**, which means light from two objects close together **can't** be told apart.
- Cones give **high visual acuity** because cones are **close together** and **one cone** joins **one neurone**. When light from two points hits two cones, two action potentials (one from **each cone**) go to the brain — so you can distinguish two points that are close together as **two separate points**.

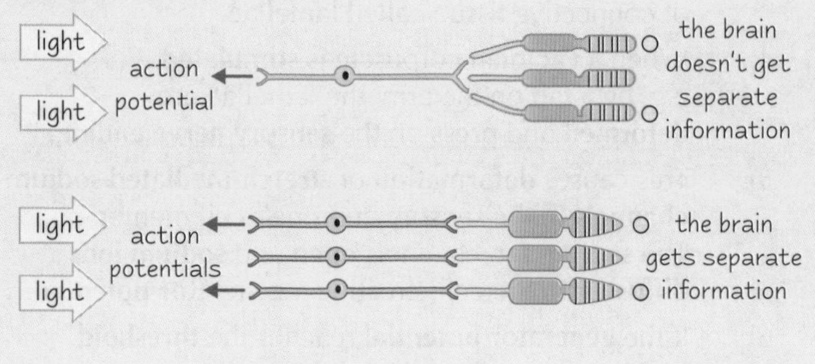

Receptors

This page is for Edexcel Unit 5 only — AQA folk, you can move along to the questions now.

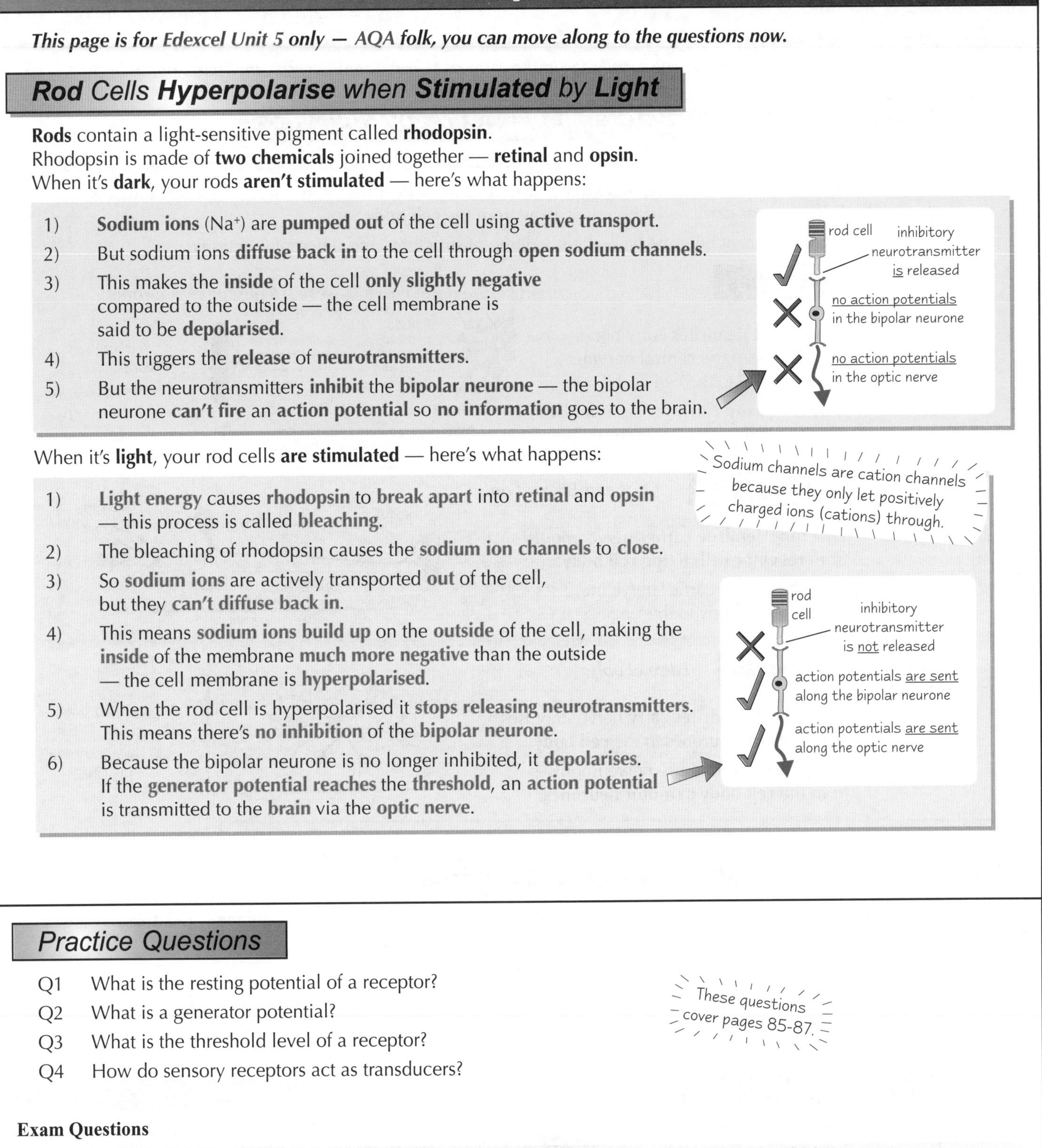

Rod Cells Hyperpolarise when Stimulated by Light

Rods contain a light-sensitive pigment called **rhodopsin**.
Rhodopsin is made of **two chemicals** joined together — **retinal** and **opsin**.
When it's **dark**, your rods **aren't stimulated** — here's what happens:

1) **Sodium ions** (Na^+) are **pumped out** of the cell using **active transport**.
2) But sodium ions **diffuse back in** to the cell through **open sodium channels**.
3) This makes the **inside** of the cell **only slightly negative** compared to the outside — the cell membrane is said to be **depolarised.**
4) This triggers the **release** of **neurotransmitters.**
5) But the neurotransmitters **inhibit** the **bipolar neurone** — the bipolar neurone **can't fire** an **action potential** so **no information** goes to the brain.

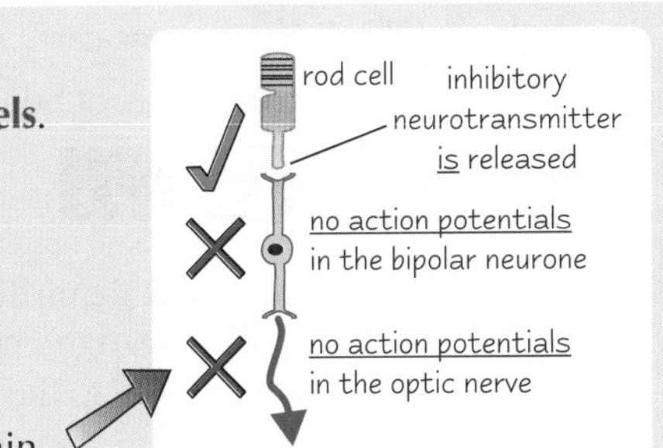

When it's **light**, your rod cells **are stimulated** — here's what happens:

Sodium channels are cation channels because they only let positively charged ions (cations) through.

1) **Light energy** causes **rhodopsin** to **break apart** into **retinal** and **opsin** — this process is called **bleaching.**
2) The bleaching of rhodopsin causes the **sodium ion channels** to **close.**
3) So **sodium ions** are actively transported **out** of the cell, but they **can't diffuse back in.**
4) This means **sodium ions build up** on the **outside** of the cell, making the **inside** of the membrane **much more negative** than the outside — the cell membrane is **hyperpolarised.**
5) When the rod cell is hyperpolarised it **stops releasing neurotransmitters.** This means there's **no inhibition** of the **bipolar neurone.**
6) Because the bipolar neurone is no longer inhibited, it **depolarises.** If the **generator potential reaches** the **threshold**, an **action potential** is transmitted to the **brain** via the **optic nerve.**

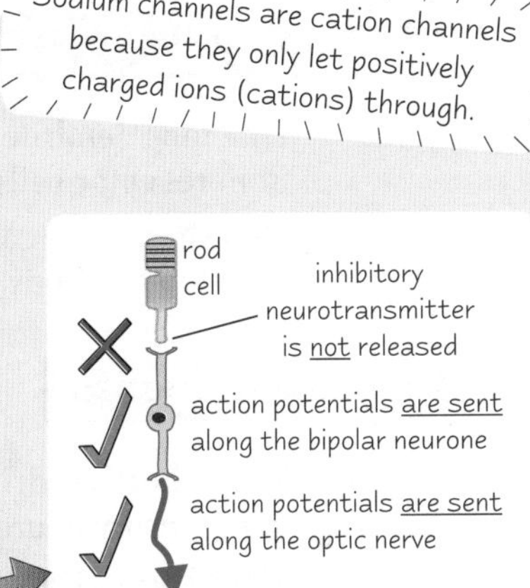

Practice Questions

Q1 What is the resting potential of a receptor?

Q2 What is a generator potential?

Q3 What is the threshold level of a receptor?

Q4 How do sensory receptors act as transducers?

Exam Questions

Q1 Explain how a tap on the arm is converted into a nerve impulse. [6 marks]

Q2 Explain how the human eye has both high sensitivity and high acuity. [5 marks]

Q3 Explain how light falling on a rod cell triggers an action potential to be transmitted to the brain. [7 marks]

Pacinian corpuscles love deadlines — they work best under pressure...

Wow, loads of stuff here, so cone-gratulations if you manage to remember it. In fact, get someone to test you, just to make sure it's well and truly fixed in that big grey blob you call your brain. Remember, receptors are really important because without them you wouldn't be able to see this book, and without this book revision would be way trickier.

Nervous System — Neurones

These pages are for AQA Unit 5, OCR Unit 4 and Edexcel Unit 5.

Ah, on to the good stuff — how neurones carry info (in the form of action potentials) to other parts of the body...

You Need to Learn the Structure and Function of Neurones

1) All neurones have a **cell body** with a **nucleus** (plus **cytoplasm** and all the other **organelles** you usually get in a cell).
2) The cell body has **extensions** that **connect** to **other neurones** — **dendrites** carry nerve impulses **towards** the **cell body**, and **axons** carry nerve impulses **away** from the **cell body**.
3) The three different types of neurone have slightly different structures and different functions:

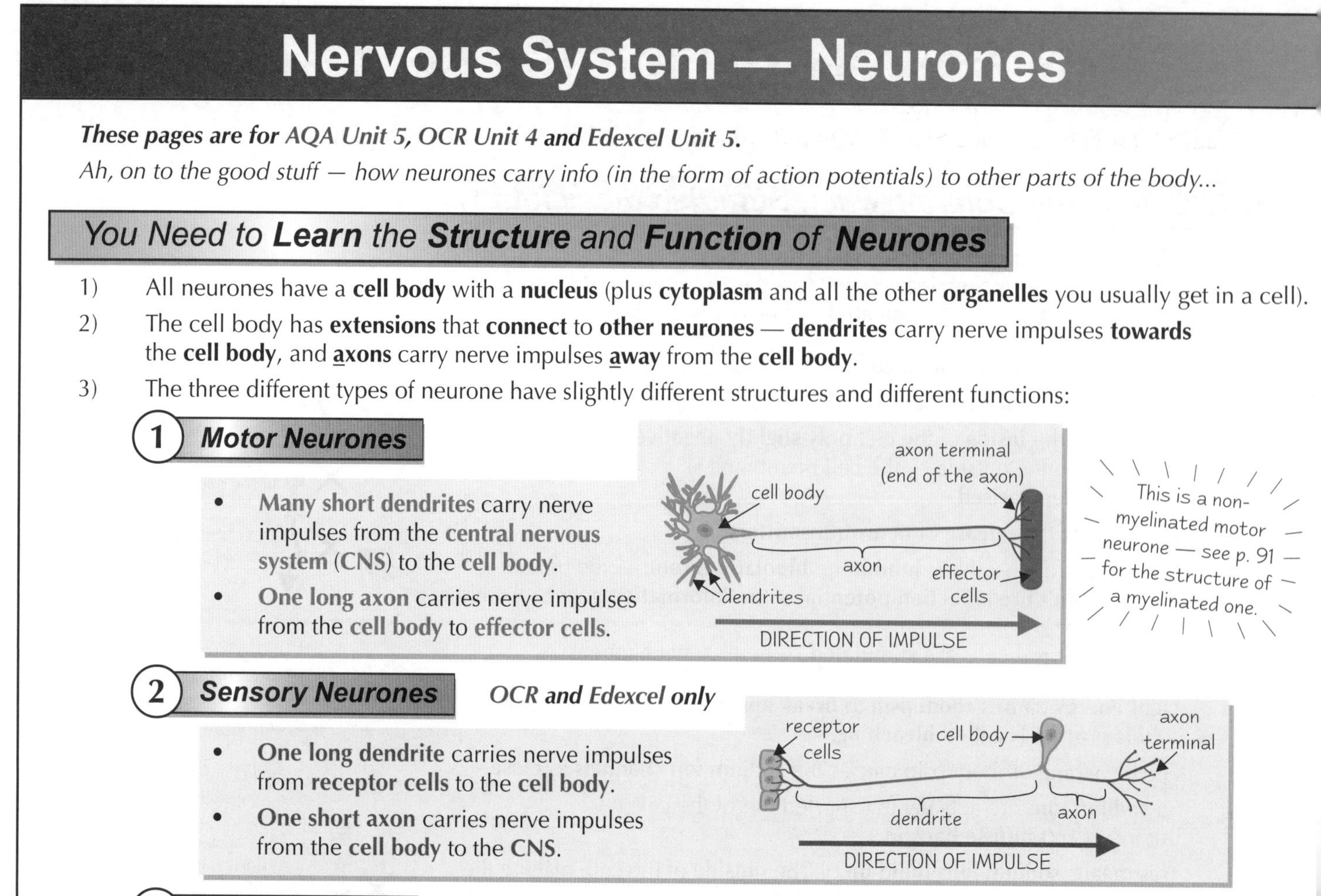

1 Motor Neurones

- **Many short dendrites** carry nerve impulses from the **central nervous system (CNS)** to the **cell body**.
- **One long axon** carries nerve impulses from the **cell body** to **effector cells.**

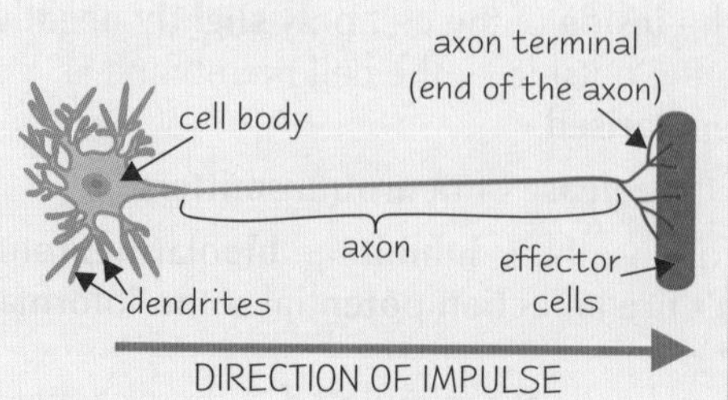

This is a non-myelinated motor neurone — see p. 91 for the structure of a myelinated one.

2 Sensory Neurones

OCR and Edexcel only

- **One long dendrite** carries nerve impulses from **receptor cells** to the **cell body**.
- **One short axon** carries nerve impulses from the **cell body** to the **CNS**.

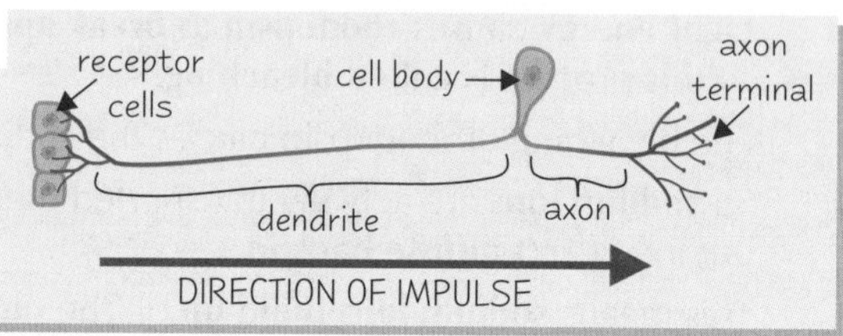

3 Relay Neurones

Edexcel only

- **Many short dendrites** carry nerve impulses from **sensory neurones** to the **cell body**.
- **Many short axons** carry nerve impulses from the **cell body** to **motor neurones**.

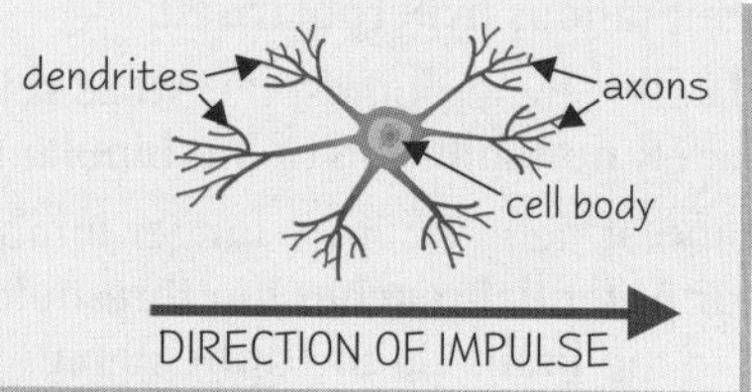

Relay neurones transmit action potentials through the CNS.

Neurone Cell Membranes are Polarised at Rest

1) In a neurone's **resting state** (when it's not being stimulated), the **outside** of the membrane is **positively charged** compared to the **inside**. This is because there are **more positive ions outside** the cell than inside.
2) So the membrane is **polarised** — there's a **difference in voltage** across it.
3) The difference in voltage across the membrane when it's at rest is called the **resting potential** — it's about **–70 mV**.
4) The resting potential is created and maintained by **sodium-potassium pumps** and **potassium ion channels** in a neurone's membrane:

Sodium-potassium pump

These pumps use **active transport** to move **three sodium ions** (Na^+) **out** of the neurone for every **two potassium ions** (K^+) moved **in**. **ATP** is needed to do this.

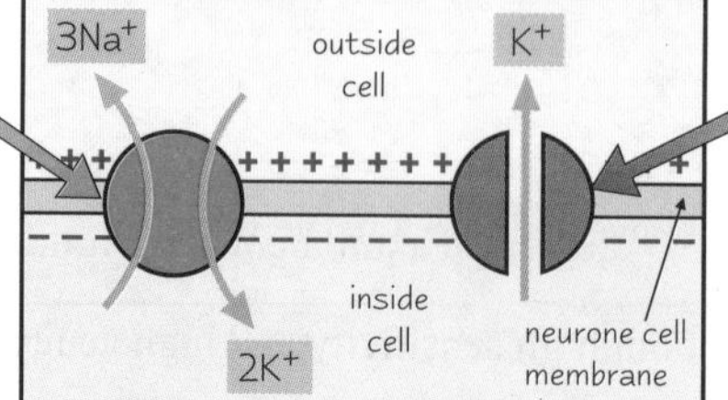

Potassium ion channel

These channels allow **facilitated diffusion** of **potassium ions** (K^+) **out** of the neurone, down their **concentration gradient.**

- The sodium-potassium pumps move **sodium ions out** of the neurone, but the membrane **isn't permeable** to **sodium ions**, so they **can't diffuse back in**. This creates a **sodium ion electrochemical gradient** (a **concentration gradient** of **ions**) because there are **more** positive sodium ions **outside** the cell than inside.
- The sodium-potassium pumps also move **potassium ions in** to the neurone, but the membrane **is permeable** to **potassium ions** so they **diffuse back out** through potassium ion channels.
- This makes the **outside** of the cell **positively charged** compared to the inside.

Nervous System — Neurones

Neurone **Cell Membranes** *Become* **Depolarised** *when they're* **Stimulated**

A **stimulus** triggers other ion channels, called **sodium ion channels**, to **open**. If the stimulus is big enough, it'll trigger a **rapid change** in **potential difference**. The sequence of events that happen are known as an **action potential**:

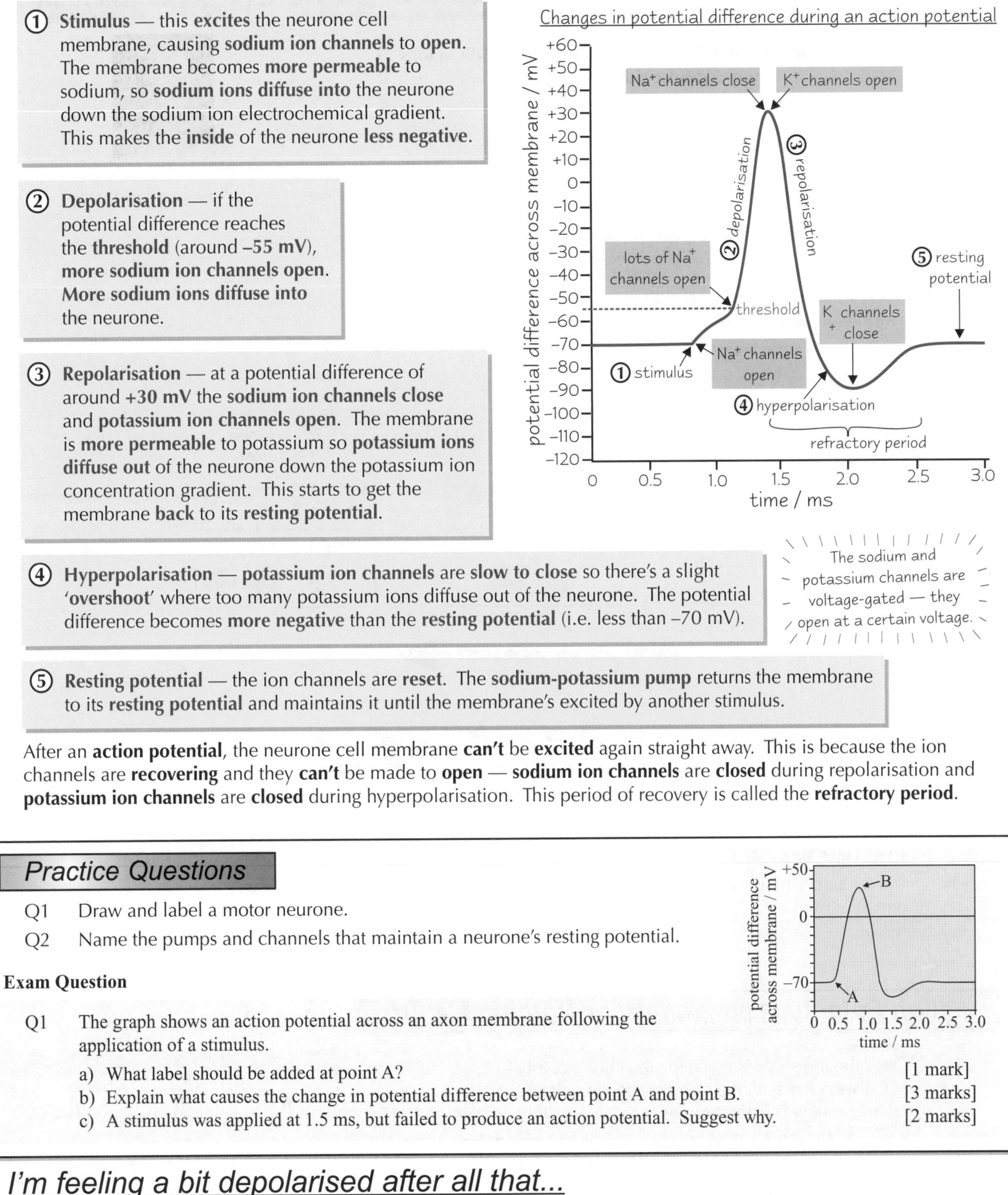

① **Stimulus** — this **excites** the neurone cell membrane, causing **sodium ion channels** to **open**. The membrane becomes **more permeable** to sodium, so **sodium ions diffuse into** the neurone down the sodium ion electrochemical gradient. This makes the **inside** of the neurone **less negative**.

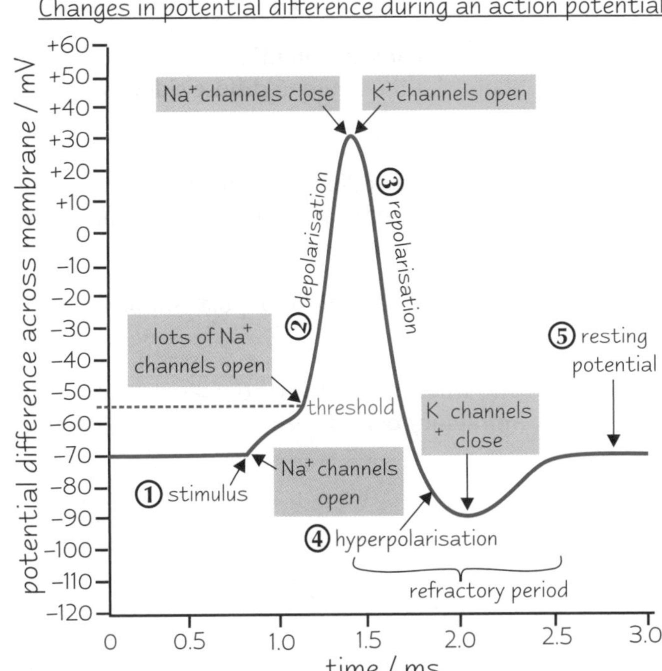

② **Depolarisation** — if the potential difference reaches the **threshold** (around **–55 mV**), **more sodium ion channels open. More sodium ions diffuse into** the neurone.

③ **Repolarisation** — at a potential difference of around **+30 mV** the **sodium ion channels close** and **potassium ion channels open**. The membrane is **more permeable** to potassium so **potassium ions diffuse out** of the neurone down the potassium ion concentration gradient. This starts to get the membrane **back** to its **resting potential**.

④ **Hyperpolarisation** — **potassium ion channels** are **slow to close** so there's a slight **'overshoot'** where too many potassium ions diffuse out of the neurone. The potential difference becomes **more negative** than the **resting potential** (i.e. less than –70 mV).

The sodium and potassium channels are voltage-gated — they open at a certain voltage.

⑤ **Resting potential** — the ion channels are **reset**. The **sodium-potassium pump** returns the membrane to its **resting potential** and maintains it until the membrane's excited by another stimulus.

After an **action potential**, the neurone cell membrane **can't** be **excited** again straight away. This is because the ion channels are **recovering** and they **can't** be made to **open** — **sodium ion channels** are **closed** during repolarisation and **potassium ion channels** are **closed** during hyperpolarisation. This period of recovery is called the **refractory period**.

Practice Questions

Q1 Draw and label a motor neurone.

Q2 Name the pumps and channels that maintain a neurone's resting potential.

Exam Question

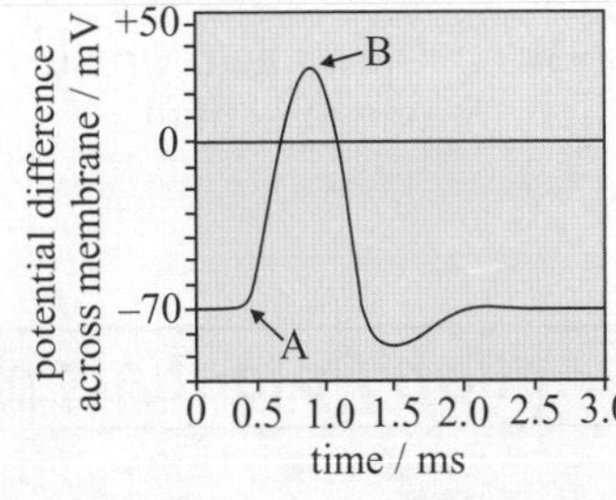

Q1 The graph shows an action potential across an axon membrane following the application of a stimulus.

a) What label should be added at point A? [1 mark]

b) Explain what causes the change in potential difference between point A and point B. [3 marks]

c) A stimulus was applied at 1.5 ms, but failed to produce an action potential. Suggest why. [2 marks]

I'm feeling a bit depolarised after all that...

All this stuff about neurones can be a bit tricky to get your head around at first. Take your time and try scribbling it all down a few times till it starts to make some kind of sense. Basically, neurones work because there's a voltage across their membrane, which is set up by ion pumps and ion channels. It's a change in this voltage that transmits an action potential.

Nervous System — Neurones

***These pages are for** AQA Unit 5, **OCR** Unit 4 **and** Edexcel Unit 5.*

Action potentials don't just sit there once they've been generated — they have to hotfoot it all the way down the neurone so the information can be passed on to the next cell...

The *Action Potential* Moves *Along* the *Neurone* as a *Wave* of *Depolarisation*

1) When an **action potential** happens, some of the **sodium ions** that enter the neurone **diffuse sideways**.

2) This causes **sodium ion channels** in the **next region** of the neurone to **open** and **sodium ions diffuse into** that part.

3) This causes a **wave of depolarisation** to travel along the neurone.

4) The **wave** moves **away** from the parts of the membrane in the **refractory period** because these parts **can't fire** an action potential.

Cindy's wave activity was looking good...

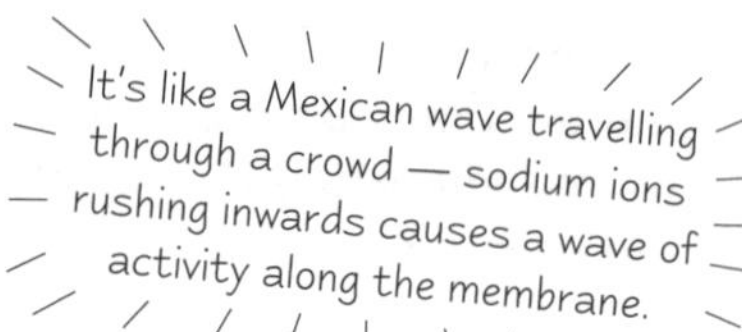

It's like a Mexican wave travelling through a crowd — sodium ions rushing inwards causes a wave of activity along the membrane.

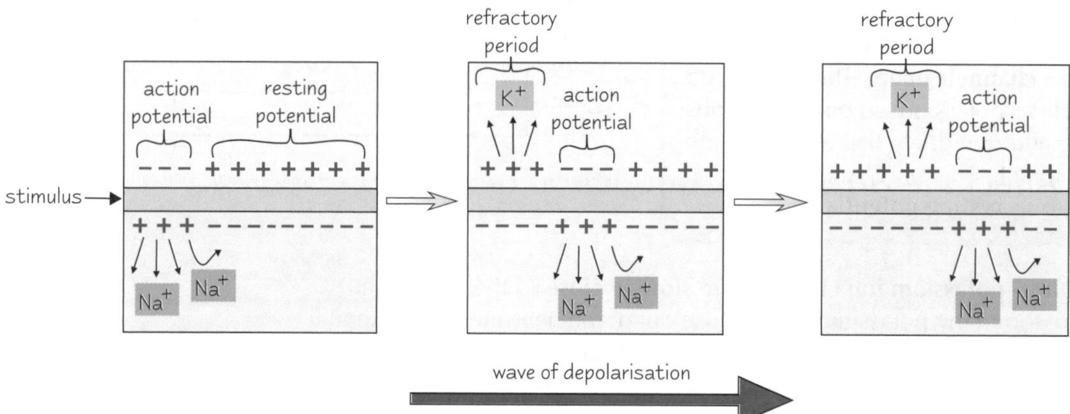

The *Refractory Period* Produces *Discrete Impulses*

1) During the **refractory period**, **ion channels** are **recovering** and **can't** be **opened**.
2) So the refractory period acts as a **time delay** between one action potential and the next. This makes sure that **action potentials don't overlap** but pass along as **discrete** (separate) **impulses**.
3) The refractory period also makes sure **action potentials** are **unidirectional** (they only travel in **one direction**).

Action Potentials have an *All-or-Nothing Nature*

1) Once the threshold is reached, an action potential will **always fire** with the **same change in voltage**, no matter how big the stimulus is.
2) If **threshold isn't reached**, an action potential **won't fire** — this is the **all-or-nothing** nature of action potentials.
3) A **bigger stimulus** won't cause a bigger action potential, but it will cause them to fire **more frequently**.

small stimulus

big stimulus

Nervous System — Neurones

Action Potentials Go Faster in Myelinated Neurones

1) Some neurones are **myelinated** — they have a **myelin sheath**.
2) The myelin sheath is an **electrical insulator**.
3) It's made of a type of cell called a **Schwann cell**.
4) Between the Schwann cells are tiny patches of **bare membrane** called the **nodes of Ranvier**. **Sodium ion channels** are **concentrated** at the nodes.
5) In a **myelinated** neurone, **depolarisation** only happens at the **nodes of Ranvier** (where sodium ions can get through the membrane).
6) The neurone's **cytoplasm conducts** enough electrical charge to **depolarise** the **next node**, so the impulse **'jumps'** from node to node.
7) This is called **saltatory conduction** and it's **really fast**.
8) In a **non-myelinated** neurone, the impulse travels as a **wave** along the **whole length** of the **axon membrane**.
9) This is **slower** than saltatory conduction (although it's still pretty quick).

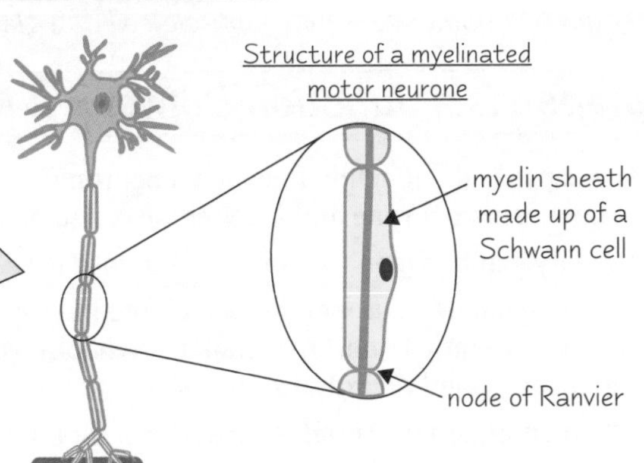

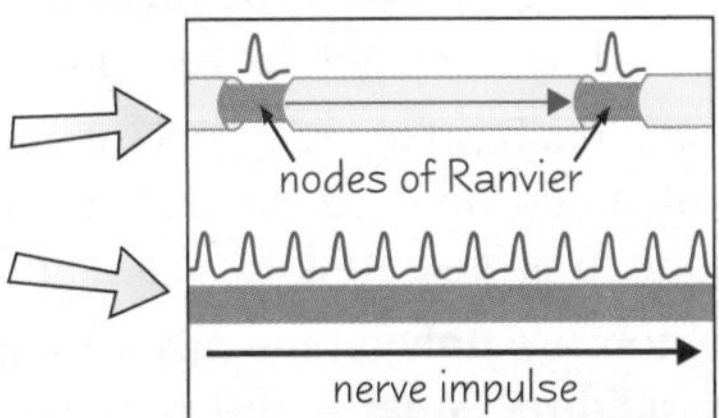

Three Factors Affect the Speed of Conduction of Action Potentials

AQA only

1) Myelination

Some neurones are **myelinated** — this **speeds up** the **conduction** of action potentials (see above).

2) Axon diameter

Action potentials are conducted **quicker** along axons with **bigger diameters** because there's **less resistance** to the **flow of ions** than in the cytoplasm of a smaller axon. With less resistance, **depolarisation reaches** other parts of the neurone cell membrane **quicker**.

3) Temperature

The speed of conduction increases as the **temperature increases** too, because **ions diffuse faster**. The speed only increases up to around **40 °C** though — after that the **proteins** begin to **denature** and the speed decreases.

Practice Questions

Q1 What is meant by the 'all-or-nothing' nature of action potentials?

Q2 How does a bigger stimulus affect the size of an action potential?

Q3 What is the function of Schwann cells on a neurone?

Q4 What are nodes of Ranvier?

Q5 Name three features of axons that speed up the conduction of action potentials.

Exam Questions

Q1 Give two functions of the refractory period in nervous transmission. [2 marks]

Q2 Multiple sclerosis is a disease of the nervous system characterised by damage to the myelin sheaths of neurones. Explain how this will affect the transmission of action potentials. [5 marks]

Never mind the ion channels, I need to recover after this lot...

I'd expect animals like cheetahs, or even humans, to have super myelinated neurones so that nerve impulses are conducted really fast. But no, apparently we've been outdone by shrimps. That's right, top of the leader board so far are shrimps with their myelinated giant nerve fibre conducting impulses faster than 200 m/s. Blimey, they must get a lot done in a day.

Nervous System — Synaptic Transmission

***These pages are for** AQA Unit 5, OCR Units 4 & 5 **and** Edexcel Unit 5.*

When an action potential arrives at the end of a neurone the information has to be passed on to the next cell — this could be another neurone, a muscle cell or a gland cell.

A **Synapse** is a **Junction** Between a **Neurone** and the **Next Cell**

1) A **synapse** is the junction between a **neurone** and another **neurone**, or between a **neurone** and an **effector cell**, e.g. a muscle or gland cell.
2) The **tiny gap** between the cells at a synapse is called the **synaptic cleft**.
3) The **presynaptic neurone** (the one before the synapse) has a **swelling** called a **synaptic knob**. This contains **synaptic vesicles** filled with **chemicals** called **neurotransmitters**.
4) When an **action potential** reaches the end of a neurone it causes **neurotransmitters** to be **released** into the synaptic cleft. They **diffuse across** to the **postsynaptic membrane** (the one after the synapse) and **bind** to **specific receptors**.
5) When neurotransmitters bind to receptors they might **trigger** an **action potential** (in a neurone), cause **muscle contraction** (in a muscle cell), or cause a **hormone** to be **secreted** (from a gland cell).
6) Because the receptors are **only** on the postsynaptic membranes, synapses make sure **impulses** are **unidirectional** — the impulse can only travel in **one direction**.
7) Neurotransmitters are **removed** from the **cleft** so the **response** doesn't keep happening, e.g. they're taken back into the **presynaptic neurone** or they're **broken down** by **enzymes** (and the products are taken into the neurone).
8) There are many **different** neurotransmitters. You need to know about one called **acetylcholine** (**ACh**), which binds to **cholinergic receptors**. Synapses that use acetylcholine are called **cholinergic synapses**.

Typical structure of a synapse

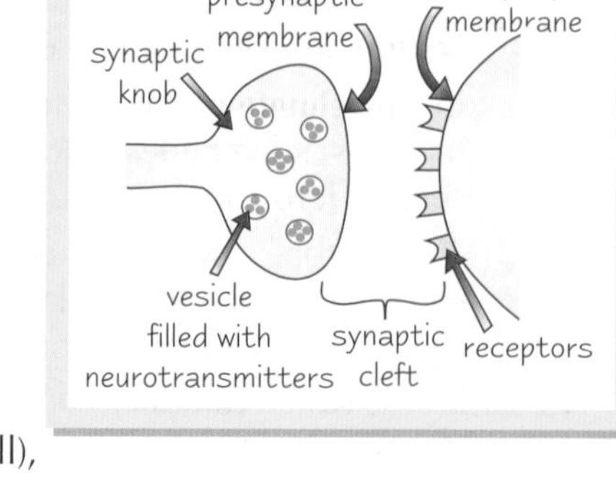

ACh Transmits the Nerve Impulse **Across** a **Cholinergic Synapse**

This is how a **nerve impulse** is transmitted across a **cholinergic synapse**:

1) An action potential (see p. 89) arrives at the **synaptic knob** of the **presynaptic neurone**.
2) The action potential stimulates **voltage-gated calcium ion channels** in the **presynaptic neurone** to **open**.
3) **Calcium ions diffuse into** the synaptic knob. (They're pumped out afterwards by active transport.)
4) The influx of **calcium ions** into the synaptic knob causes the **synaptic vesicles** to **fuse** with the **presynaptic membrane**.
5) The **vesicles release** the neurotransmitter **acetylcholine** (**ACh**) into the **synaptic cleft** — this is called **exocytosis**.

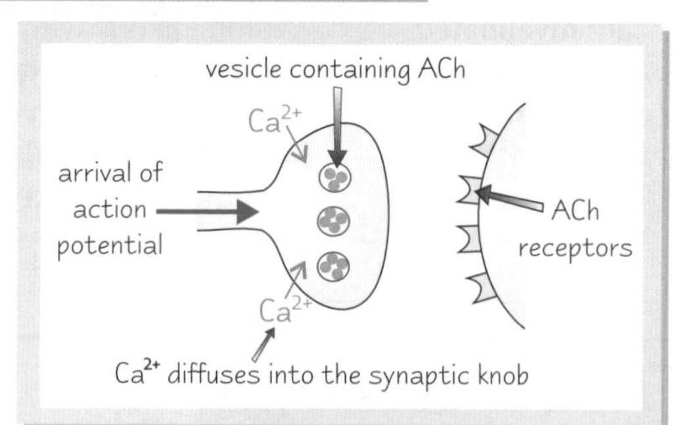

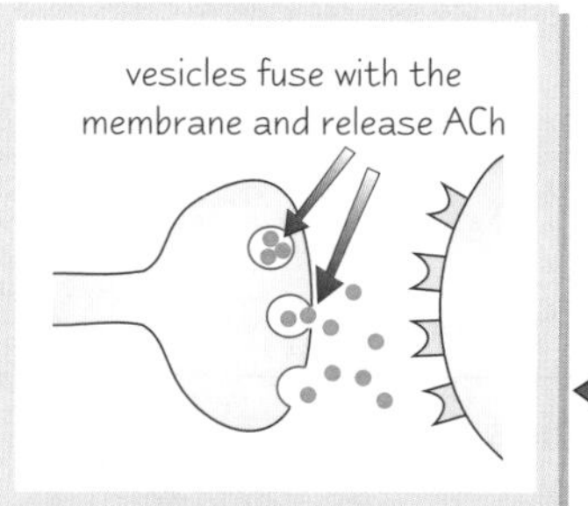

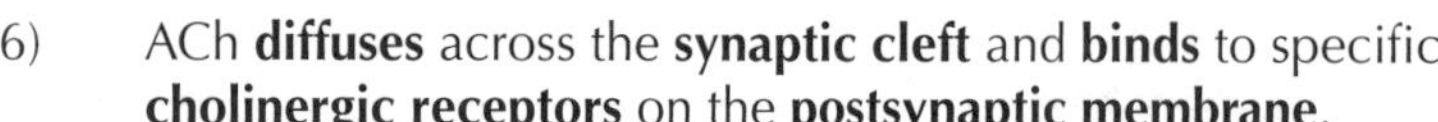

6) ACh **diffuses** across the **synaptic cleft** and **binds** to specific **cholinergic receptors** on the **postsynaptic membrane**.
7) This causes **sodium ion channels** in the **postsynaptic neurone** to **open**.
8) The **influx** of **sodium ions** into the postsynaptic membrane causes an **action potential** on the **postsynaptic membrane** (if the **threshold** is reached).
9) ACh is **removed** from the **synaptic cleft** so the **response** doesn't keep happening. It's **broken down** by an **enzyme** called **acetylcholinesterase** (**AChE**) and the products are **re-absorbed** by the **presynaptic neurone** and used to make more ACh.

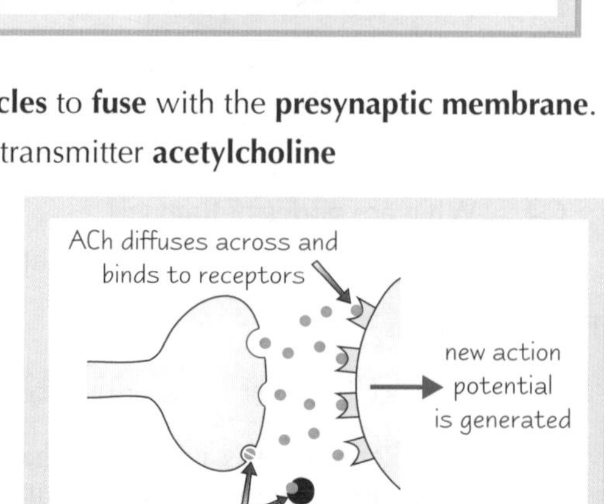

Nervous System — Synaptic Transmission

This page is for AQA and OCR Unit 5 only — everyone else, you can go straight to the questions.

Neuromuscular Junctions are Synapses Between Neurones and Muscles

1) A **neuromuscular junction** is a **synapse** between a **motor neurone** and a **muscle cell**.
2) Neuromuscular junctions use the neurotransmitter **acetylcholine** (**ACh**), which binds to cholinergic receptors called **nicotinic cholinergic receptors**.
3) Neuromuscular junctions **work** in the **same way** as the **cholinergic synapse** shown on the previous page.
4) There are a few **differences** between neuromuscular junctions and synapses where two neurones meet:

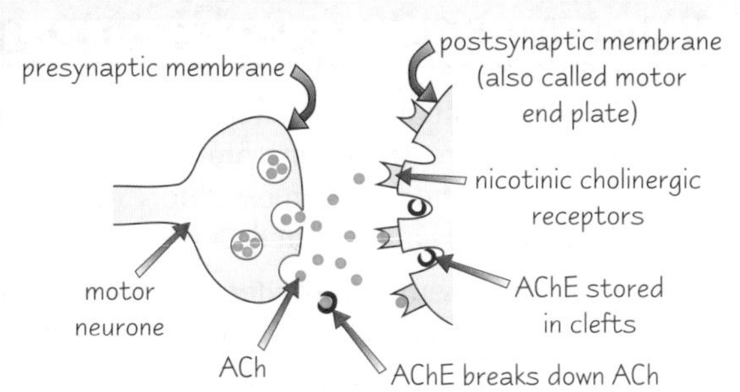

	NEUROMUSCULAR JUNCTIONS	SYNAPSES (between neurones)
Neurotransmitter	Acetylcholine	Various
Postsynaptic receptors	Nicotinic cholinergic receptors	Various
Number of postsynaptic receptors	Lots	Fewer
Postsynaptic cell	Muscle cell	Neurone
Postsynaptic membrane	Has clefts containing AChE	Smooth
Effect of neurotransmitter binding to postsynaptic receptors	Muscle cell always contracts	Action potential may or may not fire in the next neurone
Removal of neurotransmitter	Broken down by AChE	Various ways (it depends on the neurotransmitter)

Practice Questions

Q1 What neurotransmitter do you find at cholinergic synapses?

Q2 How do synapses ensure that nerve impulses are unidirectional?

Q3 Give one way that neurotransmitters are removed from the synaptic cleft.

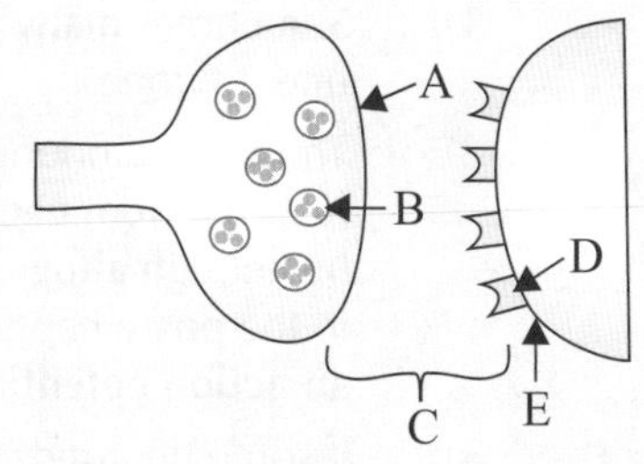

Exam Questions

Q1 The diagram on the right shows a synapse. Label parts A-E. [5 marks]

Q2 Describe the sequence of events from the arrival of an action potential at the presynaptic membrane of a cholinergic synapse to the generation of a new action potential at the postsynaptic membrane. [6 marks]

Q3 Myasthenia gravis is a disease in which the body's immune system gradually destroys receptors at neuromuscular junctions. Suggest what symptoms a sufferer might have. Explain your answer. [4 marks]

Synaptic knobs and clefts — will you stop giggling at the back...

Some more pretty tough pages here, aren't I kind to you. And lots more diagrams to have a go at drawing and re-drawing. Don't worry if you're not the world's best artist, just make sure you add labels to your drawings to explain what's happening.

Nervous System — Synaptic Transmission

***These pages are for** AQA Unit 5, OCR Unit 4 **and** Edexcel Unit 5.*

You might be wondering what the point of synapses are (or maybe you're not) — why not just have loads of neurones that go all the way from one thing to another. Well, synapses allow information to be processed and spread around...

Neurotransmitters are Excitatory or Inhibitory

1) **Excitatory** neurotransmitters **depolarise** the postsynaptic membrane, making it fire an **action potential** if the **threshold** is reached. E.g. **acetylcholine** is an excitatory neurotransmitter — it binds to cholinergic receptors to cause an **action potential** in the postsynaptic membrane.
2) **Inhibitory** neurotransmitters **hyperpolarise** the postsynaptic membrane (make the potential difference more negative), **preventing** it from firing an action potential. E.g. **GABA** is an inhibitory neurotransmitter — when it binds to its receptors it causes **potassium ion channels** to open on the postsynaptic membrane, **hyperpolarising** the neurone.

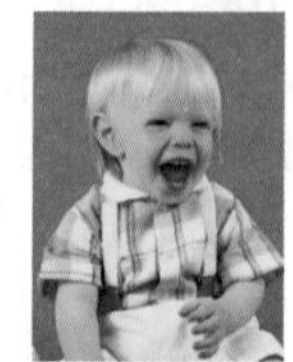

Mum couldn't help wishing Johnny had a few more inhibitory neurotransmitters.

Synapses Play Vital Roles in the Nervous System

1 Synapses allow Information to be Dispersed or Amplified

1) When **one** neurone **connects** to **many** neurones information can be **dispersed** to **different parts** of the body. This is called **synaptic divergence**.
2) When **many** neurones **connect** to **one** neurone information can be **amplified** (made stronger). This is called **synaptic convergence.**

Impulses diverge
Impulses converge

2 Summation at Synapses Finely Tunes the Nervous Response

If a stimulus is **weak**, only a **small amount** of **neurotransmitter** will be released from a neurone into the synaptic cleft. This might not be enough to **excite** the postsynaptic membrane to the **threshold** level and stimulate an action potential. **Summation** is where the effect of neurotransmitter released from many neurones (or one neurone that's stimulated a lot in a short period of time) is **added together**. There are two types of summation:

Spatial summation

1) Sometimes **many** neurones **connect** to **one** neurone.
2) The small amount of **neurotransmitter** released from **each** of these neurones can be enough **altogether** to **reach** the **threshold** in the postsynaptic neurone and **trigger** an **action potential**.
3) If some neurones release an **inhibitory neurotransmitter** then the total effect of all the neurotransmitters might be **no action potential.**

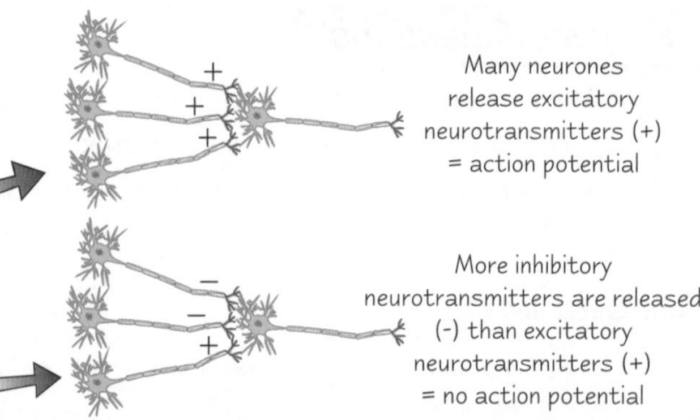
+
+
+
Many neurones release excitatory neurotransmitters (+) = action potential
–
–
+
More inhibitory neurotransmitters are released (-) than excitatory neurotransmitters (+) = no action potential

Temporal summation

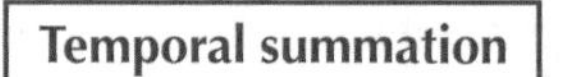

Temporal summation is where **two or more** nerve impulses arrive in **quick succession** from the **same presynaptic neurone**. This makes an action potential **more likely** because **more neurotransmitter** is released into the **synaptic cleft.**

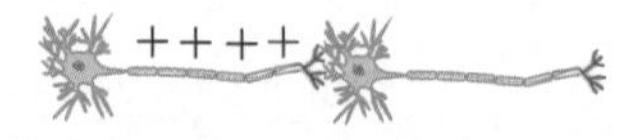
++++

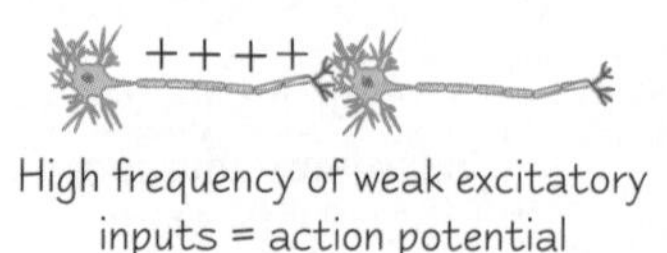

High frequency of weak excitatory inputs = action potential

Both types of **summation** mean synapses **accurately process information**, **finely tuning** the response.

Nervous System — Synaptic Transmission

This page is for AQA only — you can go straight to the questions if you're doing OCR or Edexcel.

Drugs Affect the Action of Neurotransmitters at Synapses in Various Ways

Some **drugs affect synaptic transmission**. You might have to **predict** the **effects** that a drug would have at a synapse in your exam. Here are some **examples** of how drugs can affect synaptic transmission:

1. Some drugs are the **same shape** as neurotransmitters so they **mimic** their action at receptors (these drugs are called **agonists**). This means **more receptors** are **activated**. E.g. **nicotine** mimics **acetylcholine** so binds to nicotinic cholinergic receptors in the brain.

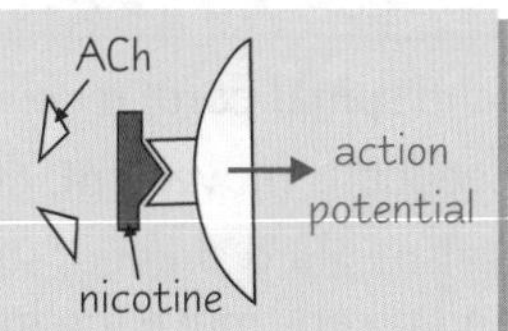

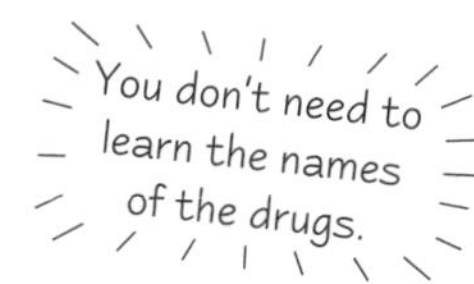

2. Some drugs **block receptors** so they **can't be activated** by neurotransmitters (these drugs are called **antagonists**). This means **fewer receptors** (if any) can be **activated**. E.g. **curare** blocks the effects of acetylcholine by blocking nicotinic cholinergic receptors at neuromuscular junctions, so muscle cells can't be stimulated. This results in the muscle being **paralysed**.

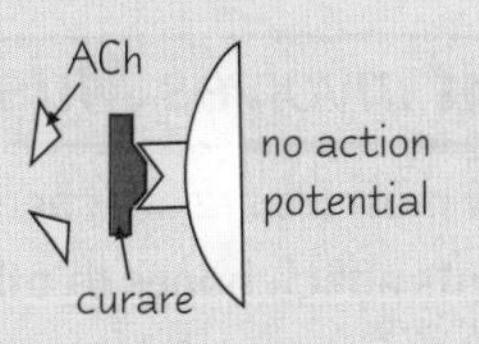

3. Some drugs **inhibit** the **enzyme** that breaks down neurotransmitters (they stop it from working). This means there are **more neurotransmitters** in the synaptic cleft to **bind** to **receptors** and they're there for **longer**. E.g. **nerve gases** stop acetylcholine from being broken down in the synaptic cleft. This can lead to **loss** of **muscle control**.

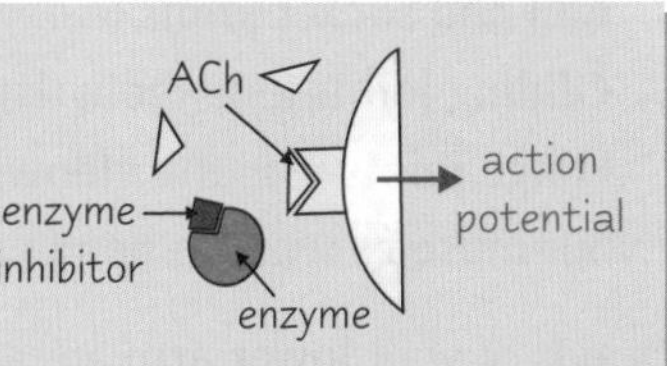

4. Some drugs **stimulate** the release of **neurotransmitter** from the presynaptic neurone so **more receptors** are activated, e.g. **amphetamines**.

5. Some drugs **inhibit** the release of neurotransmitters from the presynaptic neurone so **fewer receptors** are activated, e.g. **alcohol**.

Practice Questions

Q1 What do inhibitory neurotransmitters do at synapses?

Q2 What is synaptic divergence?

Q3 Name the two types of summation that can occur at synapses.

Q4 Would a drug that's the same shape as acetylcholine activate more or fewer acetylcholine receptors at a synapse?

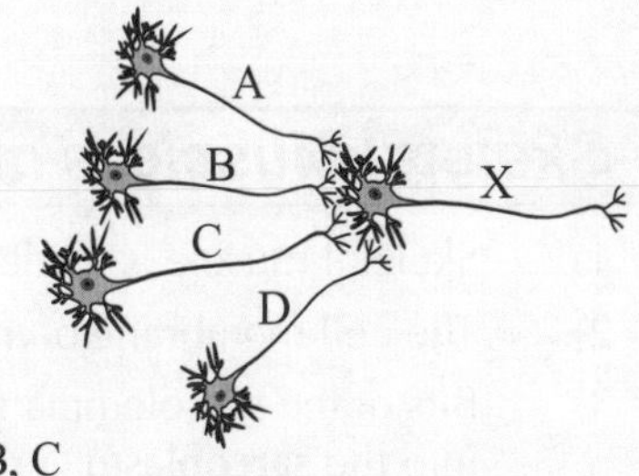

Exam Questions

Q1 The diagram on the right shows a synapse in the nervous system. In an experiment, neurone A is stimulated but neurone X doesn't fire an action potential. Neurones A, B, C and D are then stimulated at the same time and neurone X does fire an action potential.

a) Suggest why stimulating neurone A on its own didn't trigger an action potential in neurone X. [2 marks]

b) Suggest why neurone X did fire an action potential when neurones A, B, C and D were all stimulated. [2 marks]

Q2 Galantamine is a drug that inhibits the enzyme acetylcholinesterase (AChE). Predict the effect of galantamine at a neuromuscular junction and explain your answer. [3 marks]

Neurotransmitter revision inhibits any excitement...

I reckon scientists were trying to fool us all when they called neurotransmitters "excitatory" — they're a lot less exciting than they sound. Yeah they do important jobs, but it's not like they ski in the Alps, raft the Grand Canyon or do somersaults on a tightrope across Niagara Falls. Still, you've got to learn this stuff — you know the drill — close the book and start scribbling.

Effectors — Muscle Contraction

These pages are for AQA Unit 5, OCR Unit 5 and Edexcel Unit 5.

Muscles are effectors — they contract in response to nerve impulses or hormones. Luckily for you, you only need to know how they contract in response to nerve impulses — but first you need to know a bit more about them and how they're involved in movement...

The *Central Nervous System* (CNS) *Coordinates Muscular Movement*

1) The **CNS** (**brain** and **spinal cord**) receives **sensory information** and **decides** what kind of **response** is needed.
2) If the response needed is **movement**, the CNS sends signals along **neurones** to tell **skeletal muscles** to **contract**.
3) Skeletal muscle (also called striated, striped or voluntary muscle) is the type of muscle you use to **move**, e.g. the biceps and triceps move the lower arm.

Movement Involves *Muscles, Tendons, Ligaments* and *Joints*

OCR and Edexcel only

1) **Skeletal muscles** are attached to **bones** by **tendons**.
2) **Ligaments** attach **bones** to **other bones**, to hold them together.
3) The **structure** of the **joints** between your bones determines what **kind** of **movement** is possible:
   - **Ball and socket joints** (e.g. the **shoulder**) allow movement in **all directions**.
   - **Gliding joints** (e.g. the **wrist**) allow a **wide range** of movement because small bones slide over each other.
   - **Hinge joints** (e.g. the **elbow**) allow movement in **one plane only**, like up and down.

Here's how your **muscles** work to **bend** your **arm** at the **elbow**:

- The bones of your **lower arm** are attached to a **biceps** muscle and a **triceps** muscle by **tendons**.
- The biceps and triceps **work together** to move your arm — as one **contracts**, the other **relaxes**:

When your **biceps contracts** your **triceps relaxes**. This pulls the bone so your **arm bends** (**flexes**) at the elbow. A muscle that **bends** a joint when it contracts is called a **flexor**.

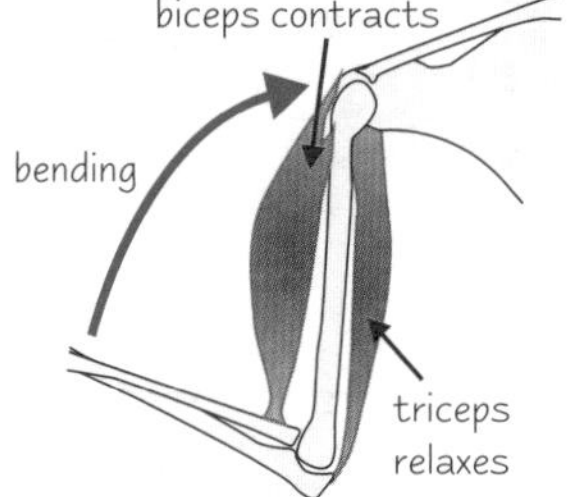

When your **triceps contracts** your **biceps relaxes**. This pulls the bone so your **arm straightens** (**extends**) at the **elbow**. A muscle that **straightens** a joint when it contracts is called an **extensor**.

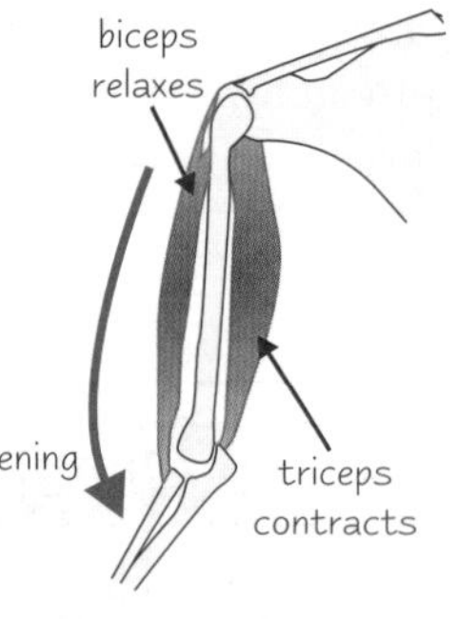

- Muscles that work together to move a bone are called **antagonistic pairs**.

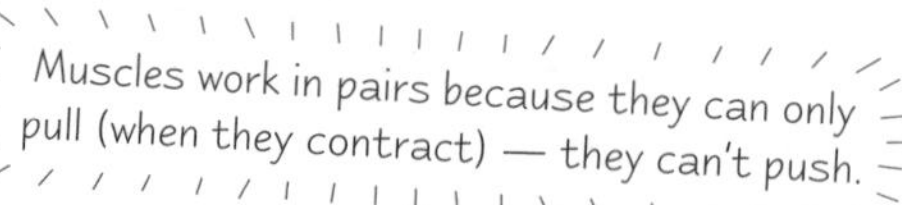

Skeletal Muscle is made up of *Long Muscle Fibres*

1) Skeletal muscle is made up of **large bundles** of **long cells**, called **muscle fibres**.
2) The cell membrane of muscle fibre cells is called the **sarcolemma**.
3) Bits of the sarcolemma **fold inwards** across the muscle fibre and stick into the **sarcoplasm** (a muscle cell's cytoplasm). These folds are called **transverse (T) tubules** and they help to **spread electrical impulses** throughout the sarcoplasm so they **reach** all parts of the **muscle fibre**.
4) A network of **internal membranes** called the **sarcoplasmic reticulum** runs through the sarcoplasm. The sarcoplasmic reticulum **stores** and **releases calcium ions** that are needed for muscle contraction (see p. 98).
5) Muscle fibres have lots of **mitochondria** to **provide** the **ATP** that's needed for **muscle contraction**.
6) They are **multinucleate** (contain many nuclei).
7) Muscle fibres have lots of **long**, **cylindrical organelles** called **myofibrils**. They're made up of proteins and are **highly specialised** for **contraction**.

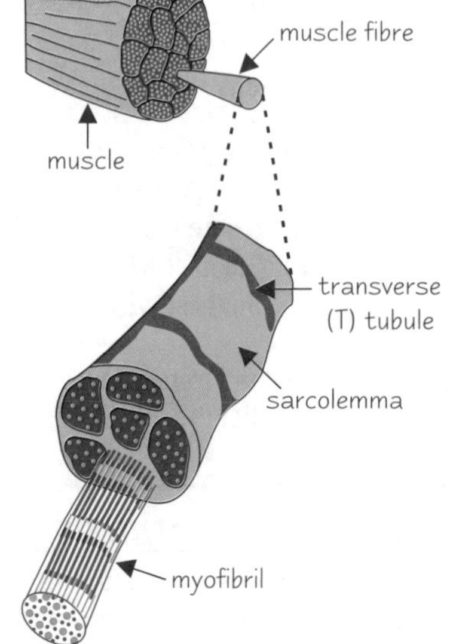

Effectors — Muscle Contraction

*Myofibrils Contain **Thick Myosin** Filaments and **Thin Actin** Filaments*

1) Myofibrils contain bundles of **thick** and **thin myofilaments** that **move past each other** to make muscles **contract**.
   - **Thick myofilaments** are made of the protein **myosin**.
   - **Thin myofilaments** are made of the protein **actin**.
2) If you look at a **myofibril** under an **electron microscope**, you'll see a pattern of alternating **dark** and **light bands**:
   - **Dark** bands contain the **thick myosin filaments** and some overlapping thin actin filaments — these are called **A-bands**.
   - **Light** bands contain **thin actin filaments** only — these are called **I-bands**.
3) A myofibril is made up of many short units called **sarcomeres**.
4) The **ends** of each **sarcomere** are marked with a **Z-line**.
5) In the **middle** of each sarcomere is an **M-line**. The M-line is the **middle** of the **myosin** filaments.
6) **Around** the M-line is the **H-zone**. The H-zone **only** contains **myosin** filaments.

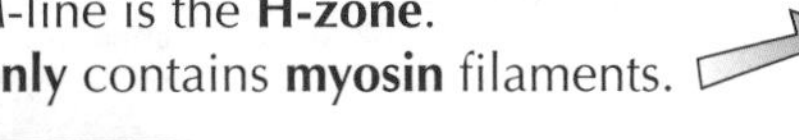

There's more detail on actin and myosin on p. 98.

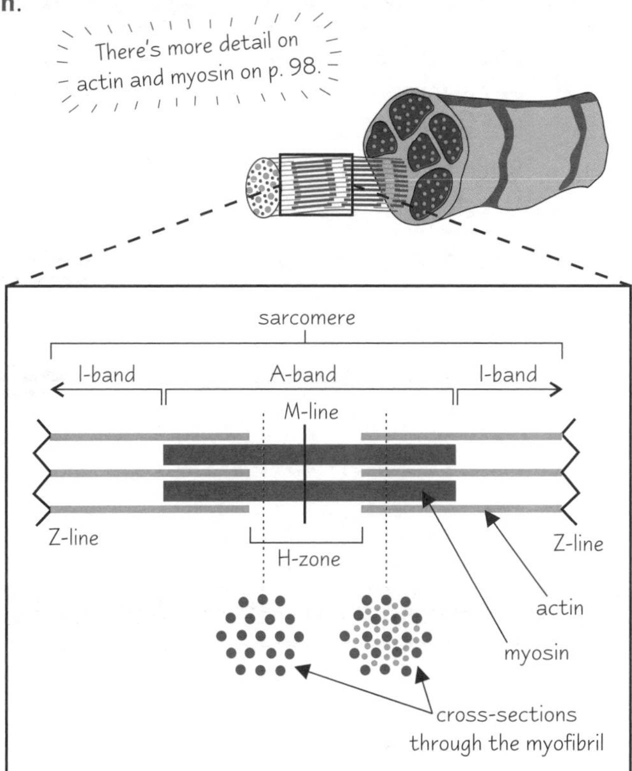

Derek was the proud winner of the biggest muscles AND the smallest pants.

Practice Questions

Q1 What is skeletal muscle?

Q2 What's meant by muscles working as an antagonistic pair?

Q3 What are transverse (T) tubules?

Q4 Name the two proteins that make up myofibrils.

Q5 What are the dark bands called?

Q6 What are the light bands called?

Exam Questions

Q1 Describe how myofilaments, muscle fibres, myofibrils and muscles are related to each other. [3 marks]

Q2 A muscle myofibril was examined under an electron microscope and a sketch was drawn (Figure 1).

a) What are the correct names for labels A, B and C? [3 marks]

b) The myofibril was then cut through the M-line (Figure 2). State which of the cross-section drawings you would expect to see and explain why. [3 marks]

Figure 1

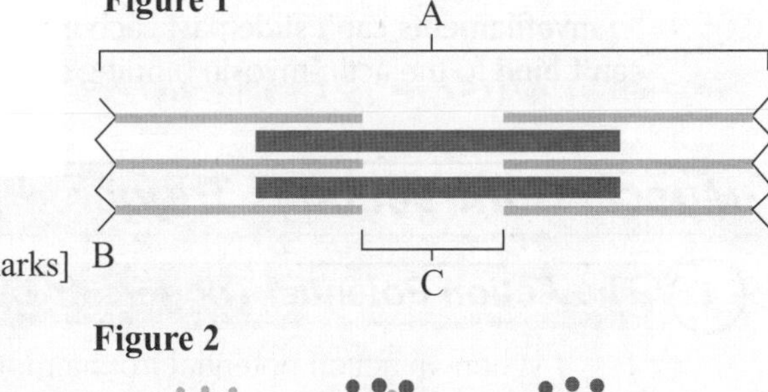

Figure 2

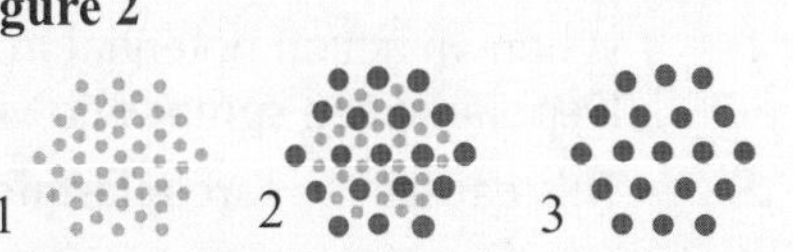

Sarcomere — a French mother with a dry sense of humour...

Blimey, there are an awful lot of similar-sounding names to learn on these pages. And then you've got your A-band, I-band, what-band, who-band to memorise too. But once you've learnt them, these are things you'll never forget — that's right, they'll take up vital brain space forever. And they'll also get you vital marks in your exam.

Effectors — Muscle Contraction

***These pages are for** AQA Unit 5, OCR Unit 5 **and** Edexcel Unit 5.*

Brace yourself — here comes the detail of muscle contraction...

Muscle Contraction is Explained by the *Sliding Filament Theory*

1) **Myosin** and **actin** filaments **slide** over one another to make the **sarcomeres contract** — the myofilaments themselves **don't** contract.
2) The **simultaneous contraction** of lots of **sarcomeres** means the **myofibrils** and **muscle fibres contract**.
3) Sarcomeres return to their **original length** as the muscle **relaxes**.

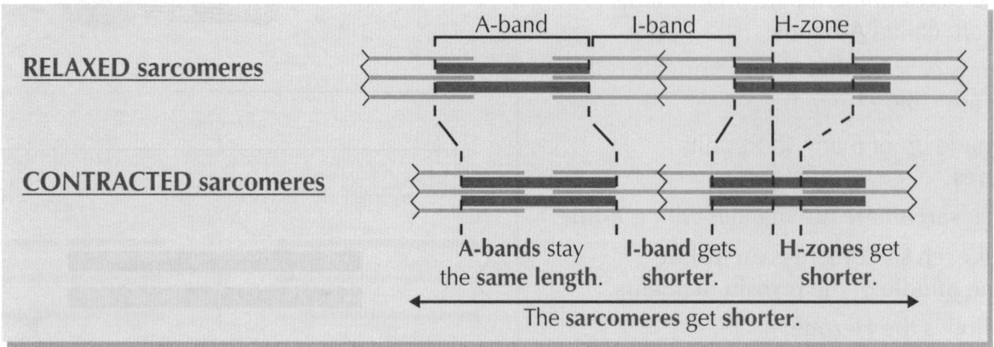

Myosin Filaments Have *Globular Heads* and *Binding Sites*

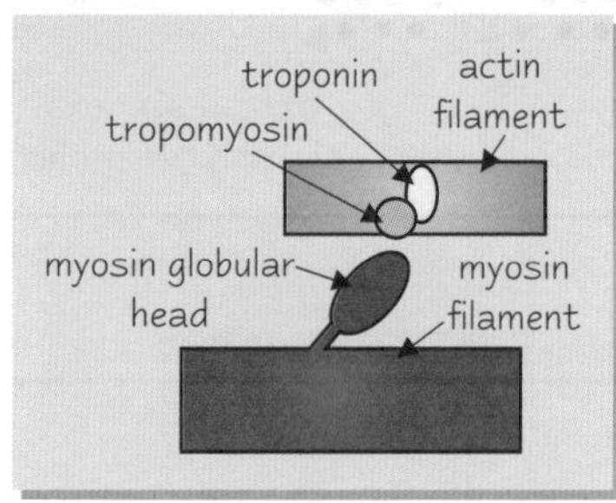

1) **Myosin filaments** have **globular heads** that are **hinged**, so they can move **back** and **forth**.
2) Each myosin head has a **binding site** for **actin** and a **binding site** for **ATP**.
3) **Actin filaments** have **binding sites** for **myosin heads**, called **actin-myosin** binding sites.
4) Two other **proteins** called **tropomyosin** and **troponin** are found between actin filaments. These proteins are **attached** to **each other** and they **help** myofilaments **move** past each other.

Binding Sites in *Resting Muscles* are *Blocked* by *Tropomyosin*

1) In a **resting** (unstimulated) muscle the **actin-myosin binding site** is **blocked** by **tropomyosin**, which is held in place by **troponin**.
2) So **myofilaments can't slide** past each other because the **myosin heads can't bind** to the actin-myosin binding site on the actin filaments.

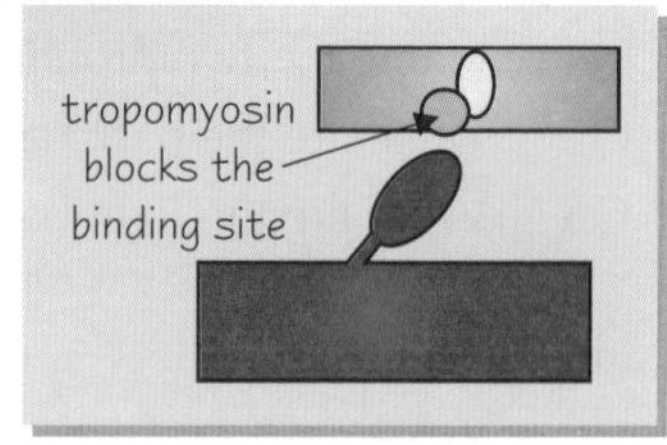

Muscle Contraction is *Triggered* by an *Action Potential*

1 The *Action Potential Triggers* an *Influx* of *Calcium Ions*

1) When an action potential from a motor neurone **stimulates** a muscle cell, it **depolarises** the **sarcolemma**. Depolarisation **spreads** down the **T-tubules** to the **sarcoplasmic reticulum** (see p. 96).
2) This causes the **sarcoplasmic reticulum** to **release** stored **calcium ions** (Ca^{2+}) into the **sarcoplasm**.

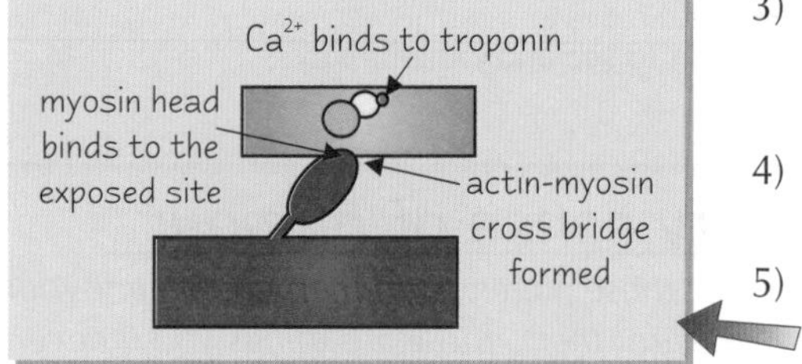

3) Calcium ions **bind** to **troponin**, causing it to **change shape**. This **pulls** the attached **tropomyosin out** of the **actin-myosin binding site** on the actin filament.
4) This **exposes** the **binding site**, which allows the **myosin head** to **bind**.
5) The bond formed when a **myosin head** binds to an **actin filament** is called an **actin-myosin cross bridge**.

Effectors — Muscle Contraction

2) ATP Provides the Energy Needed to Move the Myosin Head...

1) **Calcium** ions also **activate** the enzyme **ATPase**, which **breaks down ATP** (into ADP + P_i) to **provide** the **energy** needed for muscle contraction.
2) The **energy** released from ATP **moves** the **myosin head**, which **pulls** the **actin filament** along in a kind of **rowing action**.

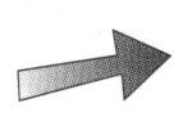

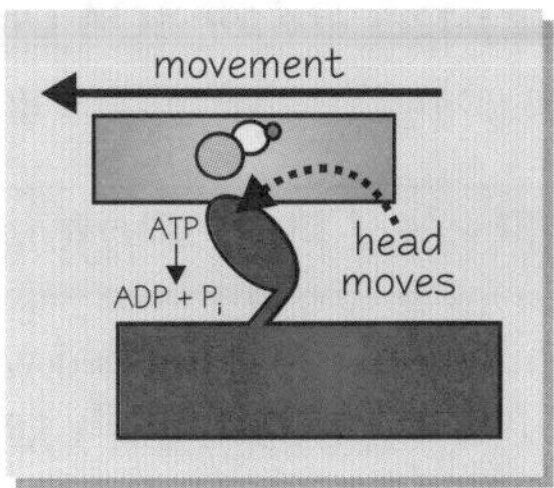

3) ...and to Break the Cross Bridge

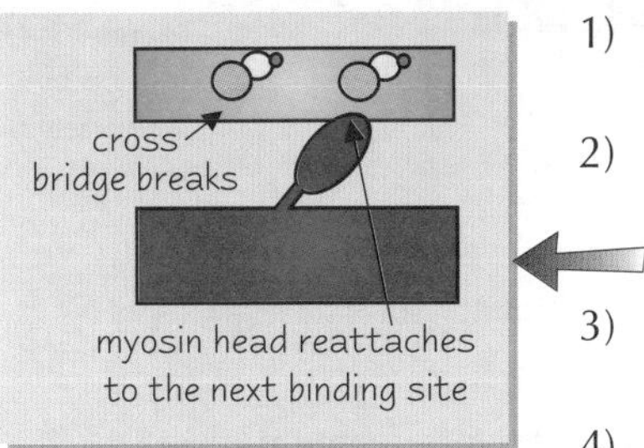

1) **ATP** also provides the **energy** to **break** the **actin-myosin cross bridge**, so the **myosin head detaches** from the actin filament **after** it's moved.
2) The **myosin head** then **reattaches** to a **different binding site** further along the actin filament. A **new actin-myosin cross bridge** is formed and the **cycle** is **repeated** (attach, move, detach, reattach to new binding site...).
3) **Many** cross bridges **form** and **break** very **rapidly**, pulling the actin filament along — which **shortens** the **sarcomere**, causing the **muscle** to **contract**.
4) The cycle will **continue** as long as **calcium ions** are **present** and **bound** to **troponin**.

When Excitation Stops, Calcium Ions Leave Troponin Molecules

1) When the muscle **stops** being **stimulated**, **calcium ions leave** their **binding sites** on the **troponin** molecules and are moved by **active transport** back into the **sarcoplasmic reticulum** (this needs **ATP** too).
2) The **troponin** molecules return to their **original shape**, pulling the attached **tropomyosin** molecules with them. This means the **tropomyosin** molecules **block** the actin-myosin **binding sites** again.
3) Muscles **aren't contracted** because **no myosin heads** are **attached** to **actin** filaments (so there are no actin-myosin cross bridges).
4) The **actin** filaments **slide back** to their **relaxed** position, which **lengthens** the **sarcomere**.

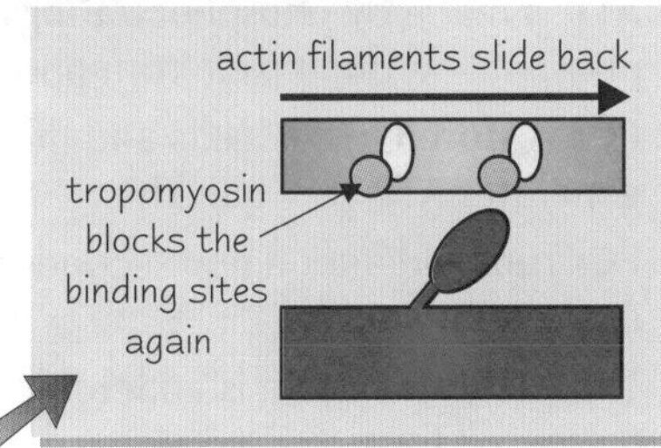

Practice Questions

Q1 What happens to sarcomeres as a muscle relaxes?

Q2 Which molecule blocks the actin-myosin binding site in resting muscles?

Q3 What's the name of the bond that's formed when a myosin head binds to an actin filament?

Exam Questions

Q1 Describe how the lengths of the different bands in a myofibril change during muscle contraction. [2 marks]

Q2 Rigor mortis is the stiffening of muscles in the body after death. It happens when ATP reserves are exhausted. Explain why a lack of ATP leads to muscles being unable to relax. [3 marks]

Q3 Bepridil is a drug that blocks calcium ion channels.
Describe and explain the effect this drug will have on muscle contraction. [3 marks]

What does muscle contraction cost? 80p...

Sorry, that's my favourite sciencey joke so I had to fit it in somewhere — a small distraction before you revisit this page. It's tough stuff but you know the best way to learn it. That's right, shut the book and scribble down what you can remember — if you can't remember much, read it again till you can (and if you can remember loads read it again anyway, just to be sure).

Effectors — Muscle Contraction

These pages are for AQA Unit 5, OCR Unit 5 and Edexcel Unit 5.

Keep going, you've almost got muscle contraction done and dusted — just a few more bits and pieces to learn...

ATP and *PCr* Provide the *Energy* for *Muscle Contraction*

AQA and OCR only

So much **energy** is **needed** when muscles contract that **ATP** gets **used up very quickly**.
ATP has to be **continually generated** so exercise can continue — this happens in **three main ways**:

① Aerobic respiration

- **Most ATP** is generated via **oxidative phosphorylation** in the cell's **mitochondria**.
- **Aerobic** respiration only works when there's **oxygen** so it's good for **long periods** of **low-intensity exercise**, e.g. walking or jogging.

See pages 14-21 for more on aerobic and anaerobic respiration.

② Anaerobic respiration

- ATP is made **rapidly** by **glycolysis**.
- The **end product** of glycolysis is **pyruvate**, which is converted to **lactate** by **lactate fermentation**.
- Lactate can **quickly build up** in the muscles and cause **muscle fatigue**.
- Anaerobic respiration is good for **short periods** of **hard exercise**, e.g. a **400 m sprint**.

③ ATP-Phosphocreatine (PCr) System

- **ATP** is made by **phosphorylating ADP** — adding a phosphate group taken from **phosphocreatine** (**PCr**).

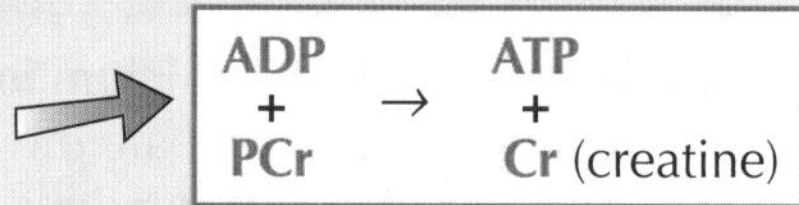

- **PCr** is **stored** inside cells and the ATP-PCr system **generates ATP** very **quickly**.
- **PCr runs out** after a few seconds so it's used during **short bursts** of **vigorous exercise**, e.g. a **tennis serve**.
- The ATP-PCr system is **anaerobic** (it doesn't need oxygen) and it's **alactic** (it doesn't form any lactate).

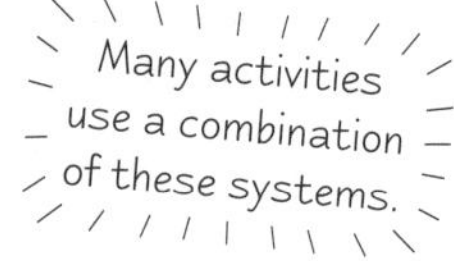

Skeletal Muscles are Made of *Slow* and *Fast Twitch Muscle Fibres*

AQA and Edexcel only

Skeletal muscles are made up of **two types** of **muscle fibres** — **slow twitch** and **fast twitch**.
Different muscles have **different proportions** of slow and fast twitch fibres. The two types have **different properties**:

SLOW TWITCH MUSCLE FIBRES	FAST TWITCH MUSCLE FIBRES
Muscle fibres that contract slowly.	Muscle fibres that contract very quickly.
Muscles you use for posture, e.g. those in the back, have a high proportion of them.	Muscles you use for fast movement, e.g. those in the eyes and legs, have a high proportion of them.
Good for endurance activities, e.g. maintaining posture, long-distance running.	Good for short bursts of speed and power, e.g. eye movement, sprinting.
Can work for a long time without getting tired.	Get tired very quickly.
Energy's released slowly through aerobic respiration. Lots of mitochondria and blood vessels supply the muscles with oxygen.	Energy's released quickly through anaerobic respiration using glycogen (stored glucose). There are few mitochondria or blood vessels.
Reddish in colour because they're rich in myoglobin — a red-coloured protein that stores oxygen.	Whitish in colour because they don't have much myoglobin (so can't store much oxygen).

Morgan used his good looks and his fast twitch muscle fibres to quickly hail a taxi.

Effectors — Muscle Contraction

This page is for OCR Unit 5 only. If you're doing AQA or Edexcel, you can go straight to the questions.

There are **Three Types** of **Muscle**

1 Voluntary muscle (skeletal muscle)

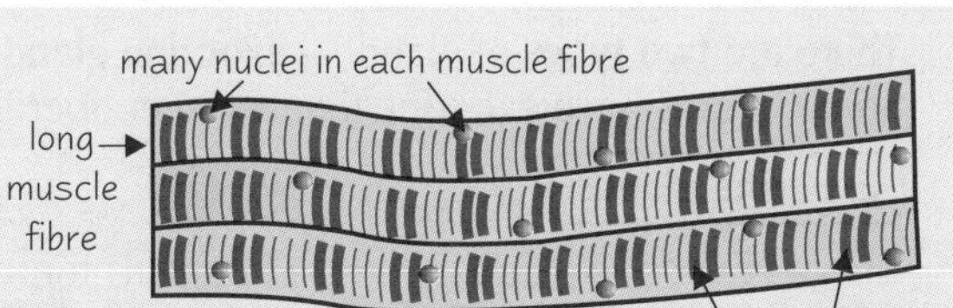

1) **Voluntary** muscle contraction is controlled **consciously** (you have to voluntarily decide to contract it).
2) It's made up of **many muscle fibres** that have **many nuclei.**
3) The muscle fibres can be **many centimetres long.**
4) You can see regular **cross-striations** (a **striped pattern**) under a **microscope.**
5) Some muscle fibres **contract very quickly** — they're used for **speed** and **strength** but **fatigue** (get tired) **quickly.**
6) Some muscle fibres **contract slowly** and **fatigue slowly** — they're used for **endurance** and **posture.**

many nuclei in each muscle fibre
long muscle fibre
cross-striations

2 Involuntary muscle (also called smooth muscle)

1) **Involuntary** muscle contraction is controlled **unconsciously** (it'll contract automatically without you deciding to).
2) It's also called **smooth muscle** because it **doesn't** have the **striped appearance** of voluntary muscle.
3) It's found in the **walls** of your **hollow internal organs**, e.g. the **gut**, the **blood vessels.** Your **gut smooth muscles contract** to **move food along** (**peristalsis**) and your **blood vessel smooth muscles contract** to **reduce** the **flow** of **blood.**
4) Each muscle fibre has **one nucleus.**
5) The muscle fibres are **spindle-shaped** with **pointed ends**, and they're only about **0.2 mm long.**
6) The muscle fibres **contract slowly** and **don't fatigue.**

nucleus
spindle-shaped muscle fibre

3 Cardiac muscle (heart muscle)

1) **Cardiac** muscle **contracts** on its **own** — it's **myogenic** (but the **rate** of contraction is controlled involuntarily by the **autonomic nervous system**).
2) It's found in the **walls** of your **heart.**
3) It's made of muscle fibres **connected** by **intercalated discs**, which have **low electrical resistance** so nerve impulses pass **easily** between cells.
4) The muscle fibres are **branched** to allow **nerve impulses** to **spread quickly** through the whole muscle.
5) Each muscle fibre has **one nucleus.**
6) The muscle fibres are shaped like **cylinders** and they're about **0.2 mm long.**
7) You can see **some cross-striations** but the striped pattern **isn't** as **strong** as it is in voluntary muscle.
8) The muscle fibres **contract rhythmically** and **don't fatigue.**

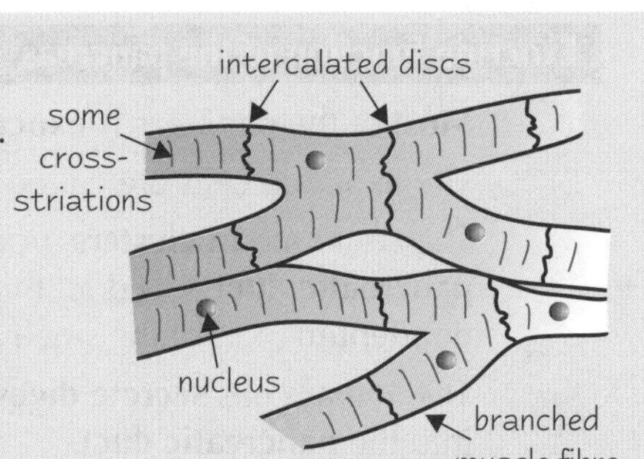

Practice Questions

Q1 State three differences between slow and fast twitch skeletal muscle fibres.

Q2 Which type of muscle has many nuclei in each muscle fibre?

Exam Questions

Q1 During short bursts of vigorous exercise, contracting muscles quickly exhaust ATP supplies. Explain how the body can quickly generate more ATP using the ATP-phosphocreatine system. [2 marks]

Q2 Compare the structure and function of involuntary muscle and cardiac muscle. [5 marks]

Smooth muscle — it has a way with the ladies...

Blimey, those biologists are an indecisive bunch. Or maybe they're not, I can't decide. They should really stick to one name per muscle type. Mind you, at least these names are logical for once — like involuntary muscle that contracts involuntarily.

Effectors — Glands

These pages are for OCR Unit 4. Go to p. 104 if you're doing AQA or OCR Unit 5, and p. 106 if you're doing Edexcel.

Muscles aren't the only effectors in town, oh no — you've got glands to learn about too, you lucky old thing...

Glands Secrete Useful Substances

There are **two types** of gland — **exocrine glands** and **endocrine glands**.
Some organs have exocrine tissue **and** endocrine tissue, so act as both types of gland.

Exocrine glands

1) Exocrine glands secrete chemicals through **ducts** (tubes) into **cavities** or onto the **surface** of the body, e.g. **sweat glands** secrete **sweat** onto the **skin surface**.
2) They usually secrete **enzymes**, e.g. **digestive glands** secrete **digestive enzymes** into the **gut**.

Endocrine glands

1) Endocrine glands secrete **hormones directly** into the **blood**.
2) Hormones are '**chemical messengers**' — they take a **message** from **one cell** to **another**.
3) Hormones are secreted when an endocrine gland's **stimulated** by a **change** in **concentration** of a **substance** (sometimes another hormone) or by **nerve impulses**.
4) Hormone secretion is the **start** of **hormonal communication** — see page 84 for more.

The Pancreas is an Exocrine and an Endocrine Gland

The pancreas is a gland that's found **below** the **stomach**.
You need to know about its exocrine function and its endocrine function:

Exocrine function of the pancreas

1) **Most** of the pancreas is **exocrine** tissue.
2) The exocrine cells are called **acinar cells**.
3) They're found in **clusters** around the **pancreatic duct** — a duct that goes to the **duodenum** (part of the small intestine).
4) The acinar cells **secrete digestive enzymes** into the **pancreatic duct**.
5) The enzymes **digest food** in the **duodenum**, e.g. **amylase** breaks down **starch** to **glucose**.

Endocrine function of the pancreas

1) The areas of **endocrine** tissue are called the **islets of Langerhans**.
2) They're found in clusters around **blood capillaries**.
3) The islets of Langerhans **secrete hormones** directly into the **blood**.
4) They're made up of **two types** of cell:
   - **Alpha** (α) **cells** secrete a **hormone** called **glucagon**.
   - **Beta** (β) **cells** secrete a **hormone** called **insulin**.
5) **Glucagon** and **insulin** help to **control blood glucose concentration** (see p. 116).

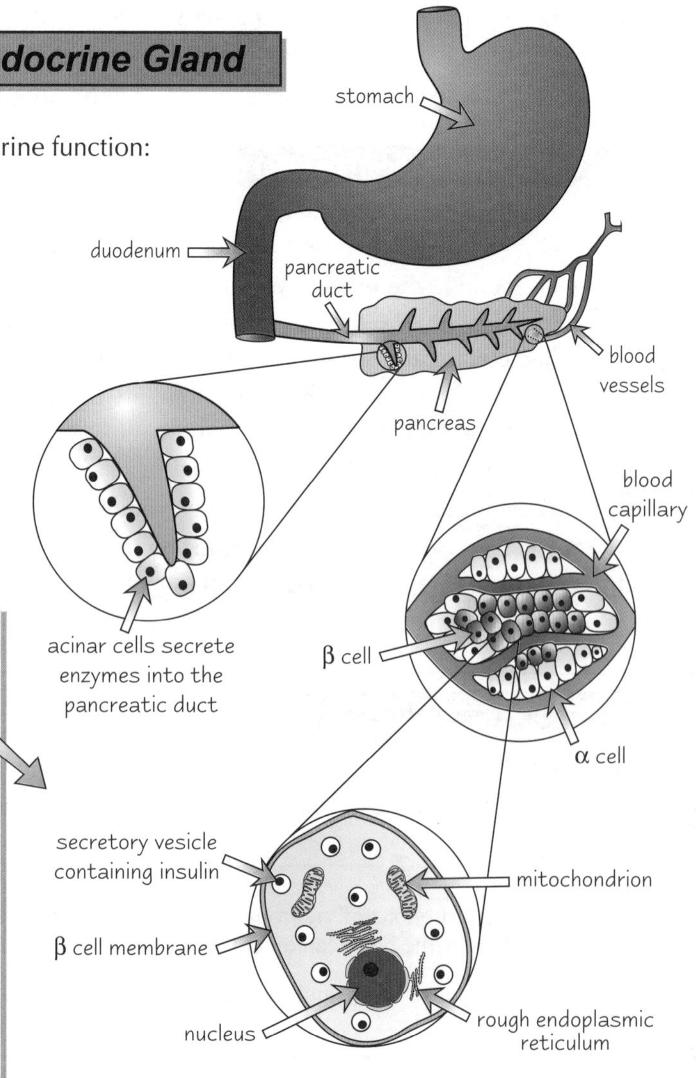

Effectors — Glands

The Adrenal Glands Secrete Hormones Including Adrenaline

1) The **adrenal glands** are **endocrine glands** that are found just **above** your **kidneys**.
2) Each adrenal gland has an **outer** part called the **cortex** and an **inner** part called the **medulla**.
3) The cortex and the medulla have **different functions**:

- The **cortex** secretes **steroid hormones**, e.g. it secretes **cortisol** when you're **stressed**.
- The **medulla** secretes **catecholamine hormones** (modified amino acids), e.g. it secretes **adrenaline** when you're **stressed**.

Cortisol and adrenaline work together to control your response to stress.

Hormones Bind to Receptors and Cause a Response via Second Messengers

1) When an **endocrine gland** is stimulated it **secretes hormones**, which is the **start** of **hormonal communication**.
2) A **hormone** is called a **first messenger** because it carries the chemical message the **first part** of the way, from the **endocrine gland** to the **receptor** on the **target cells**.
3) When a hormone **binds** to its receptor it **activates** an **enzyme** in the **cell membrane**.
4) The enzyme catalyses the **production** of a **molecule** inside the cell called a **signalling molecule** — this molecule **signals** to **other parts** of the cell to **change** how the cell **works**.
5) The **signalling molecule** is called a **second messenger** because it carries the chemical message the **second part** of the way, from the **receptor** to **other parts** of the **cell**.
6) Second messengers **activate** a **cascade** (a chain of reactions) **inside** the cell. Here's an **example** you need to **learn**:

- The hormone **adrenaline** is a **first messenger**.
- It binds to **specific receptors** in the **cell membranes** of many cells, e.g. liver cells.
- When adrenaline binds it **activates** an **enzyme** in the membrane called **adenylate cyclase**.
- **Activated adenylate cyclase** catalyses the production of a **second messenger** called **cyclic AMP** (**cAMP**).
- cAMP **activates** a **cascade**, e.g. a cascade of enzyme reactions make **more glucose available** to the cell.

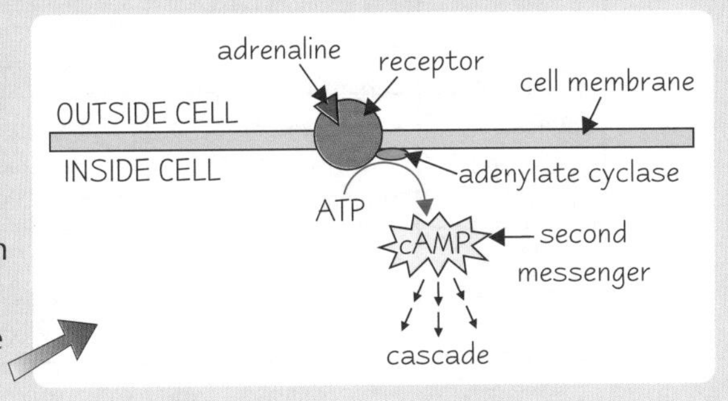

Practice Questions

Q1 What is a hormone?

Q2 Give one function of the cortex of the adrenal gland.

Exam Questions

Q1 Describe the differences between an endocrine gland and an exocrine gland. [2 marks]

Q2 In people who suffer from cystic fibrosis, the pancreatic duct may become blocked.
Suggest what effect, if any, this would have on:
a) the endocrine function of the pancreas. [2 marks]
b) the exocrine function of the pancreas. [2 marks]

Q3 In the endocrine system, explain the difference between a first messenger and a second messenger. [4 marks]

Islets of Langerhans — sounds like an exotic beach to me...

All this talk of the "islets of Langerhans" and I can think of nothing else but sun, sea and sand... but it's secretions, second messengers and cyclic AMP for you, until your exams are over and you can start planning any holidays.

Responses in Animals

This page is for AQA Unit 4 and OCR Units 4 & 5. If you're doing Edexcel, you can turn to p. 106. *And just when you thought there was no more to learn about communication systems I spring a giant example on you. (Sorry.)*

Control of Heart Rate Involves the Brain and Autonomic Nervous System

1) The **sinoatrial node** (**SAN**) generates **electrical impulses** that cause the **cardiac muscles** to **contract**.
2) The **rate** at which the SAN fires (i.e. heart rate) is **unconsciously controlled** by a part of the **brain** called the **medulla**.
3) Animals need to **alter** their **heart rate** to **respond** to **internal stimuli**, e.g. to prevent fainting due to low blood pressure or to make sure the heart rate is high enough to supply the body with enough oxygen.
4) **Stimuli** are **detected** by **pressure receptors** and **chemical receptors**:
   - There are **pressure receptors** called **baroreceptors** in the **aorta** and the **vena cava**. They're stimulated by **high** and **low blood pressure**.
   - There are **chemical receptors** called **chemoreceptors** in the **aorta**, the **carotid artery** (a major artery in the neck) and in the **medulla**. They **monitor** the **oxygen** level in the **blood** and also **carbon dioxide** and **pH** (which are indicators of O_2 level).
5) Electrical impulses from receptors are sent **to the medulla** along **sensory** neurones. The medulla processes the information and sends impulses to the SAN along **sympathetic** or **parasympathetic** neurones (which are part of the **autonomic nervous system**). Here's how it all works:

There's more about the autonomic nervous system on page 83.

Stimulus	Receptor	Neurone and neurotransmitter	Effector	Response
High blood pressure.	Baroreceptors detect high blood pressure.	Impulses are sent to the medulla, which sends impulses along parasympathetic neurones. These secrete acetylcholine, which binds to receptors on the SAN.	Cardiac muscles	Heart rate slows down to reduce blood pressure back to normal.
Low blood pressure.	Baroreceptors detect low blood pressure.	Impulses are sent to the medulla, which sends impulses along sympathetic neurones. These secrete noradrenaline (a neurotransmitter), which binds to receptors on the SAN.	Cardiac muscles	Heart rate speeds up to increase blood pressure back to normal.
High blood O_2, low CO_2 or high pH levels.	Chemoreceptors detect chemical changes in the blood.	Impulses are sent to the medulla, which sends impulses along parasympathetic neurones. These secrete acetylcholine, which binds to receptors on the SAN.	Cardiac muscles	Heart rate decreases to return O_2, CO_2 and pH levels back to normal.
Low blood O_2, high CO_2 or low pH levels.	Chemoreceptors detect chemical changes in the blood.	Impulses are sent to the medulla, which sends impulses along sympathetic neurones. These secrete noradrenaline, which binds to receptors on the SAN.	Cardiac muscles	Heart rate increases to return O_2, CO_2 and pH levels back to normal.

The Hormonal System Also Helps to Control Heart Rate

OCR Unit 4 only

1) When an organism is **threatened** (e.g. by a predator) the **adrenal glands** release **adrenaline**.
2) Adrenaline **binds** to **specific receptors** in the **heart**. This causes the cardiac muscle to **contract more frequently** and with **more force**, so **heart rate increases** and the heart **pumps more blood**.

The 'Fight or Flight' Response Gets the Body Ready for Action

OCR Unit 5 only

1) When an organism is **threatened** (e.g. by a predator) it responds by **preparing the body for action** (e.g. for fighting or running away). This **response** is called the '**fight or flight**' response.
2) The **nervous system** and **hormonal system coordinate** the fight or flight response.
3) The **sympathetic** nervous system is **activated**, which also **triggers** the **release** of **adrenaline**. The sympathetic nervous system and adrenaline have the following effects:

- **Heart rate** is **increased** — so **blood** is **pumped** around the body **faster**.
- The **muscles** around the **bronchioles relax** — so **breathing is deeper**.
- **Glycogen** is **converted** into **glucose** — so **more glucose** is **available** for **muscles** to **respire**.
- **Muscles** in the **arterioles** supplying the **skin** and **gut constrict**, and muscles in the **arterioles** supplying the **heart**, **lungs** and **skeletal muscles dilate** — so **blood** is **diverted** from the skin and gut **to the heart**, **lungs** and **skeletal muscles**.

Responses in Animals

This page is for AQA. If you're doing OCR, go straight to the questions.

Simple Responses Keep **Simple Organisms** in a **Favourable Environment**

Simple organisms, e.g. woodlice and earthworms, have **simple responses** to keep them in a **favourable environment**. Their **response** can either be **tactic** or **kinetic**:

- **Tactic responses** (**taxes**) — the organisms move towards or away from a **directional stimulus**.

  For example, **woodlice** show a **tactic** response to light (**phototaxis**) — they move **away from** a **light source**. This helps them **survive** as it keeps them **concealed** under stones during the day (where they're **safe** from predators) and keeps them in **damp conditions** (which reduces water loss).

- **Kinetic responses** (**kineses**) — the organisms' movement is affected by a **non-directional stimulus**, e.g. **intensity**.

  For example, **woodlice** show a **kinetic** response to **humidity**. In **high humidity** they move **slowly** and **turn** more often, so that they **stay where they are**. As the air gets **drier**, they move **faster** and **turn less** often, so that they move into a **new area**. This response helps woodlice **move** from **drier air** to more **humid air**, and then **stay put**. This **improves** the **survival** chances of the organism — it **reduces** their **water loss** and it helps to keep them **concealed**.

Some **Cells Communicate** with Other Cells by **Secreting Chemical Mediators**

1) A **chemical mediator** is a **chemical messenger** that acts **locally** (i.e. on **nearby cells**).
2) Communication using chemical mediators is **similar** to communication using **hormones** (see p. 84) — cells release chemicals that bind to **specific receptors** on **target cells** to cause a **response**. But there are a few **differences**:
   - Chemical mediators are secreted from **cells** that are **all over** the **body** (not just from glands).
   - Their **target cells** are right **next to** where the chemical mediator's produced. This means they stimulate a **local response** (not a widespread one).
   - They only have to travel a **short distance** to their target cells, so produce a **quicker response** than hormones (which are transported in the blood).
3) You need to **know** about **two** types of **chemical mediator** — **histamine** and **prostaglandins**:

HISTAMINE: **Histamine** is a chemical mediator that's stored in **mast cells** and **basophils** (types of immune system cell). It's **released** in response to the body being **injured** or **infected**. It **increases** the **permeability** of the **capillaries nearby** to allow more immune system cells to move out of the blood to the infected or injured area.

PROSTAGLANDINS: **Prostaglandins** are a **group** of chemical mediators that are produced by **most cells** of the **body**. They're involved in loads of things like **inflammation**, **fever**, **blood pressure regulation** and **blood clotting**. E.g. one type of prostaglandin is released from blood vessel epithelium cells and causes the muscles around them to relax.

Practice Questions

Q1 What's the 'fight or flight' response?

Q2 What's the difference between taxes and kineses?

Q3 What is a chemical mediator?

Exam Question

Q1 a) Explain how high blood pressure in the aorta causes the heart rate to slow down. [5 marks]

b) What would be the effect of severing the nerves from the medulla to the sinoatrial node (SAN)? [2 marks]

AAAAAAAAAAAAAAAAAAAAAAAAAARGH — the student's response to revision...

Right, I don't know about you, but I'm getting pretty fed up with this section now — the ways to respond to a stimulus seem endless. But, that's all the animal stuff done, just a bit more about plants to learn then you're done with it.

Responses in Plants

These pages are for AQA Unit 5, OCR Unit 5 and Edexcel Unit 5. OCR Unit 4 people, you're done for this section.
Plants also have ways of responding to stimuli — OK so they're not as quick as animals, but they're important all the same.

Plants Need to Respond to Stimuli Too

1) Plants, like animals, **increase** their chances of **survival** by **responding** to changes in their **environment**, e.g:
    - They sense the direction of **light** and **grow** towards it to **maximise** light absorption for **photosynthesis**.
    - They can sense **gravity**, so their roots and shoots **grow** in the **right direction**.
    - **Climbing** plants have a sense of **touch**, so they can find things to climb and **reach** the **sunlight**.
2) Plants are more likely to survive if they **respond** to the presence of **predators** to **avoid being eaten**, e.g. some plants produce **toxic substances**:

    White clover is a plant that can produce substances that are **toxic** to **cattle**. Cattle start to **eat** lots of white clover when fields are **overgrazed** — the white clover **responds** by **producing toxins**, to **avoid** being **eaten**.
3) Plants are more likely to survive if they **respond** to **abiotic stress** — anything **harmful** that's **natural** but **non-living**, like a **drought**. E.g. some plants respond to **extreme cold** by **producing** their own form of **antifreeze**:

    Carrots produce **antifreeze proteins** at low temperatures — the proteins **bind** to **ice crystals** and **lower** the **temperature** that water **freezes** at, **stopping** more ice crystals from **growing**.

A Tropism is a Plant's Growth Response to an External Stimulus

1) A **tropism** is the **response** of a plant to a **directional stimulus** (a stimulus coming from a particular direction).
2) Plants respond to directional stimuli by **regulating** their **growth**.
3) A **positive tropism** is growth **towards** the stimulus.
4) A **negative tropism** is growth **away** from the stimulus.

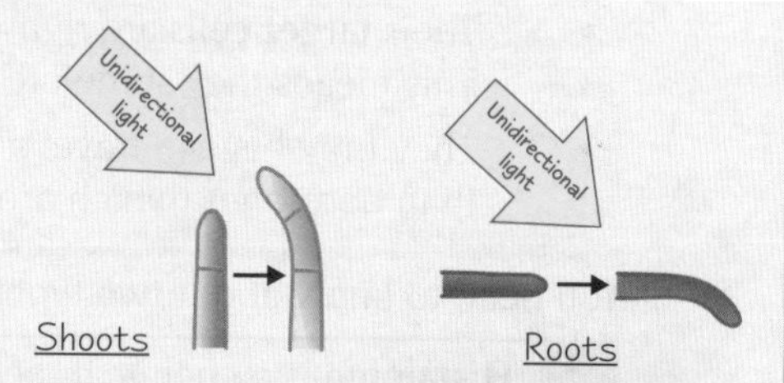

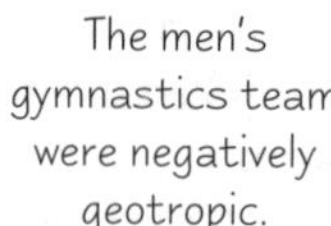

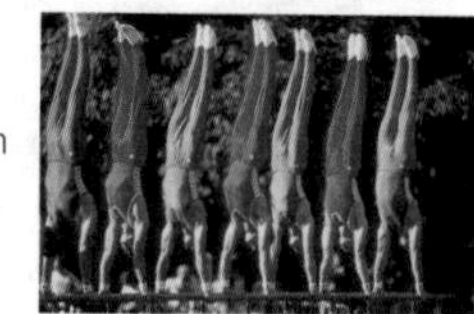

- **Phototropism** is the growth of a plant in response to **light**.
- **Shoots** are **positively phototropic** and grow **towards** light.
- **Roots** are **negatively phototropic** and grow **away** from light.

- **Geotropism** is the growth of a plant in response to **gravity**.
- **Shoots** are **negatively geotropic** and grow **upwards**.
- **Roots** are **positively geotropic** and grow **downwards**.

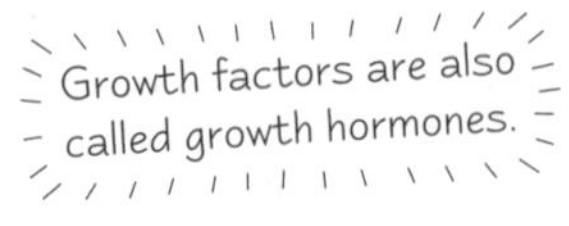
The men's gymnastics team were negatively geotropic.

Responses are Brought About by Growth Factors

1) Plants **respond** to stimuli using **growth factors** — these are chemicals that **speed up** or **slow down** plant **growth**.
2) Growth factors are **produced** in the **growing regions** of the plant (e.g. shoot tips, leaves) and they **move** to where they're needed in the **other parts** of the plant.
3) A growth factor called **gibberellin** stimulates **seed germination**, **stem elongation**, **side shoot formation** and **flowering**.
4) Growth factors called **auxins** stimulate the **growth** of shoots by **cell elongation** — this is where **cell walls** become **loose** and **stretchy**, so the cells get **longer**.
5) **High** concentrations of auxins **inhibit growth** in **roots** though.

Growth factors are also called growth hormones.

Responses in Plants

The Uneven Distribution of Auxins Causes Uneven Growth

1) **Auxins** are produced in the **tips** of **shoots** in flowering plants.
2) **Indoleacetic acid** (**IAA**) is an important **auxin** that's involved in **phototropism** and **geotropism**.
3) Auxins (including IAA) are **moved** around the plant to **control tropisms** — they move by **diffusion** and **active transport** over short distances, and via the **phloem** over longer distances.
4) This results in **different parts** of the plants having **different amounts** of auxins. The **uneven distribution** of auxins means there's **uneven growth** of the plant, e.g:

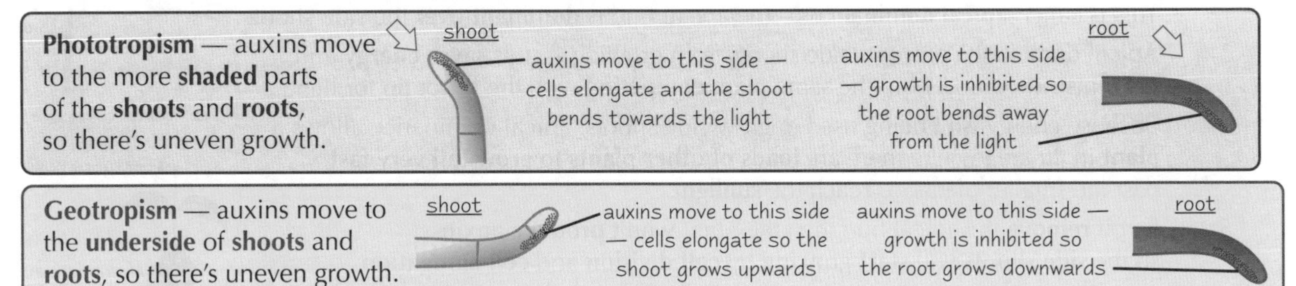

Phototropism — auxins move to the more **shaded** parts of the **shoots** and **roots**, so there's uneven growth.

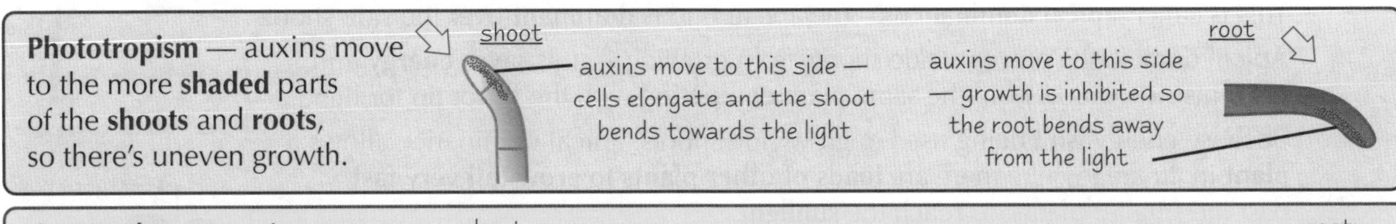

Geotropism — auxins move to the **underside** of **shoots** and **roots**, so there's uneven growth.

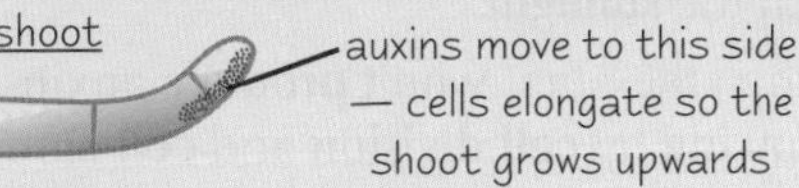

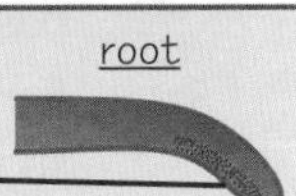

Plants Detect Light Using Photoreceptors

Edexcel only

1) Plants **detect light** using **photoreceptors** called **phytochromes**.
2) Phytochromes **control** a range of **responses**. For example, plants **flower** in **different seasons** depending on how much **daylight** there is at that time of year, e.g. some plants flower during **summer** when there are **long days**.
3) Phytochromes are **molecules** that **absorb light**. They exist in **two states** — the P_R state absorbs **red light** at a wavelength of **660 nm**, and the P_{FR} state absorbs **far-red light** at a wavelength of **730 nm**.
4) Phytochromes are **converted** from **one state** to **another** when **exposed** to **light**:
   - P_R is **quickly converted** into P_{FR} when it's exposed to **red light**.
   - P_{FR} is **quickly converted** into P_R when it's exposed to **far-red light**.
   - P_{FR} is **slowly converted** into P_R when it's in **darkness**.

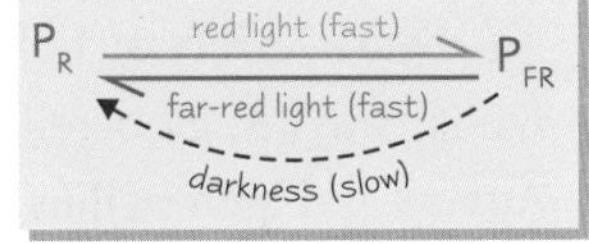

5) **Daylight** contains **more red light** than **far-red light**, so **more** P_R is converted into P_{FR} than P_{FR} is converted to P_R.
6) So the **amount** of P_R and P_{FR} **changes** depending on the **amount of light**, e.g. whether it's **day** or **night**, or **summer** or **winter**.
7) It's the **differing amounts of** P_R **and** P_{FR} that **control** the **responses** to **light**. E.g. flowering — in some plants **high levels** of P_{FR} **stimulates flowering**. When **nights** are **short** in the summer, there's not much time for P_{FR} to be converted back into P_R, so P_{FR} **builds up**. This means the plants **flower** during the **summer**.

Practice Questions

Q1 Give two reasons why plants need to respond to stimuli.

Q2 What is a tropism?

Q3 What are plant growth factors?

Exam Questions

Q1 Explain how the movement of auxins in a growing shoot enables the plant to grow towards the light. [3 marks]

Q2 Iris plants are stimulated to flower by high levels of P_{FR}.
a) What is P_{FR}? [1 mark]
b) Suggest what time of year an iris would flower and explain your answer. [4 marks]

Auxin Productions — do you have the growth factor — with Simon Trowel...

I never knew plants were so complicated — you'd never guess, just by looking at one. I thought they just grew a bit, flowered if you're lucky and then died (mine mostly seem to die — I definitely don't have green fingers). Learn these gibberish names (sorry, I mean gibberellin) and get your head around the rather bendy responses, then you'll be laughing.

Plant Hormones

These pages are for OCR Unit 5. If you're doing AQA, go to p. 111. If you're doing Edexcel you've finished this section.

Hmmm, I've decided plants are actually quite interesting. No really — you'll never look at them in the same way again...

Auxins are Involved in **Apical Dominance**

1) Auxins are **produced** in the **tips** of **shoots** (called the **apical bud**) and they're **moved** to the rest of the plant (see previous page).
2) Auxins **stimulate** the **growth** of the **apical bud** and **inhibit** the **growth** of **side shoots**. This is called **apical dominance** — the apical bud is **dominant** over the side shoots.
3) Apical dominance prevents side shoots from growing — this **saves energy** and prevents side shoots from the same plant **competing** with the shoot tip for light.
4) Because energy **isn't** being used to grow side shoots, apical dominance allows a **plant** in an area where there are **loads of other plants** to **grow tall very fast**, past the smaller plants, to **reach** the **sunlight**.
5) If you **remove** the apical bud then the plant **won't produce auxins**, so the **side shoots** will **start growing** by **cell division** and **cell elongation**.
6) Auxins become **less concentrated** as they **move away** from the apical bud to the rest of the plant. If a plant grows **very tall**, the bottom of the plant will have a **low auxin concentration** so side shoots will start to grow near the bottom.

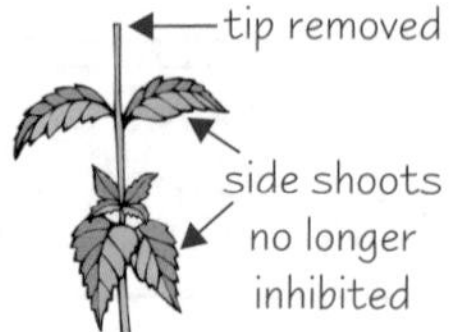

Gibberellins and **Auxins** can **Work Together**

1) **Gibberellins** are produced in **young leaves** and in **seeds**.
2) They stimulate seed germination, stem elongation, side shoot formation and flowering.
3) Gibberellins **stimulate** the **stems** of plants to **grow** by **stem elongation** — this helps plants to grow **very tall**. If a **dwarf variety** of a plant is treated with gibberellin, it will grow to the **same height** as the **tall variety**.
4) Unlike auxins, gibberellins **don't inhibit** plant growth in any way.
5) **Auxins** and **gibberellins** sometimes **work together** to affect plant growth:

Auxins and gibberellins are often **synergistic** — this means that they **work together** to have a really **big effect**. E.g. auxins and gibberellins work together to help plants grow **very tall**.

Auxins and gibberellins are sometimes **antagonistic** — this means they **oppose** each other's actions. E.g. **gibberellins stimulate** the growth of **side shoots** but **auxins inhibit** the growth of side shoots.

Hormones are Involved in **Leaf Loss** in **Deciduous Plants**

1) **Deciduous plants** are plants that **lose** their **leaves** in **winter**.
2) Losing their leaves helps plants to **conserve water** (lost from leaves) during the cold part of the year, when it might be **difficult** to **absorb water** from the **soil** (the soil water may be **frozen**), and when there's **less light** for **photosynthesis**.
3) Leaf loss is **triggered** by the **shortening day length** in the autumn.
4) Leaf loss is **controlled** by **hormones**:

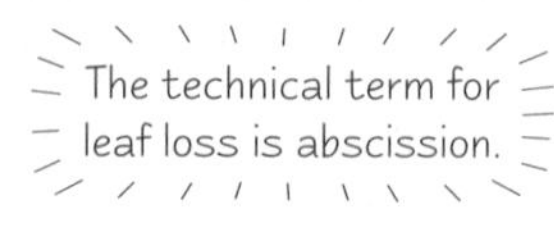

- **Auxins inhibit** leaf loss — auxins are produced by **young leaves**. As the leaf gets **older**, **less auxin** is produced, leading to **leaf loss**.
- **Ethene stimulates** leaf loss — ethene is produced by **ageing leaves**. As the leaves get **older**, **more ethene** is produced. A **layer of cells** (called the **abscission layer**) develops at the **bottom** of the **leaf stalk** (where the leaf joins the stem). The abscission layer **separates** the leaf from the rest of the plant. Ethene **stimulates** the cells in the abscission layer to **expand**, **breaking** the **cells walls** and causing the **leaf** to **fall off**.

Auxins are antagonistic to ethene.

Plant Hormones

Plant Hormones have Many Commercial Uses

The **fruit industry** uses different **plant hormones** to **control** how different fruits develop, e.g:

Ethene stimulates the ripening of fruit

Ethene stimulates enzymes that **break down cell walls**, **break down chlorophyll** and convert **starch** into **sugars**. This makes the fruit **soft**, **ripe** and **ready to eat**.

E.g. **bananas** are harvested and transported **before** they're **ripe** because they're **less likely** to be **damaged** this way. They're then **exposed** to **ethene** on arrival so they **all ripen** at the **same time** on the **shelves** and in people's **homes**.

Auxins and gibberellins make fruit develop

Auxins and gibberellins are **sprayed** onto **unpollinated flowers**, which makes the **fruit develop without fertilisation**.

E.g. **seedless grapes** can be produced using **auxins** and **gibberellins**.

Auxins can prevent or trigger fruit drop

Applying a **low concentration** of auxins in the **early stages** of fruit production **prevents** the **fruit** from **dropping** off the plant. But applying a **high concentration** of auxins at a **later stage** of fruit production **triggers** the fruit to **drop**.

E.g. **apples** can be made to **drop off** the tree at **exactly** the **right time**.

Auxins are also used **commercially** by **farmers** and **gardeners**, for example:

Auxins are used in **selective weedkillers (herbicides)** — auxins make **weeds** produce **long stems** instead of lots of **leaves**. This makes the weeds **grow too fast**, so they **can't** get enough **water** or **nutrients**, so they **die**.

Auxins are used as **rooting hormones** — auxins make a **cutting** (part of the plant, e.g. a stem cutting) **grow roots**. The **cutting** can then be **planted** and **grown** into a new plant. **Many cuttings** can be taken from **just one original plant** and **treated** with **rooting hormones**, so **lots** of the same plant can be grown **quickly** and **cheaply** from just one plant.

Practice Questions

Q1 Where are gibberellins produced in a plant?

Q2 Give one function of gibberellins in a plant.

Q3 Which hormone inhibits leaf loss in deciduous plants?

Q4 Which hormone stimulates leaf loss in deciduous plants?

Exam Questions

Q1 A gardener notices that one of the plants in his garden is showing apical dominance.

a) Name the type of plant hormone that controls apical dominance. [1 mark]
b) Give two advantages of apical dominance. [2 marks]

Q2 A tomato grower wants all her tomatoes to ripen at the same time, just before she sells them at a market.

a) Name a plant hormone she could use to make the tomatoes ripen. [1 mark]
b) Explain how the plant hormone named in part a) makes tomatoes ripen. [1 mark]
c) Suggest a commercial advantage of being able to pick and transport tomatoes before they're ripe. [1 mark]

The weeping willow — yep, that plant definitely has hormones...

See, told you plants were quite interesting — I didn't say exciting, I said interesting. Just wait till the next time you're in a supermarket — I bet you can't get round the whole shop without commenting on why the bananas are ripe...

Investigating Responses in Plants

This page is for OCR Unit 5 — if you're doing AQA Unit 5, go to the next page.

Calling all scientists — if you're anything like me, you'll be itching to get your hands on some real-life experiments to prove all the plant nonsense you've just learnt. Well, lucky you — here are some I prepared earlier...

You can **Investigate** the **Effect** of **Auxins** and **Gibberellins** in the **Lab**

You can easily do **experiments** to investigate the **effects** of **auxins** and **gibberellins** on **plant growth** — you **apply** the **hormones** to the plants and **observe** the **effects**. You need to be able to **understand** and **evaluate experiments** like the **following two examples**:

Experiment one — investigating the effect of auxins on apical dominance

1) Plant **30 plants** (e.g. **pea plants**) that are a **similar age**, **height** and **weight** in pots.
2) **Count** and **record** the number of **side shoots** growing from the main stem of **each plant**.
3) For **10 plants**, **remove** the **tip** of the **shoot** and apply a **paste containing auxins** to the **top** of the **stem.**
4) For another 10 plants, remove the tip of the shoot and apply a **paste without auxins** to the top of the stem.
5) Leave the final 10 plants as they are — these are your untouched **controls.**
6) Remember, you **always** need to have controls (e.g. without the hormone, untouched) for **comparison** — so you know the **effect** you see is **likely** to be due to the **hormone** and **not any other factor.**
7) Let each group **grow** for about **six days**. You need to keep all the plants in the **same conditions** — the same **light intensity**, **water**, etc. This makes sure any **variables** that may affect your results are **controlled**, which makes your experiment **more reliable.**
8) After six days, **count** the number of **side shoots** growing from the main stem of **each** of your **plants.**
9) You might get **results** a bit like these:
10) The results in the **table** show that **removing** the **tips** of shoots caused **extra side shoots** to **grow**, but removing tips **and** applying **auxins reduced the number** of extra side shoots.
11) The results suggest auxins **inhibit** the **growth** of side shoots — suggesting that auxins are involved in **apical dominance.**

	plants left untreated (control group)	tips removed, paste with auxins applied	tips removed, paste without auxins applied
average no. of side shoots per plant at start of experiment	4	4	4
average no. of side shoots per plant at end of experiment	5	5	9

Experiment two — investigating the effect of gibberellin on stem elongation

1) Plant **40 plants** (e.g. **dwarf pea plants**) that are a **similar age**, **height** and **mass** in pots.
2) **Leave 20** plants as they are to grow, **watering** them **all** in the **same way** and keeping them **all** in the **same conditions** — these are your **controls** (see above).
3) **Leave** the **other 20 plants** to grow in the **same conditions, except** water them with a **dilute solution** of **gibberellin** (e.g. **100 µg/ml** gibberellin).
4) Let the plants grow for about **28 days** and **measure** the **lengths** of all the **stems once each week.**
5) You might get **results** a bit like these:
6) The results in the **table** show that stems **grow more** when watered with a dilute solution of **gibberellin.**
7) The results suggest **gibberellin stimulates stem elongation.**
8) You might have to **calculate** the **rate of growth** of the plants in your exam, e.g:
   - In **28 days** the plants **watered normally** grew an **average** of **9 cm** (23 cm – 14 cm), so they grew at an average **rate** of 9 ÷ 28 = **0.32 cm/day.**
   - In **28 days** the plants **watered with gibberellin** grew an **average** of **32 cm** (46 cm – 14 cm), so they grew at an average **rate** of 32 ÷ 28 = **1.14 cm/day.**

time / days	average stem length / cm	
	plants watered normally	plants watered with gibberellin
0	14	14
7	15	17
14	18	27
21	19	38
28	23	46

Investigating Responses in Plants

This page is for AQA — if you're doing OCR, go to the questions.

*You May Have to **Interpret Experimental Data** About **Auxins***

Here's some **data** similar to what you might get in your **exam**:

1) An experiment was carried out to **investigate** the role of **auxin** in **shoot growth.**
2) Eight shoots, **equal in height and mass**, had their **tips removed**.
3) **Sponges** soaked in **glucose and either auxin** or **water** were then placed where the tip should be.
4) **Four tips** were then placed in the **dark** (**experiment A**) and the **other** four tips were exposed to a **light** source, directed at them from the **right** (**experiment B**).
5) After **two days** the **amount** of growth (in mm) and **direction** of growth was **recorded.** The results are shown in the table.

Sponge soaked in auxin and glucose
Sponge soaked in water and glucose

A B C D

Shoot minus the tip

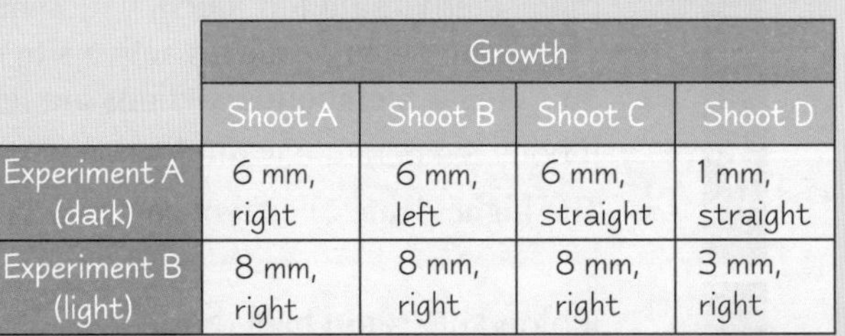

	Growth			
	Shoot A	Shoot B	Shoot C	Shoot D
Experiment A (dark)	6 mm, right	6 mm, left	6 mm, straight	1 mm, straight
Experiment B (light)	8 mm, right	8 mm, right	8 mm, right	3 mm, right

<u>You could be asked to explain the data:</u>

The results show how the **movement** of **auxin controls phototropism** in plant shoots.

In **experiment A shoot A**, the auxin **diffused** straight down from the sponge into the **left-hand side** of the shoot. This stimulated the cells on this side to **elongate**, so the shoot **grew towards the right**. In shoot B, the opposite occurred, making the shoot **grow towards the left**. In shoot C, **equal amounts** of auxin diffused down **both sides**, making **all** the cells elongate at the same rate.

In **experiment B**, the shoots were exposed to a **light source**. The auxin diffused into the shoot and **accumulated on the shaded side** (left-hand side) **regardless** of **where** the sponge was placed. All the shoots **grew towards the right**, because most auxin **accumulated** on the left, **stimulating** cell elongation there.

<u>You could be asked to comment on the experimental design:</u>

A **control** (sponge soaked in water) was included to show that it was the auxin having an **effect** and nothing else. **Glucose** was included so that the shoots would have **energy to grow in the dark** (no photosynthesis can take place).

There's more on pages 214-216 about interpreting experimental data.

Practice Questions

Q1 Why do you always need to include a control when carrying out experiments?

Q2 Why is it important to keep each group of plants in the same conditions when carrying out an experiment?

Exam Question

Q1 The table shows the results some students obtained when they investigated the effect of providing plants with auxins.

Week	Height of plant not given auxins / cm	Height of plant provided with auxins / cm
1	1	2
2	2	5
3	4	8
4	6	9
5	9	13

a) Describe and explain what the data shows. [2 marks]

b) Suggest what the students should do to increase the reliability of their results. [2 marks]

c) Why might this data be useful to a commercial tomato producer? [1 mark]

Looking at data — I'm sure there are prettier people in Star Trek...

Hoorah, you've now reached the end of this mega-section. Unless you're casually skipping through the book, in which case you've got the treat of all treats waiting for you. Just these data pages to learn — they're pretty important too, I'd bet my right trouser leg that you'll get some kind of plant experiment in your exam. But by then you'll have responses sorted.

Homeostasis Basics

These pages are for AQA Unit 5, OCR Unit 4 and Edexcel Unit 5.
Ah, there's nothing like learning a nice long word to start you off on a new section — welcome to homeostasis.

Homeostasis is the Maintenance of a Constant Internal Environment

1) **Changes** in your **external environment** can affect your **internal environment** — the blood and tissue fluid that surrounds your cells.
2) **Homeostasis** involves **control systems** that keep your **internal environment** roughly **constant** (within **certain limits**).
3) **Keeping** your internal environment **constant** is vital for cells to **function normally** and to **stop** them being **damaged**.
4) It's particularly important to **maintain** the right **core body temperature** and **blood pH**. This is because temperature and pH affect **enzyme activity**, and enzymes **control** the **rate** of **metabolic reactions**:

Temperature

- If body temperature is **too high** (e.g. 40 °C) **enzymes** may become **denatured**. The enzyme's molecules **vibrate too much**, which **breaks** the **hydrogen bonds** that hold them in their **3D shape**. The **shape** of the enzyme's **active site** is **changed** and it **no longer works** as a **catalyst**. This means **metabolic reactions** are **less efficient**.
- If body temperature is **too low enzyme activity** is **reduced**, **slowing** the rate of **metabolic reactions**.
- The **highest rate** of **enzyme activity** happens at their **optimum temperature** (about **37 °C** in humans).

pH

- If blood pH is **too high** or **too low** (highly alkaline or acidic) **enzymes** become **denatured**. The **hydrogen bonds** that hold them in their 3D shape are affected so the **shape** of the enzyme's **active site** is **changed** and it **no longer works** as a **catalyst**. This means **metabolic reactions** are **less efficient**.
- The **highest rate** of **enzyme activity** happens at their **optimum pH** — usually **around pH 7** (neutral), but some enzymes work best at other pHs, e.g. enzymes found in the stomach work best at a low pH.

5) It's important to **maintain** the right **concentration** of **glucose** in the **blood** because cells need glucose for **energy**. Blood glucose concentration also affects the **water potential** of blood — this is the potential (likelihood) of water molecules to **diffuse** out of or into a solution.

Glucose

- If blood glucose concentration is **too high** the **water potential** of blood is **reduced** to a point where **water** molecules **diffuse out** of cells into the blood by osmosis. This can cause the cells to **shrivel up** and **die**.
- If blood glucose concentration is **too low**, cells are **unable** to carry out **normal activities** because there **isn't enough glucose** for respiration to provide **energy**.

Homeostatic Systems Detect a Change and Respond by Negative Feedback

1) Homeostatic systems involve **receptors**, a **communication system** and **effectors** (see p. 82).
2) Receptors detect when a level is **too high** or **too low**, and the information's communicated via the **nervous** system or the **hormonal** system to **effectors**.
3) The effectors respond to **counteract** the change — bringing the level **back** to **normal**.
4) The mechanism that **restores** the level to **normal** is called a **negative feedback** mechanism.
5) Negative feedback **keeps** things around the **normal** level, e.g. body temperature is usually kept **within 0.5 °C** above or below **37 °C**.
6) Negative feedback only works within **certain limits** though — if the change is **too big** then the **effectors** may **not** be able to **counteract** it, e.g. a huge drop in body temperature caused by prolonged exposure to cold weather may be too large to counteract.

Control of **body temperature** by **negative feedback**:

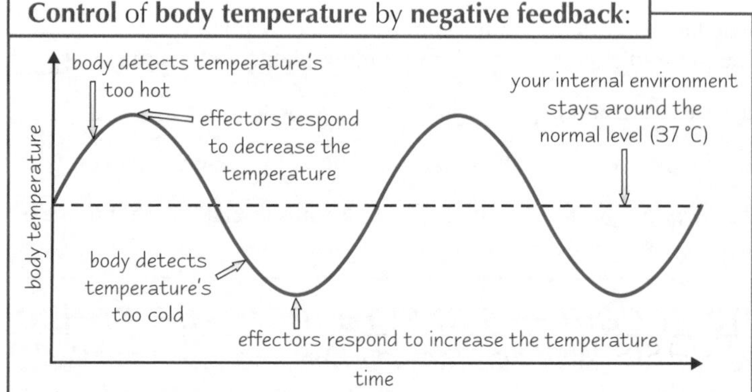

Homeostasis Basics

This page is for AQA and OCR only — if you're doing Edexcel, you can skip straight to the questions.

Multiple Negative Feedback Mechanisms Give More Control — *AQA only*

1) Homeostasis involves **multiple negative feedback mechanisms** for each thing being controlled. This is because having more than one mechanism gives **more control** over changes in your internal environment than just having one negative feedback mechanism.
2) Having multiple negative feedback mechanisms means you can **actively increase** or **decrease a level** so it returns to **normal**, e.g. you have feedback mechanisms to reduce your body temperature and mechanisms to increase it.
3) If you only had **one negative feedback mechanism**, all you could do would be **turn it on** or **turn it off**. You'd only be able to actively change a level in **one direction** so it returns to normal, e.g. it's a bit like trying to slow down a car with only an accelerator — all you can do is take your foot off the accelerator.
4) Only **one** negative feedback mechanism means a **slower response** and **less control**.

Positive Feedback Mechanisms Amplify a Change from the Normal Level

1) Some changes trigger a **positive feedback** mechanism, which **amplifies** the change.
2) The effectors respond to **further increase** the level **away** from the **normal** level.
3) Positive feedback is useful to **rapidly activate** something, e.g. a **blood clot** after an injury.

- **Platelets** become **activated** and release a **chemical** — this triggers **more platelets** to be activated, and so on.
- Platelets **very quickly** form a **blood clot** at the injury site.
- The process **ends** with **negative feedback**, when the body detects the **blood clot** has been **formed**.

4) Positive feedback can also happen when a **homeostatic system breaks down**, e.g. if you're too cold for too long:

Hypothermia involves **positive feedback**:

- **Hypothermia** is **low body temperature** (below 35 °C).
- It happens when **heat's lost** from the body **quicker** than it can be **produced**.
- As body temperature **falls** the **brain doesn't work** properly and **shivering stops** — this makes body temperature **fall even more**.
- **Positive feedback** takes body temperature **further away** from the normal level, and it continues to decrease unless action is taken.

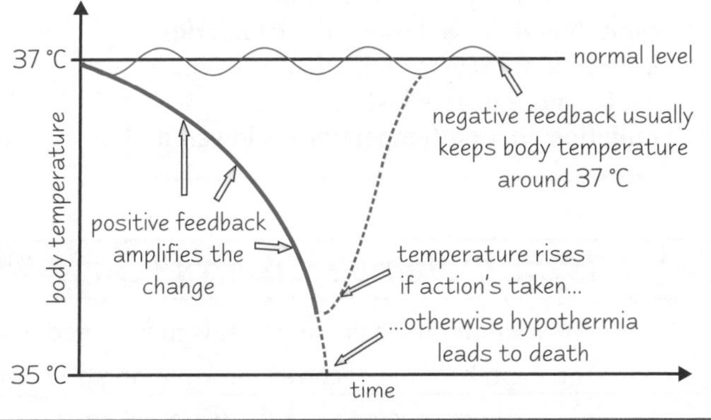

5) Positive feedback **isn't** involved in **homeostasis** because it **doesn't** keep your internal environment **constant**.

Practice Questions

Q1 What is homeostasis and why is it necessary?

Q2 Why is it important to control blood glucose concentration?

Q3 What's the advantage of having multiple negative feedback mechanisms?

Statement A: "Hyperthermia happens when the brain can't work properly and body temperature continues to increase."
Statement B: "When body temperature is low, mechanisms return the temperature to normal."

Exam Questions

Q1 Describe the role of receptors, communication systems and effectors in homeostasis. [3 marks]

Q2 Look at statements A and B in the box above.
a) Which statement is describing a positive feedback mechanism? Give a reason for your answer. [2 marks]
b) Describe and explain what effect a very high body temperature has on metabolic reactions. [2 marks]

Homeostasis works like a teacher — everything always gets corrected...

The key to understanding homeostasis is to get your head around negative feedback. Basically, if one thing goes up, the body responds to bring it down — and vice versa. And when you're ready, turn over the page for an exciting example.

Control of Body Temperature

***These pages are for** AQA Unit 5, OCR Unit 4 **and** Edexcel Unit 5.*

Your body temperature's kept around 37 °C as part of homeostasis. Mammals (including humans) have got some pretty nifty mechanisms to help them do this. Read on, oh chosen one, read on...

Mammals** have **Many Mechanisms** to **Change Body Temperature

You need to **learn** the different **mechanisms** that mammals use to **change body temperature**:

Heat loss

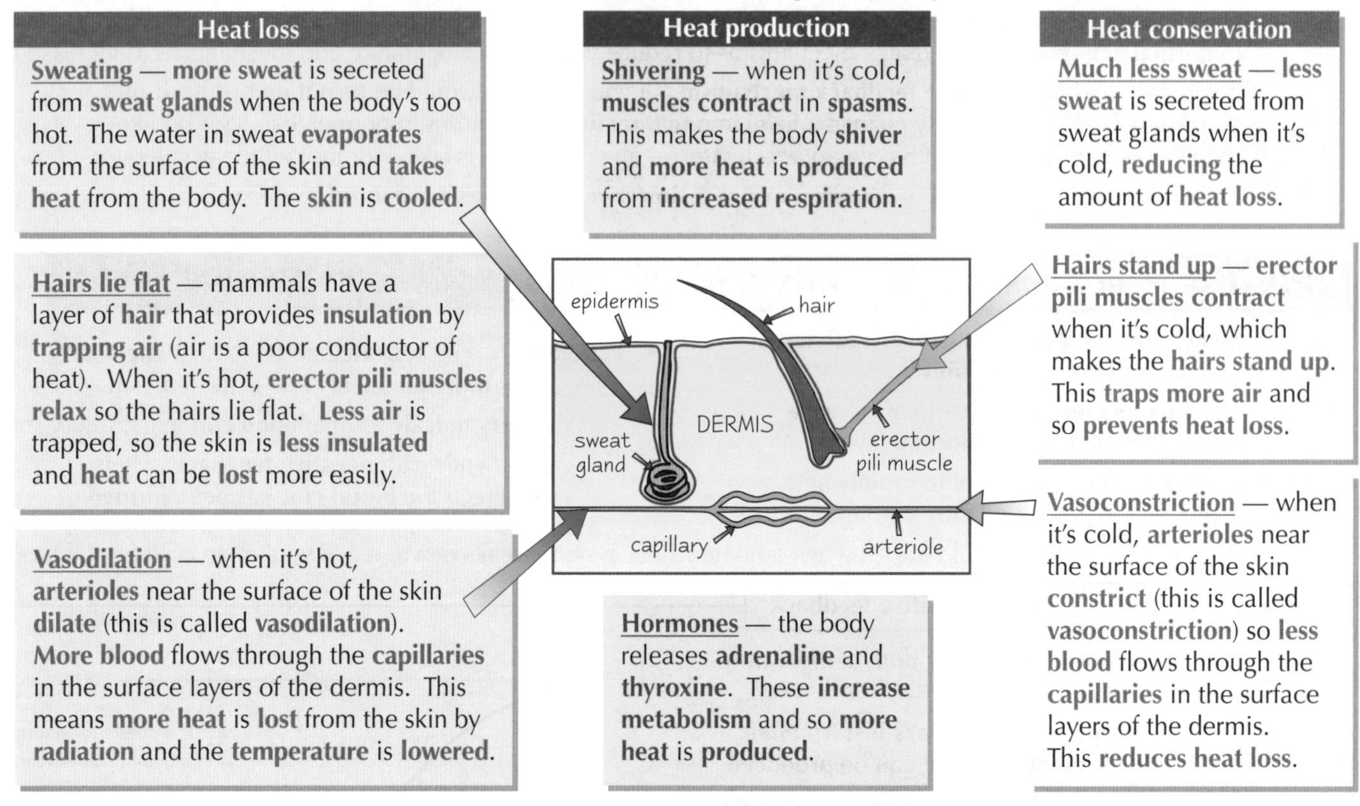

Sweating — **more sweat** is secreted from **sweat glands** when the body's too hot. The water in sweat **evaporates** from the surface of the skin and **takes heat** from the body. The **skin** is **cooled.**

Hairs lie flat — mammals have a layer of **hair** that provides **insulation** by **trapping air** (air is a poor conductor of heat). When it's hot, **erector pili muscles relax** so the hairs lie flat. **Less air** is trapped, so the skin is **less insulated** and **heat** can be **lost** more easily.

Vasodilation — when it's hot, **arterioles** near the surface of the skin **dilate** (this is called **vasodilation**). **More blood** flows through the **capillaries** in the surface layers of the dermis. This means **more heat** is **lost** from the skin by **radiation** and the **temperature** is **lowered.**

Heat production

Shivering — when it's cold, **muscles contract** in **spasms.** This makes the body **shiver** and **more heat** is **produced** from **increased respiration.**

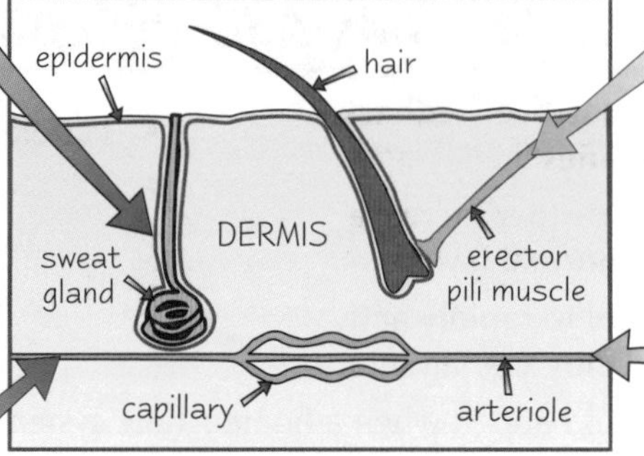

Hormones — the body releases **adrenaline** and **thyroxine.** These **increase metabolism** and so **more heat** is **produced.**

Heat conservation

Much less sweat — **less sweat** is secreted from sweat glands when it's cold, **reducing** the amount of **heat loss.**

Hairs stand up — **erector pili muscles contract** when it's cold, which makes the **hairs stand up.** This **traps more air** and so **prevents heat loss.**

Vasoconstriction — when it's cold, **arterioles** near the surface of the skin **constrict** (this is called **vasoconstriction**) so **less blood** flows through the **capillaries** in the surface layers of the dermis. This **reduces heat loss.**

*The **Hypothalamus Controls** Body Temperature in **Mammals***

1) **Body temperature** in mammals is **maintained** at a **constant level** by a part of the **brain** called the **hypothalamus**.
2) The hypothalamus **receives information** about **temperature** from **thermoreceptors** (temperature receptors):
   - Thermoreceptors in the **hypothalamus** detect **internal temperature** (the temperature of the blood).
   - Thermoreceptors in the **skin** detect **external temperature** (the temperature of the skin).
3) Thermoreceptors send **impulses** along **sensory neurones** to the **hypothalamus**, which sends **impulses** along **motor neurones** to **effectors** (muscles and glands).
4) If you're doing **AQA**, you need to know that the neurones are part of the **autonomic nervous system** (see p. 83), so it's all done **unconsciously**.
5) The effectors respond to **restore** the body temperature **back** to **normal**. Here's how it all works:

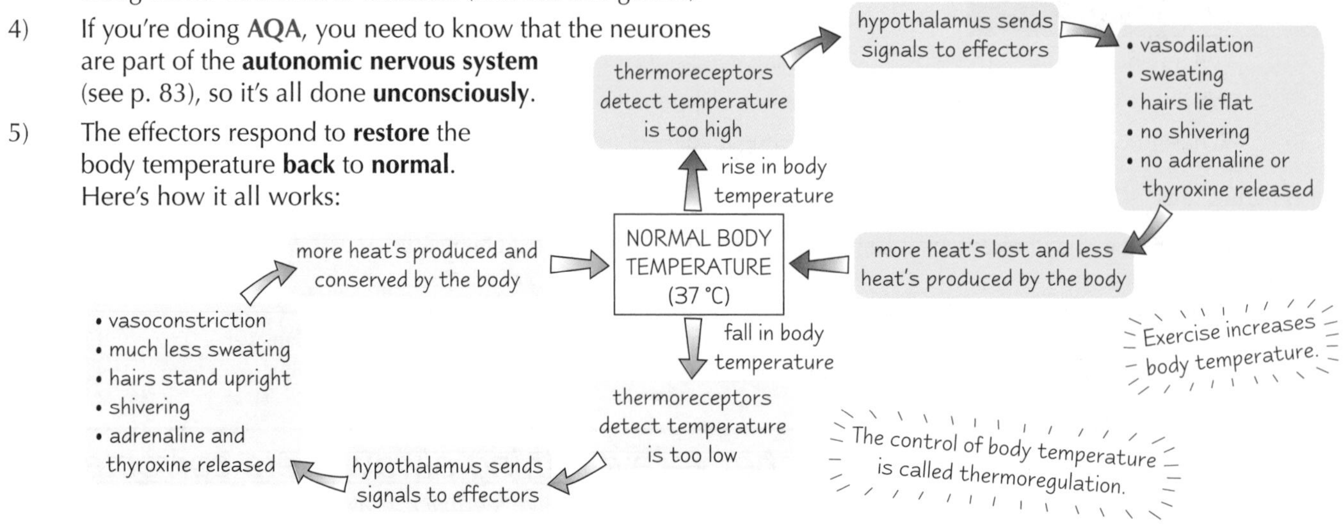

NORMAL BODY TEMPERATURE (37 °C)

rise in body temperature → thermoreceptors detect temperature is too high → hypothalamus sends signals to effectors →
- vasodilation
- sweating
- hairs lie flat
- no shivering
- no adrenaline or thyroxine released

→ more heat's lost and less heat's produced by the body → NORMAL BODY TEMPERATURE (37 °C)

fall in body temperature → thermoreceptors detect temperature is too low → hypothalamus sends signals to effectors →
- vasoconstriction
- much less sweating
- hairs stand upright
- shivering
- adrenaline and thyroxine released

→ more heat's produced and conserved by the body → NORMAL BODY TEMPERATURE (37 °C)

Exercise increases body temperature.

The control of body temperature is called thermoregulation.

Control of Body Temperature

This page is for AQA and OCR only — head straight to the questions if you're doing Edexcel.

Temperature is Controlled Differently in Ectotherms and Endotherms

Animals are classed as either **ectotherms** or **endotherms**, depending on how they **control** their body temperature:

ECTOTHERMS — e.g. reptiles, fish

Ectotherms **can't control** their body temperature **internally** — they **control** their temperature by **changing** their **behaviour**. E.g. reptiles **gain heat** by lying in the **sun**, **conserve heat** by **burrowing** down into the **sand** and **lose heat** by finding **shade**.

Their **internal** temperature **depends** on the **external temperature** (their surroundings).

Their **activity** level **depends** on the external temperature — they're **more** active at **higher** temperatures and **less** active at **lower** temperatures.

They have a **variable metabolic rate** and they **generate** very **little heat** themselves.

ENDOTHERMS — e.g. mammals, birds

Endotherms **control** their body temperature **internally** by **homeostasis**. They can also control their temperature by **behaviour**, e.g. by finding shade or lying in the sun.

Their internal temperature is **less affected** by the **external temperature** (within certain limits).

Their **activity** level is largely **independent** of the **external temperature** — they can be active at any temperature (within certain limits).

They have a constantly **high metabolic rate** and they **generate** a **lot** of **heat** from metabolic reactions.

With that beard even lying in the sun wouldn't make Chaz hot.

Practice Questions

Q1 How does sweating reduce body temperature?

Q2 How does vasodilation help the body to lose heat?

Q3 Which part of the brain is responsible for maintaining a constant body temperature in mammals?

Q4 Which part of the nervous system is involved in controlling your body temperature?

Q5 Give two differences between ectotherms and endotherms.

Exam Questions

Q1 Mammals that live in cold climates have thick fur and layers of fat beneath their skin to keep them warm. Describe and explain two other ways they maintain a constant body temperature in cold conditions. [4 marks]

Q2 Describe and explain how the body detects a high external temperature. [2 marks]

Q3 Snakes are usually found in warm climates. Suggest why they are not usually found in cold climates. Explain your answer. [4 marks]

Sweat, hormones and erector muscles — ooooh errrrrrr...

The mechanisms that change body temperature are pretty good and can cope with some extreme temperatures, but I reckon I could think up some slightly less embarrassing ways of doing it, instead of getting all red-faced and stinky. Mind you, it seems like ectotherms have got it sussed with their whole sunbathing thing — now that's definitely the life...

Control of Blood Glucose Concentration

These pages are for AQA Unit 5 and OCR Unit 4. If you're doing Edexcel, you've now finished this section.
These pages are all about how negative feedback helps you to not go totally hyper when you stuff your face with sweets.

Eating and *Exercise* Change the *Concentration* of *Glucose* in your *Blood*

1) **All cells** need a constant **energy supply** to work — so **blood glucose concentration** must be carefully **controlled**.
2) The **concentration** of **glucose** in the blood is **normally** around **90 mg per 100 cm^3** of blood. It's **monitored** by cells in the **pancreas**.
3) Blood glucose concentration **rises** after **eating food** containing **carbohydrate**.
4) Blood glucose concentration **falls** after **exercise**, as **more glucose** is used in **respiration** to **release energy**.

Insulin and *Glucagon Control* Blood Glucose Concentration

The hormonal system (see p. 84) **controls** blood glucose concentration using **two hormones** called **insulin** and **glucagon**. They're both **secreted** by clusters of cells in the **pancreas** called the **islets of Langerhans**:

Beta (β) cells secrete **insulin** into the blood.

Alpha (α) cells secrete **glucagon** into the blood.

Insulin and glucagon act on **effectors**, which respond to **restore** the blood glucose concentration to the **normal level**:

Insulin lowers blood glucose concentration when it's **too high**

1) Insulin binds to **specific receptors** on the cell membranes of **liver cells** and **muscle cells**.
2) It **increases** the **permeability** of cell membranes to glucose, so the cells **take up more glucose**.
3) Insulin also **activates enzymes** that convert **glucose** into **glycogen**.
4) Cells are able to **store glycogen** in their cytoplasm, as an **energy source**.
5) The process of **forming glycogen** from glucose is called **glycogenesis**.

Liver cells are also called hepatocytes.

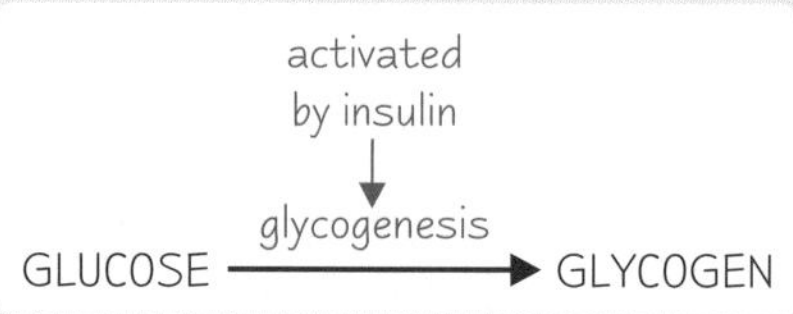

'Genesis' means 'making' — so glycogenesis means making glycogen.

6) Insulin also **increases** the **rate** of **respiration** of glucose, especially in muscle cells.

Glucagon raises blood glucose concentration when it's **too low**

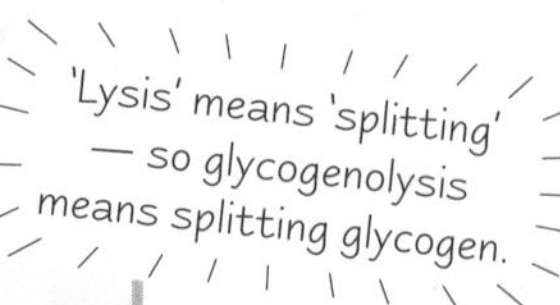

1) Glucagon binds to **specific receptors** on the cell membranes of **liver cells**.
2) Glucagon **activates enzymes** that **break down glycogen** into **glucose**.
3) The process of **breaking down glycogen** is called **glycogenolysis**.
4) Glucagon also promotes the formation of glucose from **fatty acids** and **amino acids**.
5) The process of **forming glucose** from **non-carbohydrates** is called **gluconeogenesis**.

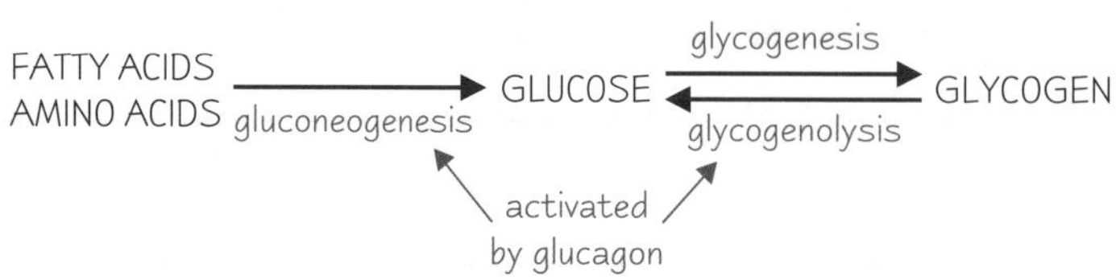

6) Glucagon **decreases** the **rate** of **respiration** of glucose in cells.

Melvin had finally mastered the ancient "chair-lysis" move.

Control of Blood Glucose Concentration

Negative Feedback Mechanisms Keep Blood Glucose Concentration Normal

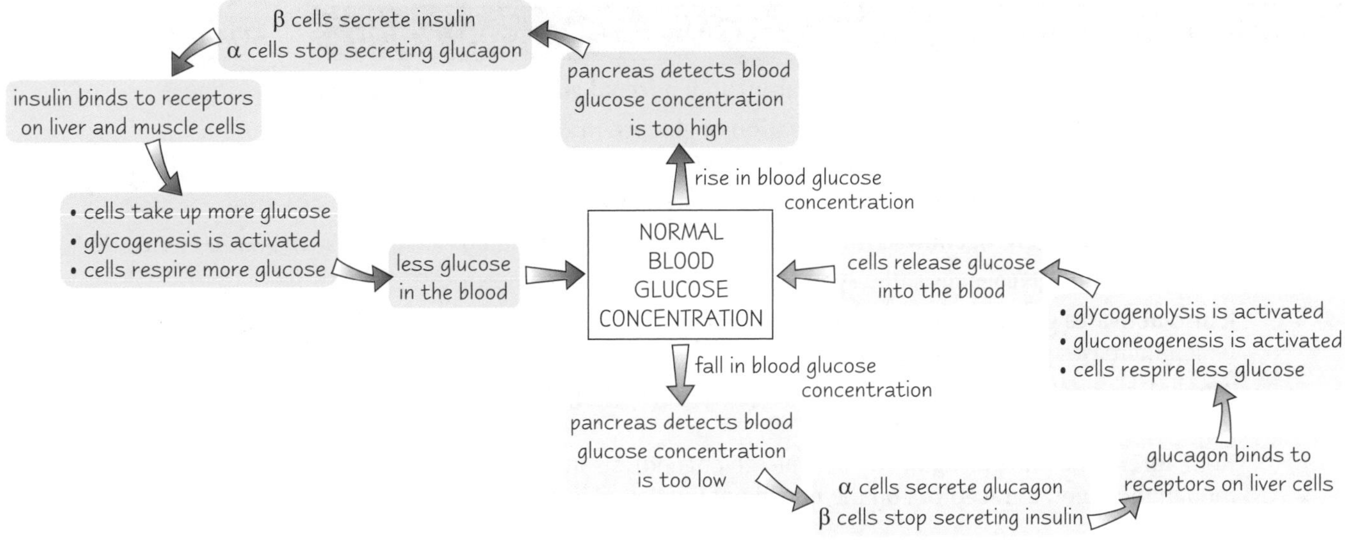

Beta (β) Cells Secrete Insulin when they're Depolarised

OCR only

β cells **contain insulin** stored in **vesicles**. β cells **secrete insulin** when they **detect high blood glucose concentration**. Here's how it happens:

1) When blood glucose concentration is **high**, **more glucose enters** the β cells by **facilitated diffusion**.
2) **More glucose** in a β cell causes the rate of **respiration** to **increase**, making **more ATP**.
3) The **rise** in **ATP** triggers the **potassium ion channels** in the β cell plasma membrane to **close**.
4) This means **potassium ions** (K^+) **can't** get through the membrane — so they **build up inside** the cell.
5) This makes the **inside** of the β cell **less negative** because there are **more positively-charged** potassium ions **inside** the cell — so the plasma membrane of the β cell is **depolarised**.
6) Depolarisation triggers **calcium ion channels** in the membrane to **open**, so **calcium ions diffuse into** the β cell.
7) This causes the **vesicles** to **fuse** with the *β* **cell plasma membrane, releasing insulin** (this is called **exocytosis**).

Practice Questions

Q1 Why does your blood glucose concentration fall after exercise?

Q2 What's the process of breaking down glycogen into glucose called?

Q3 Give two effects of glucagon on liver cells.

Exam Questions

Q1 Describe and explain how hormones return blood glucose concentration to normal after a meal. [5 marks]

Q2 Suggest the effect on a β cell of respiration being inhibited. [2 marks]

My α cells detect low glucose — urgent tea and biscuit break needed...

Aaaaargh there are so many stupidly complex names to learn and they all look and sound exactly the same to me. You can't even get away with sneakily misspelling them all in your exam — like writing 'glycusogen' or 'glucogenesisolysis'. Nope, examiners have been around for centuries, so I'm afraid old tricks like that just won't work on them. Grrrrrrrr.

Control of Blood Glucose Concentration

***These pages are for** AQA Unit 5 **and** OCR Unit 4.*

What's this — another two pages on blood glucose concentration, but still no sign of that chocolate cake I ordered...

Adrenaline Increases your Blood Glucose Concentration

AQA only

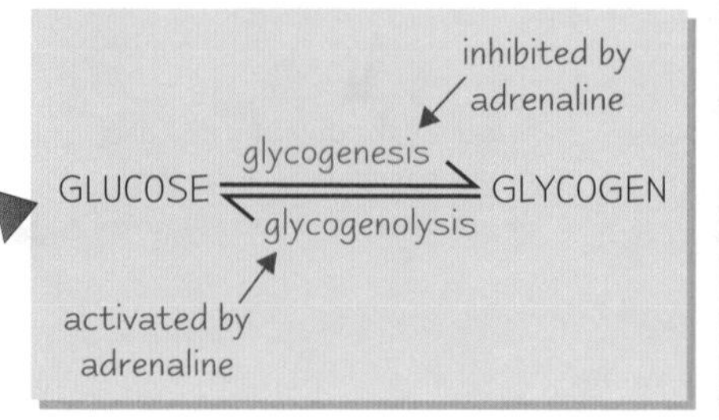

1) **Adrenaline** is a **hormone** that's secreted from your **adrenal glands** (found just above your kidneys).
2) It's secreted when there's a **low concentration** of **glucose** in your blood, when you're **stressed** and when you're **exercising**.
3) Adrenaline binds to **receptors** in the cell membrane of **liver cells**:
   - It **activates glycogenolysis** (the breakdown of glycogen to glucose).
   - It **inhibits glycogenesis** (the synthesis of glycogen from glucose).
4) It also **activates glucagon secretion** and **inhibits insulin secretion**, which increases glucose concentration.
5) Adrenaline gets the **body ready** for **action** by making **more glucose** available for **muscles** to respire.
6) Both **adrenaline** and **glucagon** can activate glycogenolysis **inside** a cell even though they bind to **receptors** on the **outside** of the cell. Here's **how** they do it:

inhibited by adrenaline

GLUCOSE ⇌ GLYCOGEN (glycogenesis →; ← glycogenolysis)

activated by adrenaline

- Adrenaline and glucagon **bind** to their specific receptors and **activate** an **enzyme** called **adenylate cyclase**.
- Activated adenylate cyclase converts **ATP** into a **chemical signal** called a **'second messenger'**.
- The second messenger is called **cyclic AMP** (**cAMP**).
- cAMP **activates** a **cascade** (a chain of reactions) that break down glycogen into glucose (**glycogenolysis**).

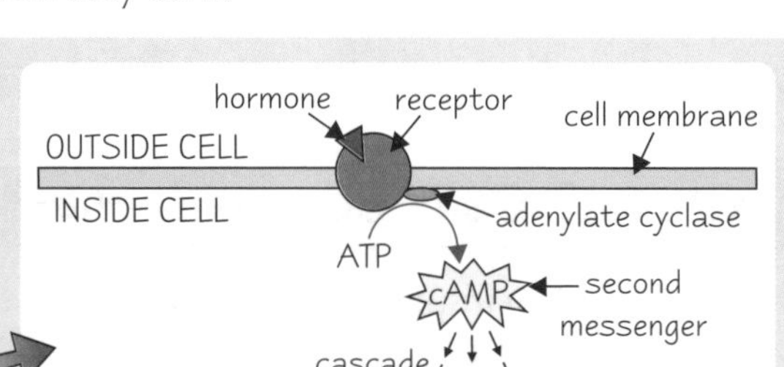

Diabetes Occurs when Blood Glucose Concentration is Not Controlled

Diabetes mellitus is a condition where **blood glucose** concentration **can't** be **controlled** properly. There are **two types**:

Type I diabetes (also called insulin-dependent diabetes)

1) In **Type I** diabetes, the **β cells** in the islets of Langerhans **don't produce** any **insulin**.
2) After **eating**, the blood glucose level **rises** and **stays high** — this is a condition called **hyperglycaemia**, which can result in **death** if left untreated. The kidneys **can't reabsorb** all this glucose, so some of it's **excreted** in the urine.
3) It can be treated by regular **injections** of **insulin**. But this has to be **carefully controlled** because too much can produce a **dangerous drop** in blood glucose levels — this is called **hypoglycaemia**.
4) **Eating regularly** and **controlling simple carbohydrate intake** (sugars) helps to **avoid** a **sudden rise** in glucose.

Type 1 diabetes usually develops in children or young adults.

Type II diabetes (also called non-insulin-dependent diabetes)

1) **Type II** diabetes is usually acquired **later** in **life** than Type I, and it's often linked with **obesity**.
2) It occurs when the β cells **don't produce enough insulin** or when the body's **cells don't respond** properly to **insulin**. Cells don't respond properly because the insulin **receptors** on their membranes **don't work** properly, so the cells **don't** take up enough glucose. This means the **blood glucose concentration** is **higher** than normal.
3) It can be treated by **controlling simple carbohydrate intake** and **losing weight**. **Glucose-lowering tablets** can be taken if diet and weight loss can't control it.

Control of Blood Glucose Concentration

This page is for OCR only — if you're doing AQA, mosey on down to the questions.

Insulin can be Produced by Genetically Modified Bacteria

1) Insulin **used** to be **extracted** from **animal pancreases** (e.g. **pigs** and **cattle**), to treat people with **Type I** diabetes.
2) But **nowadays**, **human insulin** can be made by **genetically modified** (**GM**) **bacteria** (see p. 152).
3) Using **GM bacteria** to produce insulin is **much better** for many reasons, for example:

- **Producing** insulin using GM bacteria is **cheaper** than extracting it from animal pancreases.
- **Larger quantities** of insulin can be produced using GM bacteria.
- GM bacteria make **human insulin**. This is **more effective** than using **pig** or **cattle insulin** (which is slightly different to human insulin) and it's **less likely** to trigger an **allergic response** or be **rejected** by the **immune system**.
- Some people **prefer** insulin from **GM bacteria** for **ethical** or **religious** reasons. E.g. some **vegetarians** may **object** to the **use** of **animals**, and some **religious people object** to using insulin from **pigs**.

Stem Cells Could be Used to Cure Diabetes

1) Stem cells are **unspecialised cells** — they have the **ability** to **develop** into **any type** of cell.
2) Using stem cells could **potentially cure** diabetes — here's how:

- **Stem cells** could be **grown** into **β cells**.
- The β cells would then be **implanted** into the **pancreas** of a person with **Type I diabetes**.
- This means the person would be able to **make insulin** as **normal**.
- This treatment is **still being developed**. But if it's effective, it'll **cure** people with Type I diabetes — they **won't** have to have **regular injections** of **insulin**.

Look back at your AS notes if you need to remind yourself about stem cells.

Practice Questions

Q1 Give an example of a situation when the body secretes adrenaline.
Q2 Why can't people with Type I diabetes control their blood glucose concentration?
Q3 Give two ways that Type I diabetes can be treated.
Q4 Briefly describe how stem cells could be used to cure diabetes.

Exam Questions

Q1 Describe how the hormone adrenaline activates glycogenolysis inside a liver cell. [4 marks]

Q2 Explain why someone with diabetes can produce insulin but can't control their blood glucose concentration. [3 marks]

Q3 Give two advantages of using insulin produced by genetically modified (GM) bacteria over using insulin extracted from animal pancreases. [2 marks]

I'm only passing on revision help — don't shoot the second messenger...

I don't know about you, but my ability to concentrate on blood glucose concentration is starting to waver a bit now. Still, that's not an excuse for you to sidestep (or another dance move of your choice) your revision. So, when you've got a pen and paper at the ready and a cup of tea on standby, take a deep breath and get scribbling till you know it all.

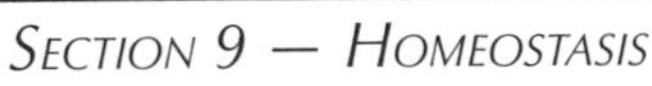

Control of the Menstrual Cycle

These pages are for AQA Unit 5 only. OCR folk, you've now finished this section.

Sorry lads — I'm afraid these two pages are pretty much devoted to the inner workings of the ladies...

The Human **Menstrual Cycle** is **Controlled** by **Hormones**

You don't need to learn this diagram (phew), but it shows you what's going on.

1) The human **menstrual cycle** (also called the **oestrous cycle**) lasts about **28 days**.
2) The menstrual cycle involves:
   - A **follicle** (an egg and its surrounding protective cells) **developing** in the **ovary**.
   - Ovulation — an **egg** being **released**.
   - The **uterus lining** becoming **thicker** so that a fertilised egg can **implant**.
   - A structure called a **corpus luteum** developing from the **remains** of the **follicle**.

3) If there's **no fertilisation**, the uterus lining **breaks down** and leaves the body through the **vagina**. This is known as **menstruation**, which marks the **end** of one cycle and the **start** of another.
4) The menstrual cycle's **controlled** by the action of **four hormones**:
   - **Follicle-stimulating hormone** (**FSH**) — does just what it says, it **stimulates** the **follicle** to develop.
   - **Luteinising hormone** (**LH**) — **stimulates ovulation** and **stimulates** the **corpus luteum** to develop.
   - **Oestrogen** — **stimulates** the **uterus lining** to **thicken**.
   - **Progesterone** — **maintains** the **thick uterus lining**, ready for implantation of an embryo.
5) **FSH** and **LH** are secreted by the **anterior pituitary gland**. **Oestrogen** and **progesterone** are secreted by the **ovaries**.

Hormone Concentrations Change During **Different Stages** of the **Cycle**

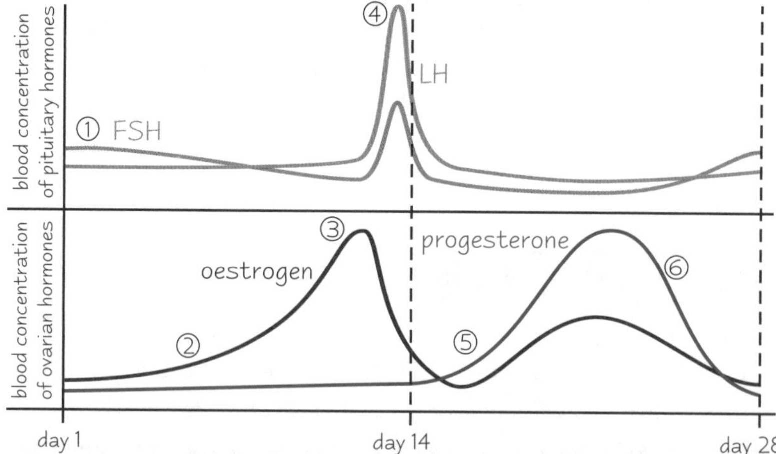

1 High FSH concentration in the blood
- FSH stimulates **follicle development**.
- The **follicle** releases **oestrogen**.
- FSH **stimulates** the **ovaries** to release **oestrogen**.

2 Rising concentration of **oestrogen**
- Oestrogen **stimulates** the **uterus lining** to **thicken**.
- Oestrogen **inhibits FSH** being released from the **pituitary** gland.

3 Oestrogen concentration **peaks**

High oestrogen concentration **stimulates** the pituitary gland to release **LH** and **FSH**.

4 LH surge (a rapid increase)
- **Ovulation** is stimulated by LH — the **follicle ruptures** and the **egg** is **released**.
- **LH** stimulates the **ruptured follicle** to turn into a **corpus luteum**.
- The **corpus luteum** releases **progesterone**.

5 Rising concentrations of **progesterone**
- Progesterone **inhibits FSH** and **LH** release from the **pituitary**.
- The **uterus lining** is **maintained** by progesterone.
- If **no embryo** implants, the **corpus luteum breaks down** and **stops releasing** progesterone.

6 Falling concentration of **progesterone**
- **FSH** and **LH** concentrations **increase** because they're **no longer inhibited** by progesterone.
- The uterus lining **isn't maintained** so it **breaks down** — **menstruation** happens and the **cycle starts again**.

Control of the Menstrual Cycle

Negative and Positive Feedback Mechanisms Control the Level of Hormones

The **different concentrations** of **hormones** in the blood during the menstrual cycle are **controlled** by **feedback loops**.

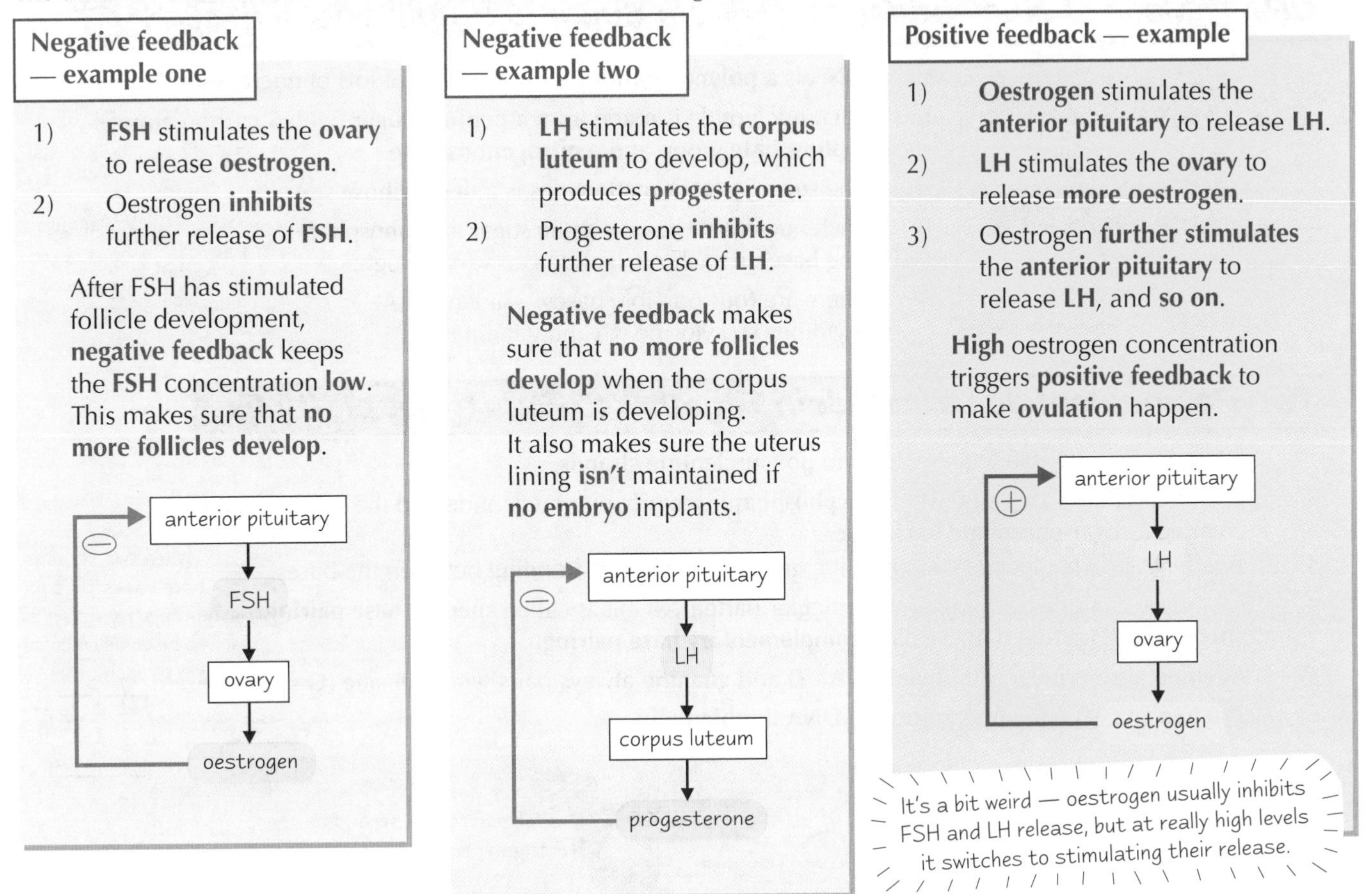

Negative feedback — example one

1) **FSH** stimulates the **ovary** to release **oestrogen**.
2) Oestrogen **inhibits** further release of **FSH**.

After FSH has stimulated follicle development, **negative feedback** keeps the **FSH** concentration **low**. This makes sure that **no more follicles develop**.

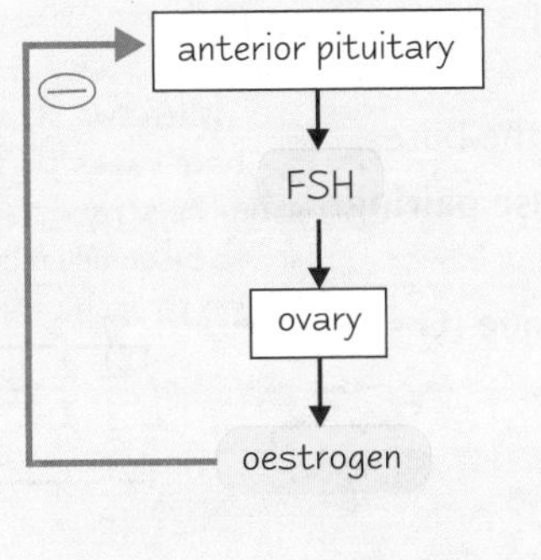

Negative feedback — example two

1) **LH** stimulates the **corpus luteum** to develop, which produces **progesterone**.
2) Progesterone **inhibits** further release of **LH**.

Negative feedback makes sure that **no more follicles develop** when the corpus luteum is developing. It also makes sure the uterus lining **isn't** maintained if **no embryo** implants.

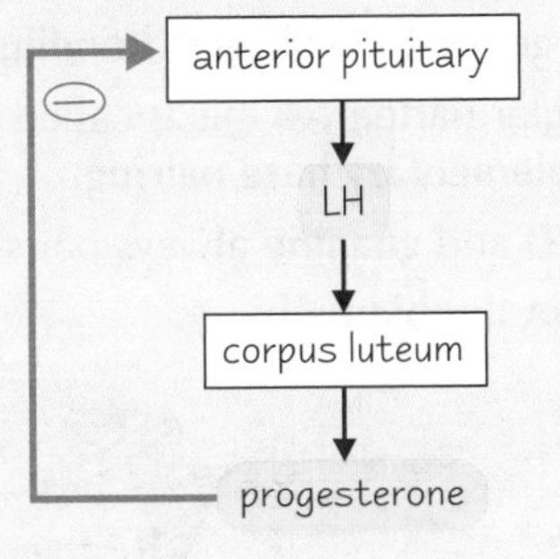

Positive feedback — example

1) **Oestrogen** stimulates the **anterior pituitary** to release **LH**.
2) **LH** stimulates the **ovary** to release **more oestrogen**.
3) Oestrogen **further stimulates** the **anterior pituitary** to release **LH**, and **so on**.

High oestrogen concentration triggers **positive feedback** to make **ovulation** happen.

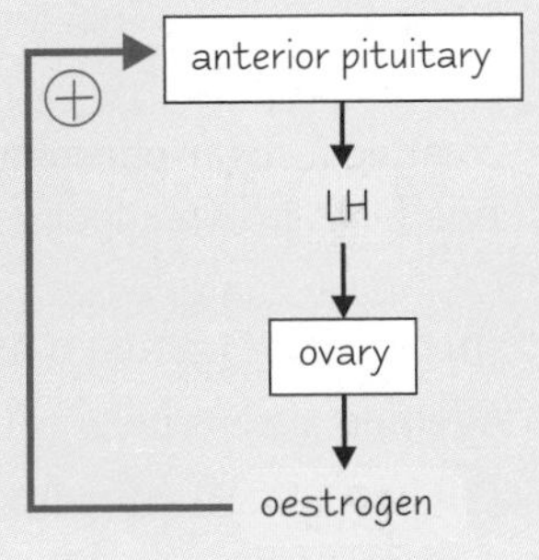

It's a bit weird — oestrogen usually inhibits FSH and LH release, but at really high levels it switches to stimulating their release.

Practice Questions

Q1 Name the hormones released by the anterior pituitary gland.

Q2 Which hormone stimulates the corpus luteum to develop?

Q3 What's the main role of oestrogen in the uterus?

Q4 What's the main role of progesterone in the uterus?

Q5 What happens to the uterus lining if no embryo implants?

Exam Questions

Q1 The human menstrual cycle is controlled by pituitary and ovarian hormones, which are present at different concentrations during the cycle.

a) Explain how negative feedback ensures only one main follicle develops. [5 marks]

b) Explain how positive feedback is involved in ovulation. [3 marks]

Q2 The contraceptive pill contains synthetic equivalents of the hormones oestrogen and progesterone. Suggest how taking the pill can prevent pregnancy. [3 marks]

Sometimes it's hard to be a woman...

What on earth... talk about women being hard to understand. Let's treat it like a good pair of shoes and take it in steps. Start by learning the four main hormones and what they do, then learn when concentrations are highest and why. Finally, get scribbling down those feedback loops. And when you know it all, it must be time for an end-of-section break.

DNA and RNA

These pages are for AQA Unit 5, OCR Unit 5 and Edexcel Unit 4.

DNA and its mysterious cousin RNA are responsible for protein production, so you need to get to know them for starters.

DNA is Made of Nucleotides that Have a Sugar, a Phosphate and a Base

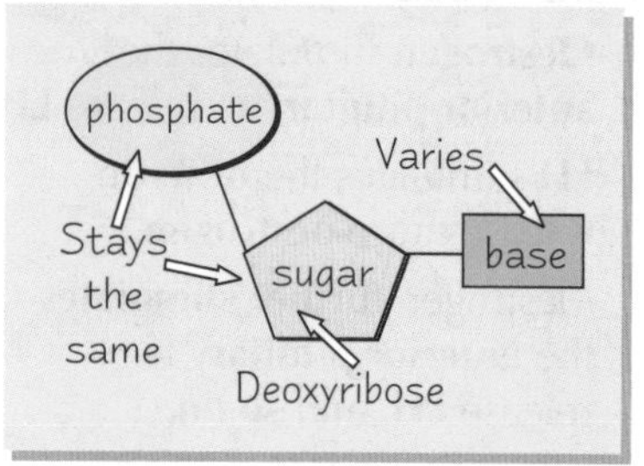

1) DNA is a **polynucleotide** — it's made up of lots of **nucleotides** joined together.
2) Each nucleotide is made from a **pentose sugar** (with 5 carbon atoms), a **phosphate** group and a **nitrogenous base**.
3) The **sugar** in DNA nucleotides is a **deoxyribose** sugar.
4) Each nucleotide has the **same sugar and phosphate**. The **base** on each nucleotide can **vary** though.
5) There are **four** possible bases — adenine (**A**), thymine (**T**), cytosine (**C**) and guanine (**G**).

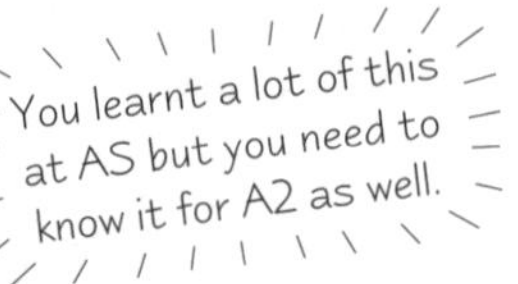

You learnt a lot of this at AS but you need to know it for A2 as well.

Two Polynucleotide Strands Join Together to Form a Double-Helix

1) DNA nucleotides join together to form **polynucleotide strands**.
2) The nucleotides join up between the **phosphate** group of one nucleotide and the **sugar** of another, creating a **sugar-phosphate backbone**.
3) **Two** DNA polynucleotide strands join together by **hydrogen bonding** between the bases.
4) Each base can only join with one particular partner — this is called **specific base pairing**. Specific base pairing is also called **complementary base pairing**.
5) **Adenine** always pairs with **thymine** (**A - T**) and **guanine** always pairs with **cytosine** (**G - C**).
6) The two strands **wind up** to form the **DNA double-helix**.

When two strands have bases that pair up the strands are said to be complementary to each other:

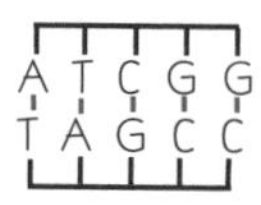

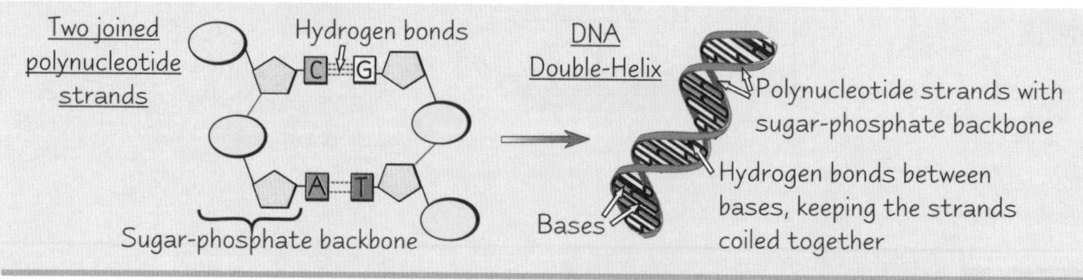

DNA Contains Genes Which are Instructions for Proteins

Polypeptide is just another word for a protein.

1) Genes are **sections of DNA**. They're found on **chromosomes**.
2) Genes **code** for **proteins** (polypeptides), including **enzymes** — they contain the **instructions** to make them.
3) Proteins are made from **amino acids**. Different proteins have a **different number** and **order** of amino acids.
4) It's the **order** of **nucleotide bases** in a gene that determines the **order of amino acids** in a particular **protein**.
5) Each amino acid is coded for by a sequence of **three bases** (called a **triplet**) in a gene. A DNA triplet is also sometimes called a **base triplet** or a **codon**.
6) **Different sequences** of bases code for different amino acids. This is the **genetic code** — see page 126 for more.
7) So the **sequence of bases** in a section of DNA is a **template** that's used to make a **protein** during **protein synthesis**.

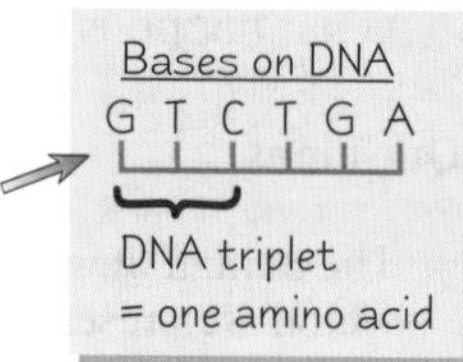

DNA is Copied into RNA for Protein Synthesis

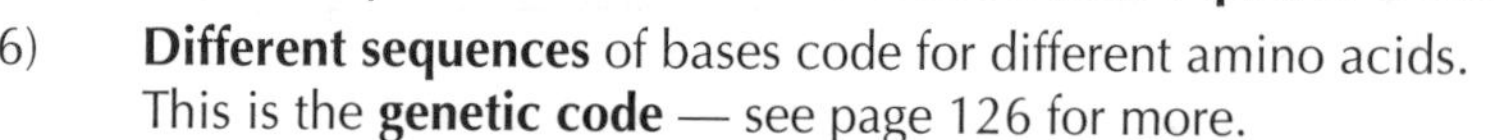

1) DNA molecules are found in the **nucleus** of the cell, but the organelles for protein synthesis (**ribosomes**) are found in the **cytoplasm**.
2) DNA is too large to move out of the nucleus, so a section is **copied** into **RNA**. This process is called **transcription** (see page 124).
3) The RNA **leaves** the nucleus and joins with a **ribosome** in the cytoplasm, where it can be used to synthesise a **protein**. This process is called **translation** (see page 125).

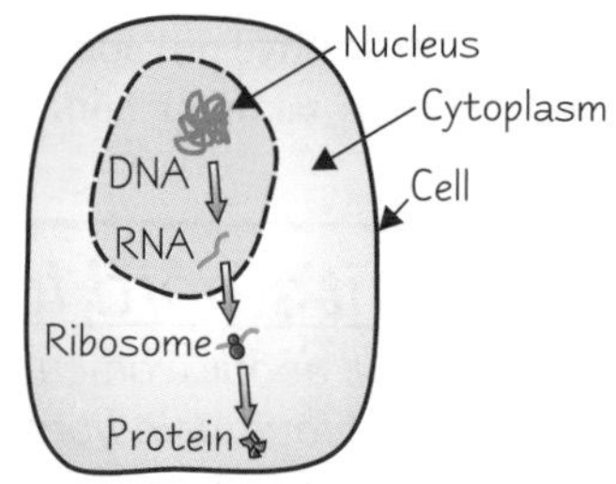

DNA and RNA

RNA is Very Similar to DNA

Like DNA, **RNA** (**ribo**nucleic **a**cid) is made of **nucleotides** that contain one of **four different bases**. The nucleotides also form a **polynucleotide strand** with a sugar-phosphate backbone. But RNA **differs** from DNA in **three** main ways:

1) The **sugar** in RNA nucleotides is a **ribose sugar** (not deoxyribose).
2) The nucleotides form a **single polynucleotide strand** (not a double one).
3) **Uracil** (**U**) replaces thymine as a base. Uracil **always pairs** with **adenine** during protein synthesis.

You need to know about **two different types** of RNA — **messenger RNA** (**mRNA**) and **transfer RNA** (**tRNA**).

Messenger RNA (mRNA)

mRNA is a **single polynucleotide strand**. In mRNA, groups of three adjacent bases are usually called **codons** (they're sometimes called **triplets** or **base triplets**). mRNA is made in the **nucleus** during **transcription**. It **carries the genetic code** from the DNA in the **nucleus** to the **cytoplasm**, where it's used to make a **protein** during **translation**.

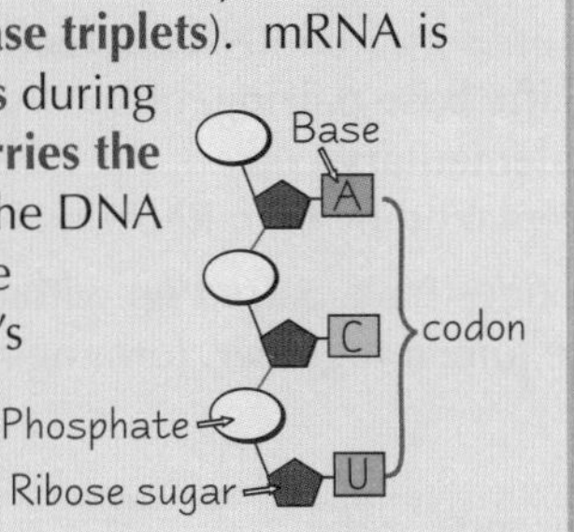

Transfer RNA (tRNA)

tRNA is a **single polynucleotide strand** that's folded into a **clover shape**. **Hydrogen bonds** between **specific base pairs** hold the molecule in this shape. Every tRNA molecule has a **specific sequence** of **three bases** at one end called an **anticodon**. They also have an **amino acid binding site** at the other end. tRNA is found in the **cytoplasm** where it's involved in **translation**. It **carries** the amino acids that are used to make **proteins** to the **ribosomes**.

You need to be able to Compare DNA, mRNA and tRNA

AQA only

You need to know the **structure** and **composition** of **DNA**, **mRNA** and **tRNA** really well — you could be asked to **compare** them in your exam. The table below outlines the **main differences** between them:

	DNA	mRNA	tRNA
Shape	Double-stranded — twisted into a double-helix and held together by hydrogen bonds	Single-stranded	Single-stranded — folded into a clover shape and held together by hydrogen bonds
Sugar	Deoxyribose sugar	Ribose sugar	Ribose sugar
Bases	A, T, C, G	A, U, C, G	A, U, C, G
Other features	Three adjacent bases are called a triplet (sometimes a base triplet or codon)	Three adjacent bases are called a codon (sometimes a triplet or base triplet)	Each tRNA molecule has a specific sequence of three bases called an anticodon and an amino acid binding site

tRNA growing in its natural environment.

Practice Questions

Q1 Name the bases found in RNA.

Q2 Describe the shape of a tRNA molecule.

Exam Questions

Q1 Name each of the following molecules:

a) A single-stranded molecule that contains ribose sugar and has an amino acid binding site. [1 mark]

b) A single-stranded molecule that contains the base uracil and has an anticodon. [1 mark]

Q2 Describe the role of mRNA and the role of tRNA. [2 marks]

Genes, genes are good for your heart, the more you eat, the more you...

An easy way to remember where mRNA and tRNA come into the whole protein synthesis game is to look at the first letters. mRNA is a messenger — it carries the code from DNA to a ribosome. tRNA transfers amino acids. Easy as that.

Protein Synthesis

***These pages are for** AQA Unit 5, OCR Unit 5 **and** Edexcel Unit 4.*

Time to find out how RNA works its magic to make proteins. It gets a bit complicated but bear with it.

First Stage of Protein Synthesis — Transcription

During transcription an **mRNA copy** of a gene (a section of DNA) is made in the **nucleus**:

1) Transcription starts when **RNA polymerase** (an **enzyme**) **attaches** to the **DNA** double-helix at the **beginning** of a **gene**.
2) The **hydrogen bonds** between the two DNA strands in the gene **break**, **separating** the strands, and the DNA molecule **uncoils** at that point.
3) One of the strands is then used as a **template** to make an **mRNA copy**.

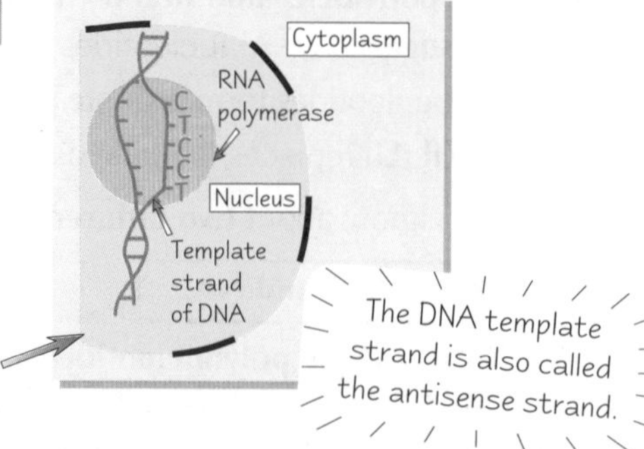

The DNA template strand is also called the antisense strand.

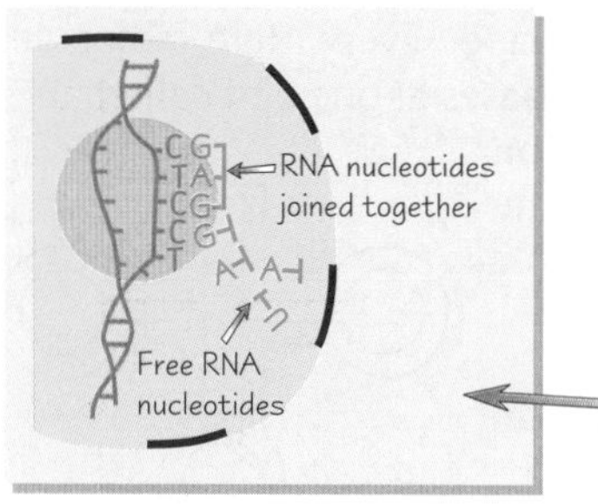

4) The RNA polymerase lines up free **RNA nucleotides** alongside the template strand. **Specific base pairing** means that the mRNA strand ends up being a **complementary copy** of the DNA template strand (except the base **T** is replaced by **U** in **RNA**).
5) Once the RNA nucleotides have **paired up** with their **specific bases** on the DNA strand they're **joined together**, forming an **mRNA** molecule.

6) The RNA polymerase moves **along** the DNA, separating the strands and **assembling** the mRNA strand.
7) The **hydrogen bonds** between the uncoiled strands of DNA **re-form** once the RNA polymerase has passed by and the strands **coil back into a double-helix**.

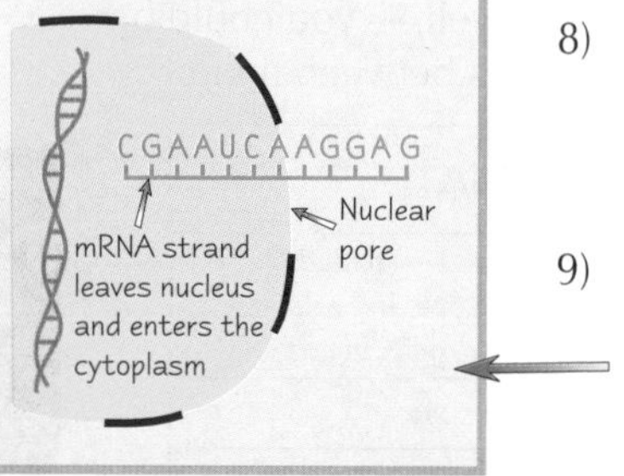

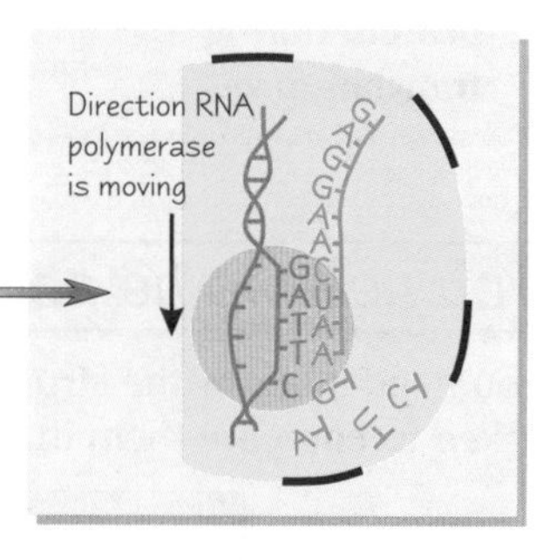

8) When RNA polymerase reaches a particular sequence of DNA called a **stop signal**, it stops making mRNA and **detaches** from the DNA.
9) The **mRNA** moves **out** of the **nucleus** through a nuclear pore and attaches to a **ribosome** in the cytoplasm, where the next stage of protein synthesis takes place (see next page).

mRNA is Modified in Eukaryotic Cells

AQA and Edexcel only

1) Genes in **eukaryotic DNA** contain sections that **don't code** for amino acids.
2) These sections of DNA are called **introns**. All the bits that **do** code for amino acids are called **exons**.
3) During transcription the introns and exons are both **copied** into mRNA. mRNA strands containing introns and exons are called **pre-mRNA**.
4) Introns are **removed** from pre-mRNA strands by a process called **splicing** — introns are removed and exons joined forming **mRNA** strands. This takes place in the **nucleus**.
5) The **exons** can be **joined together** in **different orders** to form **different mRNA strands**.
6) This means **more than one amino acid sequence** and so **more than one protein** can be produced from **one gene**.
7) After splicing the mRNA **leaves the nucleus** for the next stage of protein synthesis (**translation**).

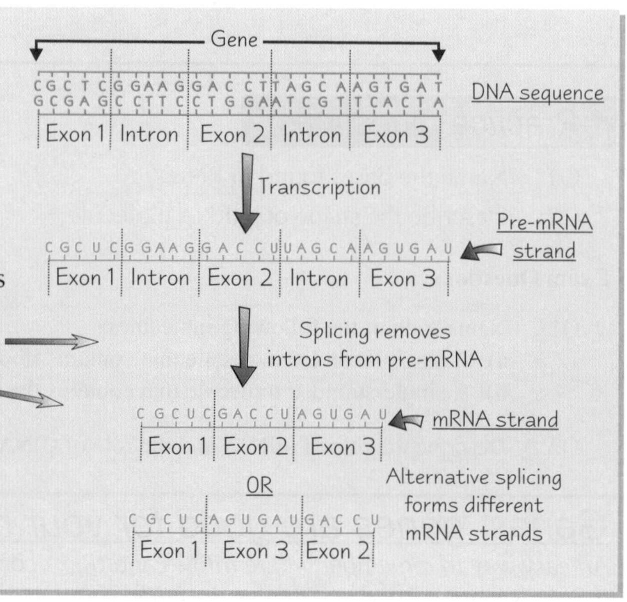

Protein Synthesis

Second Stage of Protein Synthesis — Translation

Translation occurs at the **ribosomes** in the **cytoplasm**. During **translation**, **amino acids** are **joined together** to make a **polypeptide chain** (protein), following the sequence of **codons** (triplets) carried by the mRNA.

1) The **mRNA attaches** itself to a **ribosome** and **transfer RNA** (**tRNA**) molecules **carry amino acids** to the ribosome.
2) A tRNA molecule, with an **anticodon** that's **complementary** to the **first codon** on the mRNA, attaches itself to the mRNA by **specific base pairing**. The first codon that's transcribed is called the **start codon** (see p. 126).
3) A second tRNA molecule attaches itself to the **next codon** on the mRNA in the **same way**.
4) The two amino acids attached to the tRNA molecules are **joined** by a **peptide bond**. The first tRNA molecule **moves away**, leaving its amino acid behind.
5) A third tRNA molecule binds to the **next codon** on the mRNA. Its amino acid **binds** to the first two and the second tRNA molecule **moves away**.
6) This process continues, producing a chain of linked amino acids (a **polypeptide chain**), until there's a **stop codon** on the mRNA molecule. Stop codons are codons that tell the ribosome to **stop translation** (see p. 126).
7) The polypeptide chain (**protein**) **moves away** from the ribosome and translation is complete.

anticodon on tRNA U A C
codon on mRNA A U G
(See p. 123 for more on the structure of mRNA and tRNA.)

Protein synthesis is also called polypeptide synthesis as it makes a polypeptide (protein)

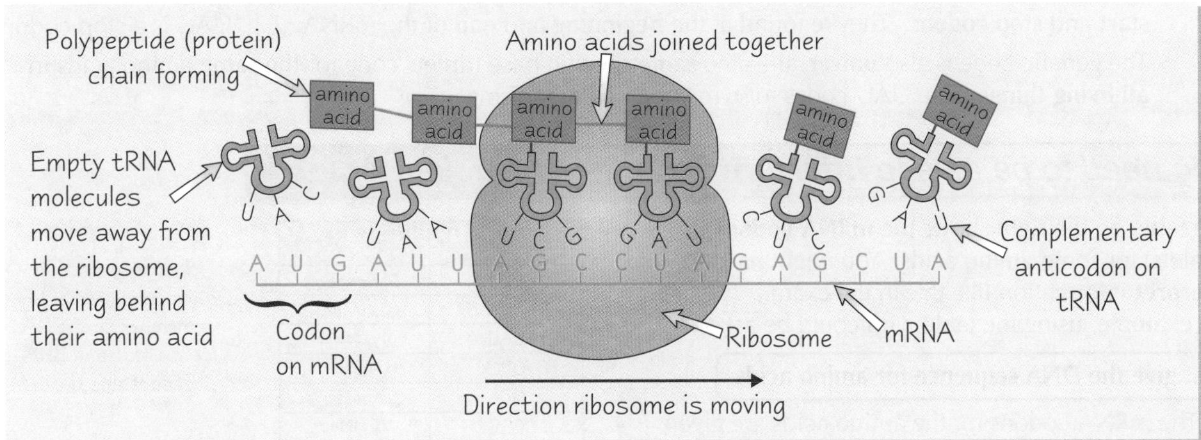

Practice Questions

Q1 What are the two stages of protein synthesis called?

Q2 Where does the first stage of protein synthesis take place?

Q3 When does RNA polymerase stop making mRNA?

Q4 What is an exon?

Q5 Where does the second stage of protein synthesis take place?

Q6 What is a stop codon?

Exam Questions

Q1 A drug that inhibits cell growth is found to be able to bind to DNA, preventing RNA polymerase from binding. Explain how this drug will affect protein synthesis. [2 marks]

Q2 A polypeptide chain (protein) from a eukaryotic cell is 10 amino acids long.

a) Predict how long the mRNA for this protein would be in nucleotides (without the start and stop codons). Explain your answer. [2 marks]

b) Describe how the mRNA is translated into the polypeptide chain. [6 marks]

The only translation I'm interested in is a translation of this page into English...

So you start off with DNA, lots of cleverness happens and bingo... you've got a protein. Only problem is you need to know the cleverness bit in quite a lot of detail. So scribble it down, recite it to yourself, explain it to your best mate or do whatever else helps you remember the joys of protein synthesis. And then think how clever you must be to know it all.

The Genetic Code and Nucleic Acids

This page is for AQA Unit 5, OCR Unit 5 and Edexcel Unit 4.

The genetic code is exactly as it sounds — a code found in your genes that tells your body how to make proteins.

The Genetic Code is **Non-Overlapping**, **Degenerate** and **Universal**

1) The genetic code is the **sequence of base triplets** (codons) in **mRNA** and **DNA** that **code** for specific **amino acids**.
2) In the genetic code, each base triplet is **read** in sequence, **separate** from the triplet **before** it and **after** it. Base triplets **don't share** their **bases** — the code is **non-overlapping**.

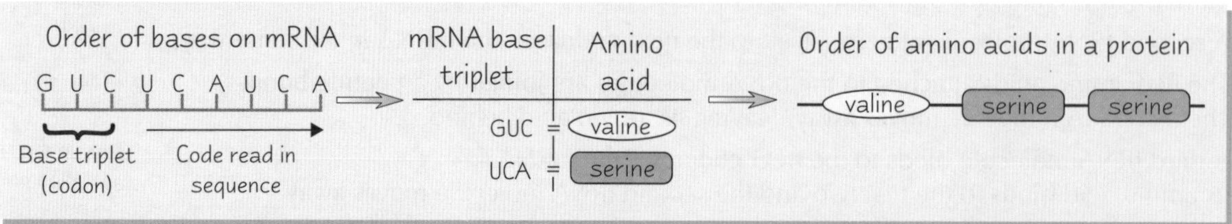

3) The genetic code is also **degenerate** — there are **more** possible combinations of **triplets** than there are amino acids (20 amino acids but 64 possible triplets). This means that some **amino acids** are coded for by **more than one** base triplet, e.g. tyrosine can be coded for by UAU or UAC.
4) Some triplets are used to tell the cell when to **start** and **stop** production of the protein — these are called **start** and **stop codons**. They're found at the **beginning** and **end** of the mRNA. E.g. UAG is a stop codon.
5) The genetic code is also **universal** — the **same** specific base triplets code for the **same** amino acids in **all living things**. E.g. UAU codes for tyrosine in all organisms.

You need to be able to **Interpret Data** about **Nucleic Acids**

The table on the right shows the **mRNA codons** (triplets) for some amino acids. You might have to **interpret** information like this in the exam. For example, using the table, you could be asked to...

mRNA codon	Amino Acid
UCU	Serine
CUA	Leucine
UAU	Tyrosine
GUG	Valine
GCA	Alanine
CGC	Arginine

When interpreting data on nucleic acids remember that DNA contains T and RNA contains U.

...give the DNA sequence for amino acids

The mRNA codons for the amino acids are given in the table. Because **mRNA is a complementary copy** of the **DNA** template, the DNA sequence for each amino acid is made up of bases that would **pair** with the mRNA sequence:

mRNA codon	Amino Acid	DNA sequence (of template strand)
UCU	Serine	AGA
CUA	Leucine	GAT
UAU	Tyrosine	ATA
GUG	Valine	CAC
GCA	Alanine	CGT
CGC	Arginine	GCG

You could also be asked to work out the amino acids from a given DNA sequence and a table.

...give the tRNA anticodons from mRNA codons

tRNA anticodons are **complementary copies** of **mRNA codons**, so you can work out the tRNA anticodon from the mRNA codon:

mRNA codon	tRNA anticodon
UCU	AGA
CUA	GAU
UAU	AUA
GUG	CAC
GCA	CGU
CGC	GCG

You might be asked to name the amino acid coded for by a tRNA anticodon using a table like the one above.

...write the amino acid sequence for a section of mRNA

To **work out** the sequence of **amino acids** from some mRNA, you need to break the genetic code into **codons** and then use the information in the table to work out what **amino acid** they code for.

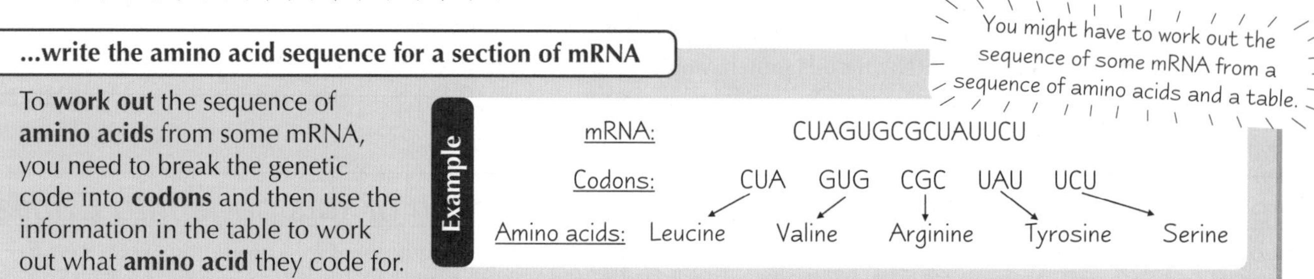

You might have to work out the sequence of some mRNA from a sequence of amino acids and a table.

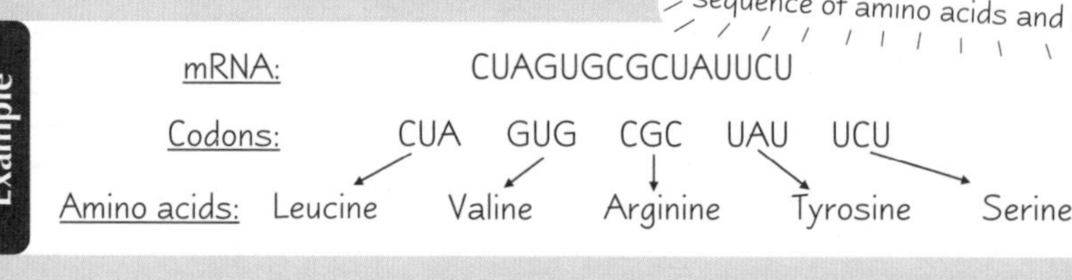

The Genetic Code and Nucleic Acids

This page is for AQA Unit 5 only. If you're doing OCR or Edexcel you can go straight to the questions.

You Might Have to Interpret Data About The Role of Nucleic Acids

In the exam you might have to **interpret data** from experiments done to **investigate nucleic acids** and their **role** in **protein synthesis**. Here's an example (you **don't** need to **learn** it):

Investigating the effect of new drugs on nucleic acids

1) To investigate **how** two new drugs affect **nucleic acids** and their **role** in protein synthesis, **bacteria** were **grown** in **normal conditions** for a few generations, then moved to media containing the drugs.

2) After a short period of time, the **concentration** of **protein** and **complete strands** of **mRNA** in the bacteria were analysed. The results are shown in the **bar graph**.

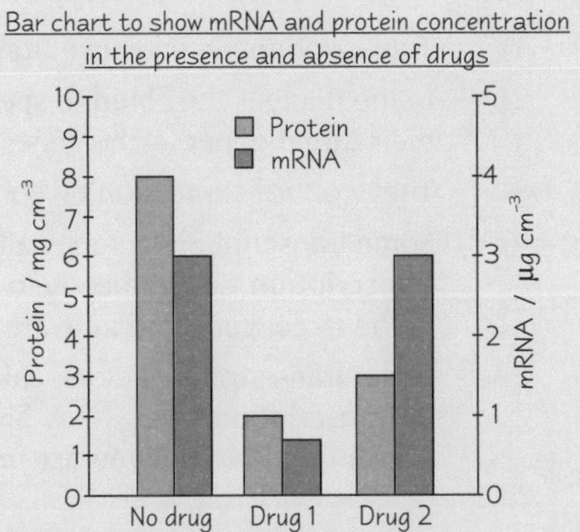

3) Both mRNA **and** protein concentration were **lower** in the presence of **drug 1 compared** to the **no-drug control**. This suggests that drug 1 **affects the production** of **full length mRNA**, so there's no mRNA for protein synthesis during **translation**.

4) **mRNA production** in the presence of **drug 2** was **unaffected**, but **less protein** was produced — **3** mg cm^{-3} compared to **8** mg cm^{-3}. This suggests that drug 2 **interferes** with **translation**. **mRNA was produced**, but **less protein** was **translated** from it.

5) **Further tests** to establish the **nature** of the two drugs were carried out.

6) **Drug 1** was found to be a **ribonuclease** (an enzyme that **digests RNA**). This could **explain** the results of the first experiment — **any strands** of **mRNA** produced by the cell would be **digested** by drug 1, so **couldn't be used** in **translation** to make proteins.

7) **Drug 2** was found to be a **single-stranded, clover-shaped** molecule capable of binding to the **ribosome**. Again, this helps to **explain** the **results** from the first experiment — drug 2 could work by **binding** to the ribosome, **blocking tRNAs** from binding to it and so **preventing translation**.

Transcription and translation are on pages 124-125.

Practice Questions

mRNA codon	amino acid
UGU	Cysteine
CGC	Arginine
GGG	Glycine
GUG	Valine
GCA	Alanine
UUG	Leucine
UUU	Phenylalanine

Q1 What is the genetic code?

Q2 Why is the genetic code described as degenerate?

Q3 Why is the genetic code described as universal?

Exam Questions

Q1 An artificial mRNA was synthesized and used in an experiment to form a protein (polypeptide). The mRNA sequence was: UUGUGUGGGUUUGCAGCA and the protein produced was: Leucine–Cysteine–Glycine–Phenylalanine–Alanine–Alanine. Use the table above to help you answer the following questions.

a) Explain how the result suggests that the genetic code is based on triplets of nucleotides in mRNA. [2 marks]

b) Explain how the result suggests that the genetic code is non-overlapping. [2 marks]

Q2 The table shows the mRNA codons for some amino acids. Show your working for the following questions.

a) Give the amino acid sequence for the mRNA sequence GUGUGUCGCGCA. [2 marks]

b) Give the mRNA sequence for the amino acid sequence arginine, alanine, leucine, phenylalanine. [2 marks]

c) Give the DNA template strand sequence that codes for the amino acid sequence valine, arginine, alanine. [3 marks]

—•• ——— ••• ——— —— • •—— ——— •—• —•—

Hurrah — a page with slightly fewer confusing terms and a lot less to remember. The key to the genetic code is to be able to interpret it, so if you know how DNA, mRNA and tRNA work together to make a protein you should be able to handle any data they can throw at you. Now repeat after me, C pairs with G, A pairs with T. Unless it's RNA — then it's U.

Regulation of Transcription and Translation

These pages are for AQA Unit 5 only. If you're doing OCR and Edexcel you can go straight to page 131.

Oh yes, you read that right — it's back to the incredibly important and immensely clever transcription and translation.

Transcription Factors Control the Transcription of Target Genes

All the **cells** in an organism carry the **same genes** (DNA) but the **structure** and **function** of different cells **varies**. This is because **not all** the **genes** in a cell are **expressed** (transcribed and used to make a protein). Because **different genes** are expressed, **different proteins** are made and these proteins modify the cell — they determine the **cell structure** and control **cell processes** (including the expression of more genes, which produce more proteins).

The **transcription** of genes is **controlled** by protein molecules called **transcription factors**:

1) Transcription factors **move** from the **cytoplasm** to the **nucleus**.
2) In the nucleus they **bind** to **specific DNA sites** near the start of their **target genes** — the genes they **control** the expression of.
3) They control expression by controlling the **rate** of transcription.
4) Some transcription factors, called **activators**, **increase** the **rate of transcription** — e.g. they help **RNA polymerase bind** to the start of the target gene and **activate** transcription.
5) Other transcription factors, called **repressors**, **decrease** the **rate of transcription** — e.g. they **bind** to the start of the target gene, **preventing RNA polymerase** from **binding**, **stopping** transcription.

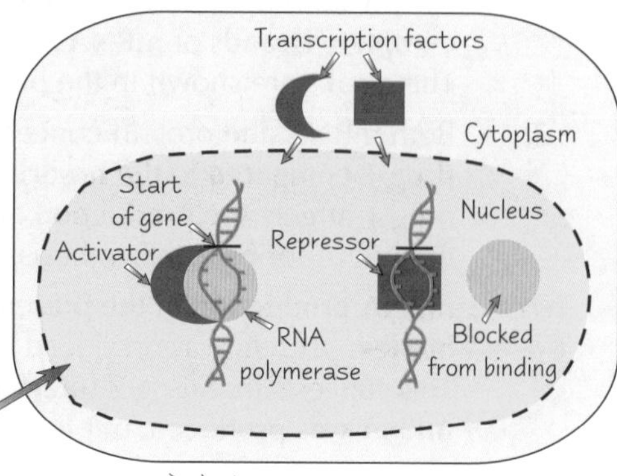

Transcription and translation are covered on pages 124-125.

Oestrogen Affects the Transcription of Target Genes

The **expression** of **genes** can also be **affected** by **other molecules** in the cell, e.g. **oestrogen**:

1) Oestrogen is a **hormone** that can affect transcription by **binding** to a **transcription factor** called an **oestrogen receptor**, forming an **oestrogen-oestrogen receptor complex**.
2) The complex moves from the **cytoplasm** into the **nucleus** where it **binds** to **specific DNA sites** near the **start** of the **target gene**.
3) The complex can **either** act as an **activator**, e.g. **helping** RNA polymerase, or as a **repressor**, e.g. **blocking** RNA polymerase.
4) Whether the complex **acts** as a repressor **or** activator **depends** on the **type of cell** and the target gene.
5) So, the **level of oestrogen** in a particular cell affects the **rate of transcription** of target genes.

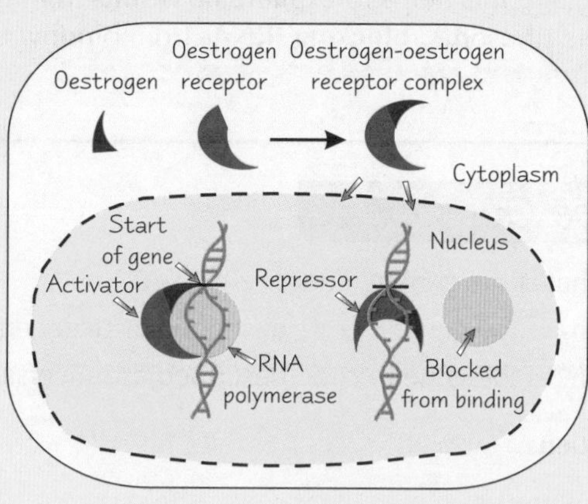

siRNA Interferes with Gene Expression

Gene expression is also affected by a **type of RNA** called **small interfering RNA** (**siRNA**):

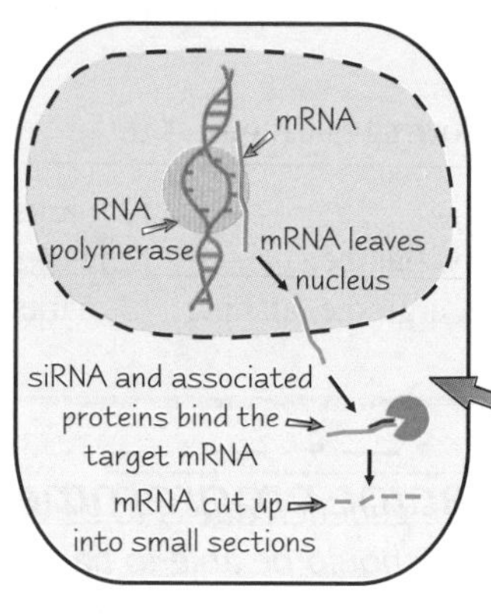

1) siRNA molecules are **short**, **double-stranded RNA** molecules that can **interfere** with the **expression** of specific genes.
2) Their bases are **complimentary** to **specific sections** of a **target gene** and the mRNA that's formed from it.
3) siRNA can **interfere** with both the **transcription** and **translation** of genes.
4) siRNA affects **translation** through a mechanism called **RNA interference**:
   - In the **cytoplasm**, siRNA and associated proteins **bind** to the **target mRNA**.
   - The proteins **cut up** the mRNA into sections so it can no longer be translated.
   - So, the siRNA **prevents the expression** of the specific gene as its protein can no longer be made during **translation**.

Regulation of Transcription and Translation

You need to be able to *Interpret Experimental Data* on *Gene Expression*

You could get a question in the exam where you have to **interpret data** about gene expression. It could be on **anything** you've learnt on the **previous page**, e.g. transcription factors, oestrogen or siRNAs. Below is an example of a **gene expression system** in bacteria and an experiment that **investigates** how it works. You **don't** need to **learn** the information, just **understand** what the results of the experiment tell you about how the expression of the gene is **controlled**.

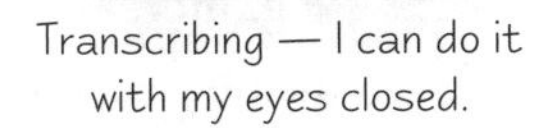

Transcribing — I can do it with my eyes closed.

The *lac* repressor:

1) ***E. coli*** is a **bacterium** that respires **glucose**, but it can use **lactose** if glucose **isn't available**.
2) If lactose is present, *E. coli* makes an **enzyme** (**β-galactosidase**) to **digest** it. But if there's **no** lactose, it doesn't **waste energy** making an enzyme it **doesn't need**. The enzyme's **gene** is **only expressed** when lactose is **present**.
3) The production of the enzyme is **controlled** by a **transcription factor** — the ***lac* repressor**.
4) When there's **no** lactose, the *lac* repressor **binds** to the **DNA** at the start of the gene, **stopping transcription**.
5) When lactose **is present** it **binds** to the *lac* repressor, **stopping** it binding to the DNA, so the gene **is transcribed**.

Experiment:

1) Different ***E. coli*** mutants were isolated and grown in **different media**, e.g. with lactose or glucose.
2) The mutants have **mutations** (**changes** in their **DNA bases**, see p. 134) that mean they **act differently** from normal *E. coli*, e.g. they **produce β-galactosidase** when grown with glucose.
3) To **detect** whether active (working) β-galactosidase was produced, a **chemical** that turns **yellow** in the presence of active β-galactosidase was **added** to the medium.
4) The production of **mRNA** that **codes for β-galactosidase** was also measured. The results are shown in the **table**.
5) In **mutant 1**, mRNA and active β-galactosidase **were produced** even when they were grown with **only glucose** — the gene is **always** being expressed.
6) This suggests that mutant 1 has a **faulty *lac* repressor**, e.g. in the **absence** of lactose the repressor **isn't able** to bind DNA, so transcription **can** occur and mRNA and active β-galactosidase **are produced**.
7) In **mutant 2**, **mRNA** is produced but **active β-galactosidase isn't** when **lactose** is present — the **gene** is being **transcribed** but it **isn't** producing **active** β-galactosidase.
8) This suggests mutant 2 is producing **faulty β-galactosidase**, e.g. because a **mutation** has affected its active site.

Medium	Mutant	mRNA	Colour
Glucose	Normal	No	No yellow
Lactose	Normal	Yes	Yellow
Glucose	Mutant 1	Yes	Yellow
Lactose	Mutant 1	Yes	Yellow
Glucose	Mutant 2	No	No yellow
Lactose	Mutant 2	Yes	No yellow

Practice Questions

Q1 Name two types of transcription factor.

Q2 Name the transcription factor that oestrogen can bind to.

Q3 What does siRNA stand for?

Exam Question

Tube	Medium	Bacteria	Full length mRNA	Protein
1	+ Oestrogen	Normal	Yes	Active
2	– Oestrogen	Normal	No	No
3	+ Oestrogen	Mutant	No	No
4	– Oestrogen	Mutant	No	No

Q1 An experiment was carried out to investigate gene expression of the Chi protein in genetically engineered bacteria. A mutant bacterium was isolated and analysed to look for mRNA coding for Chi, and active Chi protein production. The results are shown in the table above.

a) What do the results of tubes 1 and 2 suggest about the control of gene expression? Explain your answer. [4 marks]

b) What do the results of tubes 3 and 4 suggest could be wrong with the mutant? Explain your answer. [3 marks]

c) If an siRNA complimentary to the Chi gene was added to tube 1, what would you expect the results to be? Explain your answer. [3 marks]

Transcription Factor — not quite as eXciting as that other factor programme...

If it was a competition, oestrogen would totally win — it's very jazzy and awfully controlling. Flexible too — sometimes it helps to activate and other times it helps to repress. Although I'm not sure it can hold a note or wiggle in time to music.

Control of Protein Synthesis

This page is for OCR Unit 5 only. If you're doing Edexcel go to the next page, or AQA go straight to page 134.
Proteins aren't just made willy-nilly — there's some control over when they're synthesised...

Genes can be Switched On or Off

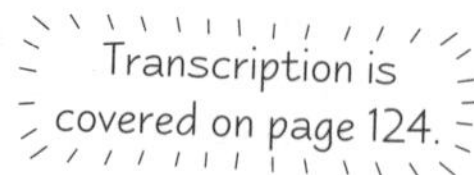

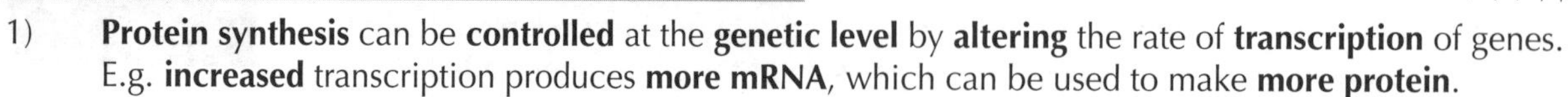

1) **Protein synthesis** can be **controlled** at the **genetic level** by **altering** the rate of **transcription** of genes. E.g. **increased** transcription produces **more mRNA**, which can be used to make **more protein**.
2) Genetic control of protein production in **prokaryotes** (e.g. bacteria) often involves **operons**.
3) An operon is a **section** of **DNA** that contains **structural genes**, **control elements** and sometimes a **regulatory gene**:
   - The structural genes code for **useful proteins**, such as **enzymes** — they're all **transcribed together**.
   - The control elements include a **promoter** (a DNA sequence located **before** the structural genes that **RNA polymerase** binds to) and an **operator** (a DNA sequence that proteins called **transcription factors** bind to).
   - The regulatory gene codes for a **transcription factor** — a protein that **binds** to **DNA** and **switches** genes **on** or **off** by **increasing** or **decreasing** the **rate** of **transcription**. Factors that **increase** the rate are called **activators** and those that **decrease** the rate are called **repressors**.
4) The **shape** of a transcription factor determines whether it **can bind to DNA** or **not**, and can be **altered** by the binding of some molecules, e.g. hormones and sugars.
5) This means the **amount** of some **molecules** in an environment or a cell can **control** the **synthesis** of some **proteins** by affecting **transcription factor binding**. You need to learn this example:

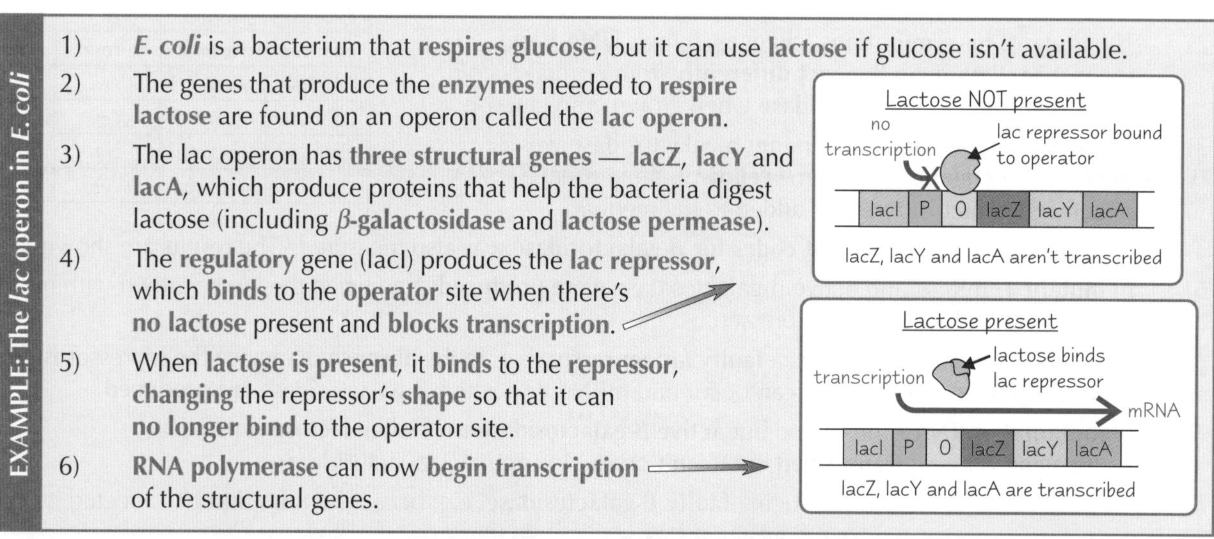
EXAMPLE: The *lac* operon in *E. coli*

1) ***E. coli*** is a bacterium that **respires glucose**, but it can use **lactose** if glucose isn't available.
2) The genes that produce the **enzymes** needed to **respire lactose** are found on an operon called the **lac operon**.
3) The lac operon has **three structural genes** — **lacZ, lacY** and **lacA**, which produce proteins that help the bacteria digest lactose (including **β-galactosidase** and **lactose permease**).
4) The **regulatory** gene (lacI) produces the **lac repressor**, which **binds** to the **operator** site when there's **no lactose** present and **blocks transcription**.
5) When **lactose is present**, it **binds** to the **repressor**, **changing** the repressor's **shape** so that it can **no longer bind** to the operator site.
6) **RNA polymerase** can now **begin transcription** of the structural genes.

Some Genes Control the Development of Body Plans

1) A **body plan** is the **general structure** of an organism, e.g. the *Drosophila* fruit fly has various **body parts** (head, abdomen, etc.) that are **arranged** in a **particular way** — this is its body plan.
2) **Proteins control** the **development** of a **body plan** — they help set up the basic body plan so that everything is in the right place, e.g. legs grow where legs should grow.
3) The proteins that control body plan development are **coded for** by genes called **homeotic genes**. E.g. two homeotic gene clusters control the development of the *Drosophila* body plan — one controls the development of the head and anterior thorax and the other controls the development of the posterior thorax and abdomen.

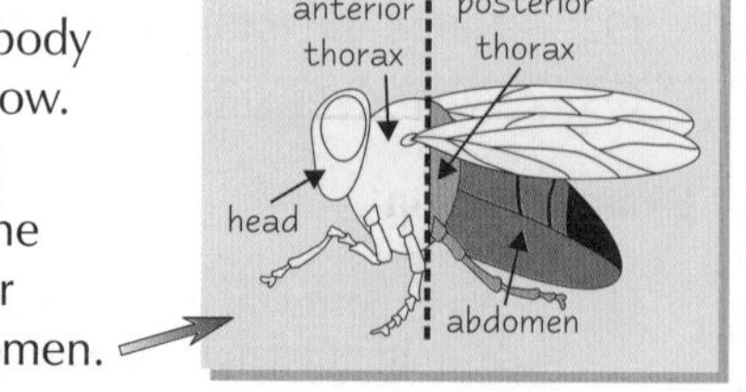

4) **Similar homeotic genes** are found in **animals**, **plants** and **fungi**, which means that body plan development is controlled in a similar way in flies, mice, humans, etc. Here's how homeotic genes control development:
   - **Homeotic genes** have **regions** called **homeobox sequences** that code for a part of the protein called the **homeodomain**.
   - The homeodomain **binds** to specific **sites** on **DNA**, enabling the protein to work as a **transcription factor** (see above).
   - The proteins bind to DNA at the **start** of **developmental genes**, **activating** or **repressing transcription** and so altering the production of proteins involved in the development of the body plan.

Control of Protein Synthesis

This page is for OCR Unit 5 and Edexcel Unit 5 only.

Programmed Cell Death is Involved in the Development of Body Plans

OCR only

1) Some cells **die** and **break down** as a **normal** part of **development**.
2) This is a **highly controlled process** called **apoptosis**, or **programmed cell death**.
3) Once **apoptosis** has been **triggered** the **cell** is **broken down** in a series of steps:
   - The cell produces **enzymes** that **break down** important cell components such as **proteins** in the cytoplasm and **DNA** in the nucleus.
   - As the cell's contents are broken down it begins to **shrink** and **breaks up** into **fragments**.
   - The **cell fragments** are **engulfed** by **phagocytes** and **digested**.
4) Apoptosis is in involved the development of **body plans** — **mitosis** and **differentiation create** the bulk of the **body parts** and then apoptosis **refines** the parts by **removing** the **unwanted structures**.
   For example:
   - When **hands** and **feet** first develop the **digits** (fingers and toes) are **connected**. They're only **separated** when cells in the **connecting tissue** undergo **apoptosis**.
   - As **tadpoles** develop into frogs their **tail cells** are **removed** by apoptosis.
   - An **excess** of **nerve cells** are produced during the development of the **nervous system**. Nerve cells that **aren't needed** undergo **apoptosis**.
5) All cells contain **genes** that code for proteins that **promote** or **inhibit apoptosis**.
6) During development, genes that **control** apoptosis are **switched on** and **off** in **appropriate** cells, so that **some die** and the **correct body plan develops**.

Hormones Switch Genes On to Help Regulate Temperature

Edexcel only

1) In a cell there are **proteins** called **transcription factors** that **control** the **transcription** of **genes**.
2) Transcription factors **bind** to **DNA sites** near the **start** of genes and **increase** or **decrease** the **rate** of transcription. Factors that **increase** the rate are called **activators** and those that **decrease** the rate are called **repressors**.
3) **Hormones** can **bind** to some **transcription factors** to **change body temperature**. Here's how:
   - At **normal** body temperature, the **thyroid hormone receptor** (a transcription factor) binds to DNA at the **start** of a gene.
   - This **decreases** the **transcription** of a gene coding for a **protein** that increases **metabolic rate**.
   - At **cold** temperatures **thyroxine** is released, which **binds** to the thyroid hormone receptor, causing it to act as an **activator**.
   - The **transcription rate increases**, producing **more protein**. The protein **increases** the **metabolic rate**, causing an increase in **body temperature** (see p.114).

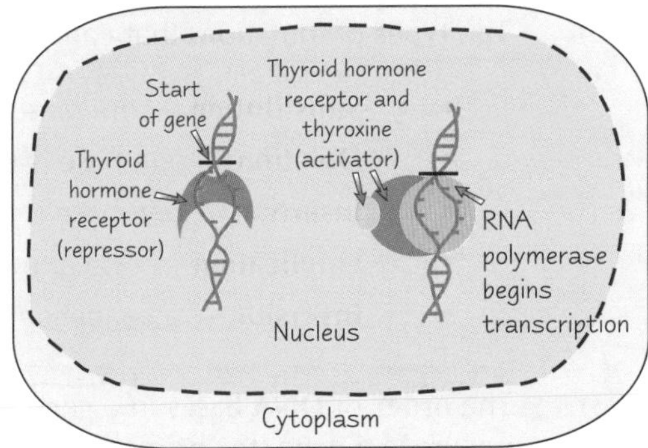

Practice Questions

Q1 What does a transcription factor do?

Q2 What is a body plan?

Q3 Give one example of how apoptosis is used during the development of a body plan.

Exam Question

Q1 Explain how the presence of lactose causes *E. coli* to produce ß-galactosidase and lactose permease. [4 marks]

Too much revision can activate programmed cell death in your brain....

OK, maybe that's not completely true. There are a lot of 'interesting' words to remember on these two pages — I bet you're glad you decided to study biology. Some of these concepts are quite hard to get your head round but keep going over it until it all makes sense — it'll click eventually. Then you can dazzle your friends with your knowledge of gene control...

Protein Activation and Gene Mutation

These pages are for OCR Unit 5 only. If you're doing Edexcel you've finished this section — have a break.

Some proteins need activating before they'll work — cyclic AMP gives them a molecular kick up their amino acid backside. Proteins can also be affected by mutations in DNA... sounds like it's a hard life being a protein...

cAMP Activates Some Proteins by Altering Their Shape

1) **Protein synthesis** can be controlled at the **genetic level** by **molecules** (see page 130).
2) Some proteins produced by protein synthesis **aren't active** though — they have to be **activated** to work.
3) **Protein activation** is also controlled by **molecules**, e.g. **hormones** and **sugars**.
4) Some of these molecules work by **binding** to **cell membranes** and **triggering** the production of cyclic **AMP** (**cAMP**) **inside** the **cell**.
5) cAMP then **activates proteins** inside the cell by **altering** their **three-dimensional** (3D) **structure**.
6) For example, altering the 3D structure can **change** the **active site** of an enzyme, making it become **more** or **less active**.
7) For example, cAMP activates **protein kinase A** (**PKA**):

cAMP is a secondary messenger — it relays the message from the control molecule, e.g. the hormone, to the inside of the cell (see page 103).

1) **PKA** is an **enzyme** made of four subunits.
2) When cAMP **isn't bound**, the four units are bound together and are **inactive**.
3) When cAMP **binds**, it causes a **change** in the enzyme's **3D structure**, releasing the active subunits — PKA is now **active**.

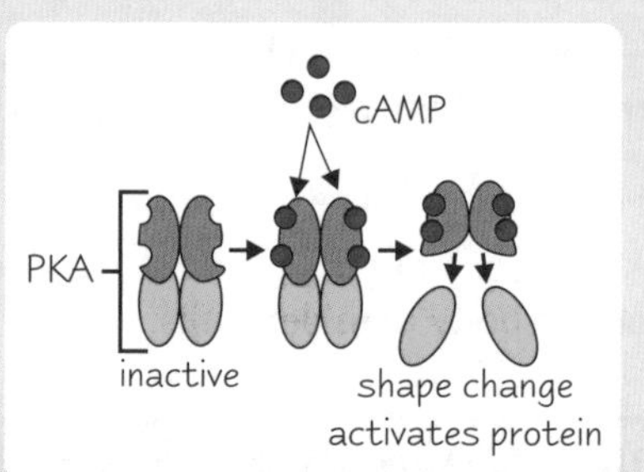

Mutations are Changes to the Base Sequence of DNA

1) Any change to the **base (nucleotide) sequence** of DNA is called a **mutation**.
2) The **types** of mutations that can occur include:

- **Substitution** — one base is swapped for another, e.g. ATGCCT becomes ATTCCT
- **Deletion** — one base is removed, e.g. ATGCCT becomes ATCCT
- **Insertion** — one base is added, e.g. ATGCCT becomes ATGACCT
- **Duplication** — one or more bases are repeated, e.g. ATGCCT becomes ATGCCCCT
- **Inversion** — a sequence of bases is reversed, e.g. ATGCCT becomes ACCGTT

3) The **order** of **DNA bases** in a gene determines the **order of amino acids** in a particular **protein**. If a mutation occurs in a gene, the **primary structure** (amino acid chain) of the protein it codes for could be **altered**:

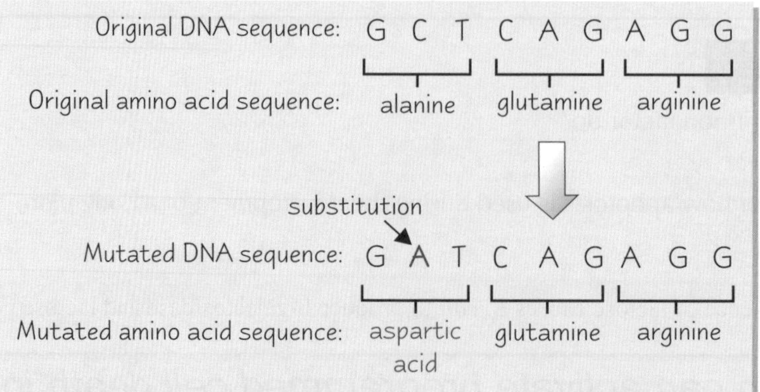

4) This may **change** the final **3D shape** of the protein so it **doesn't work properly**, e.g. **active sites** in enzymes may not form properly, meaning that **substrates can't bind** to them.

Protein Activation and Gene Mutation

Mutations can be Neutral, Beneficial or Harmful

1) Some mutations can have a **neutral effect** on a protein's **function**. They may have a neutral effect because:
   - The mutation changes a base in a triplet, but the **amino acid** that the triplet codes for **doesn't change**. This happens because **some amino acids** are coded for by **more than one triplet**. E.g. both **TAT** and **TAC** code for **tyrosine**, so if TAT is changed to TAC the **amino acid won't change**.
   - The mutation produces a triplet that codes for a **different amino acid**, but the amino acid is **chemically similar** to the original so it functions like the original amino acid. E.g. **arginine** (AGG) and **lysine** (AAG) are coded for by similar triplets — a **substitution** mutation can **swap** the amino acids. But this mutation would have a **neutral effect** on a **protein** as the amino acids are **chemically similar**.
   - The mutated triplet codes for an amino acid **not involved** with the protein's **function**, e.g. one that's located **far away** from an enzyme's **active site**, so the protein **works** as it **normally** does.
2) A **neutral effect** on protein function **won't** affect an **organism** overall.
3) However, some mutations **do** affect a protein's **function** — they can make a protein **more** or **less active**, e.g. by **changing** the **shape** of an enzyme's **active site**.
4) If protein function **is affected** it can have a **beneficial** or **harmful effect** on the **whole organism**:

Mutations with beneficial effects

- These have an **advantageous effect** on an organism, i.e. they **increase** its chance of **survival**.
- E.g. some bacterial **enzymes break down** certain **antibiotics**. **Mutations** in the genes that code for these enzymes could make them work on a **wider range** of antibiotics. This is **beneficial** to the **bacteria** because antibiotic resistance can help them to survive.

Mutations that are beneficial to the organism are passed on to future generations by the process of natural selection (see p. 74).

Mutations with harmful effects

- These have a **disadvantageous effect** on an organism, i.e. they **decrease** its chance of **survival**.
- E.g. **cystic fibrosis** (CF) can be caused by a **deletion** of three bases in the gene that codes for the **CFTR** (cystic fibrosis transmembrane conductance regulator) **protein**. The mutated CFTR protein **folds incorrectly**, so it's **broken down**. This leads to **excess mucus production**, which affects the **lungs** of CF sufferers.

Practice Questions

Q1 How does cAMP activate a protein?

Q2 Give two reasons why a mutation may have a neutral effect on a protein's function.

Amino acid	DNA triplet
Methionine	ATG
Tyrosine	TAT or TAC
Serine	TCA or TCC
Glycine	GGC or GGT
Cysteine	TGT or TGC

Exam Questions

Q1 a) Define the term mutation. [1 mark]

b) Describe two types of mutation that occur in DNA. [2 marks]

Q2 A gene begins with the following DNA sequence: ATGTATTCAGGCTGT
A mutation occurred where the ninth base was substituted by cytosine (C).

a) Write down the mutated DNA sequence. [1 mark]

b) Using the table, explain the effect that the mutation would have on the protein. [3 marks]

Mutations in adolescent turtles can enhance their ninja skills...

The important thing to remember about cAMP is that it alters the 3D structure of proteins — you can't just say that it activates proteins, you need to say how it does it as well. Don't forget that mutations can be harmless and that some can improve the way a protein functions — it's easy to associate 'mutation' with 'bad', but don't fall into that trap.

Mutations, Genetic Disorders and Cancer

The rest of this section is for AQA Unit 5 only.

Mutations — as featured in numerous superhero movies. Well, I'm sorry to be the one to break it to you, but they don't usually give you special powers or superhuman strength — in fact, they can cause a lot of problems.

Mutations are Changes to the Base Sequence of DNA

Any change to the **base sequence** of DNA is called a **mutation**:

1) They can be caused by **errors** during **DNA replication**.
2) They can also be caused by **mutagenic agents** (see below).
3) The **types** of errors that can occur include:

Errors can also be caused by insertion, duplication and inversion of bases.

- **Substitution** — one base is substituted with another, e.g. ATGCCT becomes ATTCCT (G is **swapped** for T).
- **Deletion** — one base is deleted, e.g. ATGCCT becomes ATCCT (G is **deleted**).

4) The **order** of **DNA bases** in a gene determines the **order of amino acids** in a particular **protein** (see p. 122). If a mutation occurs in a gene, the **sequence** of **amino acids** that it codes for (and the protein formed) could be **altered**:

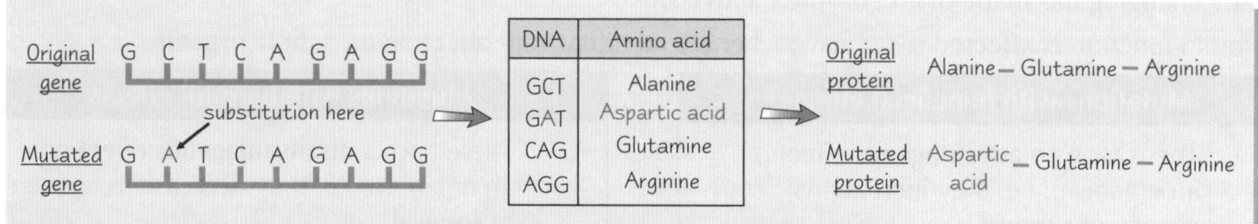

Not All Mutations Affect the Order of Amino Acids

The **degenerate nature** of the genetic code (see page 126) means that some amino acids are coded for by **more than one DNA triplet** (e.g. tyrosine can be coded for by TAT or TAC in DNA). This means that **not all** substitution mutations will result in a change to the amino acid sequence of the protein — some substitutions will still **code** for the **same amino acid**. For example:

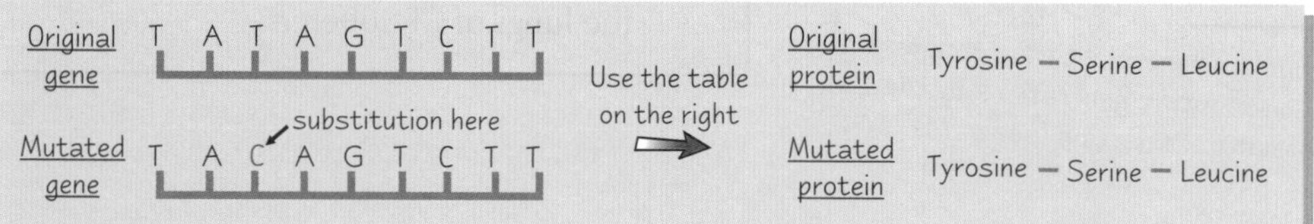

DNA	Amino acid
TAT	Tyrosine
TAC	Tyrosine
AGT	Serine
CTT	Leucine
GTC	Valine

Substitution mutations **won't always** lead to changes in the amino acid sequence, but **deletions will** — the deletion of a base will change the **number** of bases present, which will cause a **shift** in all the base triplets after it:

Mutagenic Agents Increase the Rate of Mutation

Mutations occur **spontaneously**, e.g. when DNA is **misread** during **replication**. But some things can cause an **increase** in the **rate** of **mutations** — these are called **mutagenic agents**. **Ultraviolet radiation**, **ionising radiation**, some **chemicals** and some **viruses** are examples of mutagenic agents. They can increase the rate of mutations by:

1) **Acting as a base** — chemicals called **base analogs** can **substitute** for a base during DNA replication, **changing** the **base sequence** in the new DNA. E.g. **5-bromouracil** is a base analog that can substitute for **thymine**. It can pair with **guanine** (**instead** of **adenine**), causing a **substitution mutation** in the new DNA.
2) **Altering bases** — some chemicals can **delete** or **alter bases**. E.g. **alkylating agents** can add an alkyl group to **guanine**, which **changes** the **structure** so that it pairs with **thymine** (**instead** of **cytosine**).
3) **Changing** the **structure of DNA** — some types of **radiation** can change the structure of DNA, which causes **problems** during DNA replication. E.g. **UV radiation** can cause adjacent **thymine** bases to **pair up** together.

Mutations, Genetic Disorders and Cancer

Genetic Disorders and Cancer are Caused By Mutations

Hereditary Mutations Cause Genetic Disorders and Some Cancers

Some mutations can cause **genetic disorders** — inherited disorders caused by **abnormal genes** or **chromosomes**, e.g. cystic fibrosis. Some mutations can **increase** the **likelihood** of developing certain **cancers**, e.g. mutations of the gene **BRCA1** can increase the chances of developing **breast cancer**. If a **gamete** (sex cell) containing a mutation for a genetic disorder or certain cancer is **fertilised**, the mutation will be present in the new **fetus** formed — these are called **hereditary mutations** because they are passed on to the offspring.

Acquired Mutations Can Cause Cancer

1) Mutations that occur in individual cells **after** fertilisation (e.g. in adulthood) are called **acquired mutations**.
2) If these mutations occur in the **genes** that **control** the rate of **cell division**, it can cause **uncontrolled cell division**.
3) If a cell divides uncontrollably the result is a **tumour** — a mass of abnormal cells. Tumours that **invade** and **destroy surrounding tissue** are called **cancers**.
4) There are **two types** of **gene** that control cell division — **tumour suppressor genes** and **proto-oncogenes**. Mutations in these genes can cause cancer:

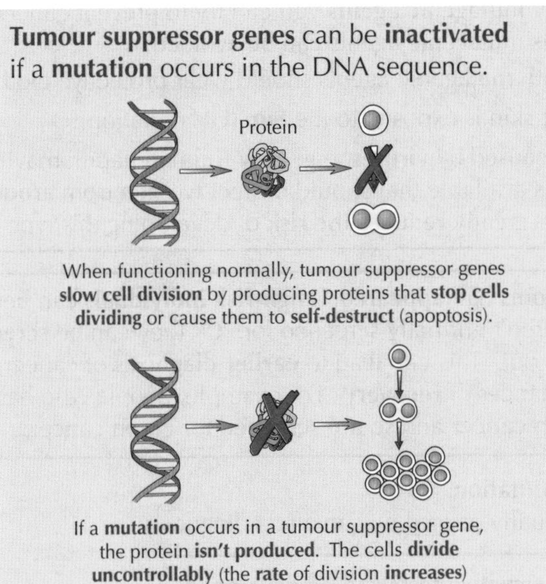

Tumour suppressor genes can be **inactivated** if a **mutation** occurs in the DNA sequence.

When functioning normally, tumour suppressor genes **slow cell division** by producing proteins that **stop cells dividing** or cause them to **self-destruct** (apoptosis).

If a **mutation** occurs in a tumour suppressor gene, the protein **isn't produced**. The cells **divide uncontrollably** (the **rate** of division **increases**) resulting in a tumour.

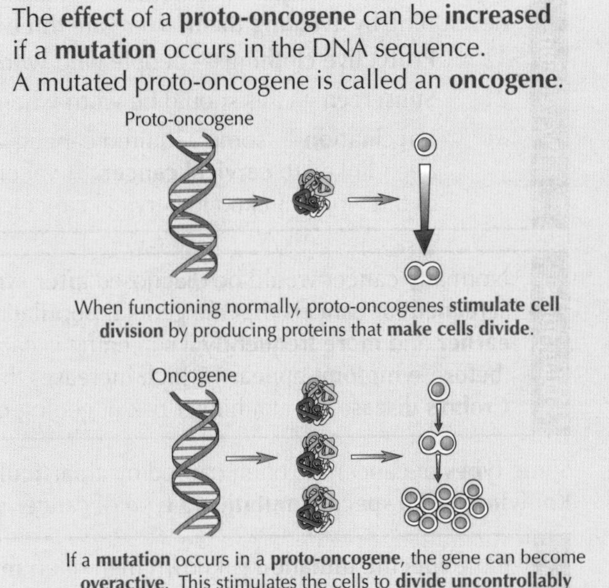

The **effect** of a **proto-oncogene** can be **increased** if a **mutation** occurs in the DNA sequence. A mutated proto-oncogene is called an **oncogene**.

When functioning normally, proto-oncogenes **stimulate cell division** by producing proteins that **make cells divide**.

If a **mutation** occurs in a **proto-oncogene**, the gene can become **overactive**. This stimulates the cells to **divide uncontrollably** (the **rate** of division **increases**) resulting in a **tumour**.

Practice Questions

Q1 What is a substitution mutation?

Q2 What are mutagenic agents?

Q3 What is a genetic disorder?

Before exposure: A G T T A T C A G G C T

After exposure: A G G T A T G A G G C C

DNA	Amino acids	DNA	Amino acids
AGT	Serine	GAG	Glutamic acid
AGG	Arginine	GCT	Alanine
TAT	Tyrosine	GCC	Alanine
CAG	Glutamine		

Exam Question

Q1 The order of bases in a liver cell's proto-oncogene before and after exposure to a mutagenic agent is shown above.

a) Underline any mutation(s) that have occurred. [1 mark]

b) Use the table to explain the changes that the mutations would cause to the sequence of amino acids. [5 marks]

c) Would you describe these mutations as acquired or inherited? Explain your answer. [2 marks]

d) Explain how the mutation(s) may lead to cancer. [3 marks]

Just hope your brain doesn't have a deletion mutation during the exam...

Right, there's plenty to learn on these pages and some of it's a bit complicated, so you know the drill. Don't read it all through at once — take the sections one by one and get all the facts straight. There could be nothing more fun...

Diagnosing and Treating Cancer and Genetic Disorders

These pages are for AQA Unit 5 only.

Before you start this page, make sure you've got one thing straight in your head from the previous page — make sure you understand the difference between acquired and hereditary disorders.

Knowing the Mutation is Useful for the Diagnosis and Treatment of Disorders

1) **Cancer** and most **genetic disorders** are caused by **mutations** (see previous page).
2) Knowing whether a disorder is caused by an **acquired** or **inherited mutation** affects the **prevention** and **diagnosis** of the disorder.
3) **Identifying** the **specific mutation** that causes a disorder in an individual affects the prevention, diagnosis and treatment.
4) Here are some **examples** for each type of **disorder**:

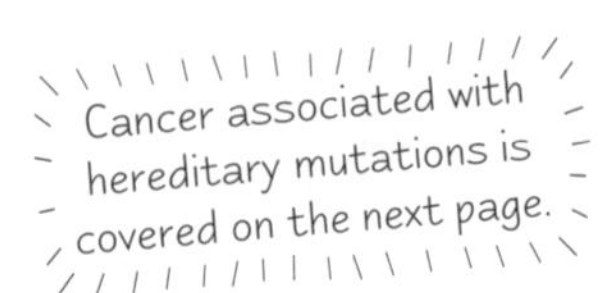

See page 158 for
how to screen
for mutations.

1 Cancer — Caused by Acquired Mutations

Cancer associated with hereditary mutations is covered on the next page.

Acquired mutations can **occur spontaneously** or be **caused by** exposure to **mutagenic agents** (see page 134). Knowing that cancer can be caused by **acquired mutations** affects the prevention and diagnosis:

Prevention

If you know that **acquired mutations** are caused by **mutagenic agents** you can try to prevent cancer developing by **avoiding them**. Here are three ways mutagenic agents can be avoided:

1) **Protective clothing** — people who **work** with mutagenic agents should wear protective clothing.
2) **Sunscreen** — this should be worn when the skin is exposed to the **Sun** (UV radiation).
3) **Vaccination** — some acquired cancers are caused by **viruses**, e.g. **HPV** (human papillomavirus) has been linked to **cervical cancer**. A vaccine is available that should protect women from **around 80%** of the viruses linked to cervical cancer. This greatly **reduces the risk** of developing this type of cancer.

Diagnosis

Normally cancer would be diagnosed **after symptoms** had **appeared**. **High-risk individuals** can be **screened** for cancers that the general population aren't normally screened for. Or they can be **screened earlier** and **more frequently** if screening is carried out. This can lead to **earlier diagnosis** of cancer (**before symptoms appear**), which **increases** the chances of **recovery**. For example, people who have **Crohn's disease** are at a higher risk of getting **colon cancer** and so are **screened** for colon cancer.

Some **types** of cancer are often caused by a **particular mutation**.
Knowing which specific mutation a type of cancer is usually caused by can affect diagnosis:

Diagnosis

If the **specific mutation** is known then often **more sensitive tests** can be **developed**, which can lead to **earlier** and **more accurate diagnosis**, improving the **chances of recovery**. For example, there's a **mutation** in the **RAS proto-oncogene** in around **half** of all **bowel cancers**. Bowel cancer can be **detected early** by looking for RAS mutations in the DNA of **bowel cells**.

Individuals diagnosed with **cancer** can also have the **DNA** from the **cancerous cells analysed** to see which mutation has caused it. Knowing **which specific mutation** the cancer is caused by affects treatment:

Treatment

1) The **treatment** can be **different** for different mutations. For example, **breast cancer** caused by mutation of the **HER2 proto-oncogene** can be treated with a drug called **Herceptin®**. This drug binds **specifically** to the altered HER2 protein receptor and **suppresses cell division and tumour growth**. Breast cancer caused by other mutations is not treated with this drug as it doesn't work.
2) The **aggressiveness** of the **treatment** can **differ** depending on the mutation. Different mutations produce **different types** of cancer, which affects the treatment. For example, if the mutation is known to cause an **aggressive** (**fast-growing**) cancer it may be treated with **higher doses** of **radiotherapy** or by **removing larger areas** of the tumour and surrounding tissue during **surgery**.
3) If the specific mutation is known, **gene therapy** (see page 160) may be able to treat it. For example, if you know it's caused by **inactivated tumour suppressor genes** (see previous page), gene therapy could be used to provide **working versions** of the genes.

Diagnosing and Treating Cancer and Genetic Disorders

2) *Cancer* — *Caused by* ***Hereditary*** *Mutations*

Cancer caused by **hereditary mutations** usually results in a **family history** of a certain type of cancer. If an individual has a family history of cancer, things can be done to **prevent** it **developing** and **diagnose it earlier** if it does:

Prevention: Most cancers are caused by mutations in **multiple genes**. So people with a family history should **avoid gaining extra acquired mutations** by **avoiding mutagenic agents**, e.g. those with a family history of **lung cancer shouldn't smoke**.

Diagnosis: **Screening**, or **increased** and **earlier screening**, if there's a **family history** can lead to **early detection** (i.e. before symptoms appear) and **increased** chances of recovery. E.g. **more frequent breast examinations** if there's a family history of **breast cancer**.

Individuals with a **family history** of cancer can have their **DNA analysed** to see if they carry the **specific mutation**. Knowing **which specific mutation** the cancer is caused by affects prevention, diagnosis and treatment:

Prevention: If the mutation causes a very high risk of cancer **preventative surgery** may be carried out — removing the **organ** the cancer is likely to affect **before cancer develops**. E.g. women with a mutation in **BRCA1** sometimes choose to have a **mastectomy** (removal of one or both breasts) to **prevent breast cancer** from developing.

Treatment: Treatment is **similar** to treating cancer caused by acquired mutations (see previous page). E.g. the treatment depends on the **particular mutation**. But cancer caused by hereditary mutations is often **diagnosed earlier**, which can **change** the treatment used.

Diagnosis: **Screening**, or **increased** and **early screening** of those with a **hereditary mutation** can lead to **early detection** and **increased** chances of recovery. E.g. **frequent colonoscopies** for those with a mutated **APC gene** to diagnose hereditary **colon cancer earlier**.

3) *Genetic Disorders* — *Caused by* ***Hereditary*** *Mutations*

Diagnosis: If a person has a **family history** of a genetic disorder they can have their **DNA analysed** to see if they have the **mutation** that causes it or if they are a **carrier** (see p. 64). If they're **tested** and **diagnosed before symptoms develop**, any **treatment** available can **begin earlier**. Also, knowing if they have the disorder or if they're a carrier can help to figure out if any **children** they have (or might have) are at **risk**.

Treatment:

1) **Gene therapy** (see page 160) — this may be able to treat **some genetic disorders**. E.g. scientists have shown it's possible to **treat** symptoms of **cystic fibrosis** by inserting a **normal copy** of the **mutated gene**.
2) The **treatment** can be **different** for different **mutations** — for example, the **exact gene mutation** for Huntington's disease affects **symptom treatment options** as it affects the **time of onset** of symptoms.
3) **Early diagnosis** can affect treatment options — for example, if **sickle cell anaemia** is diagnosed at **birth**, treatments that **relieve symptoms** and work to **avoid complications** can be given straight away.

Prevention: **Carriers** or **sufferers** of genetic disorders can undergo **preimplantation genetic diagnosis** during ***in vitro*** **fertilisation** (IVF) to **prevent** any **offspring** having the disease. **Embryos** are produced by IVF and **screened** for the mutation. Only embryos **without the mutation** are **implanted** in the womb.

Practice Questions

Q1 Describe how knowing which specific mutation a cancer is caused by affects diagnosis.

Q2 Describe how knowing which specific mutation a genetic disorder is caused by affects treatment.

Exam Question

Q1 Discuss how knowing if a disorder is caused by a hereditary or acquired mutation affects prevention and diagnosis of the disorder. [25 marks]

My genes need therapy — they've all got holes in the knees...

So whether a disorder is caused by an acquired or hereditary gene mutation has a pretty big effect on things like prevention, diagnosis and treatment. Make sure you understand the differences so you're not caught out in the exam.

Stem Cells

These pages are for AQA Unit 5 only.

Stem cells — they're the daddy of all cells, the big cheese, the top dog, and the head honcho. And here's why...

Stem Cells** are Able to **Mature** into **Any Type** of **Body Cell

1) **Multicellular organisms** are made up from many **different cell types** that are **specialised** for their function, e.g. liver cells, muscle cells, white blood cells.
2) **All** these specialised cell types originally came from **stem cells**.
3) Stem cells are **unspecialised** cells that can develop into **other types** of cell.
4) Stem cells divide to become **new** cells, which then become **specialised**.
5) All multicellular organisms have some form of stem cell.
6) Stem cells are found in the **embryo** (where they become all the **specialised cells** needed to form a **fetus**) and in **some adult tissues** (where they become **specialised** cells that need to be **replaced**, e.g. stem cells in the **bone marrow** can become **red blood cells**).
7) Stem cells that can mature (develop) into **any type** of **body cell** in an organism are called **totipotent cells**.
8) Totipotent stem cells in humans are **only present** in the **early life** of an **embryo**.
9) After this point the embryonic stem cells **lose** their ability to **specialise** into **all** types of cells, but can still become a **wide range** of cells.
10) Only a few **stem cells** remain in mature animals and they can only differentiate into a **few types** of cells.

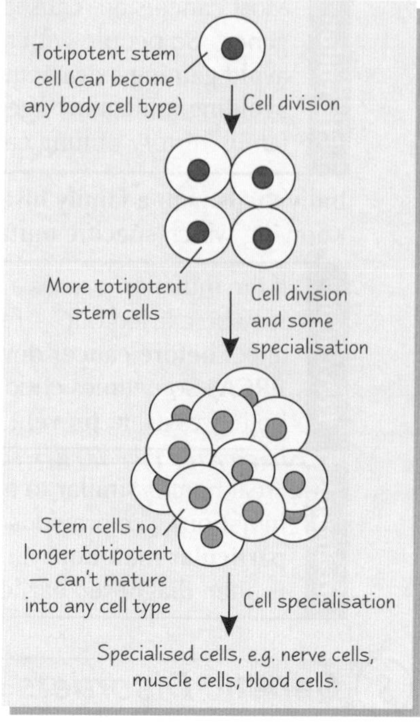

Plants** Contain **Totipotent Stem Cells

1) Mature **plants** also have **stem cells** — they're found in areas where the plant is **growing**, e.g. in roots and shoots.
2) All stem cells in plants are **totipotent** — they can mature into any **cell type**.
3) This means they can be used to grow **plant organs** (e.g. roots) or **whole new plants *in vitro*** (artificially). Growing plant tissue artificially is called **tissue culture** (see next page).

Stem Cells** Become **Specialised** Because **Different Genes** are **Expressed

Totipotent stem cells become **specialised** because during their development they only **transcribe** and **translate part** of their **DNA**:

1) **Totipotent stem cells** all contain the **same genes** — but during **development** **not all** of them are **transcribed** and **translated** (expressed).
2) Under the **right conditions**, some **genes** are **expressed** and others are switched off.
3) **mRNA** is only **transcribed** from **specific genes**.
4) The mRNA from these genes is then **translated** into **proteins**.
5) These proteins **modify** the cell — they determine the cell **structure** and **control cell processes** (including the expression of **more genes**, which produces more proteins).
6) **Changes** to the cell produced by these proteins cause the cell to become **specialised**. These changes are **difficult** to **reverse**, so once a cell has specialised it **stays** specialised.

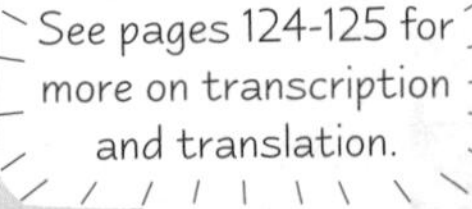

See pages 124-125 for more on transcription and translation.

All of the girls expressed different jeans.

EXAMPLE: RED BLOOD CELLS

1) **Red blood cells** are produced from a type of **stem cell** in the **bone marrow**. They contain lots of **haemoglobin** and have **no nucleus** (to make room for more haemoglobin).
2) The stem cell produces a new cell in which the genes for **haemoglobin production** are **expressed**. Other genes, such as those involved in **removing the nucleus**, are **expressed** too. Many other genes are expressed or switched off, resulting in a specialised red blood cell.

Stem Cells

Tissue Culture Can be Used to Grow Plants from a Single Totipotent Cell

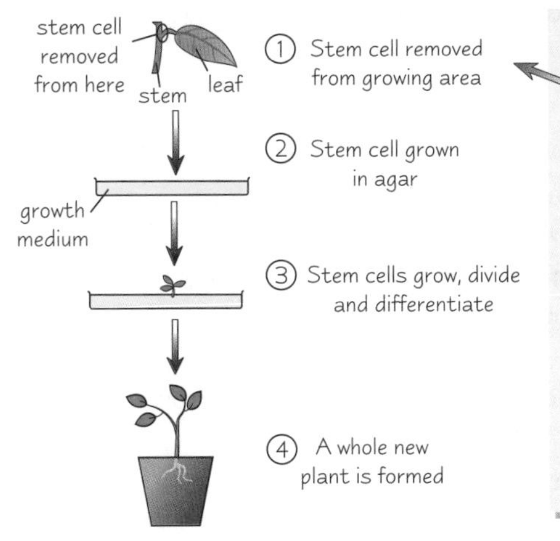

See p. 106 for more on growth factors.

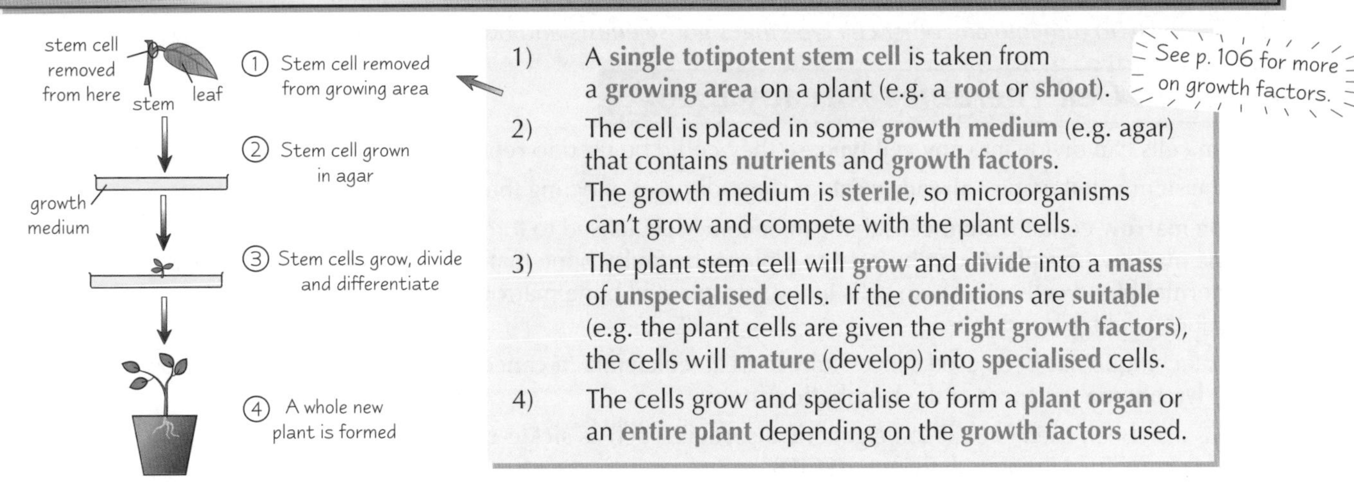

1) A **single totipotent stem cell** is taken from a **growing area** on a plant (e.g. a **root** or **shoot**).
2) The cell is placed in some **growth medium** (e.g. agar) that contains **nutrients** and **growth factors**. The growth medium is **sterile**, so microorganisms can't grow and compete with the plant cells.
3) The plant stem cell will **grow** and **divide** into a **mass** of **unspecialised** cells. If the **conditions** are **suitable** (e.g. the plant cells are given the **right growth factors**), the cells will **mature** (develop) into **specialised** cells.
4) The cells grow and specialise to form a **plant organ** or an **entire plant** depending on the **growth factors** used.

You Need to be Able to Interpret Data About Tissue Culture

Here's an example of the kind of data that might crop up in the exam:

The **table** on the right shows the results of a **tissue culture experiment** — samples of **plant tissue** were taken from a **shoot** and grown on media with **varying ratios** of the growth factors **auxin** and **cytokinin**.

Growth medium	Ratio of auxin : cytokinin	Growth after 2 months
1	1 : 1	Growth but no specialised cells
2	1 : 25	Shoot formation
3	25 : 1	Root formation

Jacob had probably overdone the growth factors for this one.

1) Growth medium **1** showed **no specialised cell growth** — so an **equal ratio** (1 : 1) of auxin : cytokinin promotes **unspecialised cell growth**.
2) Growth medium **2** showed **shoot formation** — so a **high cytokinin : auxin ratio** (25 : 1) promotes the growth of **specialised shoot cells**.
3) Growth medium **3** showed **root formation** — so a **high auxin : cytokinin ratio** (25 : 1) promotes the growth of **specialised root cells**.

The results of this experiment show that the **ratio** of these **growth factors** helps to control the **specialisation** of different tissues in this plant.

Practice Questions

Q1 What is a stem cell?

Q2 Where are stem cells found in plants?

Exam Question

Q1 Samples of plant tissue containing totipotent cells were taken from a stem and placed on growth media of different pH values. Each tissue sample weighed less than 1 g at the start of the experiment. The graph shows the results.

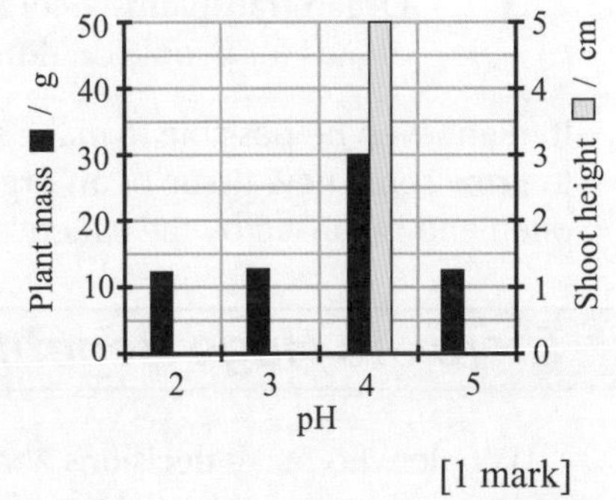

a) What are totipotent cells? [1 mark]

b) Describe and explain the results. [3 marks]

A tissue culture — what you need when you have a cold...

Jokes aside, all your biology knowledge is going to stem from some good old revision... Get it...? Stem... Sorry, I couldn't resist that one. But I mean it now — you need to know all about stem cells and how they become specialised to carry out a particular function. When you've got that straight, take a look at the tissue culture stuff. Plants are très exciting.

Stem Cells in Medicine

These pages are for AQA Unit 5 only.

Like I said before, stem cells really are the daddy of all cells because they can divide and turn into all cell types. And it's this ability to turn into any other cell type that's got scientists and doctors fairly excited...

Some *Stem Cell Therapies Already Exist*

1) Stem cells can divide into **any cell type**, so they could be used to **replace** cells **damaged** by illness or injury.
2) Some stem cell therapies **already exist** for some diseases affecting the **blood** and **immune system**.
3) **Bone marrow** contains **stem cells** that can become specialised to form **any type** of **blood cell**. **Bone marrow transplants** can be used to replace the **faulty** bone marrow in patients that produce **abnormal blood cells**. The stem cells in the transplanted bone marrow **divide** and **specialise** to produce healthy blood cells.
4) This technique has been used successfully to treat **leukaemia** (a **cancer** of the blood or bone marrow) and **lymphoma** (a cancer of the **lymphatic system**).
5) It has also been used to treat some **genetic disorders**, such as **sickle-cell anaemia** and **severe combined immunodeficiency** (**SCID**):

Example

Severe combined immunodeficiency (**SCID**) is a genetic disorder that affects the immune system. People with SCID have a **poorly functioning immune system** as their **white blood cells** (made in the bone marrow from stem cells) are **defective**. This means they **can't defend** the body against infections by identifying and destroying microorganisms. So SCID sufferers are **extremely susceptible** to **infections**. Treatment with a **bone marrow transplant** replaces the faulty bone marrow with donor bone marrow that contains **stem cells without** the **faulty genes** that cause SCID. These then **differentiate** to produce **functional** white blood cells. These cells can identify and destroy invading pathogens, so the **immune system functions properly**.

Stem Cells Could be Used to *Treat Other Diseases*

Totipotent stem cells can develop into **any** specialised cell type, so scientists think they could be used to **replace damaged tissues** in a **range** of **diseases**. Scientists are **researching** the use of stem cells as **treatment** for lots of conditions, including:

- **Spinal cord injuries** — stem cells could be used to replace damaged **nerve tissue**.
- **Heart disease** and **damage caused by heart attacks** — stem cells could be used to replace damaged **heart tissue**.
- **Bladder conditions** — stem cells could be used to grow **whole bladders**, which are then **implanted** in patients to replace diseased ones.
- **Respiratory diseases** — **donated windpipes** can be stripped down to their simple collagen structure and then covered with **tissue** generated by stem cells. This can then be **transplanted** into patients.
- **Organ transplants** — organs could be **grown** from stem cells to provide new organs for people on **donor waiting lists**.

These treatments aren't available yet but some are in the early stages of clinical trials.

It might even be possible to make **stem cells genetically identical** to a **patient's own cells**. These could then be used to **grow** some **new tissue** or **an organ** that the patient's body **wouldn't reject** (rejection of transplants occurs quite often and is caused by the patient's immune system recognising the tissue as **foreign** and **attacking it**).

There are *Huge Benefits* to Using *Stem Cells* in *Medicine*

People who make **decisions** about the **use** of stem cells to treat human disorders have to consider the **potential benefits** of stem cell therapies:

- They could **save** many **lives** — e.g. many people waiting for organ transplants **die** before a **donor organ** becomes available. Stem cells could be used to **grow organs** for those people awaiting transplants.
- They could **improve** the **quality of life** for many people — e.g. stem cells could be used to replace damaged cells in the eyes of people who are **blind**.

Stem Cells in Medicine

Human **Stem Cells** Can Come from **Adult Tissue** or **Embryos**

To **use stem cells** scientists have to get them from somewhere. There are **two** potential **sources** of human stem cells:

1 Adult stem cells

1) These are obtained from the **body tissues** of an **adult**. For example, adult stem cells are found in **bone marrow**.
2) They can be obtained in a relatively **simple operation** — with very **little risk** involved, but quite **a lot** of **discomfort**.
3) Adult stem cells **aren't** as **flexible** as embryonic stem cells — they can only specialise into a **limited** range of cells, not all body cell types. Although scientists are **trying** to find ways to make adult stem cells **specialise** into **any cell type**.

2 Embryonic stem cells

1) These are obtained from **embryos** at an **early stage of development**.
2) Embryos are created in a **laboratory** using ***in vitro* fertilisation** (IVF) — **egg cells** are **fertilised** by sperm **outside the womb**.
3) Once the embryos are approximately **4 to 5 days old, stem cells** are **removed** from them and the rest of the embryo is **destroyed**.
4) Embryonic stem cells can develop into **all types** of specialised cells.

There are **Ethical Issues Surrounding Stem Cell Use**

1) Obtaining stem cells from **embryos** created by IVF raises **ethical issues** because the procedure results in the **destruction** of an embryo that **could** become a fetus if placed in a **womb**.
2) Some people believe that at the moment of **fertilisation** an **individual** is formed that has the **right** to **life** — so they believe that it's **wrong** to **destroy** embryos.
3) Some people have **fewer objections** to stem cells being **obtained** from **unfertilised embryos** — embryos made from **egg cells** that **haven't** been fertilised by sperm. This is because the embryos **couldn't survive** past a few days and **wouldn't** produce a fetus if placed in a womb.
4) Some people think that **scientists** should **only use** adult stem cells because their production **doesn't** destroy an embryo. But adult stem cells **can't** develop into all the specialised cell types that embryonic stem cells can.
5) The decision makers in **society** have to take into account **everyone's views** when making decisions about **important scientific work** like stem cell research and its use to treat human disorders.

Practice Questions

Q1 What types of cells can bone marrow stem cells produce?

Q2 Name two conditions that stem cells could potentially be used to treat.

Q3 Describe one difference between embryonic and adult stem cells.

Exam Questions

Q1 Explain one way in which stem cell therapy is currently being used. [4 marks]

Q2 Explain why some people object to the use of embryonic stem cells in treating human disorders. [2 marks]

It's OK — you can grow yourself a new brain especially for this revision...

And that's the end of this section — whoopdidoo. It was a whopper — in size, I mean, not as in the famous 1980s chewy bars. Before you zoom off to something else (because I know you can't wait), take the time to learn all of the pages, including these last two. It might take a little while, but you'll be glad of it when it comes to the exam. Promise.

Making DNA Fragments

***These pages are for** AQA Unit 5, OCR Unit 5 **and** Edexcel Unit 4.*

You might have just done a section about genetics, but the good stuff's all in here. Three, two, one... go.

Gene Technology — Techniques Used to Study Genes

Gene technology is basically all the **techniques** used to **study genes** and their **function** — you need to learn some of these techniques for the exam. They include the **polymerase chain reaction** (**PCR**), where lots of identical copies of a DNA fragment are produced (see below) and **gel electrophoresis**, where DNA fragments are separated on a gel according to their size (see p. 144). As well as helping us to study genes, these techniques have **many** other **uses** — such as **genetic fingerprinting** (see p. 154) and **diagnosing diseases** (see p. 158).

'I know all about jean technology, baby...'

DNA Fragments can be Made Using PCR

As gene technology is all about **studying genes**, a good place to start is learning how to **get a copy** of the **DNA fragment** containing the gene you're **interested** in (the **target gene**). The **polymerase chain reaction** (PCR) can be used to make **millions of copies** of a fragment of DNA in just a few hours. PCR has **several stages** and is **repeated** over and over to make lots of copies:

1) A reaction mixture is set up that contains the **DNA sample**, **free nucleotides**, **primers** and **DNA polymerase**.
   - **Primers** are short pieces of DNA that are **complementary** to the bases at the **start** of the fragment you want.
   - **DNA polymerase** is an **enzyme** that creates new DNA strands.
2) The DNA mixture is **heated** to **95 °C** to break the **hydrogen bonds** between the two strands of DNA.
3) The mixture is then **cooled** to between **50** and **65 °C** so that the primers can **bind** (**anneal**) to the strands.

double-stranded DNA
hydrogen bonds
95 °C
single strand of DNA
primer
primer
single strand of DNA

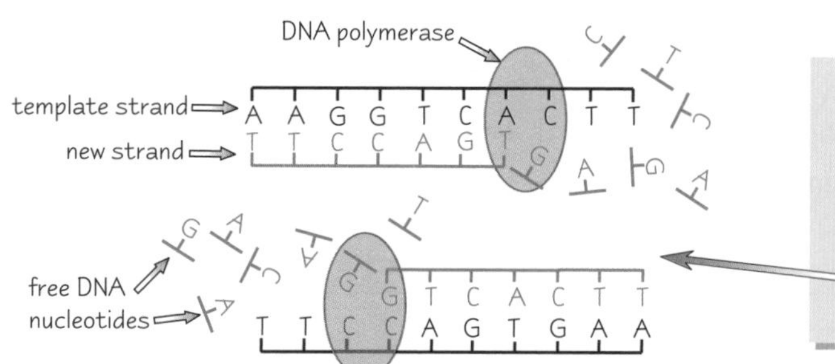

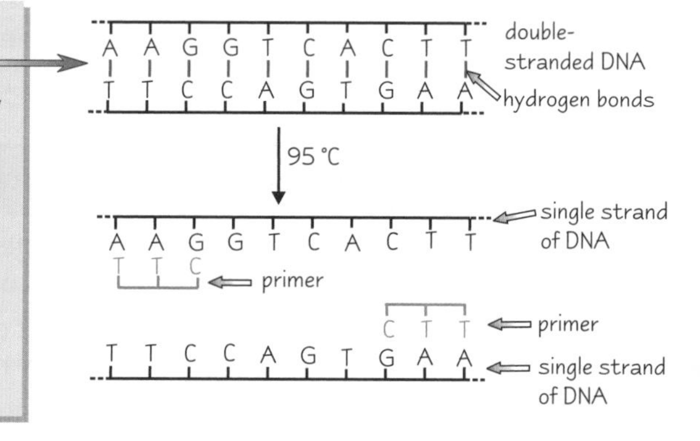

4) The reaction mixture is heated to **72 °C**, so **DNA polymerase** can **work**.
5) The DNA polymerase **lines up** free DNA nucleotides **alongside** each **template strand**. Specific **base pairing** means **new complementary strands** are formed.

6) **Two new copies** of the fragment of DNA are formed and **one cycle** of PCR is **complete**.
7) The cycle starts again, with the mixture being heated to 95 °C and this time **all four strands** (two original and two new) are used as **templates**.
8) Each PCR cycle **doubles** the amount of DNA, e.g. **1st cycle** = 2 × 2 = **4 DNA fragments**, **2nd cycle** = 4 × 2 = **8 DNA fragments**, **3rd cycle** = 8 × 2 = **16 DNA fragments**, and so on.

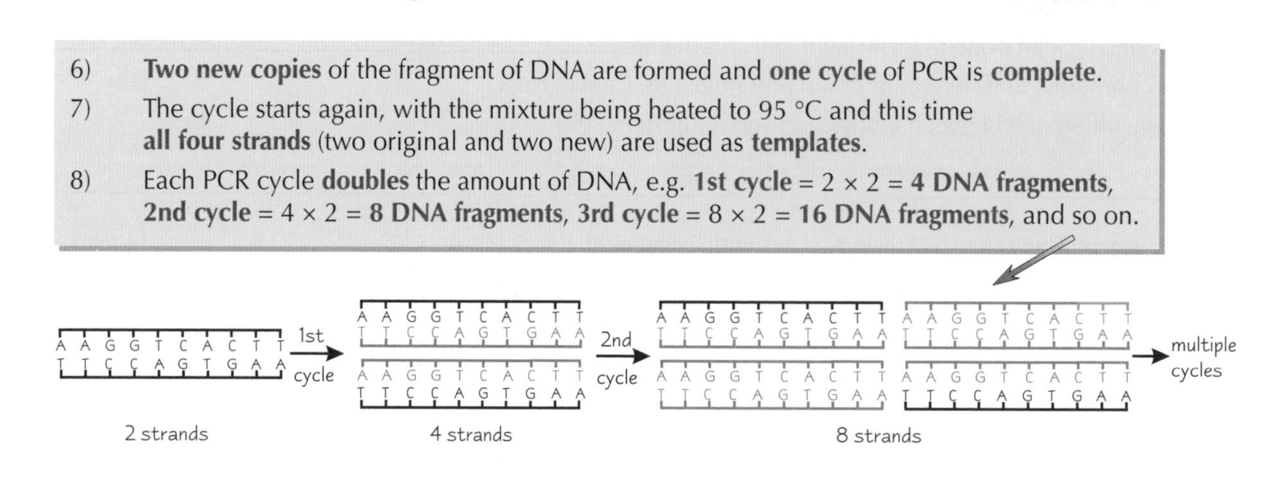

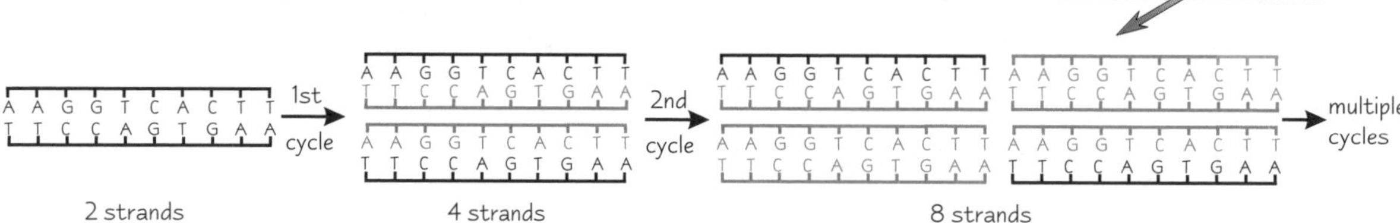

Making DNA Fragments

Restriction Endonucleases can be Used to Cut Out DNA Fragments

AQA and OCR only

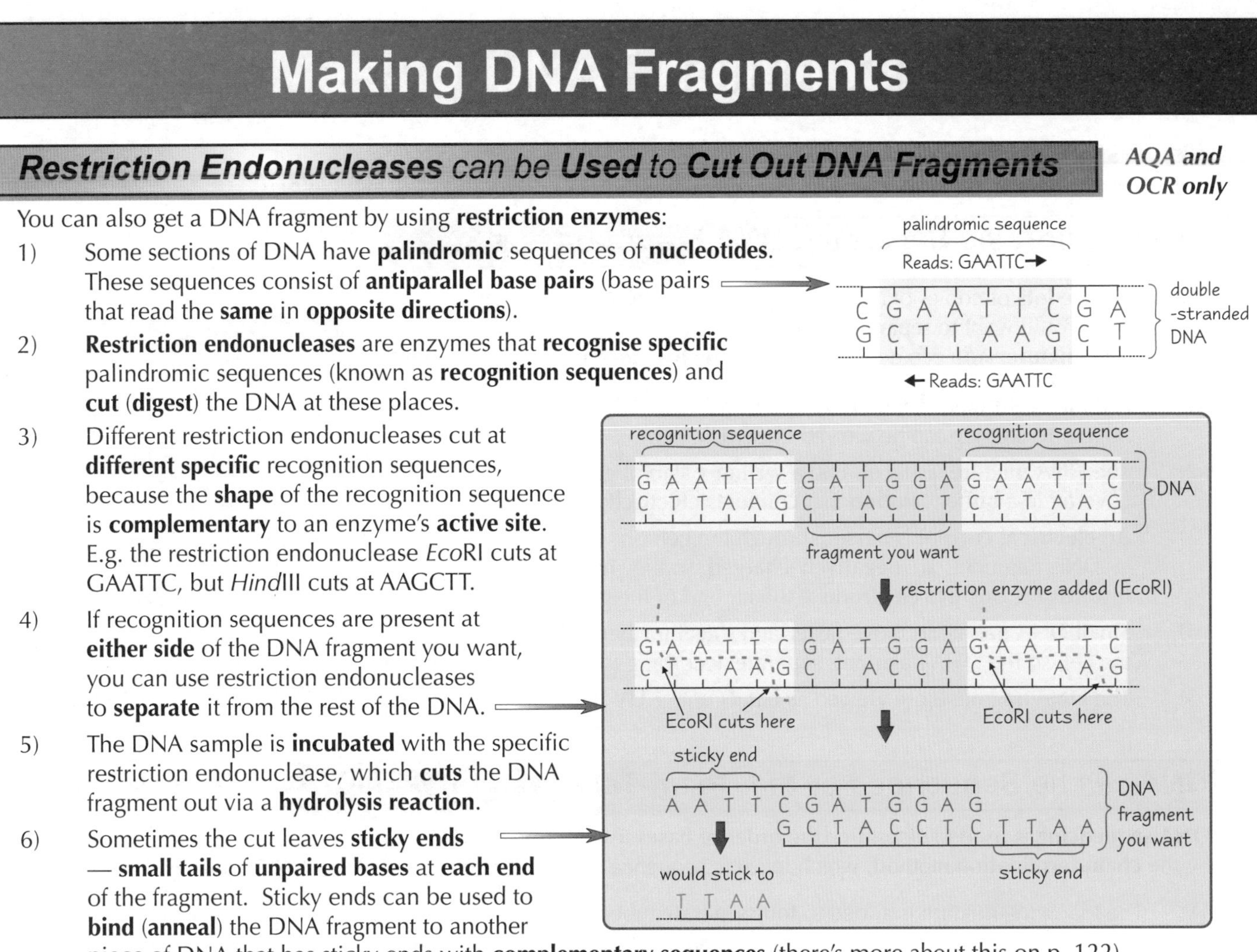

You can also get a DNA fragment by using **restriction enzymes**:

1) Some sections of DNA have **palindromic** sequences of **nucleotides**. These sequences consist of **antiparallel base pairs** (base pairs that read the **same** in **opposite directions**). 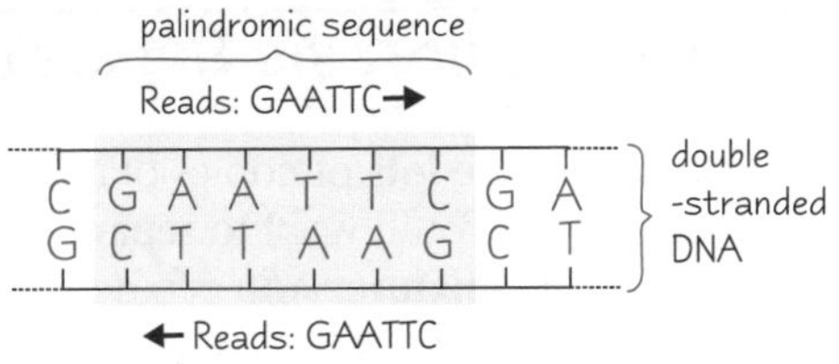
2) **Restriction endonucleases** are enzymes that **recognise specific** palindromic sequences (known as **recognition sequences**) and **cut** (**digest**) the DNA at these places.
3) Different restriction endonucleases cut at **different specific** recognition sequences, because the **shape** of the recognition sequence is **complementary** to an enzyme's **active site**. E.g. the restriction endonuclease *Eco*RI cuts at GAATTC, but *Hind*III cuts at AAGCTT.
4) If recognition sequences are present at **either side** of the DNA fragment you want, you can use restriction endonucleases to **separate** it from the rest of the DNA.
5) The DNA sample is **incubated** with the specific restriction endonuclease, which **cuts** the DNA fragment out via a **hydrolysis reaction**.
6) Sometimes the cut leaves **sticky ends** — **small tails** of **unpaired bases** at **each end** of the fragment. Sticky ends can be used to **bind** (**anneal**) the DNA fragment to another piece of DNA that has sticky ends with **complementary sequences** (there's more about this on p. 122).

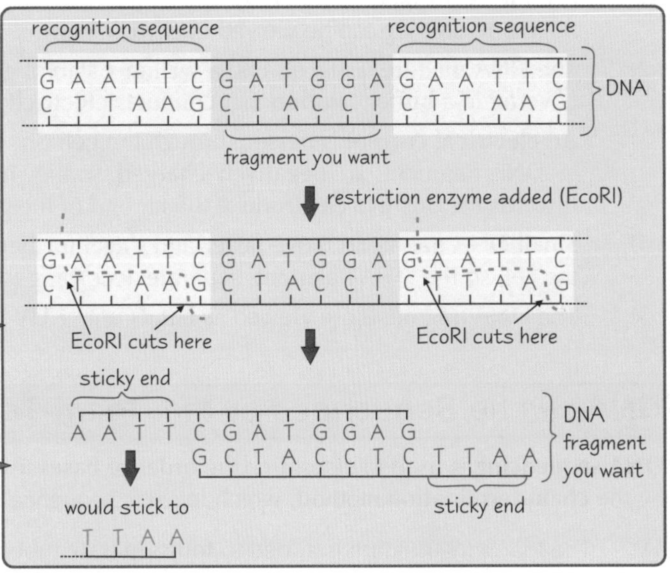

DNA Fragments can Also be Made Using Reverse Transcriptase

AQA only

1) Many **cells** only contain **two copies** of each gene, making it **difficult** to obtain a DNA fragment containing the target gene. But they can contain **many mRNA** molecules (see p. 123) complementary to the gene, so mRNA is often **easier** to obtain.
2) The mRNA molecules can be used as **templates** to **make lots of DNA**. The **enzyme reverse transcriptase makes DNA** from an RNA template. The DNA produced is called **complementary DNA (cDNA)**.
3) For example, **pancreatic cells** produce the protein **insulin**. They have loads of mRNA molecules complementary to the **insulin gene**, but only **two copies** of the gene **itself**. So reverse transcriptase could be used to **make cDNA** from the **insulin mRNA**.
4) To do this, **mRNA** is first isolated from cells. Then it's **mixed** with **free DNA nucleotides** and **reverse transcriptase**. The reverse transcriptase uses the mRNA as a **template** to synthesise a **new strand** of cDNA.

Practice Questions

Q1 What is gene technology?

Q2 What does PCR stand for?

Q3 What are sticky ends?

Exam Question

Q1 Describe and explain how to produce multiple copies of a DNA fragment using PCR. [6 marks]

Sticky ends — for once a name that actually makes sense.

Gene technology isn't the sort of technology you can buy in all good electrical stores, but it's still quite cool. Take PCR for example — you can throw a DNA fragment and some other bits into a test tube, heat it to three different temperatures and, hey presto, you've got squillions of DNA fragments. Yes, I do tend to get excited about the strangest things...

Common Techniques

These pages are for AQA Unit 5, OCR Unit 5 and Edexcel Unit 4.
I can hear you crying out for some more DNA techniques. Well, your wish is my command...

Electrophoresis Separates DNA Fragments by Size

If you've made **lots** of copies of a **DNA fragment** using **PCR** (see p. 142), you'll want to **separate** them from the rest of the DNA in the **mixture**. You can do this using **gel electrophoresis**:

1) A **fluorescent tag** is added to all the DNA fragments so they can be viewed under **UV light**.
2) The DNA mixture is placed into a **well** in a slab of **gel** and covered in a **buffer solution** that **conducts electricity**.
3) An **electrical current** is passed through the gel — DNA fragments are **negatively charged**, so they **move towards** the **positive electrode** at the far end of the gel.
4) **Small** DNA fragments move **faster** and **travel further** through the gel, so the DNA fragments **separate** according to **size**.
5) The DNA fragments are viewed as **bands** under **UV light**.

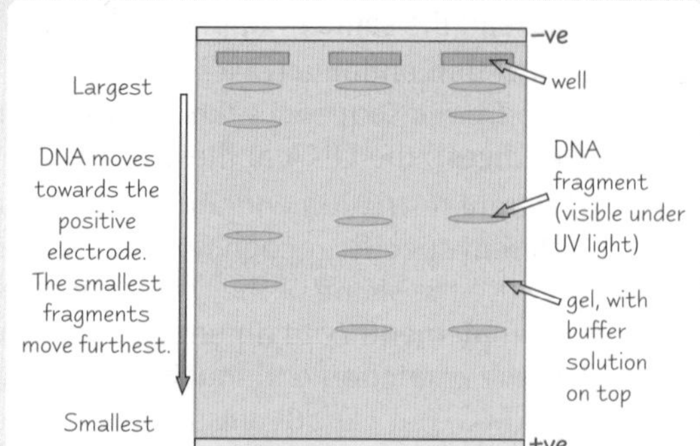

The **size** of a DNA fragment is **measured** in **bases**, e.g. ATCC = 4 bases or base pairs, **1000 bases** is **one kilobase** (1 kb).

DNA can be Sequenced by the Chain-Termination Method

AQA and OCR only

DNA sequencing is used to determine the **order** of **bases** in a section of **DNA** (gene). It can be carried out by the **chain-termination method**, which lets you sequence **small fragments** of DNA, up to **750 base pairs**:

1) The following mixture is added to **four separate** tubes:
   - A **single-stranded DNA template** — the DNA to sequence.
   - Lots of **DNA primer** — short pieces of DNA (see p. 142).
   - **DNA polymerase** — the enzyme that joins DNA nucleotides together.
   - **Free nucleotides** — lots of free A, T, C and G nucleotides.
   - **Fluorescently-labelled modified nucleotide** — like a normal nucleotide, but once it's added to a DNA strand, **no more** bases can be added after it. A **different** modified nucleotide is added to **each** tube (**A*, T*, C*, G***).
2) The tubes undergo **PCR**, which produces many **strands of DNA**. The strands are **different lengths** because each one **terminates** at a **different point** depending on where the modified nucleotide was added.
3) For example, in tube A (with the **modified adenine** nucleotide **A***) sometimes A* is **added** to the DNA at point 4 **instead** of A, **stopping** the **addition** of any more bases (the strand is **terminated**). Sometimes A is added at point 4, then **A*** is added at **point 5**. Sometimes A is added at **point 4**, A again at point 5, G at point 6 and **A*** is added at **point 7**. So strands of **three different lengths** (4 bases, 5 bases and 7 bases) all ending in **A*** are produced.
4) The DNA fragments in each tube are separated by **electrophoresis** and **visualised** under **UV light** (because of the **fluorescent label**).
5) The **complementary base sequence** can be **read** from the gel. The **smallest** nucleotide (e.g. one base) is at the **bottom** of the gel. Each band after this represents **one more base** added. So by reading the bands **from the bottom** of the gel **to the top**, you can build up the **DNA sequence** one base at a time.

Example — Tube A

DNA polymerase
gene to be sequenced
primer
free DNA nucleotides
PCR cycles
A* added at point 4
A added, then A* added at point 5
2 As added, G added, then A* added at point 7
+ all other tubes (containing C*, G*, T*)
position of nucleotide in sequence
primer
A T C G
The complementary sequence is TTCAAGA, so the original sequence is AAGTTCT.

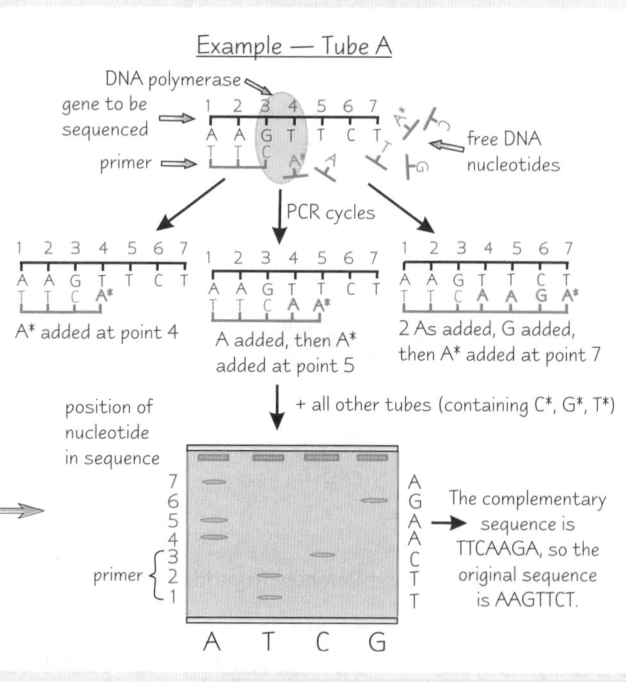

Nowadays sequencing is done **altogether** in **one tube** in an **automated DNA sequencer**. The tube contains **all** the modified nucleotides, each with a **different coloured** fluorescent label, and a machine reads the sequence for you.

Common Techniques

This page is for AQA Unit 5 and OCR Unit 5. Those of you doing Edexcel can skip straight to the questions.

DNA Probes can be used to Identify Specific Base Sequences in DNA

1) **DNA probes** (also called **gene probes**) can be used to **identify DNA fragments** that contain **specific sequences** of bases, e.g. they can be used to **locate genes** on chromosomes or see if a person's DNA **contains** a **mutated gene** (e.g. a gene that causes a genetic disorder).
2) DNA probes are **short strands** of **DNA**. They have a **specific base sequence** that's **complementary** to the target sequence — the specific sequence you're looking for.

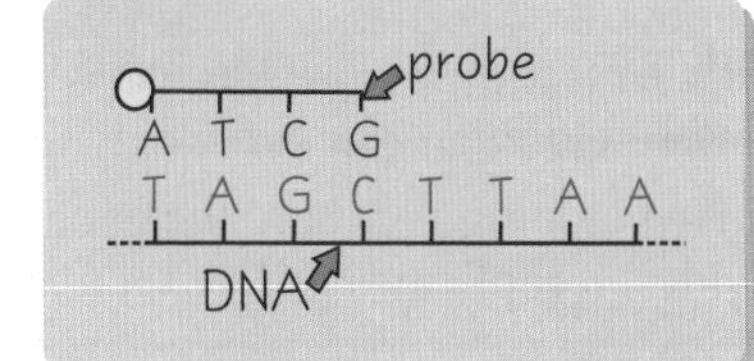

3) This means a DNA probe will **bind** (**hybridise**) to the **target sequence** if it's **present** in a **sample** of DNA.
4) A DNA probe also has a **label attached**, so that it can be **detected**. The two most common types of label are a **radioactive** label (detected using **X-ray film**) or a **fluorescent** label (detected using **UV light**).
5) For example, you can use a DNA probe to see if any members of a family have a **mutation** in a gene that causes a **genetic disorder**:
   - A **sample** of **DNA** from each family member is **digested** into fragments using **restriction enzymes** (see page 143) and **separated** using **electrophoresis** (see previous page).
   - The separated DNA fragments are then transferred to a **nylon membrane** and **incubated** with the **fluorescently labelled DNA probe**. The probe is **complementary** to the specific sequence of the mutated gene.
   - If the specific sequence **is present** in one of the DNA fragments, the DNA probe will **hybridise** (**bind**) to it.
   - The **membrane** is then **exposed** to **UV light** and if the specific sequence is present in one of the DNA fragments, then that band will **fluoresce** (**glow**).
   - For example, **person three** has a **visible band**, so that family member has the specific sequence **in** one of their DNA fragments, which means they **have** the **mutated gene**.

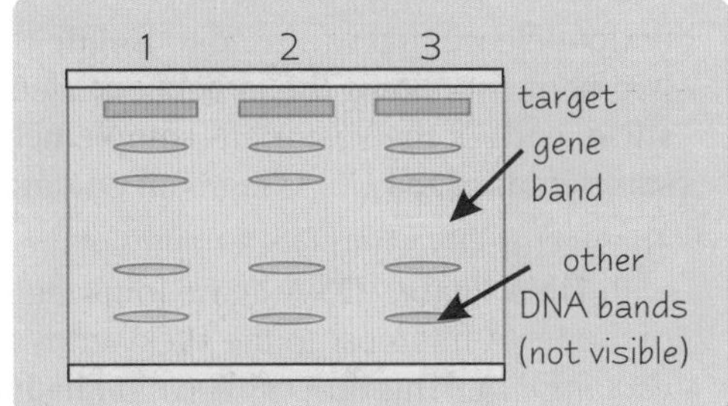

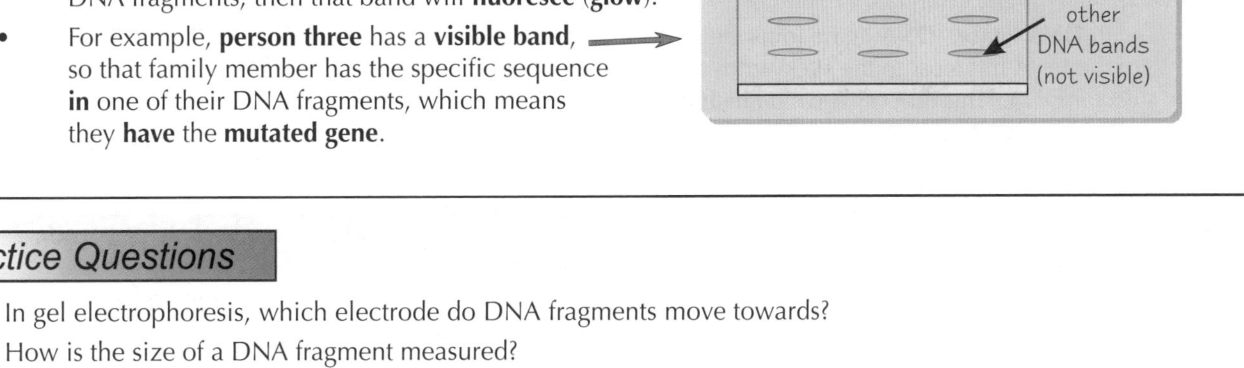

Practice Questions

Q1 In gel electrophoresis, which electrode do DNA fragments move towards?

Q2 How is the size of a DNA fragment measured?

Q3 Give the name of a method used to sequence DNA.

Q4 What is a DNA probe?

Q5 Name two types of label that can be attached to a DNA probe.

Exam Questions

Q1 Describe and explain how electrophoresis works. [5 marks]

Q2 To sequence a small DNA fragment, a single-stranded DNA template and DNA polymerase are needed.
a) Name the other three reactants needed for a sequencing reaction. [3 marks]
b) Describe and explain the process of sequencing a small DNA fragment. [6 marks]

Q3 Erin has a family history of breast cancer. She has agreed to be screened for the mutated BRCA1 gene, which can cause breast cancer. A sample of her DNA is digested and separated using gel electrophoresis. Describe how a DNA probe could be used to identify the mutated BRCA1 gene. [4 marks]

Sequincing — so 80s...

Oooh, there's nothing like revising a few lab techniques to make you feel really geeky. Jokes aside, I won't deny it, there are a couple of difficult bits on this page. But you just need to keep going over the steps of all these different techniques until they make sense. And they do make sense really, I promise. Try drawing out diagrams for each technique to help you.

Gene Cloning

These pages are for AQA Unit 5. If you're doing OCR go to page 150. If you're doing Edexcel go straight to page 154.
Hmmmmm... nope, can't think of any exciting or funny ways to start this double page. Sorry.

Gene Cloning can be done In Vitro or In Vivo

Gene cloning is all about making loads of **identical copies** of a gene. It can be done using **two** different techniques:

1) ***In vitro*** cloning — where the gene copies are made **outside** of a **living organism** using **PCR** (see page 142).
2) ***In vivo*** cloning — where the gene copies are made **within** a **living organism**. As the organism **grows** and **divides**, it **replicates** its **DNA**, creating multiple copies of the gene (see below).

In vitro is Latin for within glass. *In vivo* is Latin for within the living.

In Vivo Cloning Step 1 — The Gene is Inserted into a Vector

The **DNA fragment** containing the **target gene** has been isolated using one of the techniques on pages 142-143. The first step in *in vivo* cloning is to **stick** the fragment **into** a **vector** using **restriction endonuclease** and **ligase** (an enzyme):

1) The DNA fragment is inserted into vector DNA — a **vector** is something that's used to **transfer DNA** into a **cell**. They can be **plasmids** (**small, circular molecules** of DNA in **bacteria**) or **bacteriophages** (**viruses** that **infect** bacteria).
2) The vector DNA is **cut open** using the **same** restriction endonuclease that was used to **isolate** the DNA fragment containing the target gene (see p. 143). So the **sticky ends** of the vector are **complementary** to the sticky ends of the DNA fragment containing the gene.
3) The vector DNA and DNA fragment are **mixed together** with **DNA ligase**. DNA ligase **joins** the sticky ends of the DNA fragment to the sticky ends of the vector DNA. This process is called **ligation.**
4) The new combination of bases in the DNA (vector DNA + DNA fragment) is called **recombinant DNA.**

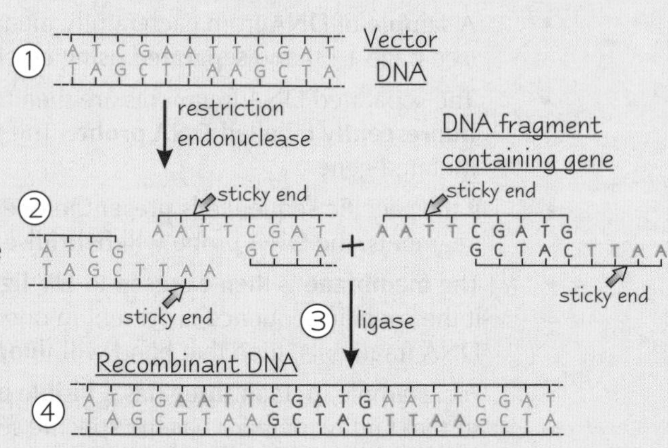

In Vivo Cloning Step 2 — The Vector Transfers the Gene into Host Cells

1) The **vector** with the **recombinant DNA** is used to **transfer** the gene into **cells** (called **host** cells).
2) If a **plasmid vector** is used, **host cells** have to be **persuaded** to **take in** the plasmid vector and its DNA. E.g. host bacterial cells are placed into ice-cold **calcium chloride** solution to make their cell walls more **permeable**. The **plasmids** are **added** and the mixture is **heat-shocked** (heated to around **42 °C** for **1-2 minutes**), which encourages the cells to take in the plasmids.
3) With a **bacteriophage** vector, the bacteriophage will **infect** the host bacterium by **injecting** its **DNA** into it. The phage DNA (with the target gene in it) then **integrates** into the bacterial DNA.
4) **Host cells** that **take up** the vectors containing the gene of interest are said to be **transformed**.

In Vivo Cloning Step 3 — Identifying Transformed Host Cells

Not all host cells will have **taken up** the vector and its DNA. **Marker genes** can be used to **identify** the **transformed** cells:

1) **Marker genes** can be inserted into vectors at the **same time** as the gene to be cloned. This means any **transformed host cells** will contain the gene to be cloned **and** the marker gene.
2) Host cells are **grown** on **agar plates** and each cell **divides** and **replicates** its DNA, creating a **colony** of **cloned cells**.
3) Transformed cells will produce colonies where **all the cells** contain the cloned gene and the marker gene.
4) The marker gene can code for **antibiotic resistance** — host cells are grown on agar plates **containing** the specific **antibiotic**, so **only** transformed cells that have the **marker gene** will **survive** and **grow**.
5) The marker gene can code for **fluorescence** — when the agar plate is placed under a **UV light only** transformed cells will **fluoresce**.
6) **Identified** transformed cells are allowed to **grow more**, producing **lots** and **lots** of **copies** of the **cloned gene**.

Gene Cloning

There are Advantages and Disadvantages to Both In Vivo and In Vitro Cloning

Depending on the **reason** why you want to clone a gene, you can choose to do it either by ***in vivo* cloning** or ***in vitro* cloning**. Both techniques have **advantages** and **disadvantages**:

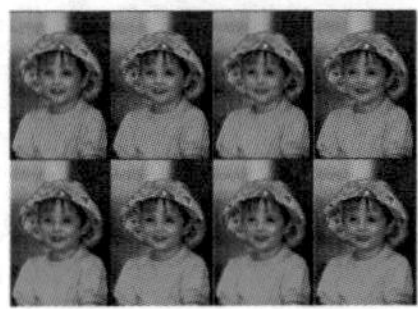

Either the cloning had worked, or Professor Dim's eyesight had gone.

In Vivo Cloning

1) Cloning *in vivo* can produce **mRNA** and **protein** as well as DNA because it's done in a **living cell** (which has the ribosomes and all the enzymes needed to produce them).
2) Cloning *in vivo* can also produce **modified DNA**, **modified mRNA** or **modified protein** — they have **modifications added** to them, e.g. **sugar** or **methyl** (-CH_3) groups.
3) **Large fragments** of DNA can be cloned using *in vivo* cloning, e.g. between **20 to 45 kilobases** of DNA can be inserted into some **plasmids** and **bacteriophages**.
4) *In vivo* cloning can be a **relatively cheap method**, depending on **how much** DNA you want to produce.

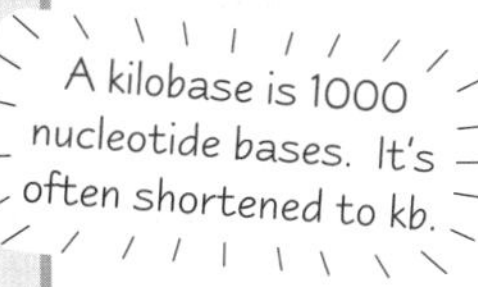
A kilobase is 1000
nucleotide bases. It's
often shortened to kb.

In vivo cloning also has **disadvantages** — the DNA fragment has to be **isolated** from other cell components, you may **not want modified** DNA, and it can be quite a **slow** process (because some types of bacteria **grow quite slowly**).

In Vitro Cloning (PCR)

1) *In vitro* cloning can be used to produce **lots of DNA** (but **not** mRNA or protein).
2) The DNA produced **isn't modified** (see above) — an advantage if you **don't want** it to be modified.
3) This technique **only** replicates the **DNA fragment** of **interest** (e.g. the target gene). This means that you **don't** have to **isolate** the DNA fragment from **host DNA** or **cell components**.
4) *In vitro* cloning is a **fast** process — PCR can clone **millions of copies** of DNA in just a **few hours**.

In vitro cloning also has **disadvantages** — it can only replicate a **small DNA fragment** (compared to *in vitro*), you may want a **modified** product, **mRNA** and **protein** aren't made as well, and it can be **expensive** if you want to produce a lot of DNA.

Practice Questions

Q1 What is *in vitro* gene cloning?

Q2 What is a vector?

Q3 Other than a plasmid, give an example of a vector.

Q4 Name the type of enzyme that can be used to cut DNA.

Q5 What is the name of the type of DNA formed from vector DNA and an inserted DNA fragment?

Q6 What is a marker gene?

Q7 Give two advantages of *in vivo* cloning.

Q8 *In vitro* cloning is a slow process — is this statement correct?

Agar plate under UV light

agar plate
colony A
colony B
colony not visible
colony visible

Exam Question

Q1 A scientist has cloned a gene by transferring a plasmid containing the target gene and a fluorescent marker gene into some bacterial cells. The cells were grown on an agar plate. The plate was then placed under UV light (see above).

a) Explain why the scientist thinks colony A contains transformed host cells, but colony B doesn't. [2 marks]

b) Explain how the scientist might have inserted the target gene into the plasmid. [3 marks]

Transformed boyfriends — made to listen, tidy up and agree with you...

These page are quite scary, I know. But don't worry, it's not as difficult as photosynthesis or respiration — you just need to keep testing yourself on all the different techniques until you get them straight in your head. I know I've said it before, but drawing out the diagram will help — then you'll know inserting DNA into vectors like a pro. Right, I need a cuppa...

Genetic Engineering

These pages are for AQA Unit 5 only.

Now that you know how to make a DNA fragment and clone a gene, it's probably a good time to tell you why you might want to. Don't worry — it's not evil stuff, but I promise to do my evil laugh. Mwah ha hah.

Genetic Engineering is the Manipulation of an Organism's DNA

1) **Genetic engineering** is also known as **recombinant DNA technology**.
2) Organisms that have had their **DNA altered** by genetic engineering are called **transformed organisms**.
3) These organisms have **recombinant DNA** — **DNA** formed by **joining together** DNA from **different sources**.
4) **Microorganisms**, **plants** and **animals** can all be **genetically engineered** to **benefit humans**.
5) **Transformed microorganisms** can be made using the same technology as ***in vivo*** cloning (see page 146). For example, **foreign DNA** can be **inserted** into **microorganisms** to produce **lots** of **useful protein**, e.g. insulin:

Transformed organisms are also known as genetically engineered or genetically modified organisms.

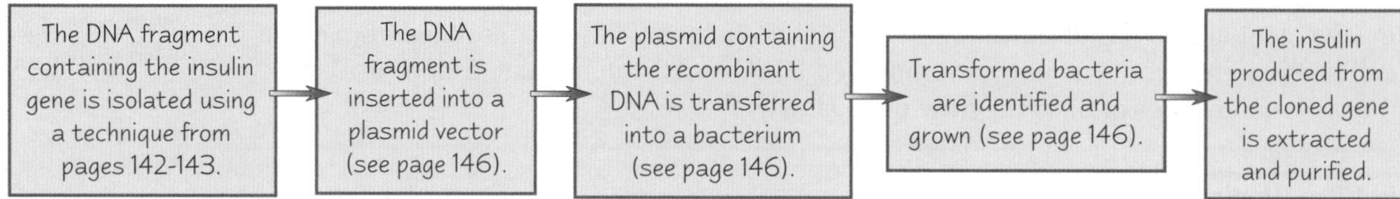

6) **Transformed plants** can also be produced — a gene that codes for a **desirable characteristic** is inserted into a **plasmid**. The plasmid is added to a **bacterium** and the bacterium is used as a **vector** to get the gene into the **plant cells**. The transformed plant will have the desirable characteristic coded for by that gene.
7) **Transformed animals** can be produced too — a gene that codes for a **desirable characteristic** is inserted into an **animal embryo**. The transformed animal will have the desirable characteristic coded for by that gene.

Transformed (Genetically Engineered) Organisms can Benefit Humans

Producing **transformed organisms** (microorganisms, plants and animals) can benefit **humans** in lots of ways:

1 Agriculture

- **Agricultural crops** can be **transformed** so that they give **higher yields** or are **more nutritious**. This means these plants can be used to reduce the risk of **famine** and **malnutrition**. Crops can also be transformed to have **pest resistance**, so that **fewer pesticides** are needed. This **reduces costs** and reduces any **environmental problems** associated with using pesticides.
- For example, ***Golden Rice*** is a variety of **transformed rice**. It contains **one gene** from a **daffodil plant** and **one gene** from a **soil bacterium**, which together enable the rice to produce **beta-carotene**. The beta-carotene is used by our bodies to produce **vitamin A**. *Golden Rice* is being developed to **reduce vitamin A deficiency** in areas where there's a **shortage** of **dietary vitamin A**, e.g. **south Asia**, **Africa**. Vitamin A deficiency is a big problem in these areas, e.g. up to **500 000 children per year** worldwide go **blind** due to vitamin A deficiency.

2 Industry

- **Industrial processes** often use **biological catalysts (enzymes)**. These enzymes can be produced from **transformed organisms**, so they can be produced in **large quantities** for **less money**, **reducing costs**.
- For example, **chymosin** (or **rennin**) is an enzyme used in **cheese-making**. It used to be made from **rennet** (a substance produced in the **stomach** of **cows**), but it can now be produced by **transformed organisms**. This means it can be made in **large quantities**, relatively **cheaply** and **without killing** any **cows**, making some cheese suitable for **vegetarians**.

3 Medicine

- Many **drugs** and **vaccines** are produced by transformed organisms, using recombinant DNA technology. They can be made **quickly**, **cheaply** and in **large quantities** using this method.
- For example, **insulin** is used to treat **Type 1 diabetes** and used to come from **animals** (cow, horse or pig pancreases). This insulin **wasn't** human insulin though, so it **didn't work quite as well**. Human insulin is now made from **transformed microorganisms**, using a **cloned human insulin gene** (see above).

Genetic Engineering

Many People are Concerned About the Use of Genetic Engineering

There are **ethical**, **moral** and **social concerns** associated with the **use** of **genetic engineering**:

1 Agriculture

- **Farmers** might plant only **one type** of transformed crop (this is called **monoculture**). This could make the **whole crop vulnerable** to **disease** because the plants are **genetically identical**.
- Some people are concerned about the possibility of **'superweeds'** — weeds that are **resistant** to **herbicides**. These could occur if transformed crops **interbreed** with **wild plants**.

2 Industry

- **Without proper labelling**, some people think they **won't** have a **choice** about whether to consume food made using genetically engineered organisms.
- Some people are worried that the process used to **purify** proteins (from genetically engineered organisms) could lead to the introduction of **toxins** into the **food industry**.

3 Medicine

- Companies who **own** genetic engineering technologies may **limit** the **use** of technologies that could be **saving lives**.
- Some people worry this technology could be used **unethically**, e.g. to make **designer babies** (babies that have characteristics **chosen** by their parents). This is currently **illegal** though.

Humanitarians Think Genetic Engineering will Benefit People

Genetic engineering has **many** potential **humanitarian benefits**:

1) **Agricultural crops** could be produced that help **reduce** the risk of **famine** and **malnutrition**, e.g. **drought-resistant** crops for **drought-prone** areas.
2) **Transformed crops** could be used to produce **useful pharmaceutical products** (e.g. **vaccines**) which could make drugs **available** to **more people**, e.g. in areas where **refrigeration** (usually needed for **storing** vaccines) **isn't available**.
3) **Medicines** could be produced more **cheaply**, so more people can **afford** them.

You need to be able to balance the humanitarian benefits with opposing views from environmentalists and anti-globalisation activists.

But some **environmentalists** and **anti-globalisation activists** have concerns:

1) **Environmentalists** — Many **oppose** recombinant DNA technology because they think it could **potentially damage** the **environment**. E.g. transformed crops could encourage **farmers** to carry out monoculture (see above), which **decreases biodiversity**. There are also fears that if **transformed crops breed** with **wild plants** there'll be **uncontrolled spread** of **recombinant DNA**, with **unknown consequences**.
2) **Anti-globalisation activists** — These are people who **oppose globalisation** (e.g. the **growth** of **large multinational companies** at the **expense** of **smaller ones**). A **few**, **large** biotechnology companies **control** some forms of genetic engineering. As the **use** of this technology **increases**, these companies get **bigger** and **more powerful**. This may **force** smaller companies **out of business**, e.g. by making it **harder** for them to **compete**.

Practice Questions

Q1 What is recombinant DNA?

Q2 Give an example of a transformed agricultural crop.

Exam Question

Q1 A large agricultural company has isolated a gene from bacteria that may increase the drought resistance of wheat plants.

a) Briefly explain how this gene could be used to make a transformed wheat plant. [3 marks]

b) Suggest how the transformed wheat plants might be beneficial to humans. [2 marks]

c) Suggest why anti-globalisation activists may be against the use of this gene. [1 mark]

Neapolitan — recombinant ice cream...

Ahhh, sitting in the sun, licking an ice cream, exams all over. That's where you'll be in a few months' time. After revising all this horrible stuff that is. As genetic engineering advances, more questions will pop up about its implications. So it's a good idea to know all sides of the argument — you need to know them for the exam anyway.

Genetic Engineering

These pages are for OCR Unit 5 only. If you're doing AQA you can go straight to page 154.

Genetic engineering — you need to know what it is and how it's done... (unlucky)...

Genetic Engineering is the Manipulation of an Organism's DNA

1) Organisms that have had their **DNA altered** by genetic engineering are called **transformed organisms**.
2) These organisms have **recombinant DNA** — **DNA** formed by **joining together** DNA from **different sources**.
3) Genetic engineering usually involves **extracting** a **gene** from **one organism** and then **inserting** it **into another organism** (often one that's a **different species**).
4) Genes can also be **manufactured** instead of extracted from an organism.
5) The organism with the inserted gene will then **produce the protein** coded for by that gene.
6) An organism that has been genetically engineered to include a **gene** from a **different species** is sometimes called a **transgenic organism**.

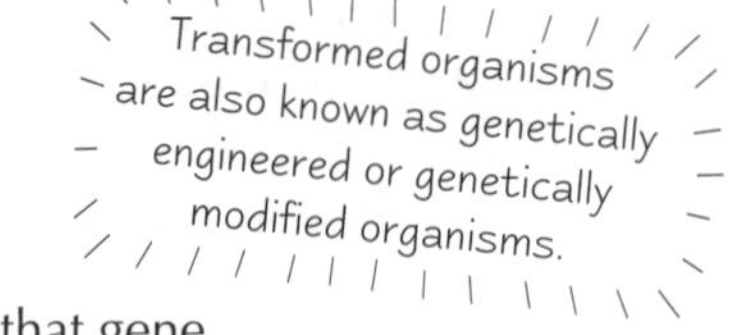

You Need to Know How to Genetically Engineer a Microorganism

1) The DNA Fragment Containing the Desired Gene is Obtained

The **DNA fragment** containing the **gene you want** can be isolated using **restriction endonucleases** (see page 143) or it can be **made using PCR** (see page 142).

2) The DNA Fragment (with the Gene in) is Inserted into a Vector

The fragment is then **inserted into** a **vector** using **restriction endonucleases** and **ligase** (an enzyme):

1) The DNA fragment is inserted into vector DNA — a **vector** is something that's used to **transfer DNA** into a **cell**. They can be **plasmids** (**small, circular molecules** of DNA in **bacteria**) or **bacteriophages** (**viruses** that **infect** bacteria).
2) The vector DNA is **cut open** using the **same** restriction endonuclease that was used to **isolate** the DNA fragment containing the target gene (see p. 143). So the **sticky ends** of the vector are **complementary** to the sticky ends of the DNA fragment containing the gene.
3) The vector DNA and DNA fragment are **mixed together** with **DNA ligase**. DNA ligase **joins** up the **sugar-phosphate backbones** of the two bits. This process is called **ligation**.

① ATCGAATTCGAT / TAGCTTAAGCTA — Vector DNA

restriction endonuclease

DNA fragment containing gene

② ATCG / TAGCTTAA — sticky end; AATTCGAT / GCTA — sticky end; + AATTCGATG / GCTACTTAA — sticky end

③ ligase

Recombinant DNA

④ ATCGAATTCGATGAATTCGAT / TAGCTTAAGCTACTTAAGCTA

4) The new combination of bases in the DNA (vector DNA + DNA fragment) is called **recombinant DNA**.

3) The Vector Transfers the Gene into the Bacteria

1) The **vector** with the **recombinant DNA** is used to **transfer** the gene into the **bacterial cells**.
2) If a **plasmid vector** is used, the bacterial cells have to be **persuaded** to **take in** the plasmid vector and its DNA. E.g. they're placed into ice-cold **calcium chloride** solution to make their cell walls more **permeable**. The **plasmids** are **added** and the mixture is **heat-shocked** (heated to around **42 °C** for **1-2 minutes**), which encourages the cells to take in the plasmids.
3) With a **bacteriophage** vector, the bacteriophage will **infect** the bacterium by **injecting** its **DNA** into it. The phage DNA (with the desired gene in it) then **integrates** into the bacterial DNA.
4) **Cells** that **take up** the vectors containing the desired gene are genetically engineered, so are called **transformed**.

Genetic Engineering

4 Identify the Transformed Bacteria

Not all the bacteria will have **taken up** the vector. **Marker genes** can be used to **identify** the ones that **have**:

1) **Marker genes** can be inserted into vectors at the **same time** as the desired gene. This means any **transformed bacterial cells** will contain the desired gene **and** the marker gene.
2) The bacteria are **grown** on **agar plates** and each cell **divides** and **replicates** its DNA, creating a **colony** of **cells**.
3) Transformed cells will produce colonies where **all the cells** contain the desired gene and the marker gene.
4) The marker gene can code for **antibiotic resistance** — bacteria are grown on agar plates **containing** the **antibiotic**, so **only** cells that have the **marker gene** will **survive** and **grow**.
5) The marker gene can code for **fluorescence** — when the agar plate is placed under a **UV light** **only** transformed cells will **fluoresce**.

It's Useful for Microorganisms to be Able to Take Up Plasmids

Microorganisms can **take up plasmids** from their surroundings, which is **beneficial** because the plasmids often contain **useful genes**. This means the microorganisms gain **useful characteristics**, so they're more likely to have an **advantage** over other microorganisms, which **increases** their **chance** of **survival**. Plasmids may contain:

- Genes that code for **resistance** to **antibiotics**, e.g. genes for enzymes that **break down antibiotics**.
- Genes that help microorganisms **invade hosts**, e.g. genes for enzymes that **break down host tissues**.
- Genes that mean microorganisms can use **different nutrients**, e.g. genes for enzymes that break down **sugars** not normally used.

Practice Questions

Q1 What is the name for an organism that has had its DNA altered?

Q2 What is a vector?

Q3 Other than a plasmid, give an example of a vector.

Q4 Name the type of enzyme that can be used to cut DNA.

Q5 What is the name of the type of DNA formed from vector DNA and an inserted DNA fragment?

Q6 What is a marker gene?

Exam Question

Q1 A scientist has genetically engineered some bacterial cells to contain a desired gene and a gene that gives resistance to penicillin. The cells were grown on an agar plate and then transferred to a plate containing penicillin. The two plates are shown above.

a) Explain why the scientist thinks colony A contains transformed bacterial cells, but colony B doesn't. [2 marks]

b) Explain how the scientist might have inserted the desired gene into the plasmid. [3 marks]

c) Explain why being able to take up plasmids is useful to bacteria. [2 marks]

Examiners — genetically engineered to contain marker genes...

This stuff might seem tricky the first time you read it, but it's not too bad really — you get the gene you want and bung it in a vector, the vector gets the gene into the cell (it's kind of like a delivery boy), then all you have to do is figure out which cells have got the gene. Easy peasy. Unfortunately you need to know each stage in detail, so get learnin'.

Genetic Engineering

These pages are for OCR Unit 5 only.

Genetic engineering can benefit humans in loads of different ways...

Transformed **Bacteria** can be used to **Produce Human Insulin**

See page 118 for more on Type 1 diabetes.

People with **Type 1 diabetes** need to **inject insulin** to **regulate** their **blood glucose concentration**. Insulin used to be obtained from the **pancreases** of dead animals, such as **pigs**. Nowadays we use **genetically engineered bacteria** to manufacture **human insulin**. Here's how the whole process works:

1) The **gene** for **human insulin** is **identified** and **isolated** using **restriction enzymes**.
2) A **plasmid** is **cut open** using the **same** restriction enzyme that was used to isolate the insulin gene.
3) The **insulin gene** is **inserted** into the **plasmid** (forming **recombinant DNA**).
4) The plasmid is **taken up** by bacteria and any **transformed** bacteria are **identified** using **marker genes**.
5) The bacteria are **grown** in a **fermenter** — **human insulin** is **produced** as the bacteria **grow** and **divide**.
6) The human insulin is **extracted** and **purified** so it can be **used in humans**.

These techniques are covered in more detail on page 150.

There are many **advantages** of using **genetically engineered human insulin** over **animal insulin**:

- It's **identical** to the insulin in our bodies, so it's **more effective** than animal insulin and there's **less risk** of an **allergic reaction**.
- It's **cheaper** and **faster** to produce than animal insulin, providing a **more reliable** and **larger supply** of insulin.
- Using genetically engineered insulin **overcomes** any **ethical** or **religious issues** arising from using animal insulin.

Transformed **Plants** can be Used to **Reduce Vitamin Deficiency**

Golden Rice is a type of **genetically engineered rice**. The rice is genetically engineered to contain a **gene** from a **daffodil plant** and a **gene** from a **soil bacterium**, which together enable the rice to produce **beta-carotene**. The beta-carotene is used by our bodies to produce **vitamin A**. *Golden Rice* is being developed to **reduce vitamin A deficiency** in areas where there's a **shortage** of **dietary vitamin A**, e.g. south Asia, parts of Africa. Here's how *Golden Rice* is produced:

1) The ***psy*** gene (from a daffodil) and the ***crtl*** gene (from the soil bacterium) are **isolated** using **restriction enzymes**.
2) A **plasmid** is **removed** from the ***Agrobacterium tumefaciens* bacterium** and **cut open** with the **same** restriction enzymes.
3) The *psy* and *crtl* genes and a **marker gene** are **inserted** into the plasmid.
4) The **recombinant plasmid** is **put back into** the bacterium.
5) **Rice plant cells** are incubated with the **transformed** *A. tumefaciens* bacteria, which **infect** the rice plant cells.
6) *A. tumefaciens* **inserts** the **genes** into the **plant cells' DNA**, creating **transformed rice plant cells**.
7) The rice plant cells are then grown on a **selective medium** — only transformed rice plants will be able to **grow** because they contain the marker gene that's needed to grow on this medium.

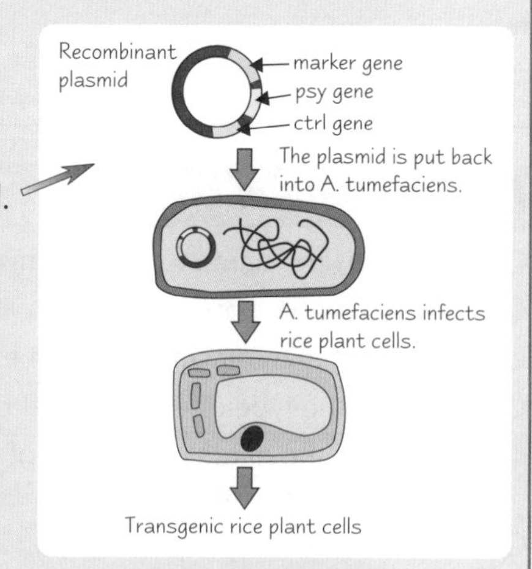

Transformed **Animals** can be Used to **Produce Organs** for **Transplant**

1) **Organ failure** (e.g. kidney or liver failure) may be **treated** with an **organ transplant**.
2) However, there's a **shortage** of **donor organs** available for transplant in the UK, which means many people **die** whilst **waiting** for a suitable donor organ.
3) **Xenotransplantation** is the **transfer** of **cells**, **tissues** or **organs** from **one species** to **another**.
4) It's hoped that xenotransplantation can be used to provide **animal donor organs** for **humans**.
5) With any form of transplantation there's a chance of **rejection** — the **immune system** of the **recipient recognises proteins** on the **surface** of the transplanted cells as **foreign** and starts an **immune response** against them.
6) Rejection is an **even greater** problem with xenotransplantation because the **genetic differences** between organisms of **different species** are even **greater** than between organisms of the same species.

Genetic Engineering

7) Scientists are trying to **genetically engineer animals** so that their **organs aren't rejected** when transplanted into humans. Here's how:

Xenotransplantation hasn't been carried out in humans yet, but there's lots of research being done on it.

1) Genes for HUMAN cell-surface proteins are INSERTED into the animal's DNA:

Human genes for **human cell-surface proteins** are **injected** into a **newly fertilised animal embryo**. The genes **integrate** into the **animal's DNA**. The animal then **produces human cell-surface proteins**, which reduces the risk of transplant rejection.

2) Genes for ANIMAL cell-surface proteins are 'KNOCKED OUT' — removed or inactivated:

- Animal genes involved in making cell-surface proteins are **removed** or **inactivated** in the **nucleus** of an **animal cell**. The nucleus is then **transferred** into an **unfertilised animal egg cell** (this is called **nuclear transfer**). The egg cell is then **stimulated** to **divide** into an embryo and the animal created **doesn't produce animal** cell-surface proteins, which reduces the risk of transplant rejection.
- For example, **pigs** have a sugar called **Gal-alpha(1,3)-Gal** attached to their cell-surface proteins, which humans don't. Scientists have developed a **knockout pig** that **doesn't produce** the **enzyme** needed to **make** this sugar.

There are Some *Ethical Issues* Surrounding *Genetic Engineering*

Genetic engineering can be used for **loads of things** other than producing insulin, reducing vitamin A deficiency and producing organs suitable for transplant from animals. For example it can be used to produce **pest-** or **herbicide-resistant crops** and **drugs** (and could even be used to genetically engineer **humans**). All these applications have ethical issues and concerns surrounding them:

1) Some people are worried that using **antibiotic-resistance** genes as **marker genes** may **increase** the number of **antibiotic-resistant**, **pathogenic** (disease-causing) **microorganisms** in our environment.
2) **Environmentalists** are worried that GM crops (like *Golden Rice*) may encourage **farmers** to carry out **monoculture** (where only one type of crop is planted). Monoculture **decreases biodiversity** and could leave the **whole crop vulnerable** to **disease**, because all the plants are **genetically identical**.
3) Some people are worried that genetically engineering **animals** for **xenotransplantation** may **cause them suffering**.
4) Some people are concerned about the possibility of '**superweeds**' — weeds that are **resistant** to **herbicides** because they've bred with **genetically engineered herbicide-resistant crops**.
5) Some people are concerned that large biotechnology companies may use GM crops to **exploit farmers** in **poor countries** — e.g. by selling them crops that they **can't** really **afford**.
6) Some people worry **humans** will be genetically engineered (e.g. to be more intelligent), creating a **genetic underclass**. This is currently **illegal** though.

Practice Questions

Q1 Give two advantages of using human insulin produced by genetic engineering compared to using animal insulin.

Q2 What is xenotransplantation?

Q3 How could xenotransplantation benefit humans?

Q4 Give two ethical issues surrounding genetic engineering.

Exam Questions

Q1 People with Type 1 diabetes need to inject insulin to regulate their blood glucose concentration.
Describe how human insulin can be made using genetically engineered bacteria. [6 marks]

Q2 *Golden Rice* is a type of transformed rice. Outline the process used to create *Golden Rice*. [7 marks]

If only they could knockout the gene for smelly feet...

...or the gene for freckles... or spots... or a big nose... or chubby ankles... the list is endless. Not that I'm vain or anything. Anyway, make sure you know all the processes in detail — it's no good just knowing that Golden Rice *is a genetically engineered crop, you need to know how it was genetically engineered too. So knuckle down and go over the page again...*

Genetic Fingerprinting

These pages are for AQA Unit 5 and Edexcel Unit 4. If you're doing OCR you can go straight to page 156.

A 100 years ago they were starting to identify people using their fingerprints, but now we can use their DNA instead.

Genomes Contain Repetitive, Non-Coding DNA Sequences

1) **Not all** of an organism's **genome** (all the genetic material in an organism) **codes** for **proteins**.
2) Some of the genome consists of **repetitive**, **non-coding base sequences** — base sequences that **don't** code for proteins and **repeat** next to each other over and over (sometimes thousands of times), e.g. CATGCATGCATGCATG is a repeat of the non-coding base sequence CATG.
3) The **number of times** these sequences are **repeated differs** from person to person, so the **length** of these sequences in nucleotides differs too. E.g. a **four** nucleotide sequence might be repeated **12 times** in one person = **48 nucleotides** (12 × 4), but repeated **16 times** in another person = **64 nucleotides** (16 × 4).
4) The repeated sequences occur in **lots of places** in the **genome**. The **number** of times a **sequence is repeated** (and so the number of nucleotides) at **different places** in their genome can be **compared** between **individuals** — this is called **genetic fingerprinting**.
5) The **probability** of **two individuals** having the **same** genetic fingerprint is **very low** because the **chance** of **two individuals** having the **same number** of sequence repeats at **each place** they're found in DNA is **very low**.

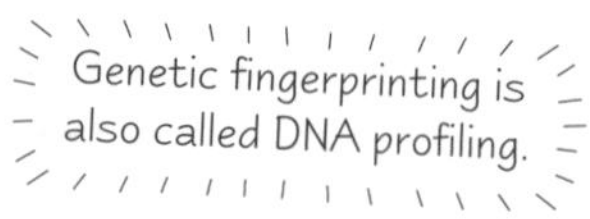

Electrophoresis Separates DNA Fragments to Make a Genetic Fingerprint

So **genetic fingerprints** can be **compared** between **different individuals**. Now you need to know how one is **made**:

1) A **sample** of **DNA** is obtained, e.g. from a person's **blood**, **saliva** etc.
2) **PCR** (see page 142) is used to make **many copies** of the **areas** of DNA that contain the repeated sequences — **primers** are used that bind to **either side** of these **repeats** and so the **whole** repeat is amplified.
3) You end up with **DNA fragments** where the **length** (in nucleotides) corresponds to the **number of repeats** the person has at that specific position, e.g. one person may have 80 nucleotides, another person 120.
4) A **fluorescent tag** is added to all the DNA fragments so they can be viewed under **UV light**.
5) The DNA fragments undergo **electrophoresis** (see page 144) to **separate** them by **size**.
6) The DNA fragments are viewed as **bands** under **UV light** — this is the **genetic fingerprint**.
7) Two genetic fingerprints can be **compared** — e.g. if both fingerprints have a band at the **same location** on the **gel** it means they have the **same number** of **nucleotides** and so the same number of **sequence repeats** at that locus — it's a **match**.

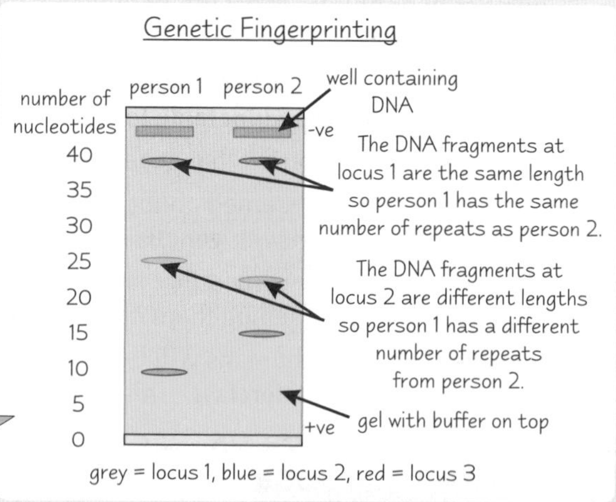

Genetic Fingerprinting is Used to Determine Relationships and Variability

Genetic fingerprinting has **many uses**, which include:

- **Determining genetic relationships** — We **inherit** the repetitive, non-coding base sequences from our **parents**. Roughly **half** of the sequences come from **each parent**. This means the **more bands** on a genetic fingerprint that match, the more **closely related** (**genetically similar**) two people are. E.g. **paternity tests** are used to determine the **biological father** of a child by comparing genetic fingerprints. If lots of bands on the fingerprint **match**, then that person is **most probably** the child's father. The **higher** the **number** of places in the genome compared, the more **accurate** the test result.
- **Determining genetic variability within a population** — The **greater** the **number of bands** that **don't** match on a genetic fingerprint, the more **genetically different** people are. This means you can **compare** the **number of repeats** at **several places** in the genome for a population to find out how **genetically varied** that population is. E.g. the **more** the **number of repeats** varies at **several places**, the **greater** the **genetic variability** within a population.

Genetic Fingerprinting

Genetic Fingerprinting can be Used in Forensic Science...

Forensic scientists use genetic fingerprinting to **compare** samples of **DNA** collected from **crime scenes** (e.g. DNA from **blood**, **semen**, **skin cells**, **saliva**, **hair** etc.) to samples of DNA from **possible suspects**, to **link them** to crime scenes.

1) The **DNA** is **isolated** from all the collected samples (from the crime scene and from the suspects).
2) Each sample is **replicated** using **PCR** (see p. 142).
3) The **PCR products** are run on an **electrophoresis gel** and the genetic fingerprints produced are **compared** to see if any match.
4) If the samples match, it **links** a **person** to the **crime scene**. E.g. this gel shows that the genetic fingerprint from **suspect C** **matches** that from the crime scene, **linking** them to the crime scene. All five bands match, so suspect C has the **same number** of repeats (nucleotides) at **five** different places.

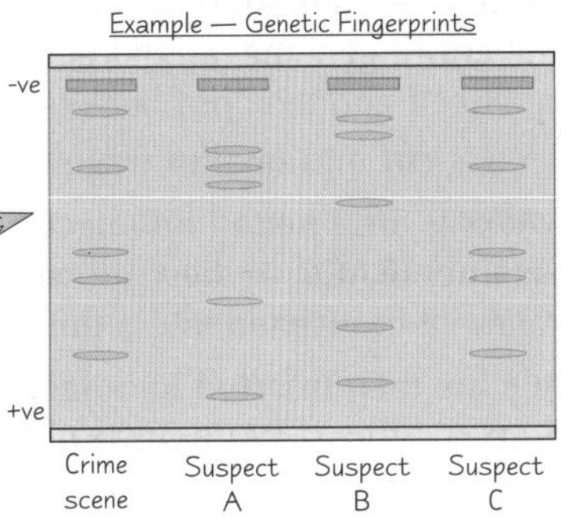

PCR amplifies the DNA, so enough is produced for it to be seen on the gel.

...and Animal and Plant Breeding

Genetic fingerprinting can be used on **animals** and **plants** to **prevent inbreeding**, which causes **health**, **productivity** and **reproductive problems**. Inbreeding **decreases** the **gene pool** (the number of **different alleles** in a population, see p. 74), which can lead to an **increased risk** of **genetic disorders**, leading to **health problems** etc. Genetic fingerprinting can be used to **identify** how **closely-related** individuals are — the **more closely-related** two individuals are, the **more similar** their genetic fingerprints will be (e.g. **more bands** will **match**). The **least related** individuals will be **bred together**.

Genetic Fingerprinting can Also be Used in Medical Diagnosis

AQA only

- In medical diagnosis, a genetic fingerprint can refer to a **unique pattern** of **several alleles**.
- It can be used to **diagnose genetic disorders** and **cancer**. It's useful when the **specific** mutation **isn't** known or where **several mutations** could have caused the disorder, because it identifies a **broader**, **altered** genetic pattern.

  EXAMPLE: **Preimplantation genetic haplotyping** (**PGH**) **screens embryos** created by **IVF** for genetic disorders **before** they're **implanted** into the uterus. The **faulty regions** of the **parents' DNA** are used to produce **genetic fingerprints**, which are **compared** to the genetic fingerprint of the **embryo**. If the fingerprints **match**, the embryo has **inherited** the **disorder**. It can be used to screen for **cystic fibrosis**, **Huntington's disease** etc.

  EXAMPLE: Genetic fingerprinting can be used to **diagnose sarcomas** (types of **tumour**). Conventional methods of identifying a tumour (e.g. biopsies) only show the **physical differences** between tumours. Now the **genetic fingerprint** of a known sarcoma (e.g. the **different mutated alleles**) can be **compared** to the genetic fingerprint of a **patient's tumour**. If there's a **match**, the sarcoma can be specifically **diagnosed** and the **treatment** can be targeted to that specific type (see page 159).

Practice Questions

Q1 Why are two people unlikely to have the same genetic fingerprint?

Q2 Why might genetic fingerprinting be used in forensic science?

Q3 How could genetic fingerprinting be used in medical diagnosis?

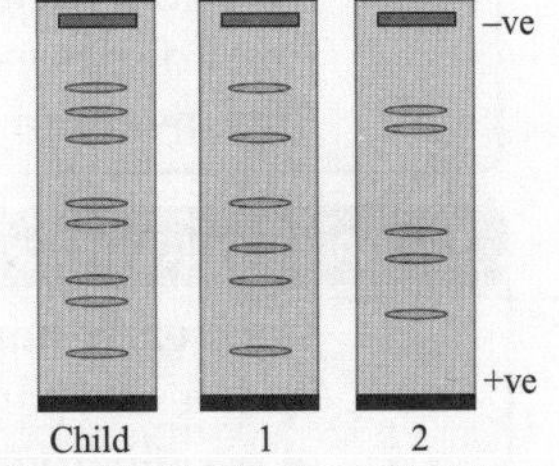

Exam Question

Q1 The diagram on the right shows three genetic fingerprints — one from a child and two from possible fathers.

a) Describe how the genetic fingerprint is made. [6 marks]

b) Which genetic fingerprint is most likely to be from the child's father? Explain your answer. [2 marks]

c) Give another use of genetic fingerprint technology. [1 mark]

Fingerprinting — in primary school it involved lots of paint and paper...

Who would have thought that tiny pieces of DNA on a gel would be that important? Well, they are and you need to know all about them. Make sure you know the theory behind fingerprinting as well as its applications. And remember, it's very unlikely that two people will have the same genetic fingerprint (except identical twins that is).

Sequencing Genomes and Restriction Mapping

This page is for OCR Unit 5 only. AQA — go to the next page, Edexcel — you've finished this section now.

We're gonna take gene technology to a whole new level now with some mighty genome sequencing...

A Whole Genome can be Sequenced Using BACs

The **chain-termination method** (see p. 144) can only be used for DNA fragments up to about **750 bp** long. So if you want to sequence the **entire genome** (all the DNA) of an organism, you need to chop it up into **smaller pieces** first. The smaller pieces are **sequenced** and then **put back in order** to give the sequence of the whole genome. Here's how it's done:

1) A genome is **cut** into **smaller fragments** (about 100 000 bp) using **restriction enzymes**.
2) The fragments are inserted into **bacterial artificial chromosomes** (**BACs**) — these are **man-made plasmids**. **Each** fragment in inserted into a **different BAC**.
3) The BACs are then **inserted** into **bacteria** — **each bacterium** contains a **BAC** with a **different DNA fragment**.
4) The bacteria **divide**, creating **colonies** of **cloned** (**identical**) cells that all contain a **specific DNA fragment**. Together the different colonies make a complete **genomic DNA library**.
5) **DNA** is **extracted** from **each colony** and **cut** up using restriction enzymes, producing **overlapping** pieces of DNA.
6) Each piece of DNA is **sequenced**, using the **chain-termination method**, and the pieces are **put back in order** to give the full sequence **from that BAC** (using **powerful computer systems**).
7) Finally the DNA fragments from **all the BACs** are **put back in order**, using computers, to **complete** the **entire genome**.

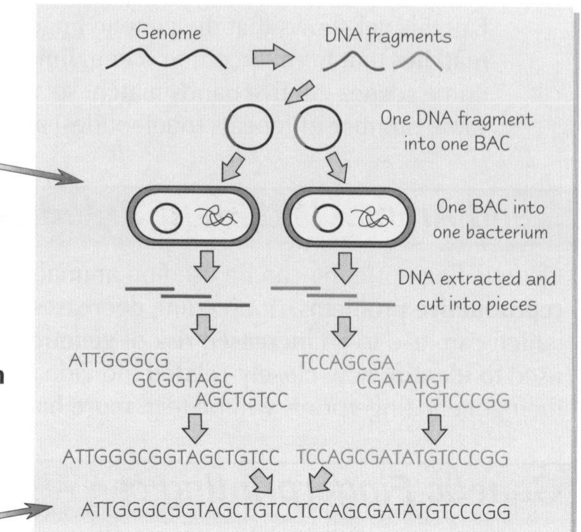

Sequenced Genomes can be Compared

Gene sequences and **whole genome** sequences can be compared **between** organisms of **different species** and between organisms of the **same species**. There are many reasons why we'd want to do this:

Comparing the genomes of DIFFERENT species can help us to:

1) Understand the **evolutionary relationships** between different species. **All** organisms **evolved** from **shared common ancestors** (relatives). **Closely related** species **evolved away** from each other more **recently** and so **share more DNA**. So DNA can tell us **how closely related** different species are. E.g. the genomes of **humans** and **chimpanzees** are about **94%** similar.
2) Understand the way in which **genes interact** during **development** and how they're **controlled**. For example, genome sequencing has shown that the **homeobox sequence** (see page 130) is the **same** in animals, plants and fungi. By studying how genes with the homeobox sequence work in the **Drosophila fruit fly** scientists can begin to piece together how they work in **humans** too.
3) Carry out **medical research**. **Human genes** that are associated with **disease**, like cancer or heart disease, can be found in the genomes of **other mammals**, such as mice and rats. This means mice or rats could be used as **animal models** for **research** into these diseases.

Comparing genomes of the SAME species can help us to:

1) Trace **early human migration**. For example, when different groups of early **humans separated** and **moved** to different parts of the world, their genomes **changed** in **slightly different ways**. By **comparing** the genomes of people from different parts of the world, it's possible to build up a picture of early human migration.
2) Study the **genetics** of **human diseases**. Some **gene mutations** have been **linked** to a **greater risk** of **disease** (e.g. mutations in the **BRCA1** gene are linked to **breast cancer**). Comparisons between the genomes of **sufferers** and **non-sufferers** can be used to **detect** particular **mutations** that could be responsible for the increased risk of disease.
3) Develop **medical treatments** for **particular genotypes**. The **same medicine** can be **more effective** in some patients than in others, which can be due to their **different genomes**. In the future, it may be possible to **sequence** a patient's genome so they can receive the **most effective medicine** for them.

Sequencing Genomes and Restriction Mapping

This page is for AQA Unit 5 only. If you're doing OCR you can go straight to the questions.

Restriction Mapping is used to put the Base Sequence of a Gene in Order

You **learnt** how to **sequence** a **gene** on page 144. But most genes are **too long** to be sequenced **all in one go**, so they're **cut** into **smaller sections** using **restriction enzymes**, then the smaller parts are sequenced. These smaller sections are then put back in the **correct order**, so the **entire gene sequence** can be **read** in the **right order** — **restriction mapping** can be used to do this:

1) **Different restriction enzymes** are used to **cut** labelled DNA into fragments (see page 143).
2) The DNA fragments are then **separated** by **electrophoresis** (see page 144).
3) The **size** of the **fragments** produced is used to **determine** the **relative locations** of **cut sites**.
4) A **restriction map** of the **original DNA** is made — a **diagram** of the piece of **DNA** showing the **different cut sites**, and so where the recognition sites of the restriction enzymes used are found.

Here's an example:

1) Some DNA, **10 kilobases** long (1 kb = **1000 nucleotides**) was **radioactively labelled**.
2) The DNA was **digested** using **two restriction enzymes**, ***Hind*III** and ***Eco*RI**, and the digested fragments were **separated** using **electrophoresis**.
3) The gel was used to build up a **restriction map** of the original DNA:

Size of fragments | Total digest | Partial digest | HindIII digest | EcoRI digest | Radioactive fragments (from total digest)
-ve
10 kb
8 kb
5 kb
3 kb
2 kb
+ve

a) The gel shows that the DNA was cut into **two fragments** by ***Hind*III**, so there's **one** *Hind*III **recognition sequence** in one of two places. But because the **2 kb piece** is **radioactive**, the label must be on the 2 kb piece. So the *Hind*III site must be 2 kb from the label.

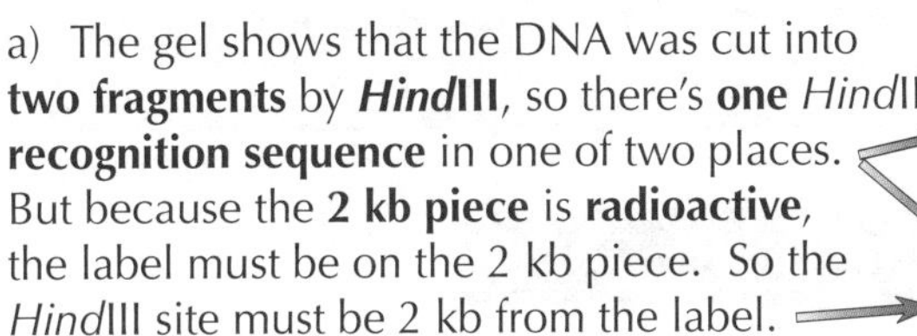

b) The gel shows that the DNA was cut into **two fragments** both **5 kb** long by ***Eco*RI**, so there's **one** *Eco*RI **recognition sequence** in the middle of the piece.

c) Finally, putting both of these together, the **complete restriction map** must be:

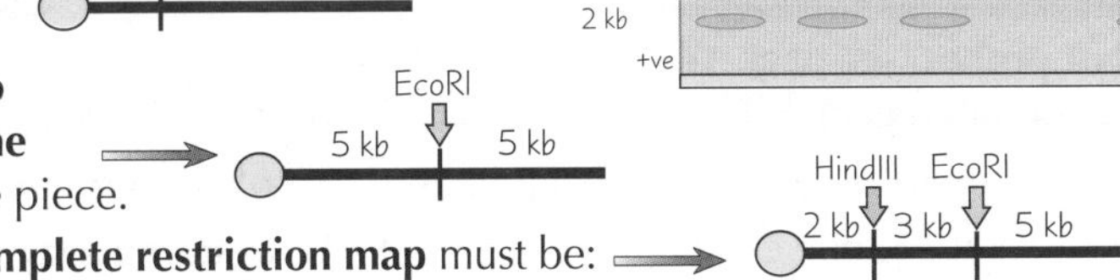

4) The restriction map matches the fragments of the **total digest** (where both enzymes are present and the DNA is cut at **all** of the **recognition sequences** present) — the **radioactive label** is on the **2 kb *Hind*III** piece.
5) A **partial digest** is where the restriction enzymes **haven't** been **left long enough** to **cut** at all of their **recognition sequences**, producing **fragments** of **other lengths**, e.g. if *Eco*RI **doesn't** cut there'll be an **8 kb** fragment produced.

Practice Questions

Q1 What is a bacterial artificial chromosome?

Q2 What is a restriction map?

Exam Questions

Q1 The genomes of over 200 different species have been sequenced.
Describe how a genome can be sequenced using BACs. [8 marks]

Size of fragments | Total digest | Partial digest | Radioactive fragments of partial digest
7 kb
6 kb
4 kb
3 kb
2 kb

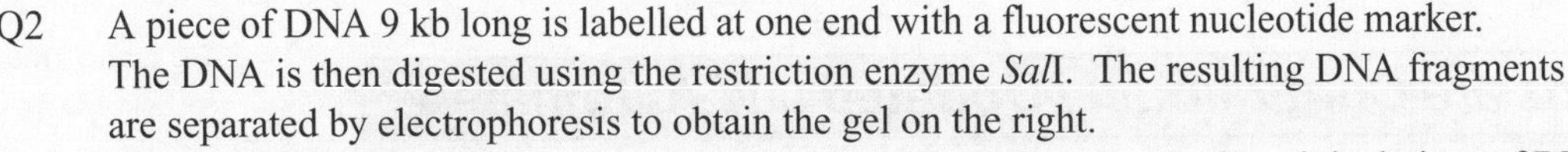

Q2 A piece of DNA 9 kb long is labelled at one end with a fluorescent nucleotide marker.
The DNA is then digested using the restriction enzyme *Sal*I. The resulting DNA fragments are separated by electrophoresis to obtain the gel on the right.

a) How many times did the recognition sequence for *Sal*I appear in the original piece of DNA? [1 mark]

b) Use the gel to produce a restriction map of the DNA piece. [3 marks]

c) Explain why there are more DNA fragments in the partial digest lane than in the total digest lane. [2 marks]

Restriction mapping — I'm not very good with coordinates...

Okay, those of you doing OCR might not get the gag, but I bet those of you doing AQA are in hysterics... Whichever board you're doing these pages are a bit tricky. But once you understand genome sequencing or restriction mapping they're the kind of techniques you won't forget. So put in the hard work now and you'll soon be thinking it's easy peasy.

DNA Probes in Medical Diagnosis

These pages are for AQA Unit 5 only. If you're doing OCR, you can go straight to page 160.

Some probes used in medical diagnosis are quite unpleasant, but luckily you only need to know about DNA probes...

Many **Human Diseases** are **Caused** by **Mutated Genes**

Some **mutated genes** can cause **diseases** such as **genetic disorders** and **cancer** (see page 135).
Other mutations can produce genes that are **useful** in **some situations** but **not** in others. For example:

Sickle-cell Anaemia

- Sickle-cell anaemia is a **recessive genetic disorder** caused by a **mutation** in the **haemoglobin gene**.
- The mutation causes an **altered haemoglobin protein** to be produced, which makes **red blood cells sickle-shaped** (**concave**).
- The **sickle** red blood cells **block capillaries** and **restrict blood flow**, causing **organ damage** and periods of **acute pain**.
- Some people are **carriers** of the disease — they have **one normal** and **one mutated copy** of the haemoglobin gene.
- Sickle-cell carriers are partially **protected** from **malaria**.
- This **advantageous** effect has caused an increase in the frequency of the sickle-cell **allele** (the mutated version of the gene) in areas where malaria is **common** (e.g. parts of Africa).
- However, this also **increases** the **likelihood** of people in those areas inheriting **two copies** of the sickle-cell allele, which means more people will **suffer** from the disease in these areas.

DNA Probes Can be Used to **Screen** for **Mutated Genes**

DNA probes (see page 145) can be used to look (**screen**) for clinically important genes (e.g. **mutated genes** that result in **genetic disorders**). There are two ways to do this:

1) The probe can be **labelled** and used to look for a **single gene** in a sample of DNA, as shown on page 145.
2) Or the probe can be used as part of a **DNA microarray**, which can screen for **lots** of **genes** at the **same time**:

- A **DNA microarray** is a **glass slide** with **microscopic spots** of **different** DNA probes **attached** to it in **rows**.
- A sample of **labelled human DNA** is washed over the array.
- If the labelled human DNA **contains** any **DNA sequences** that **match** any of the **probes**, it will **stick** to the array.
- The array is **washed**, to remove any labelled DNA that **hasn't** stuck to it.
- The array is then **visualised** under **UV light** — any **labelled DNA attached** to a probe will **show up** (fluoresce).
- Any spot that fluoresces means that the person's DNA **contains** that specific **gene**. E.g. if the probe is for a mutated gene that causes a **genetic disorder**, this person has the gene and so **has** the disorder.

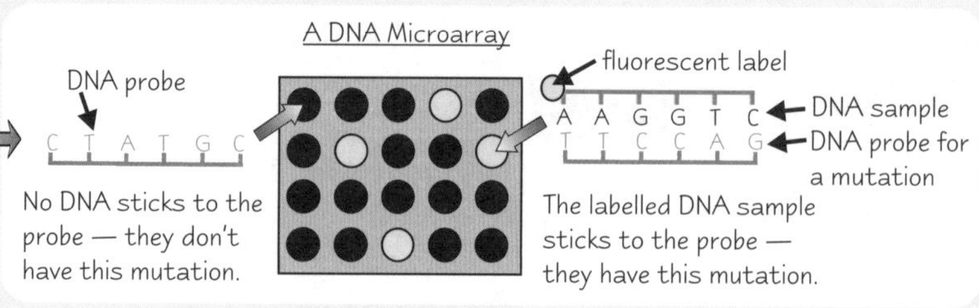

You also need to know how to **produce** a DNA probe — first the **gene** that you want to screen for is **sequenced** (see p. 144). Then **PCR** (see p. 142) is used to produce **multiple copies** of **part** of the gene — these are the **probes**.

Scientific Methods are **Continuously Updated** and **Automated**

1) In the **past**, some of the gene technologies you've learnt about on the past few pages were **labour-intensive**, **expensive** and could only be done on a **small scale**.
2) Now these techniques are often **automated**, more **cost-effective** and can be done on a **large scale**.
3) For example, using a single **probe** to **screen** for a single **mutated gene** (see page 145) is **slow** and **small-scale**. Now we have **DNA microarrays** — they're **quick** and can screen for **thousands** of genes at once (see above).
4) Scientific methods like this are **constantly** being **updated** and **automated** to be **faster**, **cheaper**, and **more accurate** (because there's less human error). This means **medical diagnoses** become **faster** and **more accurate**.

DNA Probes in Medical Diagnosis

The **Results** of **Screening** can be used for **Genetic Counselling**...

1) **Genetic counselling** is **advising patients** and their **relatives** about the **risks** of **genetic disorders**.
2) It involves **advising** people about **screening** (e.g. looking for mutated genes if there's a **history** of **cancer**) and **explaining** the **results** of a screening. Screening can help to **identify** the **carrier** of a gene, the **type of mutated gene** they're carrying (indicating the type of genetic disorder or cancer) and the **most effective treatment**.
3) If the results of a screening are **positive** (an individual **has** the mutation) then genetic counselling is used to advise the patient on the **options** of **prevention** or **treatment** available. Here are two examples:

   EXAMPLE: A **woman** with a family history of **breast cancer** may have **genetic counselling** to help her **decide** whether or not to be **screened** for **known mutations** that can lead to breast cancer, e.g. a mutation in the BRCA1 **tumour suppressor gene** (see p. 135). If she is screened and the result is **positive**, genetic counsellors might explain that a woman with the mutated BRCA1 gene has an **85%** chance of developing **breast cancer** before the age of **70**. Counselling can also help the woman to **decide** if she wants to take **preventative** measures, e.g. a **mastectomy**, to prevent breast cancer developing.

   EXAMPLE: A couple who are **both carriers** of the **sickle-cell allele** (see previous page) may **like** to have **kids**. They may undergo genetic counselling to help them **understand** the **chances** of them having a child with sickle-cell anaemia (**one in four**). Genetic counselling also provides **unbiased advice** on the possibility of having **IVF** and **screening** their **embryos** for the alleles, so embryos **without the mutation** are **implanted** in the womb. It could also provide information on the **help** and **drugs** available if they have a child with sickle-cell anaemia.

...and **Deciding Treatment**

Cancers can be caused by **mutations** in **proto-oncogenes** (forming **oncogenes**) and mutations in **tumour suppressor genes** (see page 135). **Different mutations** cause **different cancers**, which **respond** to **treatment** in **different ways**. **Screening** using DNA probes for **specific** mutations can be used to help decide the **best** course of **treatment**. For example:

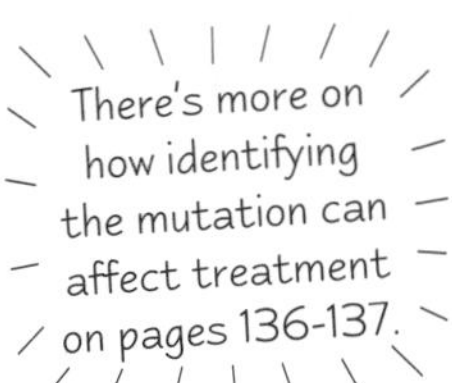

There's more on
how identifying
the mutation can
affect treatment
on pages 136-137.

Breast cancer can be caused by a **mutation** in the **HER2 proto-oncogene**. If a patient with breast cancer is screened and tests **positive** for the HER2 oncogene, they can be treated with the drug **Herceptin®**. This drug **binds** specifically to the **altered HER2 protein receptor** and **suppresses cell division**. Herceptin® is **only effective** against this type of breast cancer because it **only targets** the altered HER2 receptor. Studies have shown that **targeted treatment** like this, alongside less-specific treatment like chemotherapy, can **increase survival rate** and **decrease relapse rate** from breast cancer.

Practice Questions

Q1 Give an example of a mutation that is useful in one way but not in another.

Q2 Give an example of a scientific technique that has been automated.

Q3 What is genetic counselling?

Q4 Describe one situation where genetic counselling may be needed.

Exam Questions

Q1 a) Briefly describe how a DNA probe for a clinically important gene can be produced. [2 marks]

b) Describe how you could screen a person for this gene and many others at the same time. [4 marks]

Q2 Annette has colon cancer. A drug called Cetuximab is used to treat colon cancer caused by a mutation in the KRAS proto-oncogene. Annette is screened and tests negative for the KRAS oncogene.

a) Why is it unlikely that Annette will be treated with Cetuximab? [1 mark]

b) Suggest why Annette will undergo genetic counselling. [2 marks]

Information probes — they're called exams...

All of the techniques you've learnt earlier in this section (PCR, sequencing, DNA probes) come together nicely in this medical diagnosis stuff — it's good to know that what you've learnt has a point to it. It's also good to know that as techniques improve, better ways to diagnose some diseases are found.

Gene Therapy

These pages are for AQA Unit 5 and OCR Unit 5.

Congratulations — you've made it to the last two pages of the section. I guess I'd better make them good 'uns then...

Gene Therapy *Could be Used to* ***Treat*** *or* ***Cure Genetic Disorders*** *and* ***Cancer***

How it works:

1) Gene therapy involves **altering** the **defective genes** (mutated alleles) inside cells to treat **genetic disorders** and **cancer**.
2) How you do this depends on whether the disorder is caused by a mutated **dominant allele** or two mutated **recessive alleles** (see page 62):
   - If it's caused by two mutated **recessive** alleles you can **add** a working **dominant allele** to make up for them (you '**supplement**' the faulty ones).
   - If it's caused by a mutated **dominant** allele you can '**silence**' the **dominant allele** (e.g. by sticking a bit of DNA in the middle of the allele so it doesn't work any more).

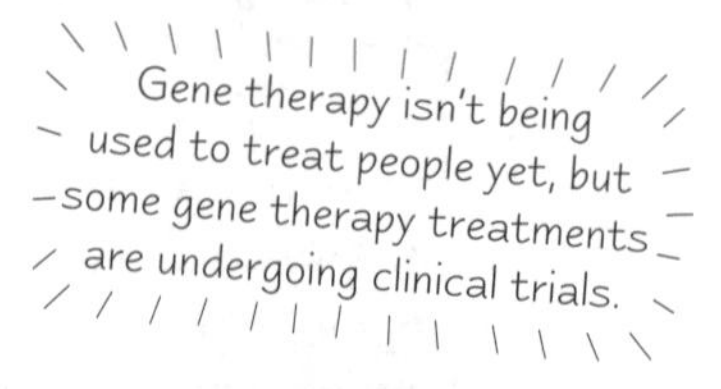

Gene therapy isn't being used to treat people yet, but some gene therapy treatments are undergoing clinical trials.

A DNA-filled doughnut — surely the best way to deliver new alleles...

How you get the 'new' allele (DNA) inside the cell:

1) The allele is **inserted into cells** using **vectors**.
2) Different **vectors** can be used, e.g. altered **viruses**, **plasmids** or **liposomes** (spheres made of lipid).

There are two types of gene therapy:

1) **Somatic therapy** — this involves **altering** the **alleles** in **body cells**, particularly the cells that are **most affected** by the disorder. For example, **cystic fibrosis** (CF) is a genetic disorder that's very **damaging** to the **respiratory system**, so somatic therapy for CF **targets** the epithelial cells lining the lungs. Somatic therapy doesn't affect the individual's **sex cells** (sperm or eggs) though, so any **offspring** could still **inherit** the disease.
2) **Germ line therapy** — this involves **altering** the **alleles** in the **sex cells**. This means that **every cell** of **any offspring** produced from these cells will be **affected** by the gene therapy and they **won't suffer from the disease**. Germ line therapy in humans is currently **illegal** though.

There are ***Advantages*** *and* ***Disadvantages*** *to* ***Gene Therapy***

You need to be able to **evaluate** the **effectiveness** of **gene therapy** — this means being able to discuss the **advantages** and **disadvantages** of the technique, some of which are given in the table below:

ADVANTAGES	DISADVANTAGES
It could prolong the lives of people with genetic disorders and cancer.	The effects of the treatment may be short-lived (only in somatic therapy).
It could give people with genetic disorders and cancer a better quality of life.	The patient might have to undergo multiple treatments (only in somatic therapy).
Carriers of genetic disorders might be able to conceive a baby without that disorder or risk of cancer (only in germ line therapy).	It might be difficult to get the allele into specific body cells.
It could decrease the number of people that suffer from genetic disorders and cancer (only in germ line therapy).	The body could identify vectors as foreign bodies and start an immune response against them.
	An allele could be inserted into the wrong place in the DNA, possibly causing more problems, e.g. cancer.
	An inserted allele could get overexpressed, producing too much of the missing protein.
	Disorders caused by multiple genes (e.g. cancer) would be difficult to treat with this technique.

There are also many **ethical issues** associated with gene therapy. For example, some people are worried that the technology could be used in ways **other** than for **medical treatment**, such as for treating the **cosmetic effects** of **aging**. Other people worry that there's the potential to do **more harm** than good by using the technology (e.g. risk of overexpression of genes — see table). There's also the concern that gene therapy is **expensive** — some people believe that **health service resources** could be **better spent** on other treatments that have passed clinical trials.

Gene Therapy

You Might have to *Interpret* Some *Data* on the *Effectiveness* of *Gene Therapy*

In the **exam**, you could get a data **question** on the **effectiveness** of **gene therapy**. So, here's one I did earlier:

Background:
X-linked severe combined immunodeficiency disease (X-linked SCID) is an **inherited disorder** affecting the **immune system**. The disorder is caused by a **mutation** in the **IL2RG gene**, located on the **X chromosome**. The IL2RG gene codes for a **protein** that's **essential** for the **development** of some immune system cells, so the sufferer is **vulnerable** to **infectious diseases** and many **die in infancy**.

The study:
A study was designed to investigate the **effectiveness** of **gene therapy** in patients with **X-linked SCID**. **Ten patients** were treated with a **virus vector** carrying a **correct version** of the **IL2RG gene**. After gene transfer, the **patient's immune system** was **monitored** for **at least three years** and noted as **functional** (good) or not. Their **health** was also **monitored** for the same time. Bar charts of the results are shown on the right.

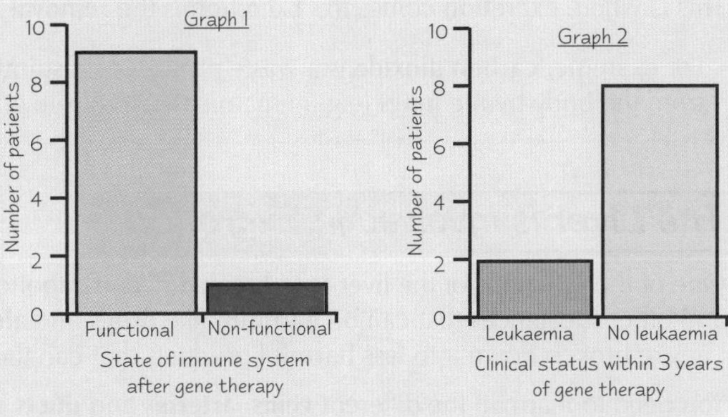

You could be asked to describe the data...

- The **first graph** shows that **nine out of the ten** patients had a **functional** immune system **after** gene therapy.
- The **second graph** shows that **two out of the ten** patients developed **leukaemia** within 3 years of the treatment.

...or draw conclusions...

- **Gene therapy** can be **used** to **correct** the symptoms of **X-linked SCID**, i.e. produce a **functioning immune system**. However, you **can't** say gene therapy can **cure** X-linked SCID as the study **doesn't** say **how long** the effects of the treatment lasted for.
- **Two out of the ten** patients developed **leukaemia** after the treatment, so there's a chance it's **linked** to the gene therapy (e.g. the vector could have **inserted** the gene into a **proto-oncogene** or **tumour suppressor**, see page 135). But you can only **suggest** a link as **other factors** may have been involved. For example, the patients could have been **more genetically predisposed** to develop cancer.

...or evaluate the methodology

- The **sample size** is **small** — only **ten** patients were treated. This makes the results **less reliable**.

Practice Questions

Q1 How could gene therapy be used to supplement mutated recessive alleles?

Q2 How are supplementary alleles added to human DNA?

Q3 What does germ line gene therapy involve?

Exam Questions

Q1 A patient suffering from cystic fibrosis was offered gene therapy targeted at his lung epithelial cells to help treat the disease.

a) What does gene therapy involve? [1 mark]

b) What type of gene therapy was the patient offered? [1 mark]

Q2 Give three possible disadvantages of somatic gene therapy. [3 marks]

Jean therapy — the washing machine is traumatic for any pair of jeans...

Now, you might think you need some therapy after this section, but don't worry, the buzzing in your head is normal — it's due to information overload. So go get yourself a cuppa and a biccie and have a break. Then go over some of the difficult bits in this section again. Believe me, the more times you go over it the more things will click into place.

The Liver and Excretion

This section is for OCR Unit 4 only.

Liver — not just what your grandparents eat with onions. The liver has loads of functions, but the one you need to know about is its job in excretion. It's great at breaking things down like excess amino acids and other harmful substances.

Excretion is the Removal of Waste Products from the Body

All the **chemical reactions** that happen in your cells make up your **metabolism**. Metabolism produces **waste products** — substances that **aren't needed** by the cells, such as **carbon dioxide** and **nitrogenous** (nitrogen-containing) **waste**. Many of these products are **toxic**, so if they were allowed to **build up** in the body they would cause **damage**. This is where **excretion** comes in. Excretion is the **removal** of the **waste products of metabolism** from the body.

For example, **carbon dioxide** is a waste product of **respiration**. **Too much** in the blood is toxic, so it's removed from the body by the **lungs** (e.g. in mammals) or **gills** (e.g. in fish). The lungs and gills act as **excretory organs**.

The Liver is Involved in Excretion

One of the functions of the **liver** is to **break down** metabolic waste products and other substances that can be **harmful**, like **drugs** and **alcohol**. They're broken down into **less harmful** products that can then be **excreted**.

You need to learn all the different **veins**, **arteries** and **ducts** connected to the liver:

1) The **hepatic artery** supplies the liver with **oxygenated blood** from the heart, so the liver has a good supply of **oxygen** for **respiration**, providing plenty of **energy**.
2) The **hepatic vein** takes **deoxygenated blood** away from the liver.
3) The **hepatic portal vein** brings blood from the **duodenum** and **ileum** (parts of the small intestine), so it's rich in the products of **digestion**. This means any ingested harmful substances are **filtered out** and **broken down straight away**.
4) The **bile duct** takes **bile** (a substance produced by the liver to **emulsify fats**) to the **gall bladder** to be **stored**.

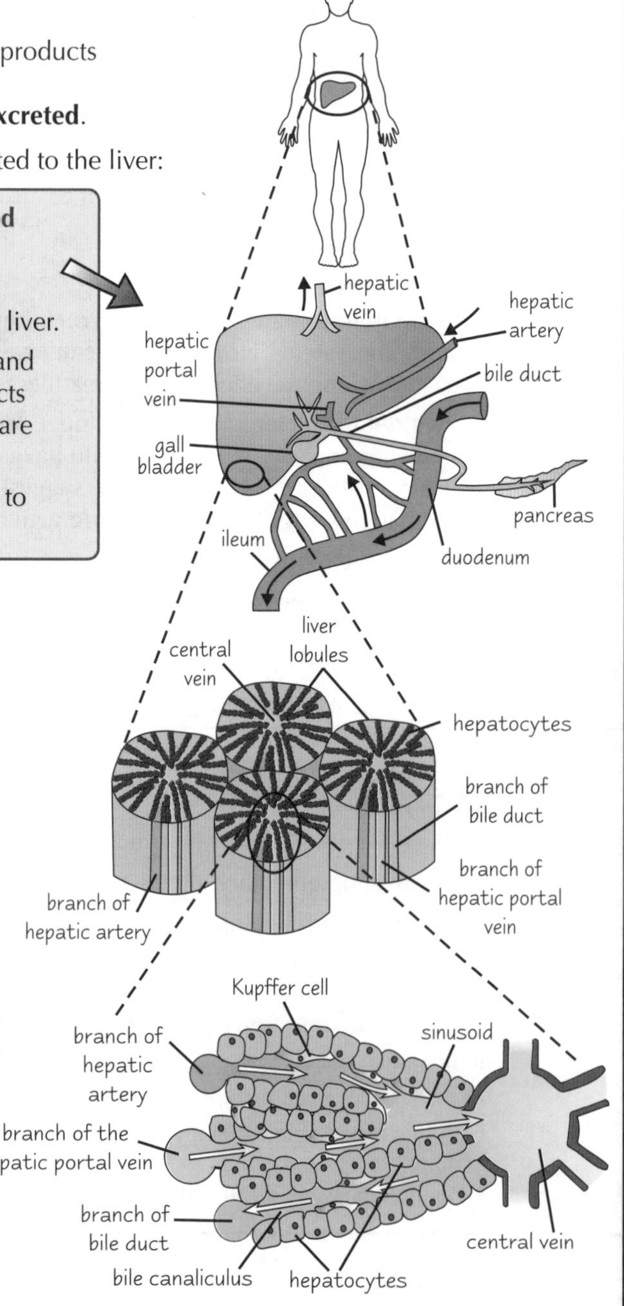

You need to learn about the **structure** of the liver too:

1) The liver is made up of **liver lobules** — **cylindrical** structures made of **cells** called **hepatocytes** that are arranged in rows **radiating** out from the centre.
2) Each lobule has a **central vein** in the middle that connects to the **hepatic vein**. **Many branches** of the **hepatic artery**, **hepatic portal vein** and **bile duct** are also found connected to each lobule (only one of each is shown in the picture).
3) The **hepatic artery** and the **hepatic portal vein** are connected to the **central vein** by **capillaries** called **sinusoids**.
4) Blood runs **through** the sinusoids, past the hepatocytes that **remove harmful substances** and **oxygen** from the blood.
5) The harmful substances are **broken down** by the hepatocytes into **less harmful** substances that then **re-enter** the blood.
6) The blood runs to the **central vein**, and the central veins from all the lobules **connect** up to form the **hepatic vein**.
7) Cells called **Kupffer cells** are also attached to the walls of the sinusoids. They **remove bacteria** and **break down** old **red blood cells**.
8) The **bile duct** is connected to the **central vein** by **tubes** called **canaliculi**. 

The Liver and Excretion

Excess Amino Acids are Broken Down by the Liver

One of the liver's most important roles is getting rid of **excess amino acids** produced by eating and **digesting protein**. **Amino acids** contain **nitrogen** in their **amino groups**. **Nitrogenous substances can't** usually be **stored** by the body. This means **excess** amino acids can be **damaging** to the body, so they must be **used** by the body (e.g. to make proteins) or be **broken down and excreted**. Here's how excess amino acids are **broken down** in the **liver**:

1) First, the nitrogen-containing **amino groups** are **removed** from any **excess** amino acids, forming **ammonia** and **organic acids** — this process is called **deamination**.
2) The organic acids can be **respired** to give **ATP** or converted to **carbohydrate** and stored as **glycogen**.
3) Ammonia is **too toxic** for mammals to excrete directly, so it's **combined** with CO_2 in the **ornithine cycle** to create **urea**.
4) The urea is **released** from the liver into the **blood**. The **kidneys** then **filter** the blood and **remove** the urea as **urine** (see p. 164-165), which is excreted from the body.

Josie felt that warm feeling that meant a little bit of urea had just slipped out.

*The **Liver** Removes Other **Harmful Substances** from the **Blood***

The **liver** also breaks down other harmful substances, like **alcohol**, **drugs** and **unwanted hormones**. They're broken down into **less harmful compounds** that can then be **excreted** from the body — this process is called **detoxification**. Some of the harmful products broken down by the liver include:

1) **Alcohol** (**ethanol**) — a **toxic** substance that can **damage** cells. It's **broken down** by the liver into **ethanal**, which is then broken down into a **less harmful** substance called **acetic acid**. **Excess** alcohol over a long period can lead to **cirrhosis** of the liver — this is when the cells of the liver **die** and **scar tissue blocks blood flow**.

2) **Paracetamol** — a common painkiller that's **broken down** by the liver. **Excess** paracetamol in the blood can lead to **liver** and **kidney failure**.

3) **Insulin** — a **hormone** that controls **blood glucose concentration** (see page 116). Insulin is also broken down by the liver as excess insulin can cause problems with blood sugar levels.

Practice Questions

Q1 Define excretion.

Q2 Why is excretion needed?

Q3 Which blood vessel brings oxygenated blood to the liver?

Q4 Name the blood vessel that brings blood to the liver from the small intestine.

Q5 What are liver lobules?

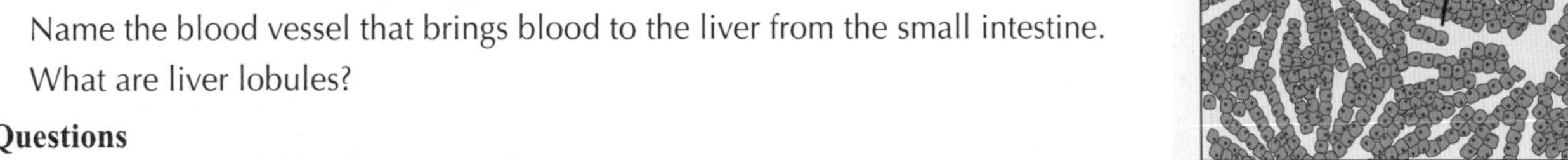

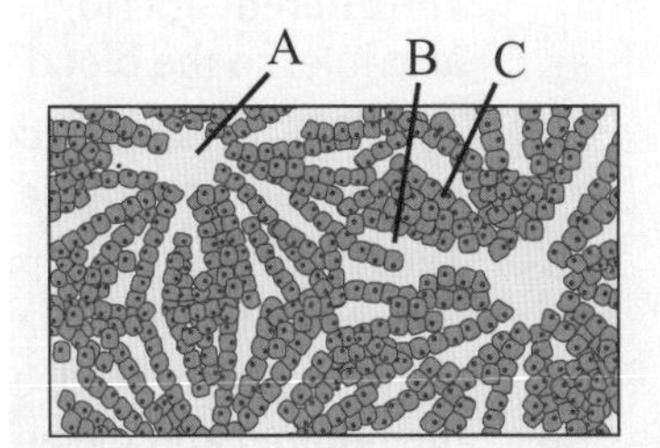

Exam Questions

Q1 Name the parts of the liver shown in the diagram on the right. [3 marks]

Q2 Explain why the concentration of urea in urine might increase after eating a meal that's rich in protein. [6 marks]

Lots of important functions — can't liver without it...

Poor little amino acids, doing no harm then suddenly they're broken down and excreted. As upsetting as it is, however, you need to learn how they're broken down in the liver. It's a heart-wrenching tale of separation — the amino group and the organic acid are torn from each other's life. Right, enough of that nonsense. Learn it and learn it good.

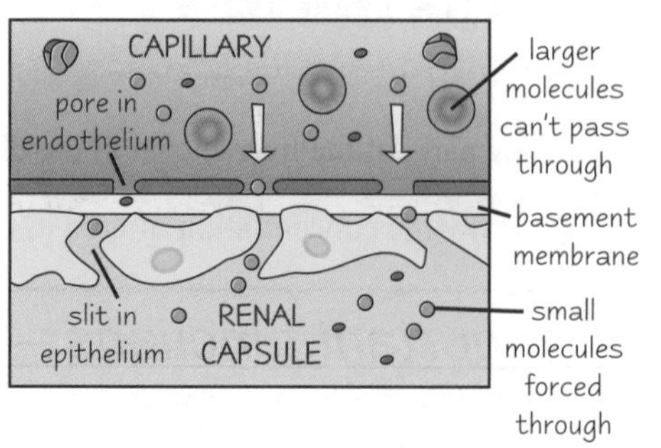

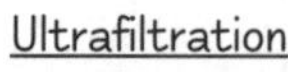

The Kidneys and Excretion

These pages are for OCR Unit 4 only.

So you've learnt about how the liver does a pretty good job at breaking down stuff for excretion. Now you get to learn that the kidneys like to play a part in this excretion malarkey too...

The Kidneys are Organs of Excretion

One of the main **functions** of the **kidneys** is to **excrete waste products**, e.g. **urea** produced by the **liver**. They also **regulate** the body's **water content** (see p. 166-167). Here's an overview of how they excrete waste products (you need to **learn** the **structure** of the kidneys too):

1) Blood **enters** the kidney through the **renal artery** and then passes through **capillaries** in the **cortex** of the kidneys.
2) As the blood passes through the capillaries, **substances** are **filtered out of the blood** and into **long tubules** that surround the capillaries. This process is called **ultrafiltration** (see below).
3) **Useful substances** (e.g. glucose) are **reabsorbed** back into the blood from the tubules in the **medulla** — this is called **selective reabsorption** (see next page).
4) The remaining **unwanted substances** (e.g. urea) pass along the tubules, then along the **ureter** to the **bladder**, where they're **expelled** as **urine**.
5) The filtered blood passes out of the kidneys through the **renal vein**.

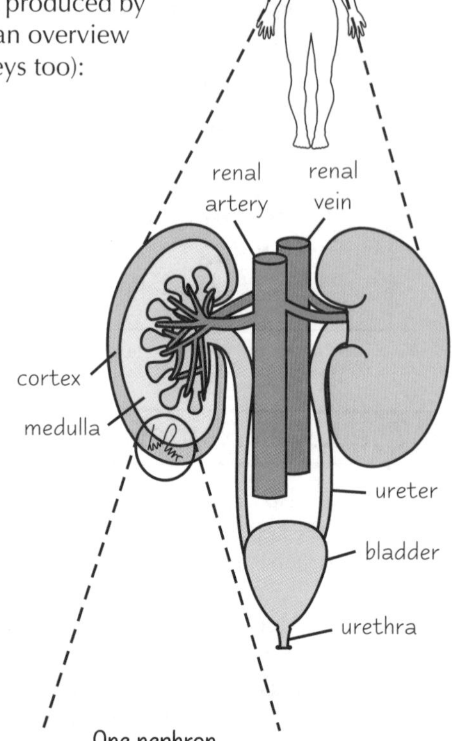

Blood is Filtered at the Start of the Nephrons

The **long tubules** along with the bundle of **capillaries** where the blood is **filtered** are called **nephrons** — there are **thousands** of nephrons in each kidney.

1) Blood from the **renal artery** enters smaller **arterioles** in the **cortex**.
2) Each arteriole splits into a structure called a **glomerulus** — a **bundle** of **capillaries** looped inside a hollow ball called a **renal capsule** (or **Bowman's capsule**).
3) This is where **ultrafiltration** takes place.
4) The **arteriole** that takes blood **into** each glomerulus is called the **afferent** arteriole, and the arteriole that takes the filtered blood **away** from the glomerulus is called the **efferent** arteriole.
5) The **efferent** arteriole is **smaller** in **diameter** than the afferent arteriole, so the blood in the glomerulus is under **high pressure**.
6) The high pressure **forces liquid** and **small molecules** in the blood **out** of the **capillary** and **into** the **renal capsule**.
7) The liquid and small molecules pass through **three** layers to get into the renal capsule and **enter** the nephron **tubules** — the **capillary wall**, a membrane (called the **basement membrane**) and the **epithelium** of the renal capsule. Larger molecules like **proteins** and **blood cells** **can't pass through** and **stay** in the blood.
8) The liquid and small molecules, now called **filtrate**, pass along the rest of the nephron and **useful substances** are **reabsorbed** along the way — see next page.
9) Finally, the filtrate flows through the **collecting duct** and passes out of the kidney along the **ureter**.

One nephron

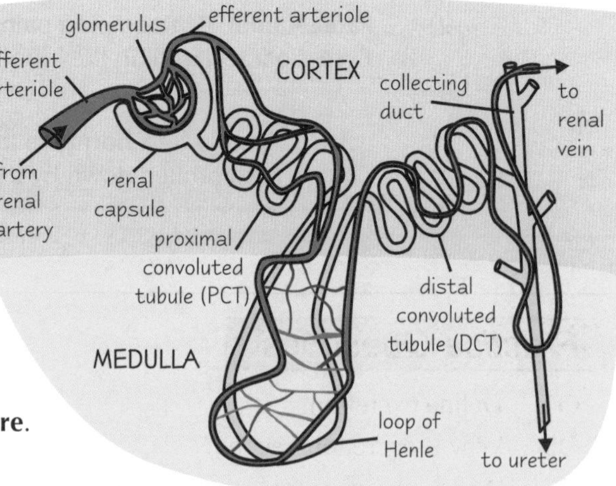

Ultrafiltration

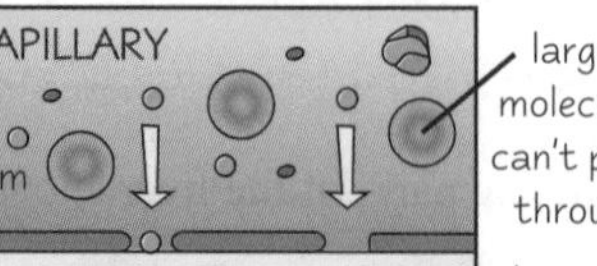

The Kidneys and Excretion

Useful Substances are Reabsorbed Along the Nephron Tubules

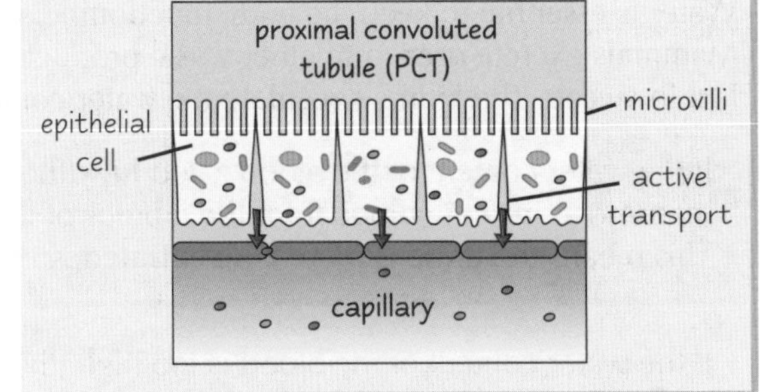

1) **Selective reabsorption** takes place as the filtrate flows along the **proximal convoluted tubule** (**PCT**), through the **loop of Henle**, and along the **distal convoluted tubule** (**DCT**).
2) **Useful substances** leave the tubules of the nephrons and **enter** the capillary network that's **wrapped** around them (see diagram on previous page).
3) The **epithelium** of the wall of the PCT has **microvilli** to provide a **large surface area** for the **reabsorption** of useful materials from the **filtrate** (in the tubules) into the **blood** (in the capillaries).
4) Useful solutes like **glucose**, **amino acids**, **vitamins** and some **salts** are reabsorbed along the PCT by **active transport** and **facilitated diffusion**.
5) Some **urea** is also reabsorbed by **diffusion**.
6) **Water** enters the blood by **osmosis** because the **water potential** of the blood is **lower** than that of the filtrate. Water is reabsorbed from the **loop of Henle**, **DCT** and the **collecting duct** (see next page).
7) The filtrate that remains is **urine**, which passes along the **ureter** to the **bladder**.

proximal convoluted tubule (PCT)
epithelial cell
microvilli
active transport
capillary

Water potential basically describes the tendency of water to move from one area to another. Water will move from an area of higher water potential to an area of lower water potential — it moves down the water potential gradient.

Urine is usually **made up of:**
- **Water** and **dissolved salts.**
- **Urea.**
- Other substances such as **hormones** and **excess vitamins.**

The volume of water in urine varies depending on how much you've drunk (see p. 166). The amount of urea also varies depending on how much protein you've eaten (see p. 163).

Urine **doesn't** usually contain:
- **Proteins** and **blood cells** — they're **too big** to be **filtered out** of the blood.
- **Glucose**, **amino acids** and **vitamins** — they're **actively reabsorbed** back into the blood (see above).

Ali was going to selectively absorb all the green jelly beans.

Practice Questions

Q1 Which blood vessel supplies the kidney with blood?

Q2 What are the bundles of capillaries found in the cortex of the kidneys called?

Q3 What is selective reabsorption?

Q4 Why aren't proteins normally found in urine?

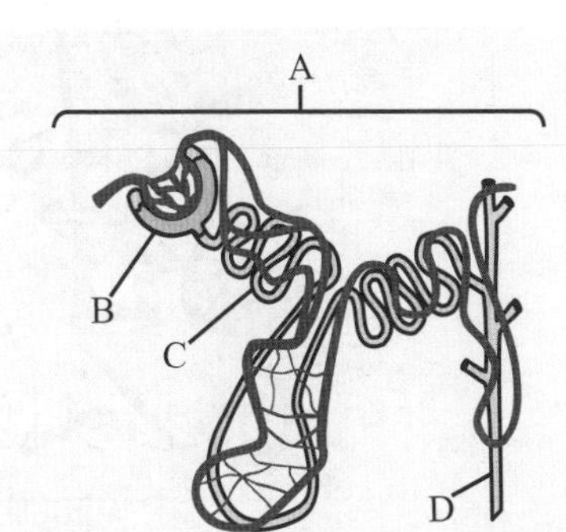

Exam Question

Q1 a) Name the structures labelled A-D shown in the diagram. [4 marks]

b) In which structure (B-D) does ultrafiltration take place? [1 mark]

c) Describe and explain the process of ultrafiltration. [5 marks]

Mmm — it's steak and excretion organ pie for dinner...

Excretion is a pretty horrible sounding word I know, but it's gotta be done. Speaking of horrible, I've never been able to eat kidney ever since I learnt all about this urine production business. Shame really because I used to love it sooooo much — I'd have kidneys on toast for breakfast, kidney sandwiches for lunch, kidney soup for tea, and kidney ice cream for pudding.

Controlling Water Content

These pages are for OCR Unit 4 only.

More lovely kidney to gobble up on these pages — this time it's their role in controlling the water content of the blood. Busy things, these kidneys.

The Kidneys Regulate the Water Content of the Blood

Water is **essential** to keep the body **functioning**, so the **amount** of water in the **blood** needs to be kept **constant**. Mammals excrete **urea** (and other waste products) in **solution**, which means **water** is **lost** during excretion. Water is also lost in **sweat**. The kidneys **regulate** the water content of the blood (and urine), so the body has just the **right amount**:

If the water content of the blood is too **low** (the body is **dehydrated**), **more** water is **reabsorbed** by osmosis **into** the blood from the tubules of the nephrons (see p. 164-165 for more). This means the urine is **more concentrated**, so **less** water is **lost** during excretion.

If the water content of the blood is too **high** (the body is too **hydrated**), **less** water is **reabsorbed** by osmosis **into** the blood from the tubules of the nephrons. This means the urine is **more dilute**, so **more** water is **lost** during excretion (see next page).

Brad liked his urine to be dilute.

Regulation of the water content of the blood takes place in the **middle** and **last parts** of the nephron — the **loop of Henle**, the **distal convoluted tubule** (DCT) and the **collecting duct** (see below). The **volume** of water reabsorbed is controlled by **hormones** (see next page).

The Loop of Henle has a Countercurrent Multiplier Mechanism

The **loop of Henle** is made up of two '**limbs**' — the **descending** limb and the **ascending** limb. They help set up a mechanism called the **countercurrent multiplier mechanism**. It's this mechanism that helps to **reabsorb water** back into the blood. Here's how it **works**:

KEY
- Water moves out by osmosis
- Na^+ and Cl^- move by diffusion
- Na^+ and Cl^- move by active transport

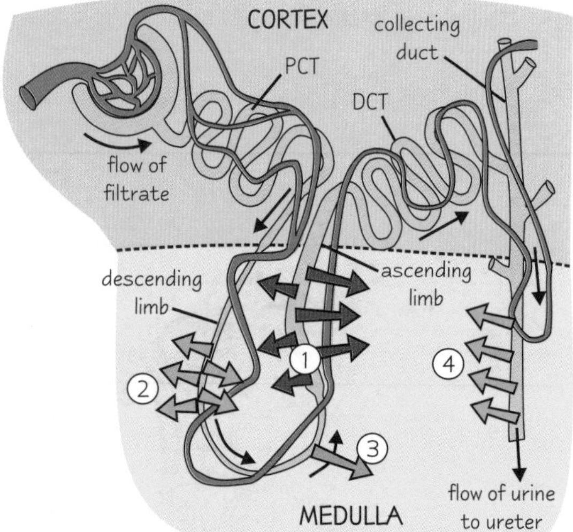

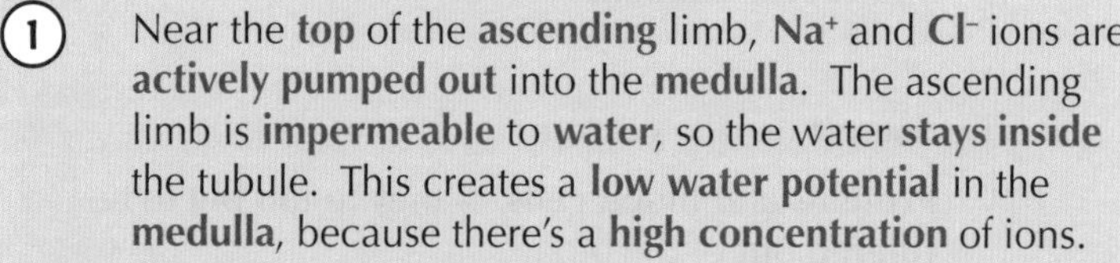

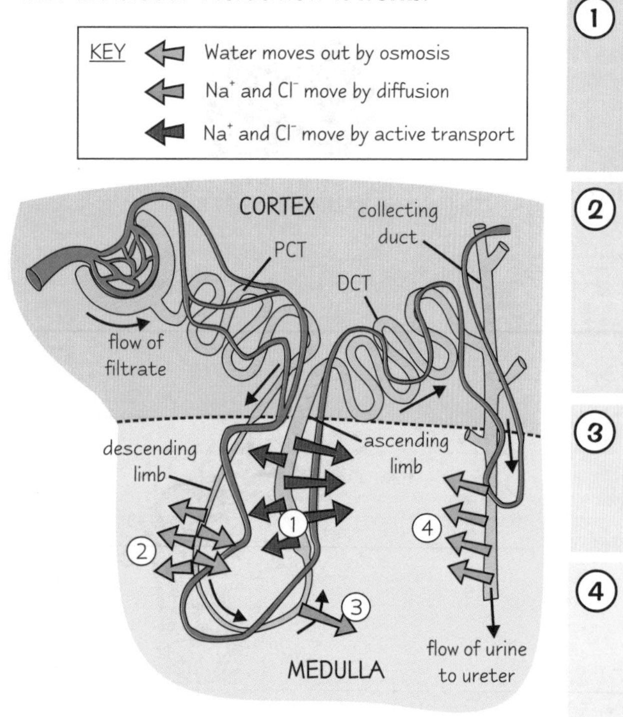

1. Near the **top** of the **ascending** limb, Na^+ and Cl^- ions are **actively pumped out** into the **medulla**. The ascending limb is **impermeable** to **water**, so the water **stays inside** the tubule. This creates a **low water potential** in the **medulla**, because there's a **high concentration** of ions.
2. Because there's a **lower** water potential in the **medulla** than in the descending limb, **water** moves **out** of the **descending limb into** the **medulla** by **osmosis**. This makes the **filtrate more concentrated** (the ions can't diffuse out — the descending limb isn't permeable to them). The water in the medulla is **reabsorbed** into the **blood** through the **capillary network**.
3. Near the **bottom** of the **ascending** limb Na^+ and Cl^- ions **diffuse out** into the **medulla**, further **lowering** the **water potential** in the medulla. (The ascending limb is **impermeable** to **water**, so it **stays in the tubule**.)
4. The first three stages massively **increase** the **ion concentration** in the **medulla**, which **lowers** the **water potential**. This causes **water** to **move out** of the **collecting duct** by **osmosis**. As before, the water in the medulla is **reabsorbed** into the **blood** through the **capillary network**.

The **volume** of water **reabsorbed** from the collecting duct into the capillaries is **controlled** by **changing the permeability** of the **collecting duct** (see next page).

Different animals have **different length loops of Henle**. The **longer** an animal's loop of Henle, the **more water they can reabsorb** from the filtrate. When there's a longer ascending limb, **more ions** are **actively pumped out** into the medulla, which creates a **really low water potential** in the medulla. This means **more water** moves **out** of the nephron and collecting duct **into** the **capillaries**, giving very **concentrated urine**. Animals that live in areas where there's **little water** usually have **long loops** to **save** as much **water** as possible.

Controlling Water Content

Water Reabsorption is Controlled by *Hormones*

1) The water content, and so water potential, of the blood is **monitored** by cells called **osmoreceptors** in a part of the **brain** called the **hypothalamus**.
2) When the osmoreceptors are **stimulated** by **low** water content in the blood, the hypothalamus sends **nerve impulses** to the **posterior pituitary gland** to release a **hormone** called **antidiuretic hormone** (ADH) into the blood.
3) ADH makes the walls of the DCT and collecting duct **more permeable** to **water**.
4) This means **more water** is **reabsorbed** from these tubules **into** the medulla and into the blood by osmosis. A **small** amount of **concentrated urine** is produced, which means **less water** is **lost** from the body.

It's called antidiuretic hormone because diuresis is when lots of dilute urine is produced, so anti means a small amount of concentrated urine is produced.

Here's how ADH changes the **water content** of the **blood** when it's too **low** or too **high**:

1 *Blood ADH* Level *Rises* When You're *Dehydrated*

Dehydration is what happens when you **lose water**, e.g. by sweating during exercise, so the **water content** of the blood needs to be **increased**:

1) The **water content** of the blood **drops**, so its **water potential drops**.
2) This is detected by **osmoreceptors** in the **hypothalamus**.
3) The **posterior pituitary gland** is stimulated to release **more ADH** into the blood.
4) **More ADH** means that the DCT and collecting duct are **more permeable**, so **more water** is **reabsorbed** into the blood by osmosis.
5) A **small amount** of **highly concentrated** urine is produced and **less water** is **lost**.

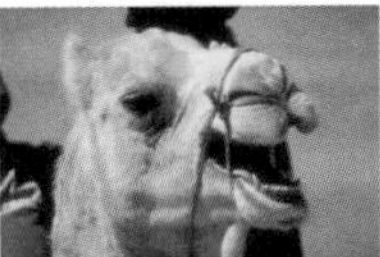

Dehydrated? Me? As if...

2 *Blood ADH* Level *Falls* When You're *Hydrated*

If you're **hydrated**, you've taken in **lots of water**, so the **water content** of the blood needs to be **reduced**:

1) The **water content** of the blood **rises**, so its **water potential rises**.
2) This is detected by the **osmoreceptors** in the **hypothalamus**.
3) The **posterior pituitary gland** releases **less ADH** into the blood.
4) **Less ADH** means that the DCT and collecting duct are **less permeable**, so **less water** is **reabsorbed** into the blood by osmosis.
5) A **large amount** of **dilute** urine is produced and **more water** is **lost**.

Practice Questions

Q1 In which parts of the nephron does water reabsorption take place?

Q2 Describe what happens along the descending limb of the loop of Henle.

Q3 Which cells monitor the water content of the blood?

Q4 Which gland releases ADH?

Exam Questions

Q1 Describe and explain how water is reabsorbed into the capillaries from the nephron. [6 marks]

Q2 The level of ADH in the blood rises during strenuous exercise.
Explain the cause of the increase and the effect it has on kidney function. [6 marks]

If you don't understand what ADH does, ur-ine trouble...

Seriously though, there are two main things to learn from these pages — how water is reabsorbed from the tubules in the kidney, and how the water content of the blood is regulated by osmoreceptors, the hypothalamus and the posterior pituitary gland. Keep writing it down until you've got it sorted in your head, and you'll be just fine. Now I need a wee.

Kidney Failure and Detecting Hormones

These pages are for OCR Unit 4 only.

Everything's fine while the kidneys are working well, but when they get damaged things don't run quite so smoothly.

Kidney Failure is When the Kidneys Stop Working Properly

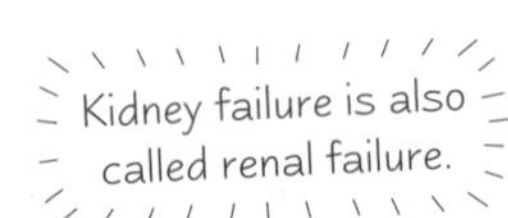
Kidney failure is also called renal failure.

Kidney failure is when the kidneys **can't** carry out their **normal functions** because they **can't work properly**. Kidney failure can be **caused** by many things including:

1) **Kidney infections** — these can cause **inflammation** (swelling) of the kidneys, which can **damage** the cells. This **interferes** with **filtering** in the renal capsules, or with **reabsorption** in the other parts of the nephrons.
2) **High blood pressure** — this can damage the **glomeruli**. The blood in the glomeruli is already under **high pressure** but the **capillaries** can be **damaged** if the blood pressure gets **too high**. This means **larger** molecules like **proteins** can get through the capillary walls and into the **urine**.

Kidney failure causes **lots of problems**, for example:

1) **Waste products** that the kidneys would normally **remove** (e.g. **urea**) begin to **build up** in the blood. **Too much** urea in the blood causes **weight loss** and **vomiting**.
2) **Fluid** starts to **accumulate** in the tissues because the kidneys **can't remove excess water** from the blood. This causes **parts of the body** to **swell**, e.g. the person's legs, face and abdomen can swell up.
3) The balance of **ions** in the body becomes, well, unbalanced. The blood may become **too acidic**, and an imbalance of calcium and phosphate can lead to **brittle bones**. **Salt build-up** may cause more **water retention**.
4) **Long-term** kidney failure causes **anaemia** — a **lack** of **haemoglobin** in the blood.

If the problems caused by kidney failure **can't be controlled**, it can eventually lead to **death**.

Renal Dialysis and Kidney Transplants can be used to Treat Kidney Failure

When the kidneys can no longer **function** (i.e. they've **totally failed**), a person is unable to **survive** without **treatment**. There are **two** main treatment options:

Renal dialysis

1) **Renal dialysis** is where a **machine** is used to **filter** a patient's blood.
   - The patient's blood is passed through a **dialysis machine** — the **blood** flows on one side of a **partially permeable membrane** and **dialysis fluid** flows on the other side.
   - **Waste products** and **excess water** and **ions** diffuse across the membrane into the dialysis fluid, **removing** them from the blood.
   - **Blood cells** and **larger** molecules like **proteins** are **prevented** from **leaving** the blood.
2) Patients can feel increasingly **unwell** between dialysis sessions because **waste products** and **fluid** starts to build up in their **blood**.
3) Each dialysis session takes **three to five hours**, and patients need **two or three sessions a week**, usually in **hospital**. This is **quite expensive** and is pretty **inconvenient** for the patient.
4) But dialysis can keep a person **alive** until a **transplant** is available (see below), and it's a lot **less risky** than having the **major surgery** involved in a transplant.

Kidney transplant

1) A **kidney transplant** is where a **new kidney** is implanted into a patient's body to **replace** a damaged kidney.
2) The new kidney has to be from a person with the **same blood** and **tissue type**. They're often donated from a **living relative**, as people can survive with **only one** kidney. They can also come from **other people** who've recently **died** — organ donors.
3) Transplants have a lot of **advantages** over dialysis:
   - It's **cheaper** to give a person a transplant than keep them on dialysis for a **long time**.
   - It's **more convenient** for a person than regular dialysis sessions.
   - Patients don't have the problem of feeling **unwell** between dialysis sessions.
4) But there are also **disadvantages** to having a kidney transplant:
   - The patient will have to undergo a **major operation**, which is **risky**.
   - The **immune system** may **reject** the transplant, so the patient has to take **drugs** to **suppress it**.

Kidney Failure and Detecting Hormones

Urine is used to Test for Pregnancy and Steroid Use

Urine is made by **filtering** the **blood**, so you can have a look at what's in a person's blood by **testing** their **urine**. For example, you can test if a woman is **pregnant** by looking for a **hormone** that only pregnant women produce, and you can test **athletes** for the presence of **banned drugs** like **steroids**:

TESTING FOR PREGNANCY

Pregnancy tests detect the hormone **human chorionic gonadotropin (hCG)** that's only found in the **urine** of **pregnant women**:

1) A **stick** is used with an **application area** that contains **antibodies for hCG** bound to a **coloured bead** (**blue**).
2) When urine is applied to the application area any hCG will **bind** to the antibody on the beads.
3) The urine **moves** up to the **test strip**, **carrying** the **beads** with it.
4) The test strip has **antibodies to hCG** stuck in place (**immobilised**).
5) If there **is hCG present** the test strip turns **blue** because the **immobilised** antibody binds to any **hCG** attached to the **blue** beads, concentrating the **blue beads** in that area. If **no hCG** is present, the beads will **pass through** the test area **without** binding to anything, and so it **won't** go blue.

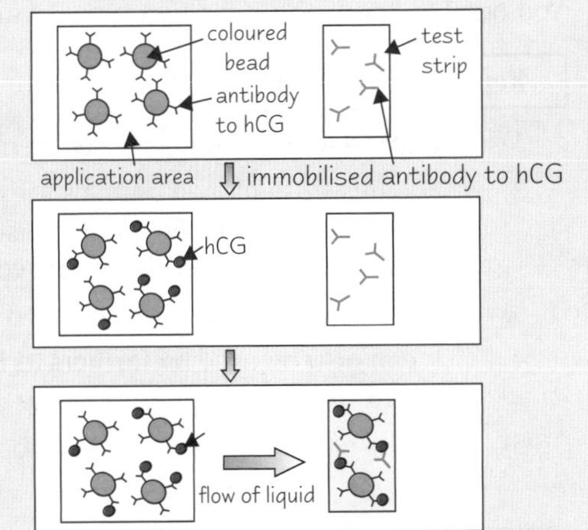

TESTING FOR STEROIDS

1) **Anabolic steroids** are **drugs** that **build up muscle tissue**.
2) **Testosterone** is an anabolic steroid, and there are other common ones such as **Nandrolone**.
3) Some **athletes** are **banned** from taking anabolic steroids. This is to try to stop the misuse of steroids that can have **dangerous side-effects**, such as **liver damage**. Also, it's considered **unfair** for some athletes to use steroids.
4) Steroids are **removed from the blood** in the **urine**, so athletes regularly have their **urine tested** for steroids.
5) Urine is tested for steroids (or the **products** made when they're **broken down**) by a technique called **gas chromatography**.
6) The urine sample is **vaporised** (turned into a **gas**) and passed through a column containing a **liquid**. **Different substances** move through the column at **different speeds**. The length of time taken for substances in the **sample** to pass through the column is **compared** to the time taken for a **steroid** to pass through the column. If the time taken is the **same** then the sample **contains the steroid**.

Practice Questions

Q1 Give one effect of kidney failure on the body.

Q2 What is renal dialysis?

Q3 What substance does a pregnancy test detect in a urine sample?

Exam Questions

Q1 Discuss the advantages and disadvantages of kidney transplants. [5 marks]

Q2 Describe and explain how urine can be used to detect steroid use. [5 marks]

Kidney failure, kidney infections, kidney transplants, kidney beans...

So you can either treat kidney failure with a kidney transplant or you can use kidney dialysis to filter the blood a few times a week. Both treatments come with their advantages and disadvantages, so make sure you can sum them both up. Here's a tip — you can usually use the disadvantages of one treatment to come up with the advantages of the other.

Cloning

This section is for OCR Unit 5 only.

Please don't try doing this at home — you'll only confuse people if there are 27 copies of you in the house...

Cloning makes Cells or Organisms Genetically Identical to Another Organism

Cloning is the process of producing **genetically identical cells** or **organisms** from the cells of an **existing organism**. Cloning can occur **naturally** in some **plants** and **animals**, but it can also be carried out **artificially**. You need to know about the **two types** of **artificial cloning** used for **animals**:

Reproductive cloning

1) **Reproductive cloning** is used to make a **complete organism** that's **genetically identical** to **another organism**.
2) Scientists use cloned animals for **research purposes**, e.g. they can **test new drugs** on cloned animals. They're all genetically identical, so the **variables** that come from **genetic differences** (e.g. the likelihood of developing cancer) are **removed**. This means the **results** are more **reliable**.
3) Reproductive cloning can be used to **save endangered animals** from **extinction** by cloning new individuals.
4) It can also be used by **farmers** to **increase** the **number** of animals with **desirable characteristics** to **breed from**, e.g. a prize-winning cow with high milk production could be cloned.
5) Loads of different animals have been cloned, e.g. **sheep**, **cattle**, **pigs** and **horses**.

Non-reproductive cloning

1) **Non-reproductive cloning** is used to make **embryonic stem cells** that are **genetically identical** to **another organism**. It's also called **therapeutic cloning**.
2) Embryonic stem cells are harvested from young **embryos**.
3) They have the **potential** to become **any cell type** in an organism, so scientists think they could be used to **replace damaged tissues** in a **range** of **diseases**, e.g. heart disease, spinal cord injuries, degenerative brain disorders like Parkinson's disease.
4) If replacement tissue is made from cloned embryonic stem cells that are **genetically identical** to the **patient's own cells** then the tissue **won't be rejected** by their immune system.

Take a look back at the stuff you learnt about stem cells at AS.

Animals are Artificially Cloned by Nuclear Transfer

Reproductive and **non-reproductive** cloning are **both carried out** using a technique called **nuclear transfer**. Here's how it's done with **sheep** (but the **principles** are the **same** for **any animal**):

1) A **body cell** is taken from sheep A. The **nucleus** is **extracted** and **kept**.
2) An **egg cell** is taken from sheep B. Its nucleus is **removed** to form an **enucleated egg cell**.
3) The nucleus from sheep A is **inserted** into the enucleated egg cell — the egg cell from **sheep B** now contains the **genetic information** from **sheep A**.
4) The egg cell is **stimulated** to **divide** and an **embryo** is formed.
5) In **reproductive cloning** the embryo is **implanted** into a **surrogate mother**. A **lamb** is produced that's a **genetically identical** copy of **sheep A**.
6) In **non-reproductive cloning stem cells** are **harvested** from the embryo. The stem cells are **genetically identical** to the cells in **sheep A**.

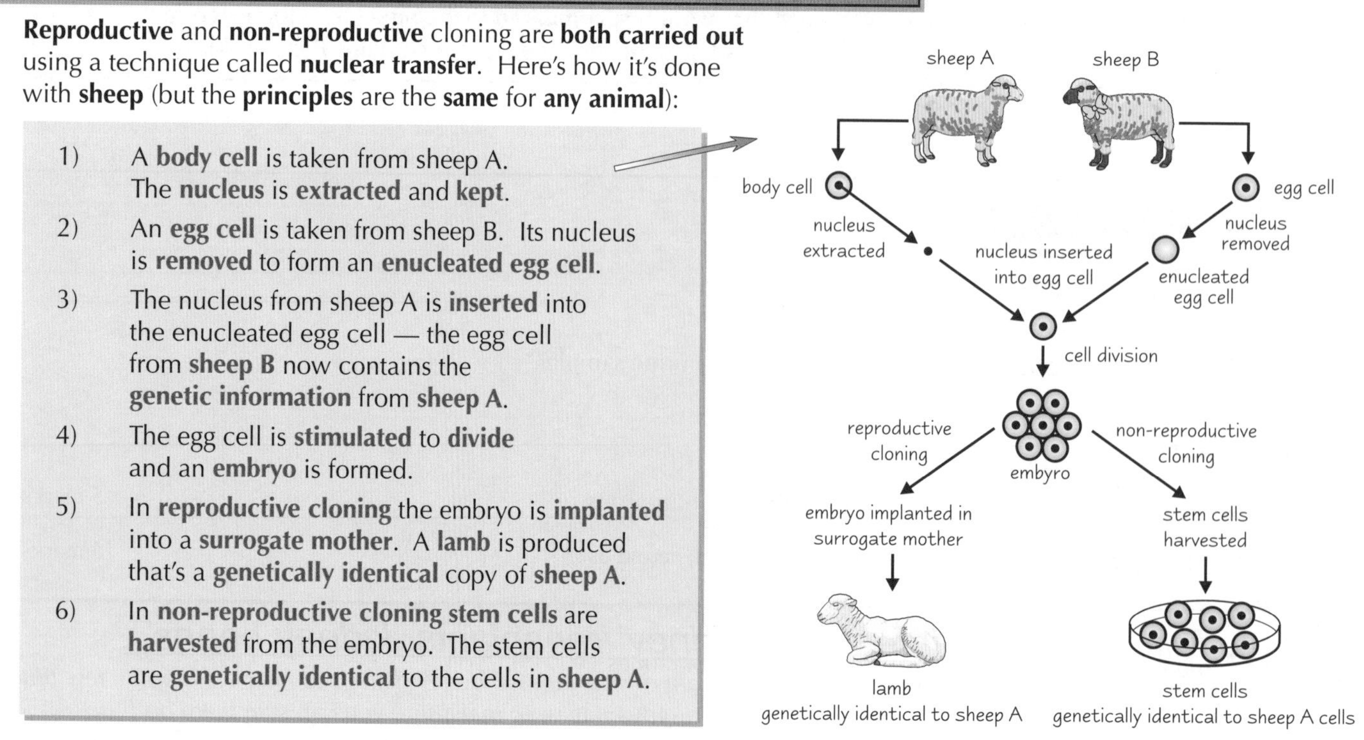

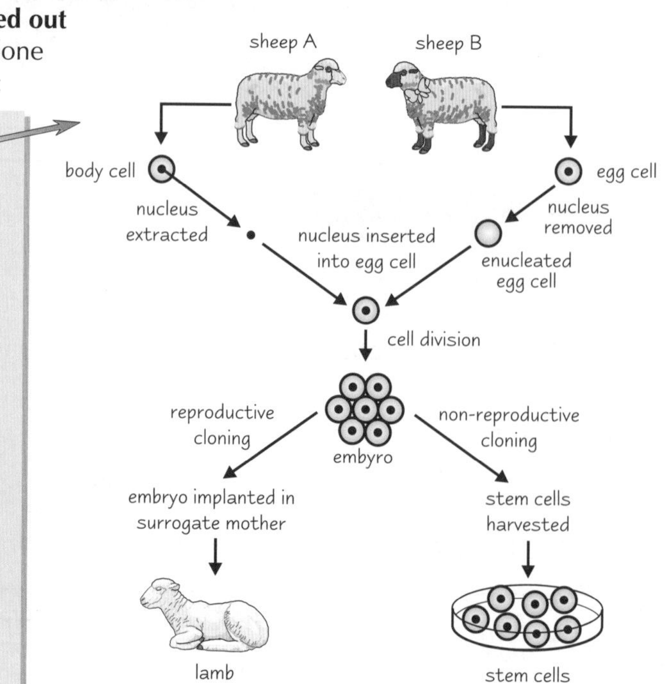

Cloning

Cloning Animals has Advantages and Disadvantages

Advantages

- **Desirable genetic characteristics** (e.g. high milk production in cows) are **always passed on** to clones — this **doesn't always** happen with **sexual reproduction**.
- **Infertile animals** can be **reproduced**.
- **Animals** can be **cloned** at **any time** — farmers wouldn't have to wait until a breeding season to produce new animals.

Disadvantages

- **Undesirable genetic characteristics** (e.g. a weak immune system) are **always passed on** to clones.
- Reproductive cloning is very **difficult, time-consuming** and **expensive** — **Dolly the sheep** was created after **277** nuclear transfer **attempts**.
- Some evidence suggests that clones **may not live as long** as natural offspring.

There are Ethical Issues to do with Human Cloning

1) The use of **human embryos** as a source of stem cells is **controversial**. The embryos are usually **destroyed** after the embryonic stem cells have been harvested — some people believe that doing this is **destroying a human life**.
2) Some people think a **cloned human** would have a **lower quality of life**, e.g. they might suffer **social exclusion** or have difficulty developing their own **personal identity**.
3) Some people think that cloning humans would be **wrong** because it **undermines** natural **sexual reproduction**, and traditional **family structures**.

Reproductive cloning of humans is currently illegal in the UK.

Plants can be Artificially Cloned using Tissue Culture

1) **Plants** can be **cloned** from existing plants using a technique called **tissue culture**. Here's how it's done:

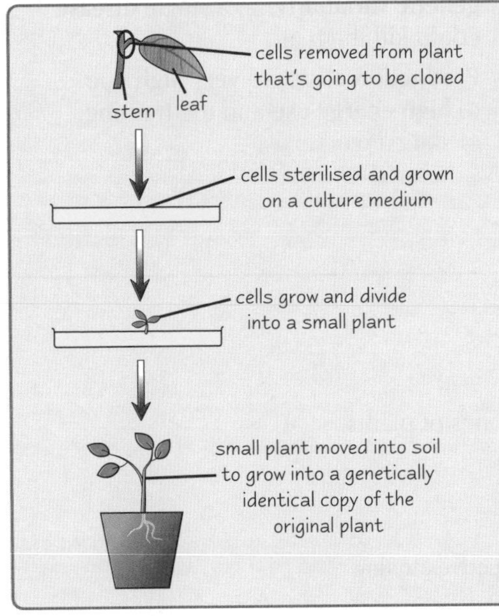

1) **Cells** are taken from the original plant that's going to be cloned.
2) Cells from the **stem** and **root tips** are used because they're **stem cells** — like in humans, plant stem cells can develop into **any type of cell**.
3) The cells are **sterilised** to kill any **microorganisms** — bacteria and fungi **compete** for nutrients with the **plant cells**, which **decreases** their **growth rate**.
4) The cells are placed on a **culture medium** containing plant **nutrients** (like **glucose** for **respiration**) and **growth factors** (such as **auxins**).
5) When the cells have **divided** and **grown** into a **small plant** they're taken out of the medium and **planted in soil** — they'll develop into plants that are **genetically identical** to the **original plant**.

2) Tissue culture is used to clone plants that **don't readily reproduce** or are **endangered** or **rare**, e.g. British orchids.
3) It's also used to grow **whole plants** from **genetically engineered plant cells**.
4) **Micropropagation** is when tissue culture is used to produce **lots** of cloned plants **very quickly**. **Cells** are taken from developing cloned plants and **subcultured** (grown on another fresh culture medium) — **repeating** this process creates **large numbers** of clones.

Cloning

These pages are for OCR Unit 5.

Some Plants can Produce Natural Clones by Vegetative Propagation

Vegetative propagation is the natural production of plant clones from **non-reproductive tissues**, e.g. roots, leaves and stems. Plants grow **structures** on roots, leaves or stems that can **grow** into an identical **new plant**. You need to know how **elm trees** produce clones from structures called **suckers**:

1) A sucker is a **shoot** that grows from the **shallow roots** of an elm tree.
2) Suckers grow from **sucker buds** (undeveloped shoots) that are scattered around the tree's **root system**. The buds are **normally dormant**.
3) During times of **stress** (e.g. drought, damage or disease) or when a tree is **dying**, the **buds** are **activated** and suckers begin to form.
4) Suckers can pop up many metres **away** from the parent tree, which can help to **avoid** the **stress** that triggered their growth.
5) They eventually form completely **separate trees** — **clones** of the tree that the suckers grew from.

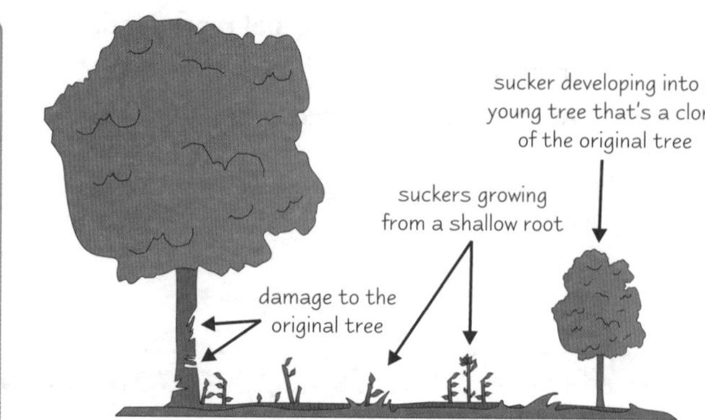

Plant Cloning in Agriculture has Advantages and Disadvantages

Advantages

- **Desirable genetic characteristics** (e.g. high fruit production) are **always passed on** to clones. This **doesn't always** happen when plants **reproduce sexually**.
- Plants can be reproduced in **any season** because tissue culture (see previous page) is carried out **indoors**.
- **Sterile plants** can be **reproduced**.
- Plants that take a **long time** to produce **seeds** can be **reproduced quickly**.

Disadvantages

- **Undesirable genetic characteristics** (e.g. producing fruit with lots of seeds) are **always passed on** to clones.
- **Cloned plant populations** have **no genetic variability**, so a **single disease** could **kill** them all.
- **Production costs** are **very high** due to **high energy use** and the **training** of skilled workers.

Practice Questions

Q1 What type of cells are made by non-reproductive cloning?

Q2 Name the technique that can be used to produce artificial clones of plants.

Q3 Give one disadvantage of plant cloning in agriculture.

Exam Question

Q1 Scientists in the UK are using stem cells produced by non-reproductive cloning to research treatments for diseases like Parkinson's disease.

a) How does reproductive cloning differ from non-reproductive cloning? [2 marks]

b) Briefly describe the technique they might use to carry out non-reproductive cloning. [6 marks]

I ain't makin' no cloned elm tree, sucker...

Although it would be nice to have lots of clones doing your revision, exams and PE lessons, it's not going to happen. Sadly there's only one of you, and you need to learn about the different types of cloning, how they're done and their advantages and disadvantages. There are ethical issues with human cloning too — it's not everyone's cup of tea...

Biotechnology

The global biotechnology industry is humongous, but fortunately you've only got to learn three pages about it...

Biotechnology is the Use of Living Organisms in Industry

1) **Biotechnology** is the **industrial use** of **living organisms** to produce **food**, **drugs** and **other products**, e.g. yeast is used to make wine.
2) The living organisms used are mostly **microorganisms** (bacteria and fungi). Here are a few reasons why:
   - Their **ideal growth conditions** can be **easily** created.
   - They grow **rapidly** under the right conditions, so **products** can be made **quickly**.
   - They can grow on a **range** of **inexpensive** materials.
   - They can be grown at **any time** of the year.
3) Biotechnology also **uses parts** of **living organisms** (such as **enzymes**) to make products, e.g. rennet (a mix of enzymes) is extracted from calf stomachs and used to make cheese.
4) Enzymes used in industry can be **contained within the cells** of organisms — these are called **intracellular enzymes**.
5) Enzymes are also used that **aren't contained within cells** — these are called **isolated enzymes**. **Some** are **secreted naturally** by microorganisms (called **extracellular enzymes**), but others have to be **extracted**.
6) **Naturally secreted** enzymes are **cheaper** to use because it can be **expensive** to **extract** enzymes from cells.

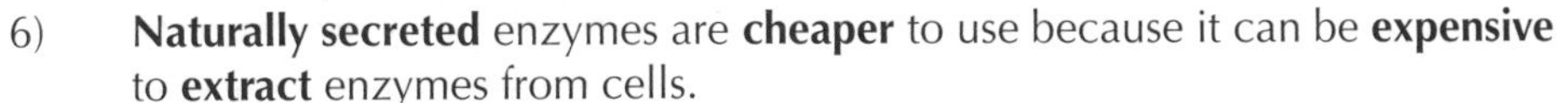

Hooray, the rennet extractor's here.

Isolated Enzymes can be Immobilised

1) **Isolated enzymes** used in industry can become **mixed in** with the **products** of a reaction.
2) The **products** then need to be **separated** from this mixture, which can be **complicated** and **costly**.
3) This is **avoided** in large-scale production by using **immobilised enzymes** — enzymes that are **attached** to an **insoluble material** so they **can't** become mixed with the products.
4) There are **three main ways** that enzymes are **immobilised**:

① **Encapsulated** in jelly-like **alginate beads**

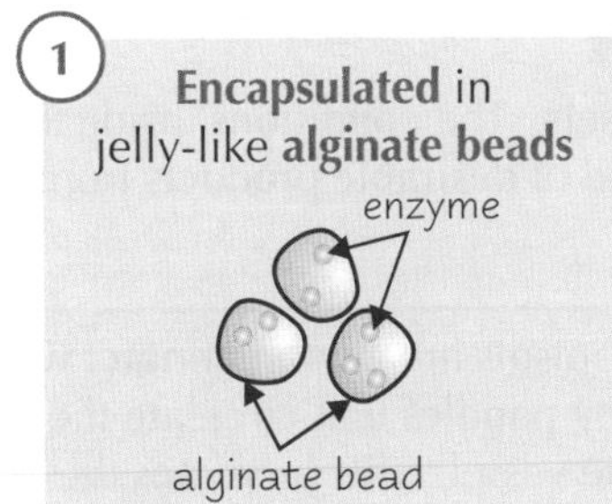

② **Trapped** in a **silica gel matrix**

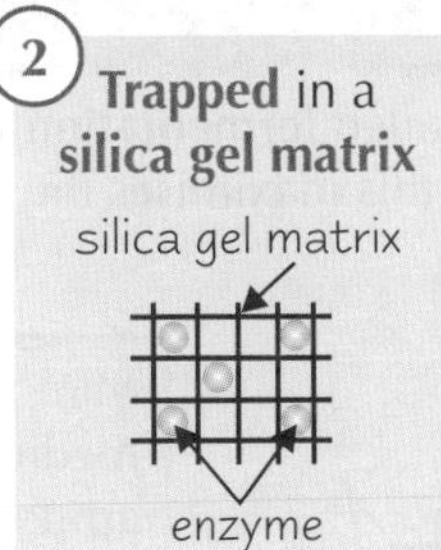

③ **Covalently bonded** to **cellulose** or **collagen fibres**

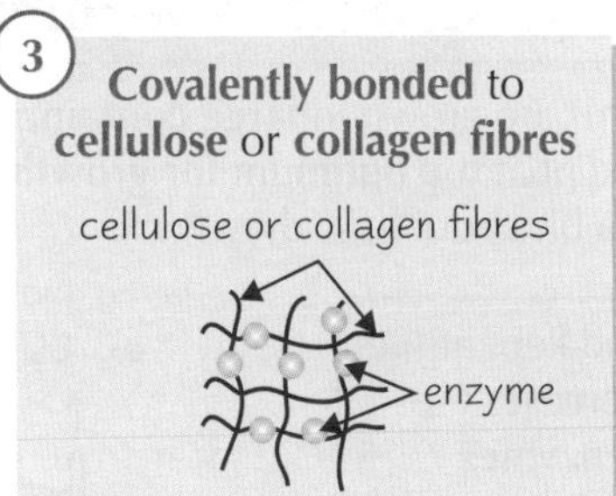

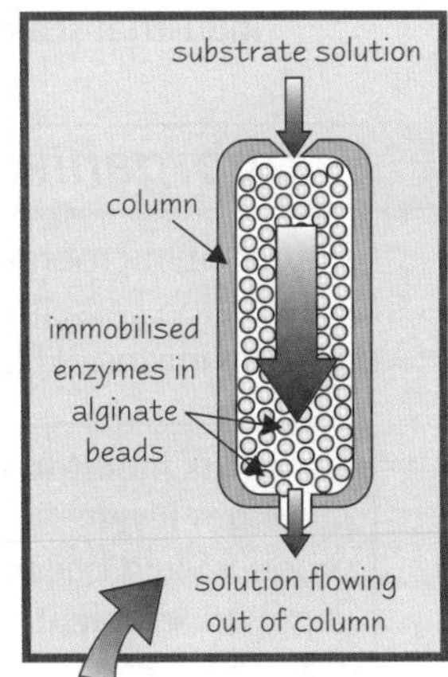

5) In industry, the **substrate solution** for a reaction is run through a **column** of **immobilised enzymes**.
6) The **active sites** of the enzymes are still **available** to **catalyse** the reaction but the solution flowing **out of** the column will **only** contain the **desired product**.
7) Here are some of the **advantages** of using **immobilised enzymes** in industry:
   - Columns of immobilised enzymes can be **washed** and **reused** — this **reduces** the **cost** of running a reaction on an **industrial scale** because you don't have to **keep buying** new enzymes.
   - The product **isn't mixed** with the enzymes — **no money** or **time** is **spent** separating them out.
   - Immobilised enzymes are **more stable** than free enzymes — they're less likely to **denature** (become inactive) in **high temperatures** or **extremes** of **pH**.

Biotechnology

These pages are for OCR Unit 5.

Closed Cultures of Microorganisms follow a Standard Growth Curve

1) A **culture** is a **population** of one type of microorganism that's been grown under **controlled conditions**.
2) A **closed culture** is when growth takes place in a vessel that's **isolated** from the **external environment** — extra nutrients **aren't added** and waste products **aren't removed** from the vessel **during growth**.
3) In a closed culture a population of microorganisms follows a **standard growth curve**:

① **Lag phase** — the population size **increases slowly** because the **microorganisms** have to make enzymes and other molecules before they can reproduce. This means the **reproduction rate** is **low**.

② **Exponential phase** — the population size **increases quickly** because the culture **conditions** are at their **most favourable** for **reproduction** (**lots of food** and **little competition**). The number of microorganisms **doubles** at **regular intervals**.

③ **Stationary phase** — the population size **stays level** because the **death rate** of the microorganisms **equals** their **reproductive rate**. Microorganisms **die** because there's **not enough food** and poisonous **waste products build up**.

④ **Decline phase** — the population size **falls** because the **death rate** is **greater** than the **reproductive rate**. This is because food is very **scarce** and waste products are at **toxic levels**.

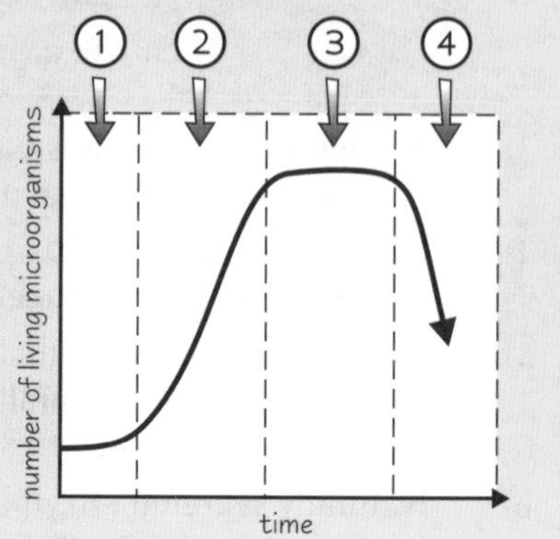

4) When growing conditions are **favourable** (e.g. during the **exponential phase**) microorganisms produce **primary metabolites** — **small molecules** that are **essential** for the **growth** of the microorganisms.
5) When growing conditions are **less favourable** (e.g. during the **stationary phase**) some microorganisms produce **secondary metabolites** — molecules that **aren't essential** for **growth** but are **useful** in **other ways**.
6) **Secondary metabolites** help microorganisms **survive**, e.g. the **antibiotic penicillin** is a secondary metabolite made by ***Penicillium*** (a fungus). It **kills bacteria** that **inhibit** its **growth**.
7) Some secondary metabolites are **desirable** to **biotechnology industries**, e.g. *Penicillium* is **cultured** on an **industrial scale** to produce lots of penicillin — it's used to treat **bacterial infections** in humans and animals.

Microorganisms are Grown in Fermentation Vessels

Cultures of microorganisms are grown in **large containers** called **fermentation vessels**. The **conditions** inside the fermentation vessels are kept at the **optimum for growth** — this **maximises** the **yield** of **desirable products** from the microorganisms. Here's a bit about how they work:

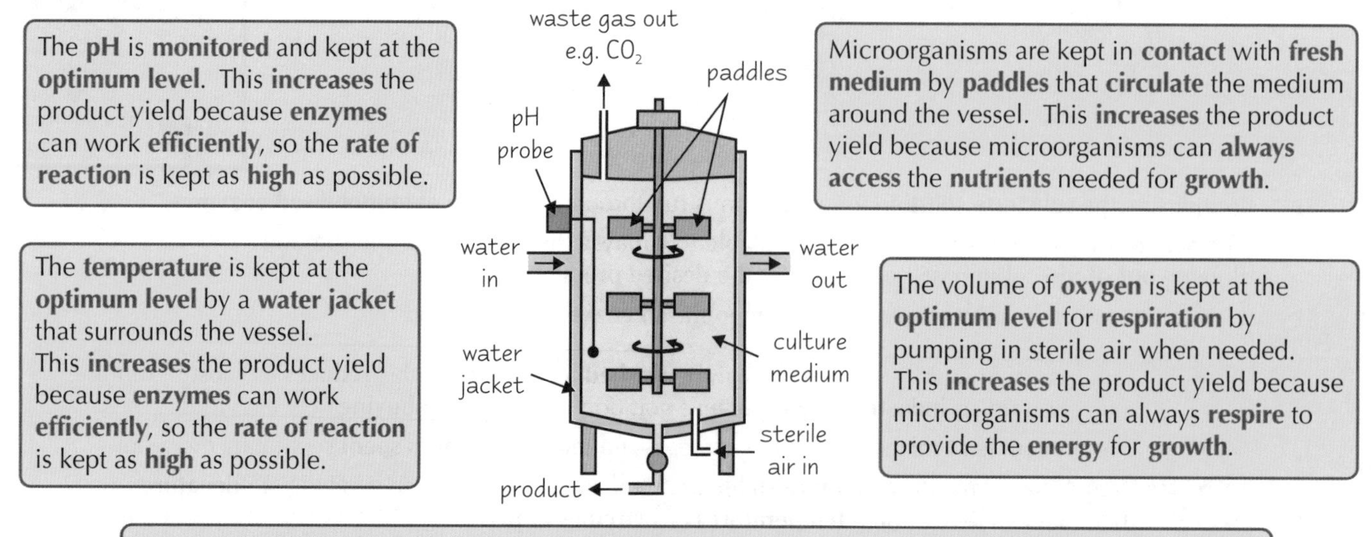

The **pH** is **monitored** and kept at the **optimum level**. This **increases** the product yield because **enzymes** can work **efficiently**, so the **rate of reaction** is kept as **high** as possible.

The **temperature** is kept at the **optimum level** by a **water jacket** that surrounds the vessel. This **increases** the product yield because **enzymes** can work **efficiently**, so the **rate of reaction** is kept as **high** as possible.

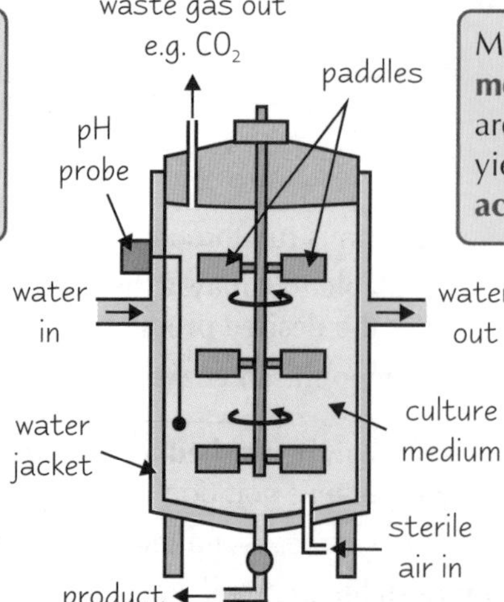

Microorganisms are kept in **contact** with **fresh medium** by **paddles** that **circulate** the medium around the vessel. This **increases** the product yield because microorganisms can **always access** the **nutrients** needed for **growth**.

The volume of **oxygen** is kept at the **optimum level** for **respiration** by pumping in sterile air when needed. This **increases** the product yield because microorganisms can always **respire** to provide the **energy** for **growth**.

Vessels are **sterilised** between uses with **superheated steam** to kill any **unwanted organisms**. This **increases** the product yield because the microorganisms **aren't competing** with other organisms.

Biotechnology

*There are **Two Main Culture Methods** — **Batch** and **Continuous***

1) **Batch culture** is where microorganisms are grown in **individual batches** in a fermentation vessel — when one culture **ends** it's **removed** and then a **different batch** of microorganisms is grown in the vessel.
2) **Continuous culture** is where microorganisms are **continually grown** in a fermentation vessel **without stopping**.
3) Here are some of the **differences** between batch culture and continuous culture:

Batch Culture	Continuous Culture
A fixed volume of growth medium (nutrients) is added to the fermentation vessel at the start of the culture and no more is added. The culture is a closed system.	Growth medium flows through the vessel at a steady rate so there's a constant supply of fresh nutrients. The culture is an open system.
Each culture goes through the lag, exponential and stationary growth phases.	The culture goes through the lag phase but is then kept at the exponential growth phase.
The product is harvested once, during the stationary phase.	The product is continuously taken out of the fermentation vessel at a steady rate.
The product yield is relatively low — stopping the reaction and sterilising the vessel between fermentations means there's a period when no product is being harvested.	The product yield is relatively high — microorganisms are constantly growing at an exponential rate.
If contamination occurs it only affects one batch. It's not very expensive to discard the contaminated batch and start a new one.	If the culture is contaminated the whole lot has to be discarded — this is very expensive when the cultures are done on an industrial scale.
Used when you want to produce secondary metabolites.	Usually used when you want primary metabolites or the microorganisms themselves as the desired product.

Asepsis** is **Important** when **Culturing Microorganisms

1) **Asepsis** is the practice of **preventing contamination** of cultures by **unwanted microorganisms**.
2) It's important when culturing microorganisms because contamination can **affect** the **growth** of the microorganism that you're **interested in**.
3) Contaminated cultures in **laboratory experiments** give **inaccurate results**.
4) Contamination on an **industrial scale** can be **very costly** because **entire cultures** may have to be **thrown away**.
5) A number of **aseptic techniques** can be used when working with microorganisms:

- **Work surfaces** are **regularly disinfected** to minimise contamination.
- **Gloves should be worn** and **long hair** is **tied back** to prevent it from falling into anything.
- The **instruments** used to **transfer** cultures are **sterilised before** and **after** each use, e.g. **inoculation loops** (small wire loops) are **heated** using a **Bunsen burner** to **kill** any microorganisms on them.
- In laboratories, the **necks** of **culture containers** are **briefly flamed** before they're **opened** or **closed** — this causes **air** to **move out** of the container, **preventing** unwanted microorganisms from **falling in**.
- **Lids** are **held over** open containers after they're removed, instead of putting them on a work surface. This **prevents** unwanted microorganisms from **falling** onto the culture.

Practice Questions

Q1 Give one way that enzymes are immobilised.

Q2 Why is it important to maintain the pH level in a fermentation vessel?

Q3 Why is asepsis important when culturing microorganisms?

Exam Question

Q1 Describe and explain the standard growth curve of microorganisms in a closed culture. [8 marks]

Calf stomachs, yeast and sterile conditions — biotechnology is sexy stuff...

Wow, biology and technology fused together — forget bionic arms, legs and eyes though, growing bacteria in a tank is where it's at. Just think of yourself like an immobilised enzyme in a column — the substrate going in is the information on these pages, then all the desired information will flow out of you (not as a liquid hopefully) onto the exam paper.

Brain Structure and Function

This section is for OCR Unit 5 and Edexcel Unit 5 — this page is for both boards.

So... the brain... a big old squelchy mass that controls all the goings on in your body, from a sniffle to your heart rate. Without your brain you wouldn't be able to see, hear, think or learn — in fact, you wouldn't be much use at all...

Different Areas of the Brain control Different Functions

Your brain **controls** the rest of your body, but **different parts** control **different functions**. You need to know the **location** and **function** of these **four** brain structures:

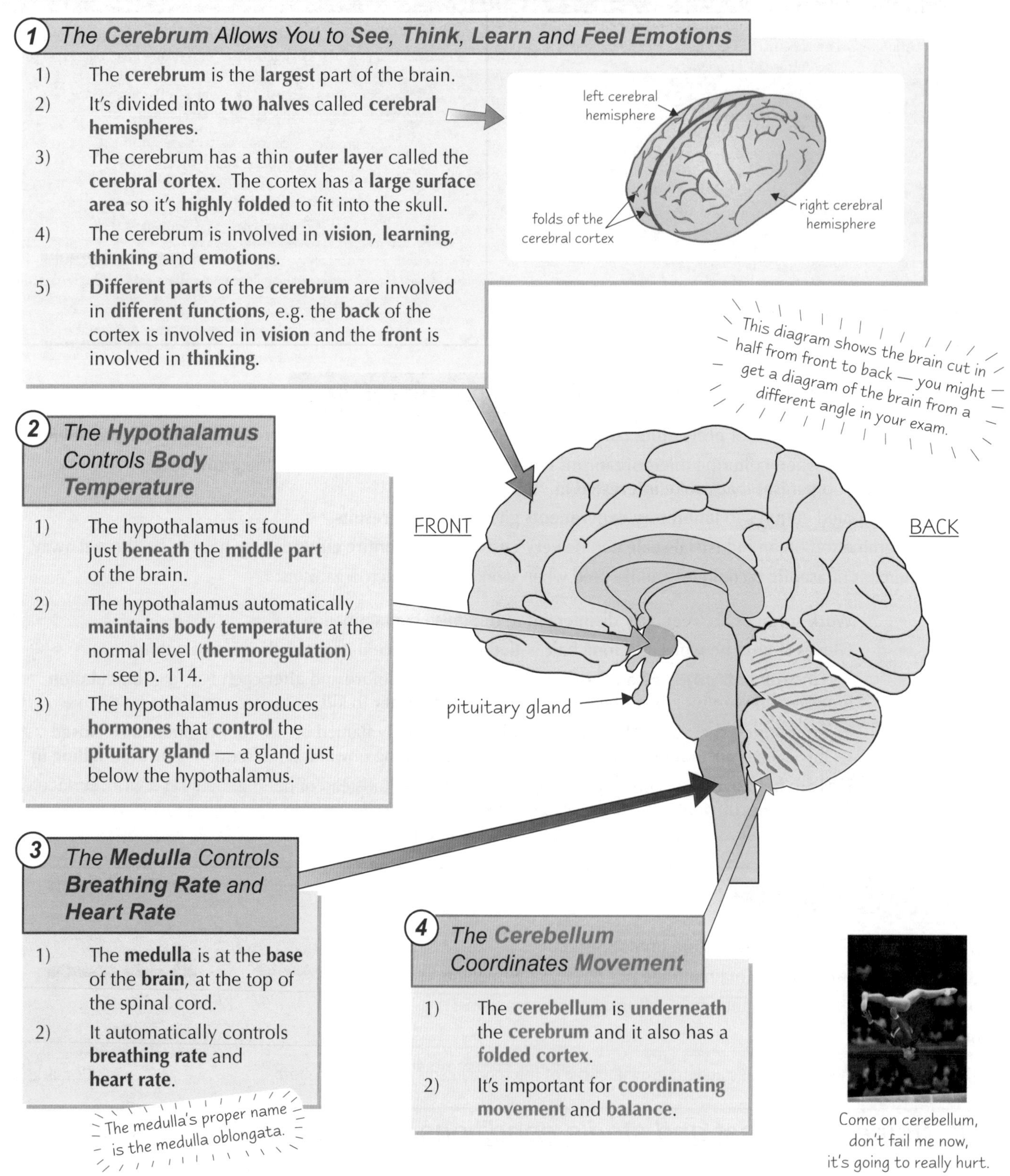

1 The Cerebrum Allows You to See, Think, Learn and Feel Emotions

1) The **cerebrum** is the **largest** part of the brain.
2) It's divided into **two halves** called **cerebral hemispheres**.
3) The cerebrum has a thin **outer layer** called the **cerebral cortex**. The cortex has a **large surface area** so it's **highly folded** to fit into the skull.
4) The cerebrum is involved in **vision**, **learning**, **thinking** and **emotions**.
5) **Different parts** of the **cerebrum** are involved in **different functions**, e.g. the **back** of the cortex is involved in **vision** and the **front** is involved in **thinking**.

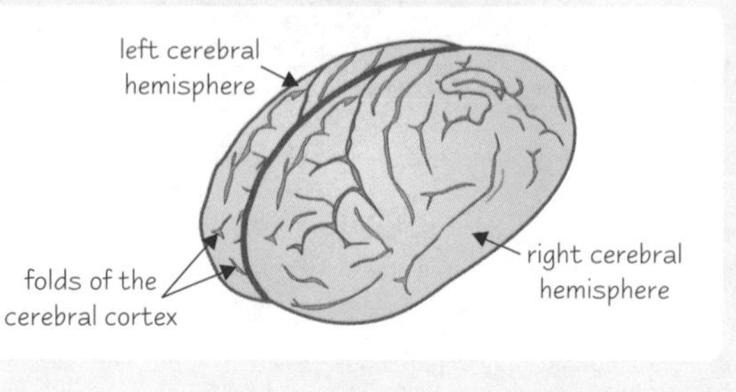

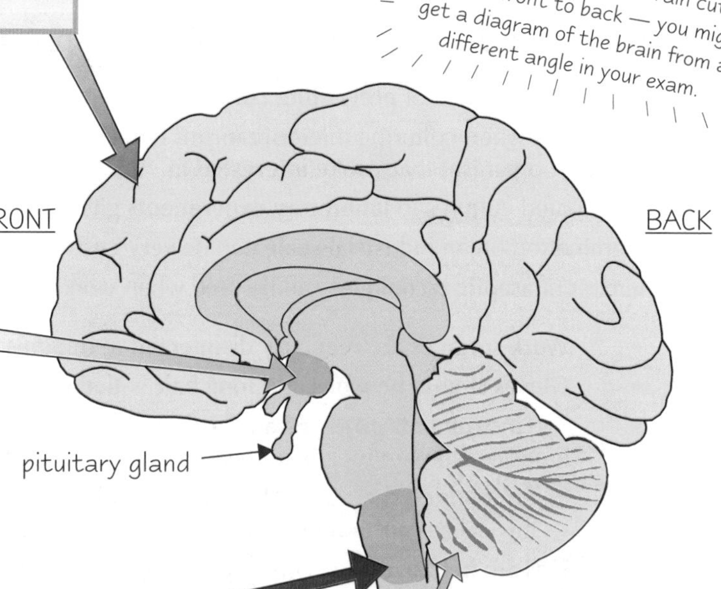

This diagram shows the brain cut in half from front to back — you might get a diagram of the brain from a different angle in your exam.

2 The Hypothalamus Controls Body Temperature

1) The hypothalamus is found just **beneath** the **middle part** of the brain.
2) The hypothalamus automatically **maintains body temperature** at the normal level (**thermoregulation**) — see p. 114.
3) The hypothalamus produces **hormones** that **control** the **pituitary gland** — a gland just below the hypothalamus.

3 The Medulla Controls Breathing Rate and Heart Rate

1) The **medulla** is at the **base** of the **brain**, at the top of the spinal cord.
2) It automatically controls **breathing rate** and **heart rate**.

The medulla's proper name is the medulla oblongata.

4 The Cerebellum Coordinates Movement

1) The **cerebellum** is **underneath** the **cerebrum** and it also has a **folded cortex**.
2) It's important for **coordinating movement** and **balance**.

Come on cerebellum, don't fail me now, it's going to really hurt.

Brain Structure and Function

This page is for** Edexcel Unit 5 **only. If you're doing** OCR, **skip to the questions on page 178.

Scanners are used to Visualise the Brain

1) To investigate the **structure** and **function** of the brain, and to **diagnose medical conditions**, you need to **look inside** the brain.
2) This can be done with **surgery**, but it's **pretty risky**.
3) The brain can be visualised without surgery using **scanners**.
4) You need to **know** about the **three** different types of scanner:

1 *Computed Tomography (CT) Scanners use Lots of X-rays*

CT scanners use **radiation** (**X-rays**) to produce **cross-section images** of the brain. **Dense structures** in the brain **absorb more radiation** than less dense structures so show up as a **lighter colour** on the scan.

Computers can build up many 2D images to produce a 3D image of the brain.

Uses of CT

Investigating brain structure — A CT scan shows the **major structures** in the brain.

Investigating brain function — A CT scan **doesn't** show brain **function** — it only shows brain structure. But if a CT scan shows a **diseased** or **damaged brain structure** and the patient has **lost some function**, the **function** of that part of the brain can be **worked out**. E.g. if an area of the brain is damaged and the patient can't see, then that area is involved in vision.

Medical diagnosis — CT scans can be used to **diagnose medical problems** because they show **damaged** or **diseased** areas of the brain, e.g. **bleeding** in the brain after a **stroke**:

- **Blood** has a **different density** from brain tissue so it shows up as a **lighter colour** on a CT scan.
- A scan will show the **extent** of the bleeding and its **location** in the brain.
- You can then work out **which blood vessels** have been damaged and what **brain functions** are likely to be **affected** by the bleeding.

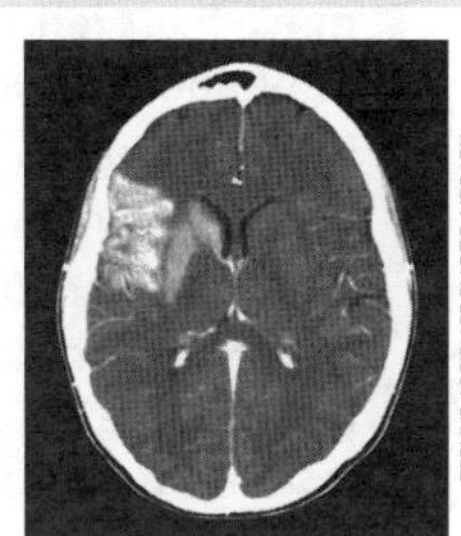

ZEPHYR / SCIENCE PHOTO LIBRARY

This CT scan looks up at the brain from below. The white area is heavy bleeding due to a blood vessel ruptured after a stroke.

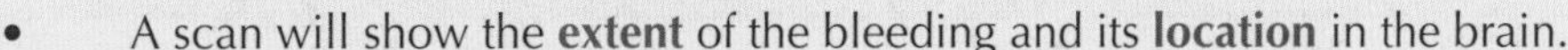

2 *Magnetic Resonance Imaging (MRI) Scanners use Magnetic Fields*

MRI scanners use a **really strong magnetic field** and **radio waves** to produce **cross-section images** of the brain.

An MRI scanner costs a lot more than a CT scanner.

Uses of MRI

Investigating brain structure — You can see the structure of the brain in a **lot more detail** using an MRI scanner than you can with a CT scanner, and you can **clearly see** the **difference** between **normal** and **abnormal** (diseased or damaged) brain tissue. For example, a scan can show diseased tissue caused by multiple sclerosis (a disease of the central nervous system).

Investigating brain function — This is done in the same way as with a CT scan.

Medical diagnosis — MRI scans can also be used to **diagnose medical problems** because they show **damaged** or **diseased** areas of the brain, e.g. a **brain tumour** (an abnormal mass of cells in the brain):

- **Tumour cells respond differently** to a **magnetic field** than healthy cells, so they show up as a **lighter colour** on an MRI scan.
- A scan will show the **exact size** of a tumour and its **location** in the brain. Doctors can then use this information to decide the most effective treatment.
- You can also work out what **brain functions** may be **affected** by the tumour.

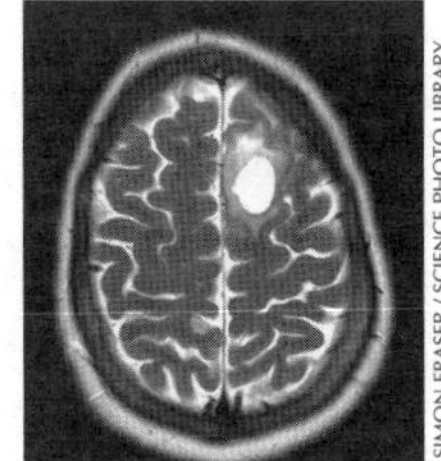

SIMON FRASER / SCIENCE PHOTO LIBRARY

This MRI scan looks down on the brain from above. The white area is a tumour.

Brain Structure and Function

This page is for Edexcel Unit 5.

3 *Functional Magnetic Resonance Imaging (fMRI) Scanners show Brain Activity*

fMRI scanners are **like MRI** scanners (see previous page), but they show **changes** in **brain activity** as they actually happen:

1) **More oxygenated blood** flows to **active areas** of the brain (to supply the neurones with oxygen and glucose).
2) Molecules in **oxygenated blood respond differently** to a **magnetic field** than those in deoxygenated blood.
3) So the **more active areas** of the brain can be **identified** on an fMRI scan.

Uses of fMRI

Investigating brain structure — An fMRI scan gives a **detailed** picture of the **brain's structure**, similar to an MRI scan.

Investigating brain function — fMRI scans are used to research the **function** of the brain as well as its structure. If a function is **carried out** whilst **in the scanner**, the **part** of the brain that's involved with that function will be **more active.** E.g. a patient might be asked to **move** their **left hand** when in the fMRI scanner. The areas of the brain involved in moving the hand will **show up** on the fMRI scan (these are often coloured by the computer so they show up more easily).

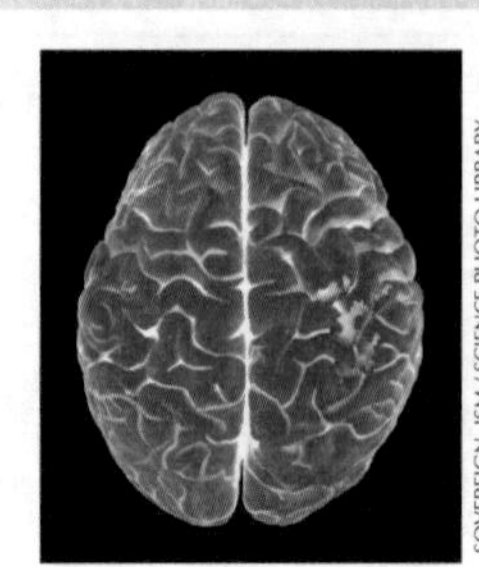

This fMRI scan looks down on the brain from above. The red area is active when the person moves their left hand. The right side's active because it controls the left side of your body.

Medical diagnosis — fMRI scans are really useful to **diagnose medical problems** because they show **damaged** or **diseased areas** of the brain and they allow you to study conditions caused by **abnormal activity** in the brain (some conditions don't have an obvious structural cause). E.g. an fMRI scan can be taken of a patient's brain **before** and **during** a **seizure.** This can help to pinpoint which part of the brain's **not working properly** and find the **cause** of the seizure. Then the patient can receive the **most effective treatment** for the seizures.

Practice Questions

These questions cover pages 176-178.

Q1 Which part of the brain is involved in learning?

Q2 Which part of the brain controls body temperature?

Q3 What type of scanner produces an image of the brain using X-rays?

Exam Questions

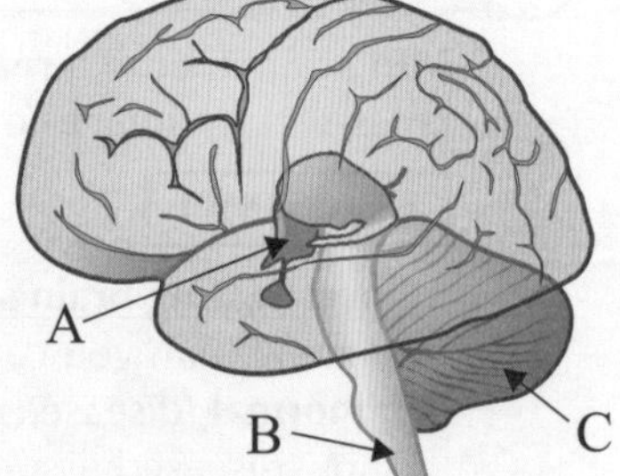

Q1 a) Name structure A on the diagram of the brain. [1 mark]

b) Give two roles of structure B. [2 marks]

c) What effect might damage to structure C have on the body? [1 mark]

Q2 A patient has fallen and hit his head. His doctor recommends an MRI (magnetic resonance imaging) scan to investigate suspected bleeding in his brain.

a) Give two pieces of information about the bleeding that the doctor would be able to get from the MRI scan. [2 marks]

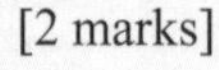

b) Before operating on him, the surgeon wants to use a different scanner to assess his brain activity. Suggest what kind of scanner should be used. [1 mark]

The cere-mum part of the brain — coordinates dirty washing and clean clothes...

Ah, the good old brain. It's a mysterious creature alright, but the invention of scanners means scientists are starting to unlock its secrets. All things in the brain have hard-to-spell names unfortunately, like hypothalamus and medulla. Annoyingly cerebrum and cerebellum sound pretty similar — so make sure you learn that the longer name is at the bottom of the brain.

Brain Development and Habituation

This page is for Edexcel Unit 5 only. If you're doing OCR, go to page 184.

I bet you've never sat down and had a good think about why your brain develops the way it does. Well now's the time...

You can Investigate the Role of **Nature and Nurture** in **Brain Development**

1) **Brain development** is how the brain **grows** and how **neurones connect together**. Measures of brain development include the **size** of the brain, the **number of neurones** it has and the **level of brain function** (e.g. speech, intelligence) a person has.
2) Your brain develops the way it does because of your **genes** (**nature**) and your **environment** (**nurture**) — your brain would **develop differently** if you had **different genes** or were brought up in a **different environment**.
3) Nature and nurture are **both involved** in controlling brain development, but **scientists disagree** about which one **influences** brain development the **most** — this **argument** is called the **nature-nurture debate**.
4) It's **really hard** to **investigate** the effects of **nature** and **nurture** because:

- Genetic and environmental factors **interact**, so it's difficult to know which one is the **main influence**.
- There are **lots** of different **genes** and **lots** of different **environmental factors** to investigate.
- To do an accurate **experiment**, you need to **cancel out** one factor to be able to **investigate** the other. This is really difficult to do — you'd need to **cut out** all **environmental** influences to **investigate** the role of a **genetics**, and vice versa.

5) You need to know the following **five methods** that are used to **investigate** the effects of **nature** and **nurture** on **brain development**:

1 *Animal Experiments*

1) Scientists study the effects of **different environments** (nurture) on the **brain development** of animals of the **same species**. Any differences in their brain development are **more likely** to be due to **nurture** than nature (if they're the same species they'll be genetically similar).
2) For example, animal experiments have shown that:

- Rats raised in a **stimulating environment** have **larger brains** and get **better scores** on **problem-solving tasks** than rats raised in boring environments (e.g. in a bare, dark cage). This suggests **nurture** plays a big role in **brain size** and the development of **problem-solving skills**.
- Rats reared in **isolation** (having no contact with other rats) have **similar brain abnormalities** to those found in **schizophrenic** patients. This suggests **nurture** plays a big role in **brain development**.

3) Scientists also study the effects of **different genes** (nature) on the **brain development** of animals raised in **similar environments**. They usually do this by **genetically engineering mice** to **lack** a particular **gene**. Any differences between the brain development of the genetically engineered mice and **normal mice** are **more likely** to be due to **nature** than nurture.
4) For example, animal experiments have shown that:

Mice engineered to **lack** the **Lgl1 gene** develop **enlarged brain regions** and **fluid builds up** in their brains. This suggests that **nature** plays a big role in **brain development**.

2 *Newborn Studies*

The brain of a newborn baby has been affected a bit by the environment in the womb.

1) The brain of a newborn baby **hasn't** really been **affected** by the **environment**.
2) Scientists study the brains of newborn babies to see what **functions** they're **born with** and **how developed different parts** of the brain are — what they're born with is **more likely** to be due to **nature** than nurture.
3) For example, newborn studies have shown that:

- Babies are **born** with a number of **abilities**, e.g. they can **cry**, **feed**, **recognise** a **human face**. This suggests that **nature** plays a big role in controlling these abilities.
- Newborn babies **don't** have the ability to **speak**, suggesting that **nurture** plays a big role in the ability to **speak**.

Brain Development and Habituation

These pages are for** Edexcel Unit 5 **only.

3 *Twin Studies*

1) **Identical twins** are **genetically identical**.
2) If identical twins are **raised separately** then they'll have **identical genes** but **different environments**.
3) Scientists can compare the brain development of **separated identical twins** — any **differences** between them **are due to nurture** not nature, and any **similarities** between them are due to **nature**, for example:

   Identical twins have **very similar IQ scores** — suggesting **nature** plays a big role in **intelligence**.

4) Scientists also **study** the brain development of **identical twins raised together**. These twins are **genetically identical** and have **similar environments**, so it's hard to tell if any differences between them are due to nature or nurture. So scientists compare them to **non-identical twins** (who are **genetically different** but have **similar environments**) — they act like a control to **cancel out** the **influence** of the **environment**. So any difference in brain development between identical and non-identical twins is **more likely** to be due to **nature** than nurture.
5) For example:
   - **Stuttering** of both twins is **more common** in **identical twins** than in non-identical twins. This suggests **nature** plays a big role in developing the **speech area** of the brain.
   - There's **no difference** in **reading ability** between pairs of identical and non-identical twins. This suggests **nurture** plays a big role in **reading ability**.

4 *Brain Damage Studies*

1) **Damage** to an adult's brain can lead to the **loss** of **brain function**, e.g. a stroke may cause loss of vision.
2) If an **adult's brain** is damaged, it can't repair itself so well because it's **already fully developed**. But a **child's brain** is **still developing** — so scientists can **study** the effects of **brain damage** on their development.
3) To do this, scientists **compare** the development of a chosen **function** (e.g. language) in children **with** brain damage to those **without** brain damage.
4) If the **characteristic still develops** in children who have brain damage, then brain development is **more likely** to be due to **nurture** than nature for that characteristic.
5) If the **characteristic doesn't develop** in children who have brain damage, then brain development is **more likely** to be due to **nature** than nurture for that characteristic (because nurture isn't having an effect).
6) For example:
   - Children aged 1-3 who were **born** with **damage** to the area of the brain associated with **language**, show a **delay** in the major language milestones (e.g. understanding words, producing sentences) when compared to children born without brain damage.
   - But by **age 5**, their **language skills** are the **same** as children with no brain damage — showing that if a young child's brain is damaged, they can **still develop language**.
   - This suggests that **nurture** plays a big role in **language development**.

5 *Cross-Cultural Studies*

1) Children brought up in **different cultures** have **different environmental influences**, e.g. social practices, beliefs, education, gender influences.
2) Scientists can study the **effects** of a different upbringing on **brain development** by comparing **large groups** of children who are the **same age** but from **different cultures**.
3) Scientists look for **major differences** in **characteristics** — any **differences** in brain development between different cultures are **more likely** to be due to **nurture** than nature. Any **similarities** in brain development between different cultures are **more likely** to be due to **nature** than nurture.
4) For example:
   - **Kenyan children** who eat **protein-rich food** (providing nutrients such as **zinc** and **iron**) have **higher IQs** than children who have a **poor diet** and **limited protein**. This suggests **nurture** plays a big role in **intelligence**.
   - The **mapping abilities** (e.g. perspective drawing) of young children are **well-developed across cultures**. This suggests that **nature** plays a big role in **mapping abilities**.

Brain Development and Habituation

Habituation is a type of *Learned Behaviour*

1) Animals (including humans) **increase** their chance of **survival** by **responding** to **stimuli** (see p. 82).
2) But if the stimulus is **unimportant** (if it's not threatening or rewarding), there's **no point** in **responding** to it.
3) If an unimportant stimulus is **repeated** over a period of **time**, an animal **learns** to **ignore it**.

E.g. you **learn** to **sleep through traffic noise** at night.

4) This **reduced response** to an **unimportant stimulus** after **repeated** exposure **over time** is called **habituation**.
5) Habituation means animals **don't waste time** and **energy** responding to unimportant stimuli.

E.g. **prairie dogs** use **alarm calls** to warn others of a predator — but they've **habituated** to **humans** because we're **not a threat**. They **no longer** send out alarm calls when they see humans, so they **don't waste time** or **energy**.

6) Animals still remain **alert** to **important stimuli** though (stimuli which might **threaten** their **survival**).

E.g. you can sleep through traffic noise, but you'll **instantly wake up** if you hear an **unfamiliar noise**.

7) You need to know how to **investigate habituation** to a stimulus:

To **investigate habituation**, you need to be able to **measure** an animal's **response** to an **unimportant stimulus**. Here's how you could do it with **tortoises**, but the principles are the **same** for any organism:

1) **Gently tap** a **tortoise** on its shell — it should cause the tortoise to **withdraw** its head, feet and tail **into its shell**.
2) **Time** how **long** it takes for the tortoise to **reappear** out of its shell after you've tapped it.
3) Tap the tortoise's shell at **regular intervals** (e.g. every minute) and **record** the **time** it takes for it to reappear.

Fearless Mr Tappy made the experiment very boring for Bob.

If **habituation has taken place** the tortoise should **reappear** out of its shell **quicker** the **more** you **repeat** the **stimulus** (or it **might not withdraw** at all eventually).
If habituation **hasn't occurred** the tortoise will take the **same length of time** to **reappear** each time.

The tortoise should still **remain alert** to an **unfamiliar stimulus**, e.g. if you **clap** your hands near the tortoise it should **cause it to withdraw** into its shell.

Practice Questions

These questions cover pages 179-181.

Q1 What is meant by the term 'brain development'?

Q2 What is meant by the term 'nurture'?

Exam Questions

Q1 A scientist carries out a series of studies on newborn babies to find out how brain development is affected by nature.
- a) Explain what is meant by 'nature'. [1 mark]
- b) What is the advantage of using newborn babies to study the effects of nature on brain development? [2 marks]
- c) Suggest two other types of study that a scientist could carry out to directly research the effect of nature and nurture on brain development in humans. [2 marks]

Q2 A birdwatcher sat in his garden quietly for an hour each day. At first this scared the birds away, but over the next few weeks he saw more and more birds. The birdwatcher predicted that this was due to habituation.
- a) Explain why the birds' behaviour could be described as habituation. [3 marks]
- b) What is the advantage of habituation to birds in the wild? [1 mark]

You say nature and I say nurture — let's call the whole thing off...

Whether a brain function is influenced by nature or nurture is pretty important — if it's heavily influenced by nurture then you can figure out how to improve that brain function by changing an organism's environment. Clever stuff.

Development of the Visual Cortex

These pages are for Edexcel Unit 5 only.

Your brain continues to develop even after you're born — good job too, you'd have trouble learning this stuff if it didn't.

Animal Models are used to *Study Brain Development*

Some animals have fairly **similar brains** to **humans**. This means scientists can do **experiments** on these animals (that would be **unethical** to do **in humans**) to **investigate brain development**, e.g. experiments have been done on cats to investigate the development of the visual cortex (see below).

The Visual Cortex *is Made Up of* Ocular Dominance Columns

1) The **visual cortex** is an area of the **cerebral cortex** (see p. 176) at the **back** of your brain.
2) The role of the visual cortex is to **receive** and **process visual information**.
3) **Neurones** in the visual cortex **receive information** from **either** your **left** or **right eye**.
4) Neurones are **grouped together** in columns called **ocular dominance columns**. If they receive information from the **right eye** they're called **right** ocular dominance columns, and if they receive information from the **left eye** they're called **left** ocular dominance columns.
5) The columns are the **same size** and they're arranged in an **alternating pattern** (left, right, left, right) across the visual cortex.

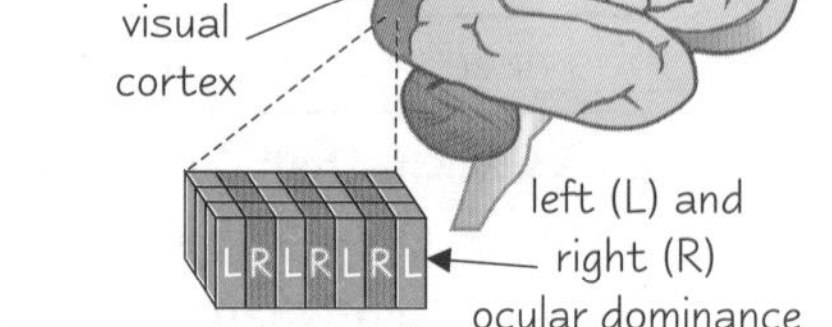

Hubel and *Wiesel* Investigated *How* the *Visual Cortex Develops*

1) The **structure** of the **visual cortex** was **discovered** by two scientists called **Hubel** and **Wiesel**.
2) They used **cats** and **monkeys** to study the **electrical activity** of **neurones** in the visual cortex.
3) They found that the **left ocular dominance columns** were **stimulated** when an animal used its **left eye**, and the **right ocular dominance columns** were **stimulated** when it used its **right eye**.
4) Hubel and Wiesel investigated **how** the **visual cortex develops** by experimenting on very **young kittens**:

   - They **stitched shut one eye** of each kitten so they could only see out of their other eye.
   - The kittens were kept like this for **several months** before their eyes were **unstitched**.
   - Hubel and Wiesel found that the kitten's **eye** that had been **stitched up** was **blind**.
   - They also found that **ocular dominance columns** for the **stitched up eye** were a **lot smaller** than normal, and the ocular dominance columns for the **open eye** were a **lot bigger** than normal.
   - The ocular dominance columns for the **open eye** had **expanded** to **take over** the other columns that **weren't** being stimulated — when this happens, the **neurones** in the visual cortex are said to have **switched dominance**.

5) Hubel and Wiesel then investigated if the **same thing** happened in an **adult's brain**, e.g. they experimented on **cats**:

   - They **stitched shut one eye** of each cat, who were kept like this for **several months**.
   - When their eyes were **unstitched**, Hubel and Wiesel found that these eyes **hadn't gone blind**.
   - The cats **fully recovered** their **vision** and their **ocular dominance columns** remained the **same**.

6) Hubel and Wiesel **repeated** the experiments on **young monkeys** and **adult monkeys** — they saw the **same results**.
7) Hubel and Wiesel's experiments showed that the **visual cortex only develops** into normal **left** and **right** ocular dominance columns if **both eyes** are **visually stimulated** in the **very early stages** of **life**.

Their Experiments Provide Evidence *for a* Critical 'Window' *in* Humans

1) Hubel and Wiesel's experiments on cats show there's a **period** in **early life** when it's **critical** that a kitten is **exposed** to **visual stimuli** for its visual cortex to **develop properly**. This period of time is called the **critical 'window'**.
2) The **human visual cortex** is **similar** to a **cat's visual cortex** (the human visual cortex has **ocular dominance columns too**) so Hubel and Wiesel's experiments provide **evidence** for a **critical 'window'** in **humans**.
3) There's **other evidence** for a visual cortex critical 'window' in humans as well (see next page).

Development of the Visual Cortex

Evidence from Human Studies Suggests a Critical 'Window' Does Exist

Scientists have **investigated** how the **visual system develops** in **humans**, e.g. by looking at **cataracts** in the **eye**:

- A **cataract** makes the **lens** in the eye go **cloudy**, causing **blurry vision**.
- If a **baby** has a **cataract**, it's **important** to **remove** the cataract within the **first few months** of the baby's life — otherwise their visual system **won't develop properly** and their vision will be **damaged for life**.
- If an **adult** has a **cataract** then it's not so serious — when the cataract is **removed, normal vision** comes back **straight away**. This is because the visual system is **already developed** in an adult.

Using Animals in Medical Research Raises Ethical Issues

Hubel and Wiesel used **animals** in their experiments, which is **common** in **medical research**.
This raises some **ethical issues** — you need to **know** a range of **arguments** from **both sides**:

Arguments AGAINST using animals in medical research	Arguments FOR using animals in medical research
Animals are different from humans, so drugs tested on animals may have different effects in humans.	Animals are similar to humans, so research has led to loads of medical breakthroughs, e.g. antibiotics, insulin for diabetics, organ transplants.
Experiments can cause pain and distress to animals.	Animal experiments are only done when it's absolutely necessary and scientists follow strict rules, e.g. animals must be properly looked after, painkillers and anaesthetics must be used to minimise pain.
There are alternatives to using animals in research, e.g. using cultures of human cells or using computer models to predict the effects of experiments.	Using animals is currently the only way to study how a drug affects the whole body — cell cultures and computers aren't a true representation of how cells may react when surrounded by other body tissues. It's also the only way to study behaviour.
Some people think that animals have the right to not be experimented on, e.g. animal rights activists.	Other people think that humans have a greater right to life than animals because we have more complex brains, e.g. compared to rats, fish, fruit flies (which are commonly used in experiments).

Practice Questions

Q1 Where in the brain are ocular dominance columns found?

Q2 What kind of pattern are ocular dominance columns arranged in?

Q3 Describe one piece of evidence that suggests a critical window exists for human visual system development.

Q4 Give two arguments against the use of animals in medical research.

Exam Question

Q1 Hubel and Wiesel conducted experiments on animals to investigate the structure and development of the visual cortex.

a) Describe their experiment on kittens and explain what this showed about how the visual cortex develops. [4 marks]

b) Do their experiments give evidence for a critical 'window' in the development of the human visual system? Explain your answer, with reference to what is meant by a critical 'window'. [2 marks]

c) Give two arguments for using animals in medical research. [2 marks]

Hubel and Wiesel — weren't they a '60s pop duo...

The experiments that Hubel and Wiesel did on cats and monkeys may have been a bit gross, but they did provide us with good knowledge of how the visual cortex develops. There's evidence to suggest a critical 'window' for other things too, like language development. Interesting stuff, so make sure you learn it and then you can impress others (like the examiners).

Behaviour

These pages are for OCR Unit 5 only. If you're doing Edexcel Unit 5, you can now put your feet up.

I thought there were only two types of behaviour — behaving and misbehaving. Turns out there are a few more...

Behaviour** is an Organism's **Response** to **Changes** in its External **Environment

1) Responding in the **right way** helps organisms **survive** and **reproduce** (e.g. by finding food and a mate).
2) An organism's behaviour is influenced by both its **genes** and its **environment**.

Innate Behaviour** is **Instinctive** and **Inherited

1) **Innate** behaviour is **behaviour** that organisms do **instinctively**.
2) It's **genetically determined** — it's **inherited** from parents and it's **not** influenced by the **environment**.
3) It's also **stereotyped** — it's always carried out in the **same way** and by **all** the **individuals** in a species.
4) The **advantage** of innate behaviour is that organisms **respond** in the **right way** to the stimulus **straight away** because **no learning** is needed, e.g. newborn babies instinctively suckle from their mothers.
5) You need to know **three examples** of **innate behaviours**:

① **Escape reflexes** — the organisms **move away** from **potential danger**.

E.g. **cockroaches run away** when your foot's about to squash them.

Dave's escape reflex was about to kick in.

Taxes (**tactic** responses) and **kineses** (**kinetic** responses) allow simple organisms to move **away from unpleasant stimuli** and into **more favourable** environments:

② **Taxes** — the organisms move **towards** or **away from** a **directional stimulus.**

E.g. **woodlice** move **away from** a **light source**. This helps keep them **concealed** under stones where they're **safe** from predators, so it helps them **survive**.

③ **Kineses** — the organisms' **movement** response is affected by a **non-directional stimulus**, e.g. **intensity**.

E.g. **woodlice** show a **kinetic** response to **low humidity**. This helps them **move** from **drier air** to more **humid air**, and then **stay put**. This **reduces** their **water loss** so **improves** their **survival** chances.

Learned Behaviour** is **Behaviour** that's **Modified** as a Result of **Experience

Learned behaviour is **influenced** by the **environment**. It allows animals to **respond** to **changing conditions**, e.g. they **learn** to **avoid harmful food**. Here are some **examples** that you need to **learn**:

❶ Habituation

- **Habituation** is a **reduced response** to an **unimportant stimulus** after **repeated** exposure **over time**.
- An **unimportant stimulus** is a change that **isn't threatening or rewarding**. An animal quickly learns to **ignore** it so it **doesn't waste time** and **energy** responding to unimportant things, e.g. you **learn** to **sleep through traffic noise** at night.
- Animals **remain alert** to **unfamiliar stimuli** though, e.g. you instantly **wake up** if you hear an **unfamiliar noise**.

❷ Classical Conditioning

- **Classical conditioning** is **learning** to **respond naturally** to a **stimulus** that **doesn't normally** cause that response.
- A **natural stimulus** (called the **unconditioned** stimulus) can cause a **natural response** (called the unconditioned response). E.g. in dogs, **food** (an **unconditioned** stimulus) causes **salivation** (an unconditioned response).
- If another stimulus **coincides** with an **unconditioned** stimulus **enough times**, eventually this other stimulus will cause the **same response**. E.g. if a **bell** is rung **immediately before** dogs are given **food**, after a **time** the dogs will learn to **salivate** in **response** to the **bell only**.

The behaviour is automatic.

Behaviour

3 Operant Conditioning

- **Operant conditioning** is **learning** to **associate** a particular **response** with a **reward** or a **punishment**.
- When put in the **same situation lots of times**, an animal will work out **which response** gets a **reward** (e.g. pressing the right lever gets food) or a **punishment** (e.g. pressing the left lever gives a shock).
- The response must be **rewarded** (or punished) **straight away** — this **reinforces** the animal's behaviour so it's **more likely** to respond in the **same way** to get the **reward again** (or less likely to do it to be punished again).

E.g. a **rat** was put in a cage with a **choice** of **levers**. Pressing one of the levers **rewarded** the rat with food **straight away**. The rat was **repeatedly** put in the **same cage**, so learned which **lever** to **press** to get the **reward**.

- Lots of mistakes are made at first, but animals **quickly learn** to make **fewer mistakes** by using **trial and error**.

4 Latent Learning

- **Latent** learning is **hidden** learning — an animal **doesn't immediately show** it's learned something.
- It involves **learning** through **repeatedly** doing the **same task**.
- The animal only **shows** it's learned something when it's given a **reward** or a **punishment**.

E.g. **three groups** of **rats** were **repeatedly** put in the **same maze**:

1) The first group of rats were **reinforced** with a **reward** each time they reached the **end** of the **maze** — they **quickly learned** their way around the maze.
2) The second group of rats were **not reinforced** (they didn't receive a reward) — they continued to plod about the maze and **took ages** to reach the end.
3) The third group of rats were **only rewarded** from the **11th time** they did the maze — after this they **very quickly reached the end**, with **hardly any errors**. The rats had been **learning** the maze all along **without reinforcement**, but they **didn't show** their learning **until** there **was** a **reward**.

5 Insight Learning

- **Insight** is learning to **solve** a **problem** by **working out** a **solution** using **previous experience**.
- Solving problems by **insight** is **quicker** than by trial and error because actions are **planned** and **worked out**.

E.g. **chimpanzees** were put in a play area with **sticks**, **clubs** and **boxes**. Bunches of **bananas** were hung just **out of reach**. The chimps used their **previous experience** of playing with the objects to **work out** a **solution** — they **piled up** the boxes to reach the bananas, and used the sticks and clubs to **knock them down**.

Practice Questions

Q1 What is an escape reflex?

Q2 What's the difference between taxes and kineses?

Q3 What is habituation?

Exam Questions

Q1 When a postman puts a letter through Number 10, a dog barks loudly causing the postman's heart rate to increase. This happens for the next few days until the postman's heart rate increases as he approaches Number 10, even if the dog is not there. State what type of learning has occurred and give a reason for your answer. [2 marks]

Q2 Give an example of how operant conditioning could be used in dog training. [2 marks]

My hair's so shiny — it's classically conditioned...

Behaviour is a bit of a weird topic, really. It seems you can only find anything out by doing lots of strange experiments. Might be fun to work in this area of research though — it's certainly not your average day at the office. "How was your day, dear?" "Great thanks — the rats sped around the maze to get their chocolate." And who said science was boring...

Behaviour

These pages are for OCR Unit 5 only.

All this studying of animal behaviour helps us to understand human behaviour. But animals are a bit different from us. Not many dogs study for exams, for a start. Mind you, they don't need A2 levels — they've already got a pe-degree...

Imprinting is a Combination of *Innate* and *Learned* Behaviour

1) **Imprinting** is a combination of a **learned** behaviour and an **innate** behaviour — e.g. an animal **learns to recognise its parents**, and **instinctively follows them**.
2) Imprinting occurs in several species, mainly **birds**, which are **able to move** very **soon** after they're **born**. A newly-born animal has an **innate instinct** to **follow** the **first moving object** it sees — usually this would be its **mother** or **father**, who would **provide warmth**, **shelter** and **food** (helping it to survive).
3) But the animal has **no innate instinct** of what its parents **look like** — they have to **learn** this.
4) Imprinting **only happens** during a certain period of time **soon after** the animal is **born**. This period of time is called the **critical period**.

E.g. ducklings usually imprint on their parent ducks. But if ducklings are **reared from birth** (during the **critical period**) by a **human**, then the human is the **first moving object** the ducklings **see** — so the ducklings **imprint** on the human (they follow them).

"Who's your daddy..."

5) Once learned, imprinting is **fixed** and **irreversible**. Animals use imprinting later in life to **identify mates** from the **same species**.

The Dopamine Receptor D_4 *is* Linked *to* Human Behaviour

1) An animal's **behaviour** depends on the **structure** and **function** of its **brain** (e.g. neurotransmitters, synapses, receptors, etc.).
2) Even fairly **small differences** in the brain can produce **big differences** in **behaviour**.
3) Much of our **understanding** of **human behaviour** comes from **studying** people with **abnormal behaviour**, to see how their **brains** are **different** from the brains of people who behave 'normally'.
4) Any **differences** in the brain give scientists **clues** to understanding how normal behaviour is **controlled**. For example:

- The **D_4 receptor** is a receptor in the brain for a neurotransmitter called **dopamine**.
- Having **too many D_4 receptors** in the brain has been **linked** to **abnormal behaviour**, e.g. the abnormal behaviour seen in **schizophrenia** — a disorder that affects **thinking**, **perception**, **memory** and **emotions**.
- The **evidence** for this **link** includes:

  1. If **drugs** that **stimulate** dopamine receptors are given to **healthy people**, it **causes** the **abnormal behaviour** seen in **schizophrenia**.
  2. **Drugs** that **block** D_4 receptors **reduce symptoms** in people with schizophrenia.
  3. People with **schizophrenia** have a **higher density** of **D_4 receptors** in their brain.
  4. One of the drugs that's used to **treat** schizophrenia **binds** to D_4 receptors **better** than it binds to other dopamine receptors.

- The **link** between the **D_4 receptor** and **abnormal behaviour** helps us to understand the **role** that the **D_4 receptor** plays in **normal behaviour**, e.g. it's involved in **thinking**, **perception**, **memory** and **emotions**.

See p. 92 if you can't remember about neurotransmitters and their receptors.

The D4 receptor protein is made by the DRD4 gene.

Behaviour

Social Behaviour in Primates has Many Advantages

Many animals **live together** in large **groups**. Behaviour that involves members of the group **interacting** with each other is called **social behaviour**. The **primates** (e.g. baboons, apes, humans) have more developed **social behaviour** than other animals. Social behaviour has many **advantages**. Here are some **examples** of **social behaviour** in **baboons** and the **advantages** of the behaviours:

1) **Baboons** live in groups, with about 50 baboons in each group.

   A large group like this is **more efficient** at **hunting** for **food** — together the baboons can **search** a **large area** and **communicate** back to the group where there's a good source of **food**.

2) Within each group there's a **clear-cut hierarchy** of **adult males**.

   This helps to **prevent fighting** (which **wastes energy**) because the males already **know** their **rank order** in the group.

The kids knew they'd have to move up the hierarchy before they could enjoy their go on the slide.

3) As each group **moves through** its own **territory** hunting for food, baboons **cooperate** with each other — **infant baboons** stay with their **mother** in the **middle** of the group and the **adult males** stay on the **outside** of the group.

   Infants and the **females** are **protected** if they're on the **inside** of the group. The young baboons need to be kept **safe** and there needs to be **enough female baboons** for the males to **mate** with, to make sure that **reproduction** is **successful** and the group continues.

4) Members of the group **groom** each other (they **pick out small insects** and **dirt** from each other's **fur**).

   Grooming is **hygienic** and helps to **reinforce** the **social bonds** within the group.

Practice Questions

Q1 Give an example of a species that shows imprinting behaviour.

Q2 Can imprinting be reversed?

Q3 Name one behaviour that's linked to the D_4 receptor.

Q4 Give one example of social behaviour in primates.

Exam Questions

Q1 Goslings usually imprint on their parent geese. However, it's possible for a gosling to imprint on a human.

a) Explain what is meant by imprinting. [1 mark]

b) Explain how a gosling can imprint on a human. [2 marks]

Q2 Gorillas eat leaves, fruits and bark. They usually live in groups of 8-12 individuals. They exhibit many social behaviours that have many advantages.

a) Describe what is meant by the term 'social behaviour'. [1 mark]

b) Suggest a possible advantage to gorillas of:

i) working together to look for food. [1 mark]

ii) grooming each other. [1 mark]

Dopamine — wasn't he one of the seven dwarfs...

More crazy behaviour stuff. Apparently, a duckling will imprint to almost anything that moves about and makes a noise — try it with a football next time you're near a pond. No, don't really. It'll only go and disturb every match you play. So forget the practical, concentrate on the theory — learn these pages and then you've finished the book. Give me a whoop whoop...

Viral and Bacterial Infections

This section is for Edexcel Unit 4 only.

Unfortunately, some microorganisms have got nothing better to do than cause disease. The little blighters...

Pathogens Cause *Infectious Diseases*

1) A **pathogen** is **any organism** that **causes disease**.
2) Diseases caused by pathogens are called **infectious diseases**.
3) Pathogenic microorganisms include **some bacteria**, **some fungi** and **all viruses**.
4) As an infectious disease **develops** in an organism it causes a **sequence of symptoms**, which **may** lead to **death**.
5) You need to know the sequence of symptoms for the diseases caused by the **human immunodeficiency virus** (HIV) and ***Mycobacterium tuberculosis***:

1 The *Human Immunodeficiency Virus (HIV)* Causes *AIDS*

1) The **human immunodeficiency virus** (**HIV**) infects and destroys **immune system cells**.
2) HIV infection eventually leads to **acquired immune deficiency syndrome** (**AIDS**).
3) AIDS is a condition where the immune system **deteriorates** and eventually **fails**.
4) People with HIV are classed as having AIDS when **symptoms** of their **failing immune system** start to **appear**.
5) AIDS sufferers generally develop diseases and infections that **wouldn't** cause serious problems in people with a **healthy** immune system — these are called **opportunistic infections**.
6) The length of **time** between **infection** with HIV and the **development** of AIDS **varies** between individuals but it's usually **8-10 years**. The disease then progresses through a **sequence of symptoms**:

> 1) The **initial symptoms** of AIDS include **minor infections** of mucous membranes (e.g. the inside of the nose, ears and genitals), and recurring respiratory infections. These are caused by a **lower than normal** number of **immune system cells**.
> 2) As AIDS **progresses** the number of **immune system cells decreases** further. Patients become susceptible to **more serious infections** including chronic diarrhoea, severe bacterial infections and tuberculosis (see below).
> 3) During the **late stages** of AIDS, patients have a **very low number** of immune system cells and suffer from a **range of serious infections** such as toxoplasmosis of the brain (a parasite infection) and candidiasis of the respiratory system (fungal infection). It's these serious infections that kill AIDS patients, not HIV itself.

The infections become more and more serious as there are fewer and fewer immune system cells to fight them.

2 The *Bacterium* Mycobacterium tuberculosis *Causes Tuberculosis (TB)*

1) ***Mycobacterium tuberculosis*** infects **phagocytes** (see p. 191) in the **lungs**.
2) It causes the lung disease **tuberculosis** (**TB**).
3) Most people **don't develop TB straight away** — their immune system **seals off** the **infected phagocytes** in structures in the lungs called **tubercles**.
4) The bacteria become **dormant** inside the tubercles and the infected person shows **no obvious symptoms**.
5) Later on, the **dormant bacteria** can become **reactivated** and **overcome** the **immune system**, causing TB.
6) **Reactivation** is **more likely** in people with **weakened immune systems**, e.g. people with **AIDS** (see above).
7) The length of **time** between the **infection** with *Mycobacterium tuberculosis* and the **development** of TB **varies** between individuals — it can be **weeks** to **years**. TB then progresses through a **sequence of symptoms**:

> 1) The **initial symptoms** of TB include **fever**, **general weakness** and **severe coughing**, caused by **inflammation** of the lungs.
> 2) As TB **progresses** it **damages** the **lungs** and if it's left **untreated** it can cause **respiratory failure**, which can lead to **death**.
> 3) TB can also **spread** from the lungs to **other parts** of the body, e.g. the **brain** and **kidneys**. If it's left **untreated** it can cause **organ failure**, which can lead to **death**.

See p. 190 for more on inflammation.

Viral and Bacterial Infections

You need to Know the Structure of Bacteria...

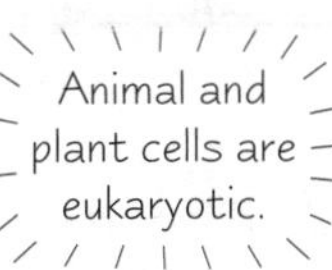

1) Bacteria are **single-celled**, **prokaryotic** microorganisms (prokaryotic means they have **no nucleus**).
2) Most bacteria are only a **few micrometers** (μm) long, e.g. the TB bacterium is about 1 μm.
3) Bacterial cells have a **plasma membrane**, **cytoplasm**, **ribosomes** and other features:

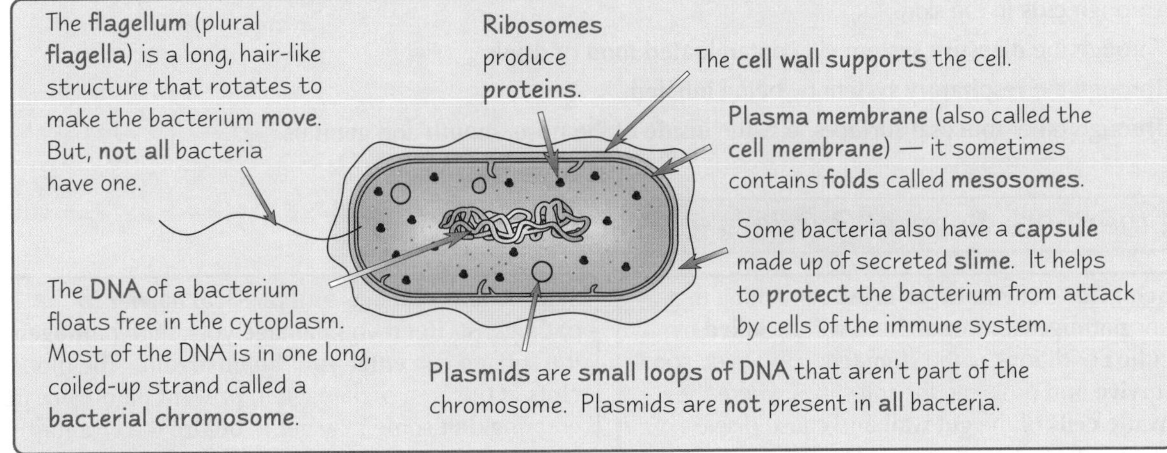

...And Viruses

1) Viruses are microorganisms but they're **not cells** — they're just **nucleic acids** surrounded by **protein**.
2) They're **tiny**, even **smaller** than bacteria, e.g. HIV is about 0.1 μm across.
3) **Unlike** bacteria, viruses have **no** plasma membrane, **no** cytoplasm and **no** ribosomes.
4) But they do have **nucleic acids** (like bacteria) and some other features:

Remember — DNA and RNA are nucleic acids.

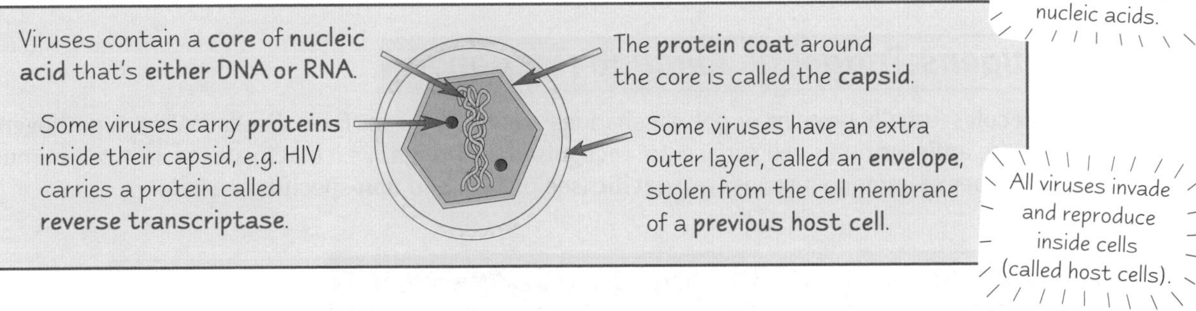

All viruses invade and reproduce inside cells (called host cells).

Practice Questions

Q1 What is the name of the bacterium that causes tuberculosis?

Q2 Describe two ways in which tuberculosis can cause death.

Exam Questions

Q1 People infected with HIV eventually develop AIDS.
Describe and explain the sequence of symptoms for AIDS, from infection to death. [7 marks]

Q2 All viruses are pathogens and some bacteria are pathogens.
Give three differences between the structure of bacteria and the structure of viruses. [3 marks]

My computer has a virus — I knew I shouldn't have sneezed on it...

Not the nicest of topics, but one you have to learn about — you need to know the sequence of symptoms, from infection to death, that are caused by AIDS and TB. I always find it weird just how simple a virus is. For something so basic, it can do a lot of damage to humans. I'm sure there's a moral in there somewhere, but don't worry, I won't bore you with it.

Infection and The Non-Specific Immune Response

These pages are for Edexcel Unit 4 only.

If you're a bit worried about getting an infection, fear not — your body can stop those pesky pathogens in their tracks...

Pathogens Need to Enter the Body to Cause Disease...

Pathogens can **enter** the body via four major routes:

1) Through **cuts** in the **skin**.
2) Through the **digestive system** via **contaminated food** or **drink**.
3) Through the **respiratory system** by being **inhaled**.
4) Through other **mucosal surfaces**, e.g. the **inside** of the **nose**, **mouth** and **genitals**.

...but there are Several Barriers to Prevent Infection

Stomach acid — If you **eat** or **drink** something that contains **pathogens**, most of them will be **killed** by the **acidic** conditions of the **stomach**. However, some may **survive** and pass into the intestines where they can **invade cells** of the **gut wall** and cause disease.

Skin — Your skin acts as a **physical barrier** to pathogens. But if you **damage** your skin, **pathogens** on the surface can **enter** your **bloodstream**. The blood **clots** at the area of damage to **prevent** pathogens from entering, but some may get in **before** the clot forms.

Gut and skin flora — Your **intestines** and **skin** are **naturally covered** in billions of **harmless microorganisms** (called **flora**). They **compete** with **pathogens** for **nutrients** and **space**. This **limits** the **number** of **pathogens** living in the gut and on the skin and makes it **harder** for them to **infect** the body.

Lysozyme — **Mucosal surfaces** (e.g. eyes, mouth and nose) produce **secretions** (e.g. tears, saliva and mucus). These secretions all **contain** an **enzyme** called **lysozyme**. Lysozyme **kills bacteria** by **damaging** their **cell walls** — it makes the bacteria **burst open** (**lyse**).

Foreign Antigens Trigger an Immune Response

Antigens are **molecules** (usually proteins or polysaccharides) found on the **surface** of **cells**. When a **pathogen invades** the body, the **antigens** on its cell surface are **recognised as foreign**, which **activates** cells in the **immune system**. The body has **two types** of immune response — **specific** (see p. 192) and **non-specific** (see below).

The Non-Specific Immune Response Happens First

The non-specific response happens in the **same** way for **all microorganisms** (regardless of the foreign antigen they have) — it's **not** antigen-specific. It starts attacking the microorganisms **straight away**. You need to know about **three mechanisms** that are part of the non-specific immune response:

1 Inflammation at the Site of Infection

The **site** where a **pathogen enters** the body (the **site of infection**) usually becomes **red**, **warm**, **swollen** and **painful** — this is called **inflammation**. Here's how it happens:

1) Immune system cells **recognise foreign antigens** on the surface of a pathogen and **release molecules** that trigger inflammation.
2) The molecules cause **vasodilation** (**widening** of the blood vessels) around the site of infection, **increasing** the **blood flow** to it.
3) The molecules also **increase** the **permeability** of the **blood vessels**.
4) The increased blood flow brings **loads** of **immune system cells** to the **site of infection** and the increased permeability allows those cells to **move out** of the blood vessels and **into** the infected tissue.
5) The immune system cells can then start to **destroy** the **pathogen**.

Trevor's throat infection triggered a small amount of inflammation.

Infection and The Non-Specific Immune Response

2) Production of *Anti-Viral Proteins* Called *Interferons*

1) When cells are **infected** with **viruses**, they produce **proteins** called **interferons**.
2) Interferons help to **prevent** viruses **spreading** to **uninfected cells**.
3) They do this in several ways:

- They **prevent** viral **replication** by **inhibiting** the production of **viral proteins**.
- They **activate** cells involved in the **specific** immune response (see p. 192) to **kill** infected cells.
- They **activate** other mechanisms of the **non-specific** immune response, e.g. they **promote inflammation** to bring immune system cells to the **site of infection** (see previous page).

3) *Phagocytosis*

A **phagocyte** (e.g. a macrophage) is a type of **white blood cell** that carries out **phagocytosis** (**engulfment** of pathogens). They're found in the **blood** and in **tissues** and are the **first cells** to **respond** to a pathogen inside the body.
Here's how they work:

1) A phagocyte **recognises** the **antigens** on a pathogen.
2) The cytoplasm of the phagocyte moves round the pathogen, **engulfing** it.
3) The pathogen is now contained in a **phagocytic vacuole** (a bubble) in the cytoplasm of the phagocyte.
4) A **lysosome** (an organelle that contains **digestive enzymes**) **fuses** with the phagocytic vacuole. The enzymes **break down** the pathogen.
5) The phagocyte then **presents** the pathogen's **antigens**. It sticks the antigens on its **surface** to **activate** other immune system cells (see p. 192) — so it's also called an **antigen-presenting cell**.

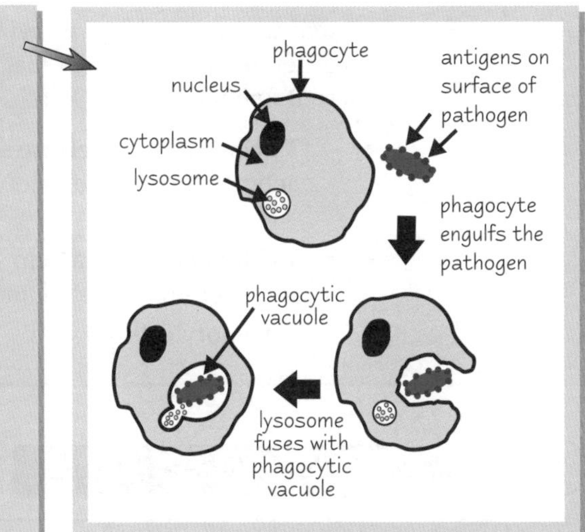

Practice Questions

Q1 State four ways in which pathogens can enter the body.

Q2 State two barriers that prevent infection.

Q3 What are antigens?

Q4 Describe two ways in which interferons prevent viruses from spreading to uninfected cells.

Exam Questions

Q1 Inflammation is part of the non-specific immune response.
Describe and explain how it occurs. [6 marks]

Q2 Describe how a phagocyte responds to an invading pathogen. [6 marks]

Studying for exams interferons with your social life...

There's a lot to remember here, but fear not, you just need to know about your body's barriers to pathogens and how the non-specific immune response works. It's also important that you know what an antigen is — it'll help you to understand how the specific immune response works. Next stop, T and B cells — and no, you can't get off...

The Specific Immune Response

These pages are for Edexcel Unit 4 only.

Most pathogens will regret even thinking about sneaking into your body when the specific response gets going...

The *Specific Immune Response* Involves *T and B Cells*

The **specific immune response** is **antigen-specific** — it produces responses that are **aimed** at **specific pathogens**. It involves white blood cells called **T** and **B cells**:

1 Phagocytes *Activate T Cells*

1) A **T cell** is a type of **white blood cell**.
2) Their surface is covered with **receptors**.
3) The receptors **bind** to **antigens** presented by the phagocytes.
4) Each T cell has a **different shaped receptor** on its surface.
5) When the receptor on the surface of a T cell meets a **complementary antigen**, it binds to it — so each T cell will bind to a **different antigen**.
6) This **activates** the T cell — it **divides** and **differentiates** into **different types** of T cells that carry out **different functions**:
   - **T helper cells** — **release substances** to **activate B cells** (see below), **T killer cells** and macrophages.
   - **T killer cells** — **attach** to antigens on a pathogen-infected cell and **kill** the cell.
   - **T memory cells** — see p. 194.

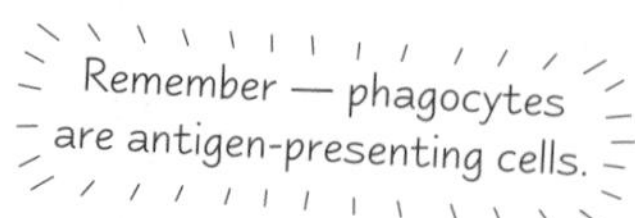

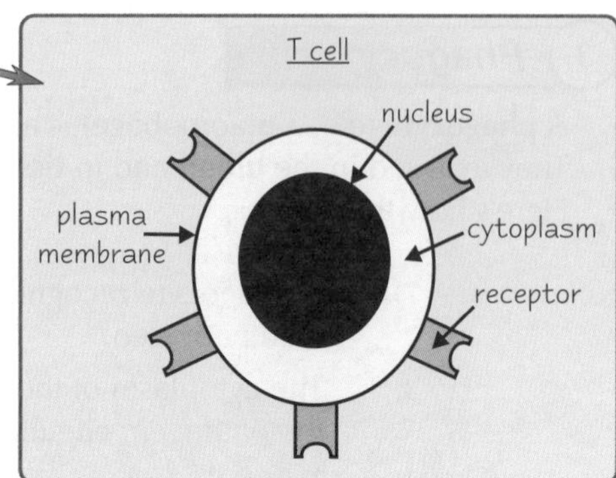

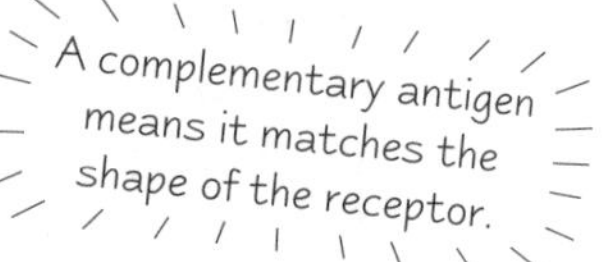

2 T Helper Cells Activate B Cells

1) **B cells** are another type of **white blood cell**.
2) They're covered with proteins called **antibodies**.
3) Antibodies **bind to antigens** to form an **antigen-antibody complex**.
4) Each B cell has a **different shaped antibody** on its surface.
5) When the antibody on the surface of a B cell meets a **complementary antigen**, it binds to it — so each B cell will bind to a **different antigen**.
6) This, together with substances **released** from the T cell, **activates** the B cell.
7) The activated B cell **divides**, by **mitosis**, into **plasma cells** (also called **B effector cells**) and **B memory cells** (see p. 194).

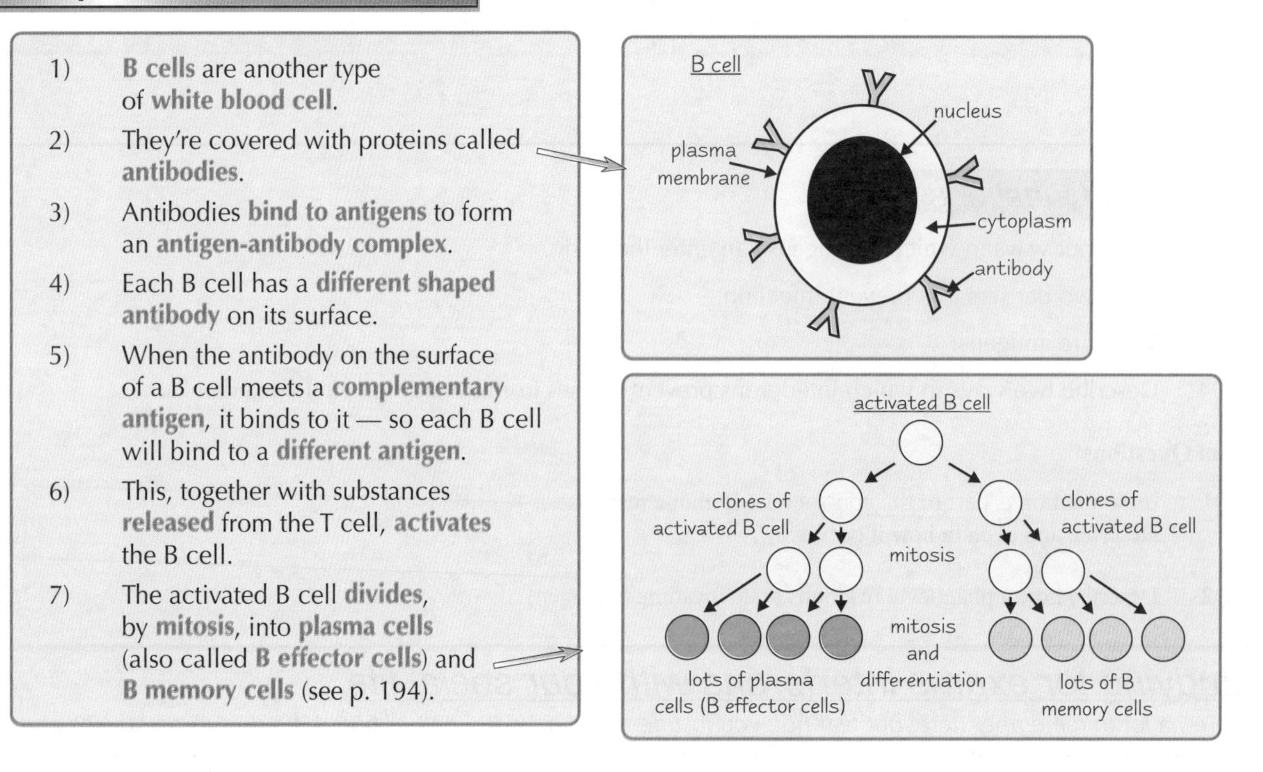

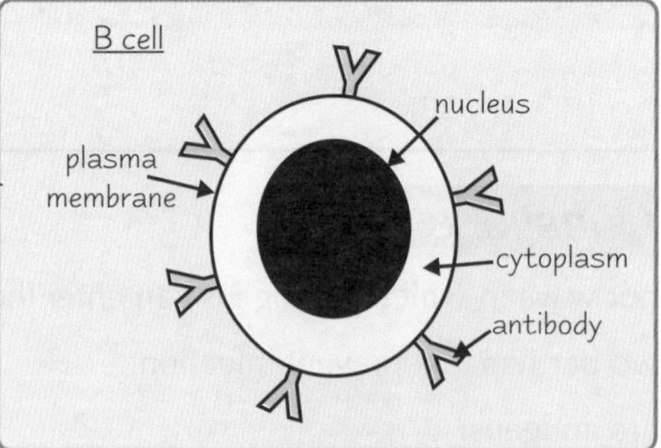

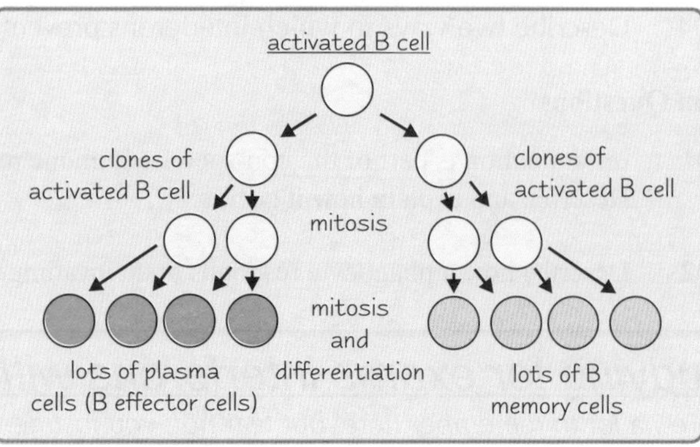

The Specific Immune Response

Plasma cells Make Antibodies to a Specific Antigen

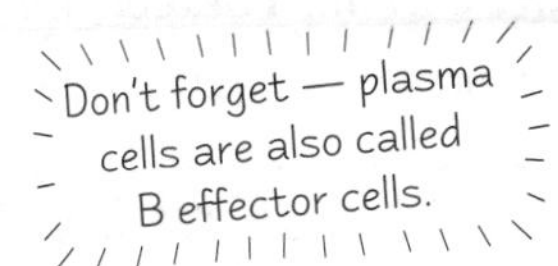

1) **Plasma cells** are **clones** of the **B cells** (they're **identical** to the B cells).
2) They secrete **loads** of the **antibody**, specific to the antigen, into the blood.
3) These antibodies will bind to the antigens on the surface of the pathogen to form **lots** of **antigen-antibody complexes**:

- The **variable regions** of the antibody form the **antigen binding sites.** The **shape** of the variable region is **complementary** to a particular antigen. The variable regions **differ** between antibodies.
- The **hinge region** allows **flexibility** when the antibody binds to the antigen.
- The **constant regions** allow binding to **receptors** on **immune system cells**, e.g. phagocytes. The constant region is the **same in all** antibodies.
- **Disulfide bridges** (a type of bond) hold the polypeptide chains together.

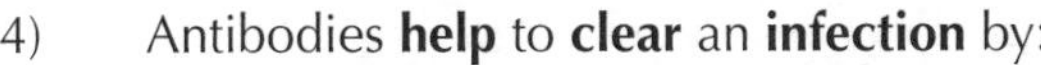

An Antigen-Antibody Complex

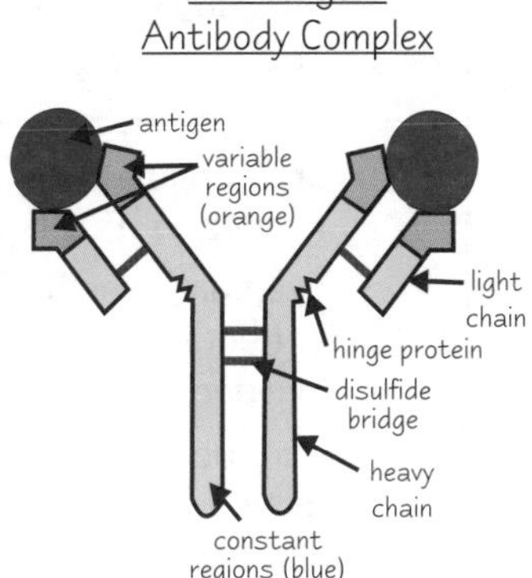

4) Antibodies **help** to **clear** an **infection** by:

1) **Agglutinating pathogens** — each antibody has **two binding sites**, so an antibody can **bind** to **two pathogens** at the **same time** — the pathogens become **clumped together**. Phagocytes then bind to the antibodies and phagocytose a lot of pathogens **all at once**.
2) **Neutralising toxins** — antibodies can **bind** to the **toxins** produced by pathogens. This **prevents** the toxins from **affecting human cells**, so the toxins are **neutralised** (inactivated). The toxin-antibody complexes are also phagocytosed.
3) **Preventing the pathogen binding to human cells** — when antibodies bind to the antigens on pathogens, they may **block** the cell surface **receptors** that the pathogens need to **bind to the host cells**. This means the pathogen **can't attach to** or **infect** the host cells.

Agglutination

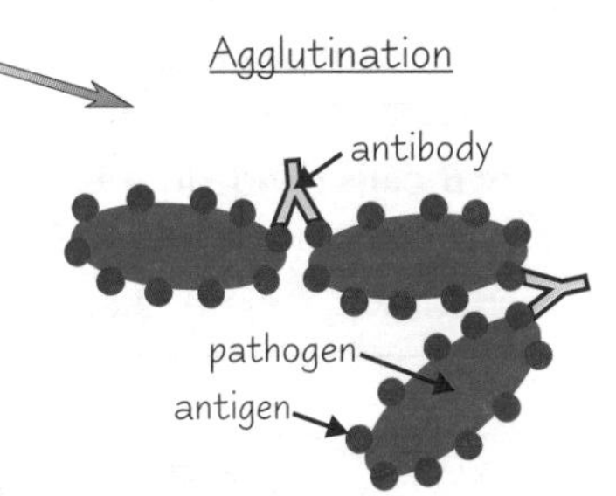

Practice Questions

Q1 What structures are found on the surface of T cells?

Q2 Briefly describe how a B cell is activated.

Q3 What cells do activated B cells divide into?

Exam Questions

Q1 a) Describe how a T cell is activated. [2 marks]

b) Name the cells that activated T cells divide and differentiate into. [3 marks]

Q2 Describe the function of antibodies. [3 marks]

The student-revision complex — only present the night before an exam...

Memory cells are still types of B and T cells. Luckily for you, you get to learn all about them on the next page. Basically they're the elephants of the immune system — they have a really long memory. They stick around in the body waiting for the same pathogen to strike again, and if it does, these cells can immediately get to work destroying it. Ha ha (evil laugh).

Developing Immunity

These pages are for Edexcel Unit 4 only.

Your immune system has a memory — it's handy for fighting infections, not for remembering biology notes...

The Production of **Memory Cells** Gives **Immunity**

1) When a **pathogen** enters the body for the **first time** the **antigens** on its surface **activate** the **immune system**. This is called the **primary response**.
2) The primary response is **slow** because there **aren't many B cells** that can make the antibody needed to bind to the antigen.
3) Eventually the body will produce **enough** of the right antibody to overcome the infection. Meanwhile the infected person will show **symptoms** of the disease.

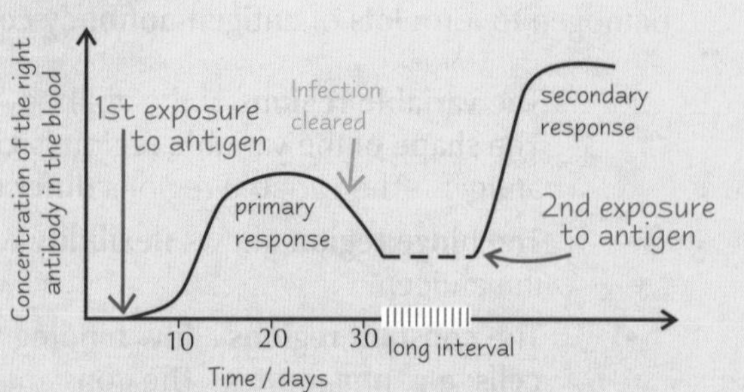

4) After being exposed to an antigen, both T and B cells produce **memory cells**. These memory cells **remain in the body** for a **long** time. Memory T cells remember the **specific antigen** and will recognise it a second time round. Memory B cells record the specific **antibodies** needed to bind to the antigen.
5) The person is now **immune** — their immune system has the **ability** to respond **quickly** to a second infection.
6) If the **same pathogen** enters the body again, the immune system will produce a **quicker**, **stronger** immune response — the **secondary response**.
7) **Memory B cells** divide into **plasma cells** that produce the right antibody to the antigen. **Memory T cells** divide into the **correct type** of **T cells** to kill the cell carrying the antigen.
8) The secondary response often gets rid of the pathogen **before** you begin to show any **symptoms**.

Immunity can be **Active** or **Passive**

ACTIVE IMMUNITY

This is the type of immunity you get when **your immune system makes its own antibodies** after being **stimulated** by an **antigen**. There are **two** different types of active immunity:

1) **Natural** — this is when you become immune after **catching a disease**.
2) **Artificial** — this is when you become immune after you've been given a **vaccine** containing a harmless dose of antigen (see below).

Active immunity gives you long-term protection, but it takes a while for the protection to develop.

PASSIVE IMMUNITY

This is the type of immunity you get from being **given antibodies made by a different organism** — your immune system **doesn't** produce any antibodies of its own. Again, there are **two** types:

1) **Natural** — this is when a **baby** becomes immune due to the antibodies it receives from its **mother**, through the **placenta** and in **breast milk**.
2) **Artificial** — this is when you become immune after being **injected** with **antibodies**, e.g. if you contract tetanus you can be injected with antibodies against the tetanus toxin.

Passive immunity gives you short-term protection, but the protection is immediate.

Vaccines Give You **Immunity Without Getting** the **Disease**

1) While your B cells are busy **dividing** to build up their numbers to deal with a pathogen (i.e. the **primary response** — see above), you **suffer** from the disease. **Vaccination** can help avoid this.
2) Vaccines **contain antigens** that cause your body to **produce memory cells** against a particular pathogen, **without** the pathogen **causing disease**. This means you become **immune** without getting any **symptoms**.
3) Some vaccines contain **many different antigens** to protect against **different strains** of pathogens. Different strains of pathogens are created by **antigenic variation** (see next page).

Developing Immunity

Pathogens Evolve Mechanisms to Evade the Immune System

1) Over **millions** of **years** vertebrates (e.g. humans) have **evolved** better and better **immune systems** — ones that **fight** a **greater variety** of pathogens in lots of **different ways**.
2) At the same time, **pathogens** have **evolved** better and better ways to **evade** (**avoid**) the immune systems of their **hosts** (the **organisms** that they **infect**).
3) This struggle between **pathogens** and their **hosts** to outdo each other is known as an **evolutionary race**.
4) An evolutionary race is **similar** to an **arms race** — where **two countries** constantly **develop better weapons** in an attempt to **overpower** each other.
5) **Evidence** to **support** the **theory** of an **evolutionary race** comes from the **evasion mechanisms** that pathogens have **developed**. For example:

HIV's evasion mechanisms

- HIV **kills** the **immune systems cells** that it **infects**. This **reduces** the overall **number** of immune system cells in the body, which **reduces** the **chance** of HIV being **detected**.
- HIV has a **high rate of mutation** in the **genes** that code for **antigen proteins**. The mutations **change** the **structure** of the antigens and this forms **new strains** of the virus — this process is called **antigenic variation**. The **memory cells** produced for **one strain** of HIV **won't recognise** other strains with **different antigens**, so the immune system has to produce a **primary response** against **each new strain**.
- HIV **disrupts antigen presentation** in infected cells. This **prevents** immune system cells from **recognising** and **killing** the **infected cells**.

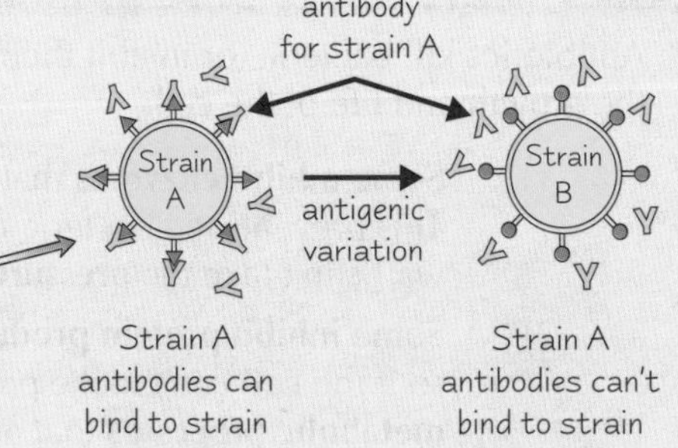

***Mycobacterium tuberculosis'* evasion mechanisms**

- When *M. tuberculosis* bacteria **infect** the **lungs** they're **engulfed** by **phagocytes**. Here, they **produce substances** that **prevent** the **lysosome fusing** with the **phagocytic vacuole** (see p. 191). This means the bacteria **aren't broken down** and they can **multiply undetected** inside phagocytes.
- This bacterium also **disrupts antigen presentation** in infected cells, which **prevents** immune system cells from **recognising** and **killing** the **infected phagocytes**.

Practice Questions

Q1 Describe the role of memory B cells in the secondary response.

Q2 Describe the role of memory T cells in the secondary response.

Q3 State two differences between active and passive immunity.

Q4 Describe how a vaccine gives immunity to a pathogen.

Q5 Describe one way in which *Mycobacterium tuberculosis* evades the immune system.

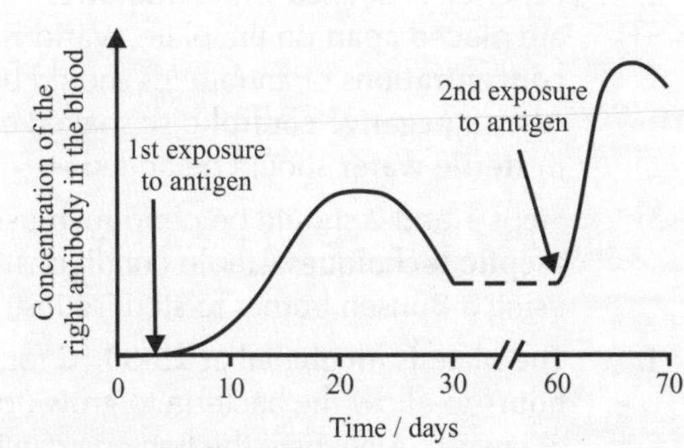

Exam Questions

Q1 The graph shows the concentration of antibody in a person's bloodstream following two separate exposures to the same antigen. Describe and explain the changes in the concentration of antibody. [10 marks]

Q2 Describe and explain three mechanisms that HIV has evolved to evade immune systems. [6 marks]

Even if you could change your antigens, you can't evade your exams...

Who would've thought there were so many ways to become immune to a microorganism. I don't like the sound of being injected with antibodies (I'm not keen on needles), but if it stops me dying from something like tetanus, then I'll give it a go. We're in an evolutionary race with loads of nasty microbes at the minute... let's hope we're the first past the post...

Antibiotics

These pages are for Edexcel Unit 4 only.

You've probably taken antibiotics at some point. Now you get to learn about how they work — the fun never ends...

Antibiotics Kill or Prevent the Growth of Microorganisms

1) **Antibiotics** are **chemicals** that **kill** or **inhibit** the **growth** of microorganisms.
2) There are **two different types** of antibiotics:
   - **Bacteriocidal** antibiotics **kill** bacteria.
   - **Bacteriostatic** antibiotics **prevent** bacteria **growing**.
3) Antibiotics are **used** by humans as **drugs** to **treat bacterial infections**.

Antibiotics Work by Inhibiting Bacterial Metabolism

Antibiotics kill bacteria (or inhibit their growth) because they **interfere** with **metabolic reactions** that are **crucial** for the growth and life of the cell:

1) Some **inhibit enzymes** that are needed to make the chemical **bonds** in bacterial **cell walls**. This prevents the bacteria from **growing** properly. It can also lead to **cell death** — the weakened cell wall can't take the **pressure** as water moves into the cell by **osmosis**. This can cause the cell to **burst**.
2) Some **inhibit protein production** by binding to bacterial **ribosomes**. All **enzymes** are proteins, so if the cell can't make proteins, it can't make enzymes. This means it can't carry out important **metabolic processes** that are needed for growth and development.

Bacterial cells are **different** from **mammalian** cells (e.g. human cells) — mammalian cells are **eukaryotic**, they **don't** have **cell walls**, they have **different enzymes** and they have different, **larger ribosomes**. This means antibiotics can be designed to **only target** the **bacterial cells**, so they don't damage mammalian cells. **Viruses don't** have their **own** enzymes and ribosomes — they use the ones in the host's cell, so antibiotics **don't affect them**.

You Can Test the Effect of Different Antibiotics on Bacteria

There are lots of reasons why you might want to **test** the **effect** of **different antibiotics** on **different strains** of bacteria. For example, doctors need to find out **which** antibiotics will **treat** a **patient's bacterial** infection.

Here's one way to do it:

1) The bacteria to be tested are **spread** onto an **agar plate**.
2) Paper discs **soaked** with **antibiotics** are placed apart on the plate. Various **concentrations** of antibiotics should be used. Also, a **negative control** disc soaked only in **sterile water** should be added.
3) Steps 1 and 2 should be performed using **aseptic techniques** (sterile conditions), e.g. using a Bunsen burner to sterilise instruments.
4) The plate is **incubated** at **25-30 °C** for **24-36 hours** to allow the bacteria to **grow** (forming a '**lawn**'). Anywhere the bacteria **can't grow** can be seen as a **clear patch** in the lawn of bacteria. This is called an **inhibition zone**.
5) The size of an **inhibition zone** tells you how well an antibiotic works. The **larger** the zone, the **more** the bacteria were inhibited from growing.
6) Bacteria that are **unaffected** by antibiotics are said to be **antibiotic resistant**.

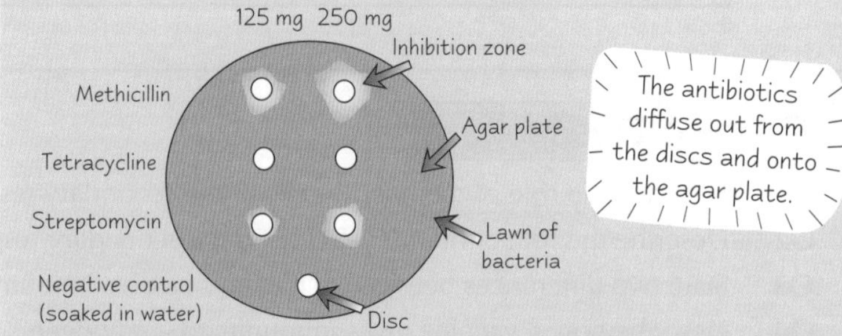

This diagram shows an agar plate with **methicillin**, **tetracycline** and **streptomycin** discs **after** it has been **incubated**.

- The **tetracycline** discs have **no** inhibition zones, so the bacteria are **resistant** to tetracycline up to 250 mg.
- The **streptomycin** discs have **small** inhibition zones, with the zone at 250 mg slightly larger than the one at 125 mg. So streptomycin has **some effect** on the bacteria.
- The **methicillin** discs have the **largest** inhibition zones, especially at 250 mg, so methicillin has the **strongest effect** on these bacteria.

Antibiotics

Hospital Acquired Infections (HAIs) can be Transmitted by Poor Hygiene

1) **Hospital acquired infections** (HAIs) are **infections** that are **caught** while a patient is being treated **in hospital**.
2) HAIs are **transmitted** by poor hygiene, such as:

- Hospital **staff** and **visitors not washing** their **hands** before and after visiting a patient.
- **Coughs** and **sneezes not** being **contained**, e.g. in a tissue.
- **Equipment** (e.g. beds or surgical instruments) and **surfaces not** being **disinfected** after they're used.

3) People are **more likely** to catch infections in hospital because many patients are ill, so have **weakened immune systems**, and they're **around** other **ill people**.
4) **Codes of practice** have been developed to **prevent** and **control HAIs**. They include:

- Hospital **staff** and **visitors** should be **encouraged** to **wash** their **hands**, **before** and **after** they've been with a patient.
- **Equipment** and **surfaces** should be **disinfected** after they're used.
- **People with HAIs** should be **moved** to an **isolation ward** so they're **less likely** to **transmit** the infection to **other patients**.

Despite his protests, Huxley wasn't allowed in the ward with his mucky trotters.

Some HAIs are Antibiotic-Resistant

1) Some **HAIs** are caused by bacteria that are **resistant** to **antibiotics**, e.g. MRSA.
2) These HAIs are **difficult** to **treat** because antibiotics **don't** get rid of the infection. This means these HAIs can lead to **serious health problems** or even **death**.
3) Infections caused by antibiotic-resistant bacteria are **more common** in **hospitals** because **more** antibiotics are used there, so bacteria in hospitals are **more likely** to have **evolved resistance** against them.
4) **Codes of practice** have also been developed to **prevent** and **control** HAIs caused by antibiotic-resistant bacteria:

- Doctors **shouldn't** prescribe antibiotics for **minor** bacterial infections or **viral** infections.
- Doctors **shouldn't** prescribe antibiotics to **prevent** infections.
- Doctors **should** use **narrow-spectrum antibiotics** (which only affect a specific bacterium) if possible, e.g. when the **strain** of bacteria the person has is **identified**.
- Doctors **should rotate** the **use** of **different** antibiotics.
- Patients **should** take **all** of the antibiotics that they're **prescribed** so infections are **fully cleared**.

These codes reduce the likelihood that bacteria will evolve antibiotic resistance.

Practice Questions

Q1 What are bacteriostatic antibiotics?

Q2 Name two processes in a bacterial cell that antibiotics can inhibit.

Q3 What is a hospital acquired infection?

Exam Questions

Negative control (N)
Penicillin 125 mg (A)
Amoxicillin 125 mg (B)
Erythromycin 125 mg (C)
Streptomycin 125 mg (D)

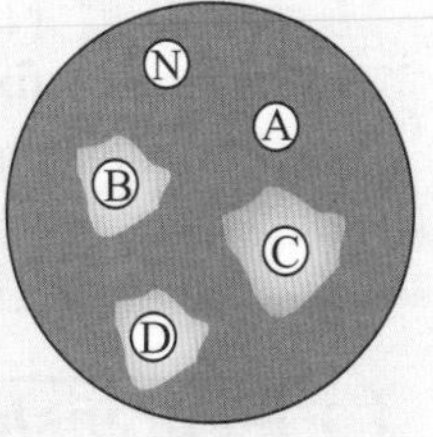

Q1 The agar plate on the right shows the effects of different antibiotics on a strain of bacteria.
a) Describe how the plate could have been prepared. [4 marks]
b) Which antibiotic is most effective against this strain of bacterium? Explain your answer. [2 marks]

Q2 Describe one way in which poor hygiene can cause HAIs, and a code of practice designed to prevent it. [2 marks]

The Market Research Society of Australia — not a deadly bacterium...

It's just typical, scientists discover all these fancy chemicals that get rid of bacteria and then some of the swines decide they won't be affected by them. Spoilsports. So, if you ever visit a hospital, make sure you use the alcohol gel to clean your hands. Not only will you be preventing those pesky infections, but you'll experience a cooling sensation like no other. Lovely...

Microbial Decomposition and Time of Death

These pages are for Edexcel Unit 4 only.

Wowsers, this is pretty morbid stuff. Make sure you read these pages before you've had your lunch...

Microorganisms **Decompose Organic Matter**

1) **Microorganisms**, e.g. **bacteria** and **fungi**, are an important part of the **carbon cycle** (see p. 52).
2) When plants and animals die, microorganisms on and in them **secrete enzymes** that decompose the **dead organic matter** into **small molecules** that they can **respire**.
3) When the microorganisms respire these small molecules, **methane** and CO_2 are released — this **recycles** carbon back into the atmosphere.

Scientists can **Estimate** the **Time of Death** of a **Body**

Police and **forensic scientists** often need to establish a body's **time of death** (**TOD**). This can give them a lot of information about the **circumstances** of the death, e.g. if they know when someone died they might be able to figure out who was present. The TOD can be established by looking at **several different factors** together — on their **own** these factors **aren't accurate** enough to give a reliable time of death. The **five** factors you need to know about are:

1 Body Temperature

1) All mammals **produce heat** from metabolic **reactions** like **respiration**, e.g. the **human body** has an internal temperature of around **37 °C**.
2) From the TOD the metabolic reactions **slow down** and eventually **stop**, causing **body temperature** to **fall** until it **equals** the temperature of its **surroundings** — this process is called ***algor mortis***.
3) Forensic scientists know that **human bodies** cool at a rate of around **1.5 °C** to **2.0 °C per hour**, so from the temperature of a dead body they can **work out** the approximate TOD. E.g. a dead body with a temperature of **35 °C** might have been **dead** for about an **hour**.
4) Conditions such as **air temperature**, **clothing** and **body weight** can **affect** the **cooling rate** of a body. E.g. the cooling rate of a **clothed body** will be **slower** than one without clothing, because it's **insulated**.

2 **Degree** of **Muscle Contraction**

About **4-6 hours** after death, the **muscles** in a dead body **start** to **contract** and become **stiff** — this is called ***rigor mortis***:

1) *Rigor mortis* begins when **muscle cells** become **deprived** of **oxygen**.
2) **Respiration** still takes place in the muscle cells, but it's **anaerobic**, which causes a build-up of **lactic acid** in the muscle.
3) The **pH** of the cells **decreases** due to the lactic acid, **inhibiting enzymes** that produce ATP.
4) **No ATP** means the **bonds** between the **myosin** and **actin** in the muscle cells (see p. 98) become **fixed** and the body **stiffens**.

It usually takes around 12-18 hours after the TOD for every muscle in the body to contract.

Smaller muscles in the head **contract first**, with **larger muscles** in the lower body being the **last** to contract. *Rigor mortis* is affected by **degree of muscle development** and **temperature**. E.g. *rigor mortis* occurs **more quickly** at **higher** temperatures because the chemical reactions in the body are **faster**.

Rigor mortis wears off around 24-36 hours after the TOD.

3 Forensic Entomology

1) When somebody dies the body is quickly **colonised** by a **variety** of **different insects** — the study of this is called **forensic entomology**.
2) TOD can be estimated by identifying the **type of insect** present on the body — e.g. **flies** are often the **first** insects to appear, usually a **few hours** after death. Other insects, like **beetles**, colonise a body at **later** stages.
3) TOD can also be estimated by identifying the **stage of lifecycle** the insect is in — e.g. **blowfly larvae hatch** from eggs about **24 hours** after they're **laid**. If **only** blowfly **eggs** are found on a body you could estimate that the TOD was **no more** than **24 hours ago**.
4) Different conditions will **affect** an insect's **lifecycle**, such as **drugs**, **humidity**, **oxygen** and **temperature**. E.g. the **higher** the temperature, the **faster** the **metabolic rate** and the **shorter** the **lifecycle**.

Microbial Decomposition and Time of Death

4 Extent of Decomposition

1) **Immediately** after death **bacteria** and **enzymes** begin to **decompose** the **body**.
2) Forensic scientists can use the **extent** of decomposition to establish a **TOD**:

Approximate time since TOD	Extent of decomposition
Hours to a few days	Cells and tissues are being broken down by the body's own enzymes and bacteria that were present before death. The skin on the body begins to turn a greenish colour.
A few days to a few weeks	Microorganisms decompose tissues and organs. This produces gases (e.g. methane), which cause the body to become bloated. The skin begins to blister and fall off.
A few weeks	Tissues begin to liquefy and seep out into the area around the body.
A few months to a few years	Only a skeleton remains.
Decades to centuries	The skeleton begins to disintegrate until there's nothing left of the body.

3) Different conditions **affect** the **rate** of decomposition, such as **temperature** and **oxygen availability**. E.g. **aerobic** microorganisms need **oxygen**, so decomposition could be **slower** if there's a **lack** of oxygen.

5 Stage of Succession

1) The **types of organism** found in a dead body **change over time**, going through a number of **stages** — this is called **succession**.
2) Forensic scientists can establish a **TOD** from the particular **stage of succession** that the body's in.
3) If a dead body is left to decompose **above ground** succession will usually follow these stages:

- **Immediately after** the TOD conditions in a dead body are **most favourable** for **bacteria**.
- As bacteria **decompose tissues**, conditions in a dead body become favourable for **flies** and their **larvae**.
- When fly larvae **feed** on a dead body they make conditions favourable for **beetles**, so beetles move in.
- As a dead **body dries out** conditions become **less favourable** for **flies** — they leave the body. Beetles **remain** as they can decompose **dry tissue**.
- When **no tissues** remain, conditions are **no longer favourable** for **most organisms**.

4) Succession in a dead body is **similar** to plant succession (see p. 54) — the **only difference** is that most of the **early insects** (e.g. beetles) **remain** on the body as other insects colonise it.
5) The stage of succession of a dead body (and so the type of organism that's present) is affected by many things including the **location** of the body, such as above ground, under ground, in water or sealed away. E.g. a body that's been **sealed away won't** be **colonised** by any species of insect.

Practice Questions

Q1 Describe how microorganisms recycle carbon from dead plants and animals.

Q2 What is forensic entomology?

Exam Question

Q1 A human body with a temperature of 29 °C was found at 22:45. *Rigor mortis* was only present in the face and shoulders. There was no visible decomposition of the body. Blowfly eggs, but no larvae, were found on the body.
For each piece of evidence above estimate the person's time of death and explain your answer. [8 marks]

CSI: Cumbria — it doesn't really have the same ring to it, does it...

Well, that was just lovely. Admittedly, I found it all rather interesting 'cause I like to watch repeats of Quincy on the telly. The main aim of all this grim stuff is to estimate the time of death of a body — remember that it's only an estimate, and lots of things (especially temperature) affect a dead body. Maybe if I pickle myself now, I'll never decompose. Hum...

Electrical Activity in the Heart

This section is for Edexcel Unit 5 only.

The heart beats all day every day. Doctors can monitor the heart with a fancy machine to check if it's beating OK and to check for signs of disease. Beep, beep, beep, beep, beep, beeeeeeeeeeep.... uh oh...

Cardiac Muscle *Controls the* ***Regular Beating*** *of the Heart*

Cardiac (heart) muscle is '**myogenic**' — it can contract and relax without receiving signals from neurones. This pattern of contractions controls the **regular heartbeat**.

This is a view of the heart from the front so the left and right sides are reversed.

1) The process starts in the **sinoatrial node (SAN)**, which is in the wall of the **right atrium**.
2) The SAN is like a pacemaker — it sets the **rhythm** of the heartbeat by sending out regular **waves of electrical activity** to the **atrial walls**.
3) This causes the right and left **atria** to **contract at the same time**.
4) A band of non-conducting **collagen tissue** prevents the waves of electrical activity from being passed directly from the atria to the ventricles.
5) Instead, these waves of electrical activity are transferred from the SAN to the **atrioventricular node (AVN)**.

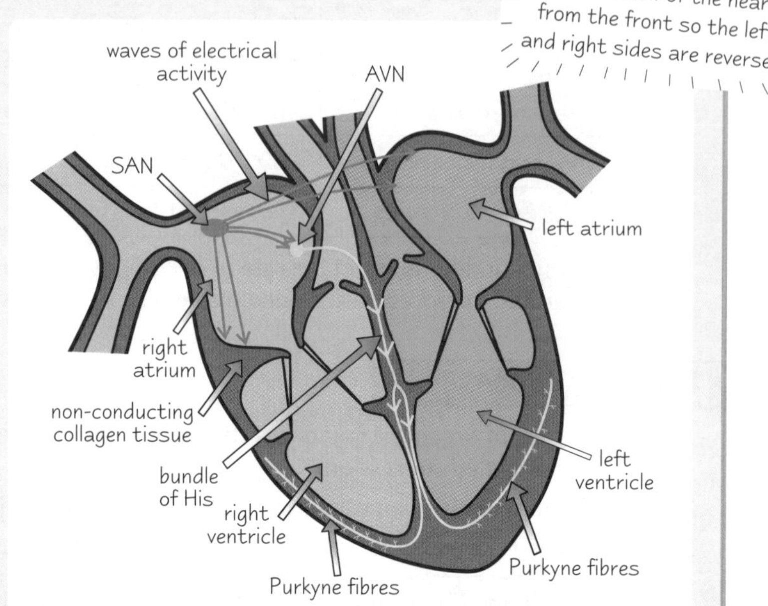

6) The AVN is responsible for passing the waves of electrical activity onto the bundle of His. But, there's a **slight delay** before the AVN reacts, to make sure the ventricles contract **after** the atria have emptied.
7) The **bundle of His** is a group of muscle fibres responsible for conducting the waves of electrical activity to the finer muscle fibres in the right and left **ventricle walls**, called the **Purkyne fibres**.
8) The Purkyne fibres carry the waves of electrical activity into the muscular walls of the right and left ventricles, causing them to **contract simultaneously**, from the bottom up.

An ***Electrocardiograph*** *Records the* ***Electrical Activity*** *of the Heart*

A doctor can check someone's **heart function** using an **electrocardiograph** — a machine that **records** the **electrical activity** of the heart. The heart muscle **depolarises** (loses electrical charge) when it **contracts**, and **repolarises** (regains charge) when it **relaxes**. An electrocardiograph records changes in electrical charge using **electrodes** placed on the chest.

The trace produced by an electrocardiograph is called an **electrocardiogram**, or **ECG**. A **normal** ECG looks like this:

1) The **P wave** is caused by **contraction** (depolarisation) of the **atria**.
2) The main peak of the heartbeat, together with the dips at either side, is called the **QRS complex** — it's caused by **contraction** (depolarisation) of the **ventricles**.
3) The **T wave** is due to **relaxation** (repolarisation) of the **ventricles**.

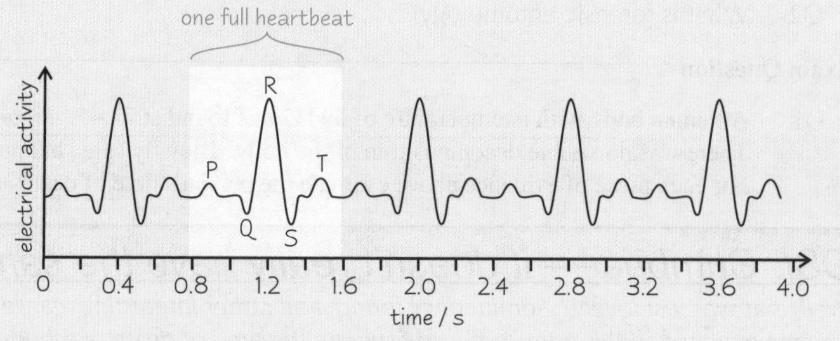

Electrical Activity in the Heart

Doctors use ECGs to Diagnose Heart Problems

Doctors **compare** their patients' ECGs with a **normal trace**. This helps them to **diagnose** any **problems** with the heart's **rhythm**, which may indicate **cardiovascular disease** (heart and circulatory disease).
Here are some examples of abnormal traces:

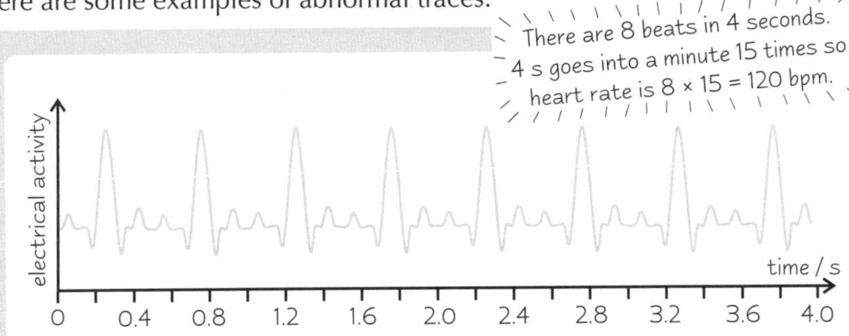

Tachycardia — increased heart rate

Here, the heartbeat is **too fast**, around **120** beats per minute. This could be a sign of **heart failure** — a problem with the heart means that it **can't pump blood efficiently**, so heart rate **increases** to ensure **enough blood** is pumped around the body. Tachycardia can also increase the **risk of a heart attack.**

Problem with the AVN

Here, the **atria** are contracting but the **ventricles** are **not** (some **P waves** aren't followed by a **QRS complex**). This might mean there's a problem with the **AVN** — impulses aren't travelling from the atria through to the ventricles.

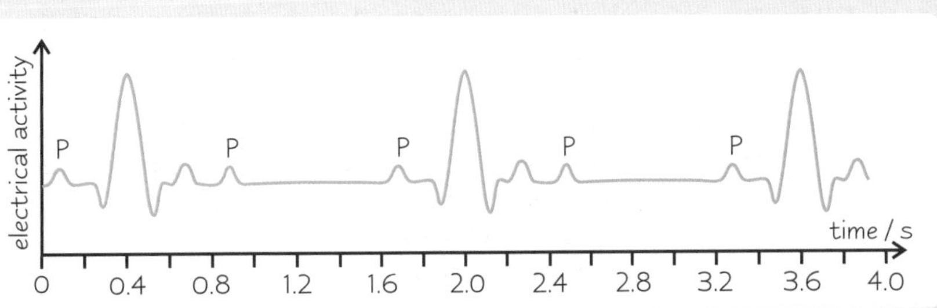

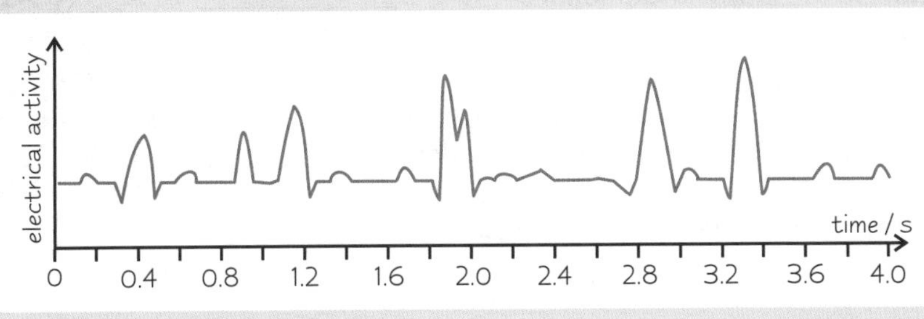

Fibrillation — irregular heart beat

The heart beat in this ECG is **irregular**. Both the **atria** or **ventricles** have lost their rhythm and **stopped contracting properly**. Atrial fibrillation can lead to **chest pains**, **fainting** and an **increased risk** of **stroke**. Ventricular fibrillation can quickly **cause death.** It may be **caused** by a **heart attack.**

Practice Questions

Q1 What prevents impulses from the atria travelling straight into the ventricles?

Q2 What is the name of the structure that picks up impulses from the atria and passes them on to the ventricles?

Q3 What causes the QRS part of the ECG trace?

Exam Questions

Q1 Describe the function of:

a) the sinoatrial node. [1 mark]

b) the bundle of His. [1 mark]

Q2 Suggest the cause of an ECG which has a QRS complex that is smaller than normal. [2 marks]

Perhaps if I plug myself into the mains, my heart'll be supercharged...

It's pretty incredible that your heart manages to go through all those stages in the right order, at exactly the right time, without getting it even slightly wrong. It does it perfectly, about 70 times every minute. That's about 100 800 times a day. If only my brain was that efficient. I'd have all this revision done in five minutes, then I could go and watch TV...

Variations in Heart Rate and Breathing Rate

These pages are for Edexcel Unit 5 only.

Your heart doesn't beat away steadily all day — heart rate increases and decreases, depending on what you're doing. What you're doing also affects your breathing rate...

Breathing Rate and Heart Rate Increase When you Exercise

When you exercise your **muscles contract more frequently**, which means they use **more energy**. To replace this energy your body needs to do **more aerobic respiration**, so it needs to **take in more oxygen** and **breathe out more carbon dioxide**. The body does this by:

1) **Increasing breathing rate** and **depth** — to **obtain more oxygen** and to **get rid** of **more carbon dioxide**.
2) **Increasing heart rate** — to **deliver oxygen** (and glucose) to the muscles **faster** and **remove extra carbon dioxide** produced by the increased rate of **respiration** in muscle cells.

The Medulla Controls Breathing Rate

The ventilation centre is also called the respiratory centre.

The **medulla** (a part of the **brain** — see p. 176) has areas called **ventilation centres**. There are two ventilation centres — the **inspiratory** centre and the **expiratory** centre. They control the **rate of breathing**:

1) The **inspiratory centre** in the **medulla** sends nerve impulses to the **intercostal** and **diaphragm** muscles to make them **contract**. This **increases** the **volume** of the lungs, which **lowers** the **pressure** in the lungs. (The inspiratory centre also sends nerve impulses to the **expiratory centre**. These impulses **inhibit** the action of the **expiratory centre**.)
2) **Air enters** the lungs due to the **pressure difference** between the lungs and the air outside.
3) As the **lungs inflate**, **stretch receptors** in the lungs are **stimulated**. The stretch receptors send nerve impulses back to the **medulla**. These impulses **inhibit** the action of the **inspiratory centre**.
4) The expiratory centre (no longer inhibited) then sends nerve impulses to the **diaphragm** and **intercostal muscles** to **relax**. This causes the **lungs to deflate**, expelling air. As the lungs deflate, the **stretch receptors** become **inactive**. The inspiratory centre is no longer inhibited and the cycle starts again.

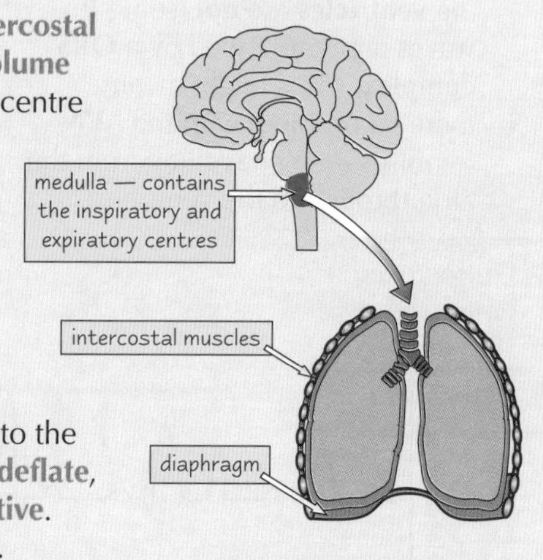

Exercise Triggers an Increase in Breathing Rate by Decreasing Blood pH

1) During exercise, the level of **carbon dioxide** (CO_2) in the blood **increases**. This **decreases** the **pH** of the blood.
2) There are **chemoreceptors** (receptors that sense chemicals) in the **medulla, aortic bodies** (in the aorta) and **carotid bodies** (in the carotid arteries carrying blood to the brain) that are **sensitive** to **changes in blood pH**.
3) If the chemoreceptors **detect** a **decrease** in blood **pH**, they send nerve impulses to the **medulla**, which sends **more frequent** nerve impulses to the **intercostal muscles** and **diaphragm**. This **increases** the **rate** and **depth** of breathing.
4) This causes **gaseous exchange** to **speed up** — the CO_2 level drops and extra O_2 is supplied for the muscles.

Ventilation Rate Increases with Exercise

1) Ventilation rate is the **volume** of air **breathed in or out** in a **period of time**, e.g. a minute.
2) It increases during exercise because **breathing rate** and **depth increase**.

Variations in Heart Rate and Breathing Rate

The Medulla Controls Heart Rate Too

Heart rate is **controlled** by the **cardiovascular control centre** in the **medulla** of the brain:

<u>Decreased blood pH causes an increase in heart rate</u>

1) A decrease in **blood pH** (caused by an increase in CO_2) is detected by **chemoreceptors**.
2) The chemoreceptors send **nerve impulses** to the medulla.
3) The medulla sends nerve impulses to the **SAN** to **increase the heart rate**.

Derek wasn't sure if his heart rate had increased because of running or the fact that Janice was wearing a thong.

<u>Increased blood pressure causes a decrease in heart rate</u>

1) **Pressure receptors** in the **aorta wall** and in the **carotid sinuses** (at the start of the carotid arteries carrying blood to the brain) **detect changes** in **blood pressure**.
2) If the pressure is **too high**, the pressure receptors send **nerve impulses** to the cardiovascular centre, which sends nerve impulses to the **SAN**, to **slow down the heart rate**.
3) If the **pressure is too low**, pressure receptors send nerve impulses to the cardiovascular centre, which sends nerve impulses to, yep you guessed it, the **SAN**, to **speed up** the **heart rate**.

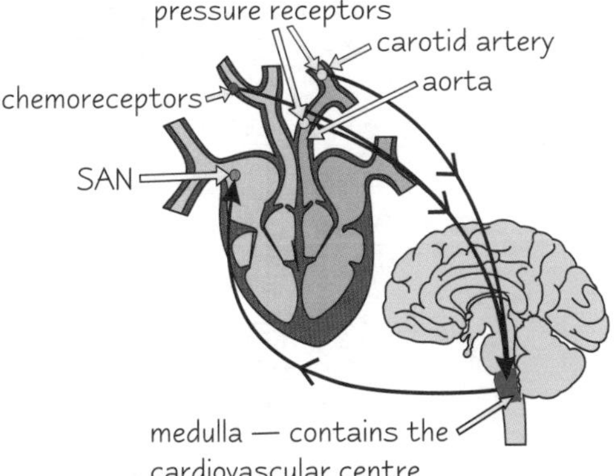

Exercise Triggers an Increase in Heart Rate by Decreasing Blood pH

During exercise, the level of **carbon dioxide** (CO_2) in the blood **increases**. This **decreases** the **pH** of the blood, which the chemoreceptors detect. The leads to an increase in heart rate (see above).

Cardiac Output Increases with Exercise

1) Cardiac output is the **total volume** of blood pumped by a **ventricle** every **minute**.
2) The **equation** for working out cardiac output is:

   Cardiac output (cm^3/min) = **heart rate** (beats per minute) × **stroke volume** (cm^3)

3) So cardiac output increases during exercise because **heart rate increases** (stroke volume also increases because the heart pumps harder as well).

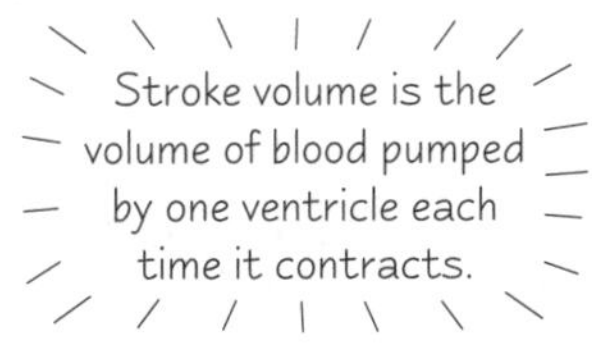

Practice Questions

Q1 Which part of the brain controls breathing rate and heart rate?

Q2 What effect does exercise have on cardiac output?

Exam Question

Q1 In a laboratory experiment, an animal was anaesthetised and dilute carbonic acid (carbon dioxide in solution) was added to the blood in the coronary artery.

a) What effect would you expect this to have on the animal's breathing rate? Explain your answer. [5 marks]

b) Cardiac output increased in response to the addition of carbonic acid. How is cardiac output calculated? [1 mark]

My heart rate increases when Ronaldo exercises...

...because I find him incredibly annoying, not because he's attractive in any way. Breathing rate and heart rate can be increased to supply more oxygen for aerobic respiration, which releases the energy the body needs during exercise. Cardiac output increases as heart rate (and stroke volume) increases, so a greater volume of blood is pumped around the body.

Investigating Ventilation

These pages are for Edexcel Unit 5 only.

You need to know how to investigate the effects of exercise on all things to do with ventilation. And I'm not talking about the ventilation that Bruce Willis is fond of climbing through in films...

Tidal Volume *is the* ***Volume of Air*** *in a* ***Normal Breath***

Here are some terms that you need to know about breathing:

1) **Tidal volume** — the **volume** of air in **each breath**, usually about **0.4 dm^3**.
2) **Breathing rate** — **how many breaths** are taken, usually in a **minute**.
3) **Ventilation rate** — the **volume** of air **breathed in or out**, usually in a **minute**. Here's how it's calculated:

ventilation rate = tidal volume × breathing rate

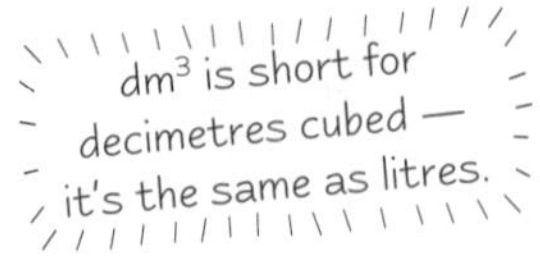

dm^3 is short for decimetres cubed — it's the same as litres.

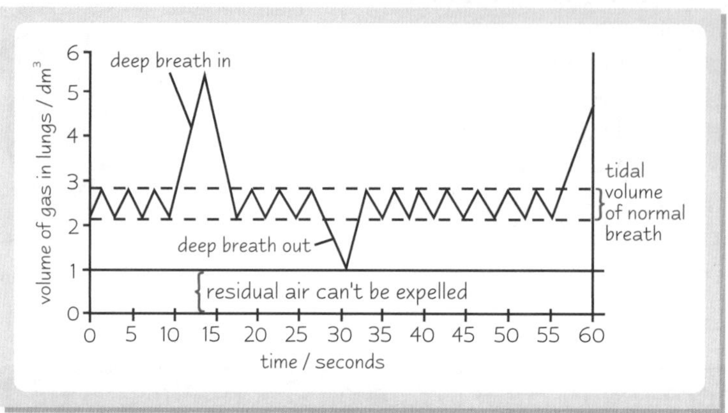

Jane couldn't maintain her breathing rate when she saw all those TVs.

Spirometers *Can be Used to* ***Measure Tidal Volume*** *and* ***Breathing rate***

A spirometer is a machine that can give readings of **tidal volume** and **breathing rate**.

1) A spirometer has an **oxygen-filled** chamber with a **movable lid**.
2) A person breathes through a **tube** connected to the oxygen chamber.
3) As the person breathes **in** the lid of the chamber moves **down**. When they breathe **out** it moves **up**.
4) These movements are recorded by a **pen** attached to the lid of the chamber — this writes on a **rotating drum**, creating a **spirometer trace**.
5) The **soda lime** in the tube the person breathes into absorbs **carbon dioxide**.

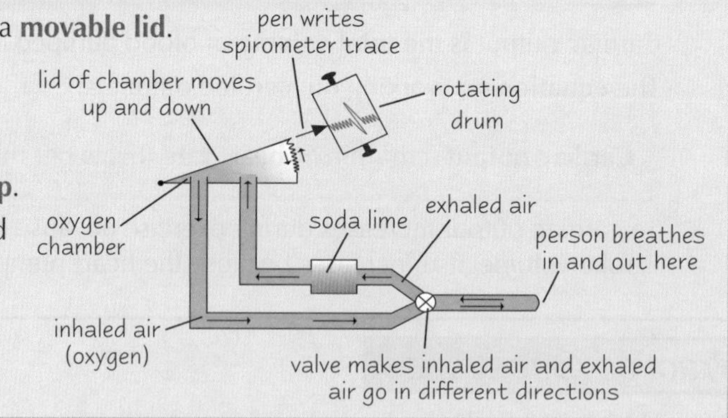

The **total volume of gas** in the chamber **decreases** over time. This is because the air that's breathed out is a **mixture** of oxygen and carbon dioxide. The carbon dioxide is absorbed by the **soda lime** — so there's **only oxygen** in the chamber which the person inhales from. As this oxygen gets used up by respiration, the total volume decreases.

Spirometers *Can be Used to* ***Investigate the Effects of Exercise***

Exercise causes an increase in **breathing rate** (see page 202) and **tidal volume**. You can use a spirometer to measure the change in breathing rate and tidal volume at **rest**, **during exercise** and **after exercise**. For example:

1) A person is connected to a spirometer using a **mask** so that continuous readings can be recorded.
2) Readings are recorded for **one minute** at **rest**.
3) The person then begins to **exercise**, e.g. running on a **treadmill**, for **two minutes**.
4) The person then **stops exercising** and readings are continued for **one minute** at **rest**.

Investigating Ventilation

You Need to be Able to Analyse Data from a Spirometer

You can use spirometer traces to look at the **effect** of **exercise** on breathing rate and tidal volume. Here's an example:

EXAMPLE 1

1) A person's **breathing rate** and **tidal volume** were measured using a **spirometer** and the method described on the previous page. The spirometer trace is shown in the **graph** below.
2) At **rest**, tidal volume is about **0.4 dm³** and breathing rate is **12 breaths per minute**.
3) During **exercise**, the body needs **more oxygen** for muscle contraction and it needs to **remove more carbon dioxide** (see page 202). So breathing rate and tidal volume **increase** — to around **20 breaths per minute** and **3.2 dm³**.
4) During **recovery**, the body still needs to keep **breathing hard** (to get oxygen to remove any lactate that's built up), but eventually breathing rate and tidal volume return to **rest levels**.

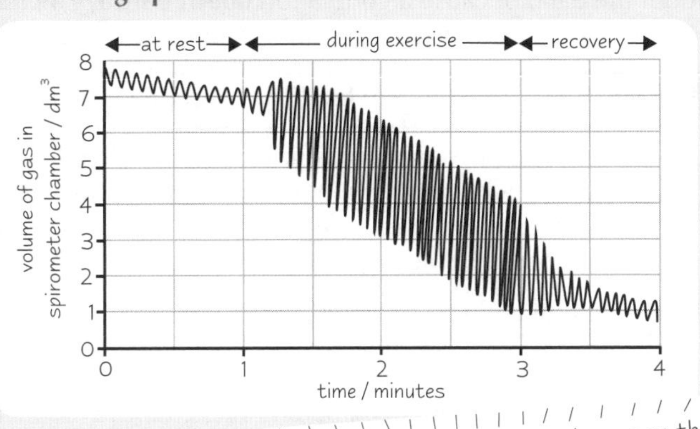

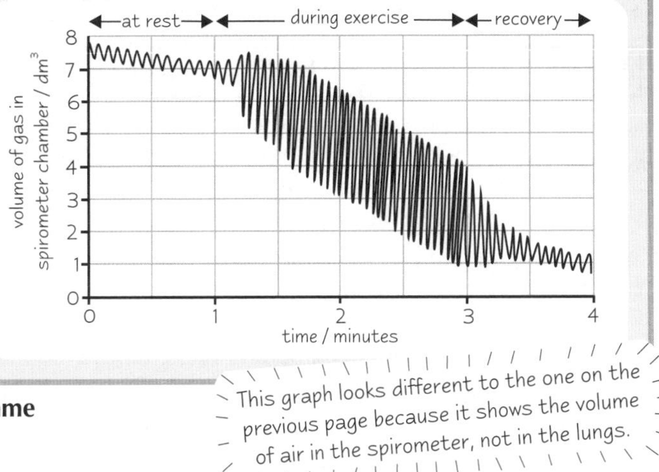

This graph looks different to the one on the previous page because it shows the volume of air in the spirometer, not in the lungs.

You can also look at the **effect** of a **fitness training programme** on breathing rate and tidal volume:

EXAMPLE 2

A person's **breathing rate** and **tidal volume** were measured using a **spirometer** and the method described on the previous page. The results are shown in the **graph** below. The **same test** was repeated three months later on the **same person**, after they'd gone through a **fitness training programme**. The spirometer results were plotted to **compare** the **effects** of training:

1) Training **decreased breathing rate at all stages** because the lung muscles are **strengthened**, so **more air** is taken in with each breath, meaning **fewer breaths** are needed.

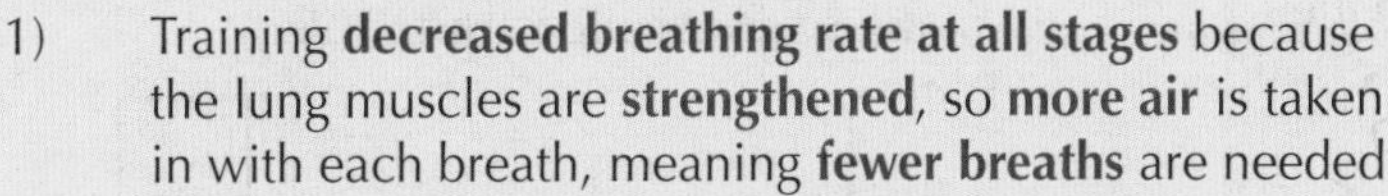

2) Training **increased tidal volume at all stages**, again because muscles are strengthened, so more air is taken in with each breath.
3) During **recovery**, **breathing rate** and **tidal volume decreased faster** due to training because the muscles are strengthened, so the lungs can get oxygen and carbon dioxide supplies back to normal quicker.

Practice Questions

Q1 What is tidal volume?

Q2 What's the purpose of soda lime in a spirometer?

Exam Question

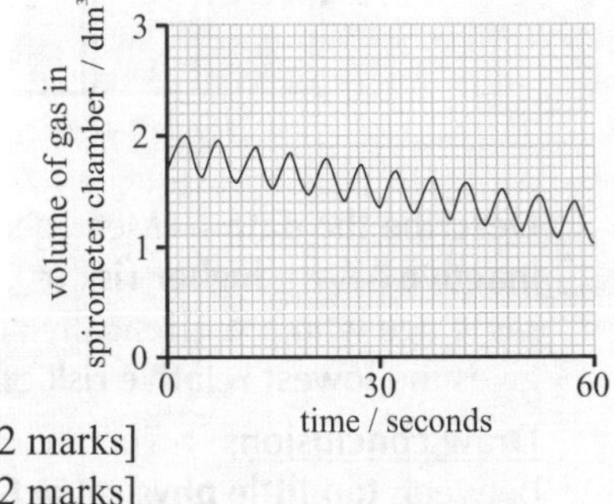

Q1 The graph on the right shows a spirometer trace taken from a student at rest.

a) Calculate the tidal volume and breathing rate. [2 marks]

b) How would you expect the trace to differ if the student was exercising? [2 marks]

Investigate someone's breathing — make sure they've had a mint first...

I thought spirometers were those circular plastic things you draw crazy patterns with... apparently not. I know the graphs don't look that approachable, but it's important you understand what the squiggly lines show, and the meaning of terms used when investigating breathing — I'd bet my right lung there'll be a question on spirometer graphs in the exam.

Exercise and Health

These pages are for Edexcel Unit 5 only. *More data analysis now, but don't worry, it's not as scary as it looks...*

Not Doing *Enough Exercise* can be *Unhealthy...*

There are loads of **studies** out there that have looked at the **effects** of **not doing enough exercise**. You might have to analyse some **data** from a study in your exam — so you need to be able to **describe** what the data shows and say if there's a **correlation** (a relationship between two variables, see p. 214). You need to be careful what you **conclude** about the data, because a correlation **doesn't** always mean that one thing **causes** another. Here are some **examples** of the kind of things you might get in your **exam**:

See pages 214-216 for more on analysing data.

EXAMPLE 1 — The effect of **too little exercise** on **obesity**

15 239 men and **women** across the EU were asked to **estimate** how many **hours** they spent **sitting during their leisure time** each week. The **body mass index** (**BMI**) of each individual was also calculated as a measure of **obesity** (BMI greater than 30 kg / m^2). The **table** below shows the **percentage** of men and women who are **obese compared** to the amount of time spent sitting in their leisure time each week.

Hours sitting down per week	% obese men	% obese women
<15	7.6	9.2
15-20	7.3	6.5
21-25	8.3	10.3
26-35	9.0	11.8
>35	13.3	12.4

Describe the data — The table shows that **overall**, the **percentage** of men and women who are **obese increases** as the number of hours spent sitting down during leisure time per week **increases**. But the table shows a **slight decrease** for people who spent **15-20 hours** compared to less than 15 hours a week sitting down.

Draw conclusions — The table shows there's a **correlation** (link) between sitting down for a long time each week and being **obese**, for both men and women. But you **can't** say that long periods of sitting **causes** obesity. There could be **other reasons** for the trend, e.g. people may sit down more **because** they're obese.

EXAMPLE 2 — The effect of **too little exercise** on **coronary heart disease** (**CHD**)

5159 men aged **40 to 59** with **no history of CHD** were asked about their **exercise habits**. The health of the men was then followed for an **average time** of **16.8 years**, and the results were used to assess the **risk of CHD** according to the level of **physical activity** carried out.

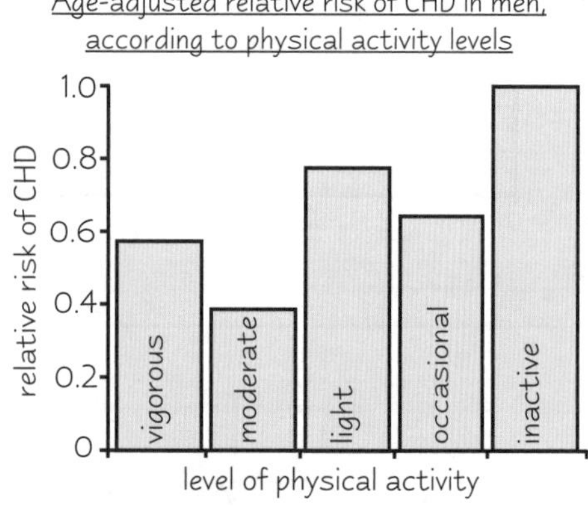

Describe the data — Men aged 40 to 59 who are **inactive** have a **higher risk** of CHD than men of the same age who are physically active. **Moderate exercise** gives the **lowest relative risk** of CHD, at about 0.39.

Draw conclusions — There's a **weak correlation** between **too little physical activity** and an **increased risk** of CHD in **men** aged 40 to 59, but it **doesn't** steadily increase with the amount of exercise. The study **only** involved **men**, so you **can't say** that the same risk occurs in **women**.

EXAMPLE 3 — The effect of **too little exercise** on **Type 2 diabetes**

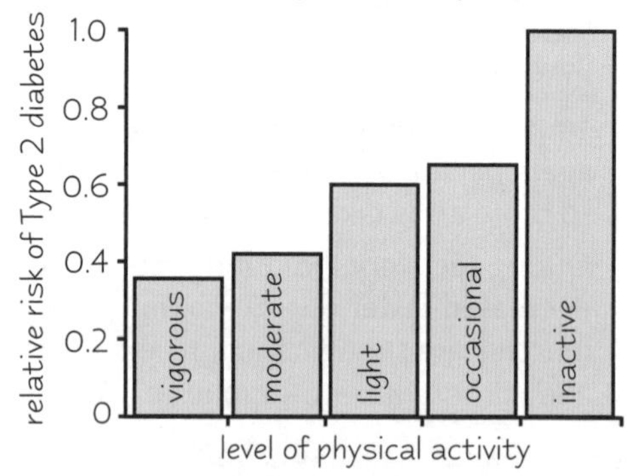

5159 men aged **40 to 59** with **no history of Type 2 diabetes** were asked about their **exercise habits**. The health of the men was then followed for an **average time** of **16.8 years**, and the results were used to assess the **risk of developing Type 2 diabetes** according to the level of **physical activity** carried out.

Describe the data — The **relative risk** of Type 2 diabetes **increases** with **more inactivity** in men aged 40 to 59.

Draw conclusions — There's a **correlation** between **too little physical activity** and an **increased risk** of Type 2 diabetes in **men** aged 40-59. But, you can't say one **causes** the other as there could be **other reasons**, e.g. diet.

Exercise and Health

... But Doing Too Much Exercise can be Unhealthy Too

It's thought that **too much exercise** could possibly cause some problems too, e.g. wear and tear of the joints. So you might get asked to interpret **data** looking at the **effects** of **too much exercise** as well...

EXAMPLE 1 — The effect of **too much exercise** on **wear** and **tear** of **joints**

The number of **hospital admissions** for **osteoarthritis** of the **hip, knee** and **ankle joints** for **2049** former **male elite athletes** and **1403 healthy, fit controls** were compared by analysing **hospital records** from 1970 to 1990.

<u>Describe the data</u> — The percentage of **male former elite athletes** admitted to hospital for **osteoarthritis** (wear and tear of the joints) is **more than twice** that of healthy men (5.9% compared to 2.6%). The table shows similar high percentages for different kinds of athletes, with **endurance athletes** being **worse affected** (6.8% admitted to hospital).

<u>Draw conclusions</u> — The table shows a **correlation** between being an **elite male athlete** of any kind and having **osteoarthritis** of the hip, knee or ankle. But you can't say that doing a lot of exercise **causes** osteoarthritis — there may be **other reasons** for the trend, e.g. elite athletes may be more likely to injure themselves in competitions, which could lead to arthritis.

Group of men	% of men admitted to hospital for osteoarthritis of the hip, knee or ankle
Healthy men, fully fit for military service	2.6
Former elite athletes (all)	5.9
• Endurance athletes (e.g. long-distance runners)	6.8
• Mixed sports athletes (e.g. footballers)	5.0
• Power sports athletes (e.g. boxers)	6.6

EXAMPLE 2 — The effect of **too much exercise** on the **immune system**

The number of people showing **two or more** symptoms of **upper respiratory tract infection** was recorded for a group containing **32 elite** and **31 recreational triathletes** and **cyclists** during a **five-month** period in their training season. It was also recorded for **20 control** subjects during a five-month period.

<u>Describe the data</u> — There was a **much higher** number of cases of respiratory illnesses in **elite athletes** (21) than in athletes (7) and sedentary controls (9) over the five-month period.

<u>Draw conclusions</u> — There's a **correlation** between doing **a lot of exercise** (being an elite athlete) and getting **more** cases of **respiratory illnesses**. But there's also a **correlation** between doing **some exercise** (recreationally competitive athletes) and getting **fewer** cases of respiratory illnesses. You **can't** say doing a lot of exercise **causes** more respiratory illnesses — there could be **other reasons** for the trend, e.g. elite athletes might be exposed to lots of infections when competing a lot.

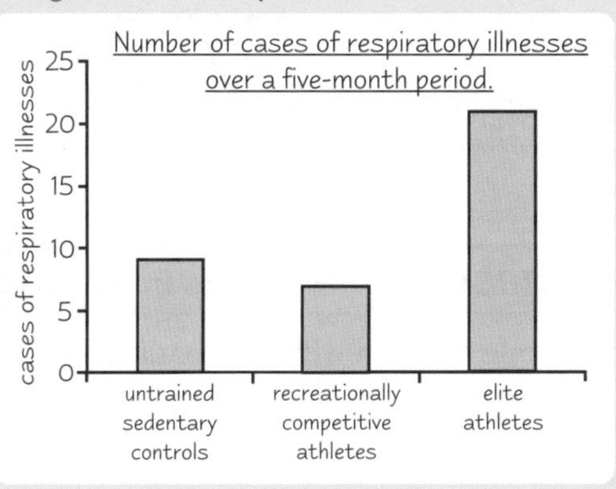

Practice Questions

Q1 What is a correlation?

Exam Question

Physical activity (hours/week)	≥3.5	1–3.5	<1
Relative risk of CHD	1.0	1.32	1.48

Q1 A study investigated the effects of physical activity on the risk of coronary heart disease (CHD) in 88 393 women. The results are shown in the table.

a) Describe the results. [1 mark]

b) From the table, what can you conclude about the effects of physical activity on CHD? [1 mark]

Drawing conclusions — you'll need wax crayons and some paper...

These pages give you some examples to help you deal with what the examiners are sure to hurl at you — they really love throwing data around. There's some important advice here (even if I say so myself) — it's easy to leap to a conclusion that isn't really justified. So make sure you know that just because things are correlated, it doesn't mean one causes the other.

Exercise and Health

These pages are for Edexcel Unit 5 only.

Yoga on an exotic beach somewhere, that's my kind of exercise. Apparently not all sports are as calm or as stress-free. There are injuries galore...

Surgery can Help People with Injuries to Play Sports

Some injuries can cause **permanent damage** to the body, e.g. head or spinal injuries. But people can **recover** from some injuries if they're **treated correctly**. A lot of injuries happen when **playing sports** because the body's put under a lot of **stress** (fast running, hard tackles, etc.). Advances in **medical technology** can help people with an injury to **recover** and **participate in sports**. One advance you need to know about is **keyhole surgery**:

1) Keyhole surgery is a way of doing surgery **without** making a **large incision** (cut) in the skin.
2) Surgeons make a much **smaller incision** in the patient, and they insert a tiny **video camera** and **specialised medical instruments** through the incision into the body.
3) There are many **advantages** of keyhole surgery over regular surgery:
   - Operations don't involve opening up the patient as much, so patients **lose less blood** and have **less scarring** of the skin.
   - Patients are usually in **less pain** after their operation and they **recover more quickly**, because less damage is done to the body.
   - This makes it **easier** for the patient to return to **normal activities** and their **hospital stay** is **shorter**.

Debra knew she'd have to do some serious keyhole surgery to see what was in the case.

For example, damaged **cruciate ligaments** can be fixed by **keyhole surgery**:

1) A **common sports injury** is damage to the cruciate ligaments — **ligaments** found in the middle of your **knee**, connecting your **thigh bone** to your **lower leg bone**.
2) **Damaged** cruciate ligament can be **removed** and **replaced** with a **graft** of ligament through a small incision in the knee.

Injuries aren't usually fixed by surgery alone — other treatments (e.g. physiotherapy, anti-inflammatory drugs) are needed too for a full recovery.

Prostheses can Replace Damaged Body Parts

Some people are **born without** a particular **body part**, e.g. without a leg. Other people suffer **injuries** that result in them **losing** or **badly damaging** a **body part**, e.g. tennis players can damage their knees so much that they can no longer play sports. Sometimes it's possible to **replace** damaged or missing body parts with an **artificial device** called a **prosthetic**:

1) Prostheses can be used to **replace whole limbs** (e.g. an artificial leg can replace a missing leg) or **parts of limbs** (e.g. artificial hip joints can replace damaged hip joints).
2) Some prostheses include **electronic devices** that **operate** the prosthesis by picking up information sent by the **nervous system** (e.g. artificial hand prostheses with an electronic device allow the user to move the fingers).
3) So prostheses make it possible for people with some **disabilities** to **participate** in **sport**, e.g. prosthetic 'legs' (called blades) allow people without legs to run.
4) They also make it possible for people who have certain **injuries** to **play sport again**.

For example, damaged **knee joints** can be replaced by **prosthetic joints**:

1) A **metal device** is inserted into the knee to **replace damaged cartilage** and **bone**.
2) The knee joint and the ends of the leg bones are replaced to provide a **smooth knee joint**. **Cushioning** in the new joint helps to **reduce** the **impact** on the knee.
3) A knee joint replacement allows people with serious knee problems to **move around** and participate in **low-impact sports**, such as walking and swimming.

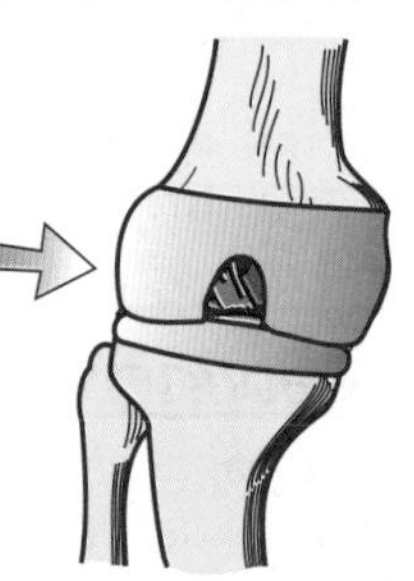

Exercise and Health

Some Athletes Use Performance-Enhancing Drugs

When involved in a very **competitive sport**, some people choose to take **performance-enhancing drugs** — these are drugs that will **improve** a person's **performance**. There are various kinds of performance-enhancing drugs that have different effects on the body, for example:

- **Anabolic steroids** — these drugs **increase strength**, **speed** and **stamina** by increasing **muscle size** and allowing athletes to train harder. They also **increase aggression**.
- **Stimulants** — these drugs **speed up reactions**, **reduce fatigue** and **increase aggression**.
- **Narcotic analgesics** — these drugs **reduce pain**, so **injuries don't affect performance**.

Performance-enhancing drugs are **banned** in most sports. Athletes can be **tested** for drugs at any time and if they're **caught**, they can be **banned** from **competing** and stripped of any medals.

The Use of Performance-Enhancing Drugs is Controversial

There are many reasons why **performance-enhancing drugs** are **banned** but some people think they should be **allowed** in sport. You need to know **both sides** of the argument:

Arguments AGAINST using performance-enhancing drugs

- Some performance-enhancing drugs are **illegal**.
- Competitions become **unfair** if some people take drugs — people gain an advantage by taking drugs, not through training or hard work.
- There are some **serious health risks** associated with the drugs used, such as high blood pressure and heart problems.
- Athletes may **not** be **fully informed** of the health risks of the drugs they take.

Arguments FOR using performance-enhancing drugs

- It's up to each individual — athletes have the right to make their **own decision** about taking drugs and whether they're worth the risk or not.
- Drug-free sport **isn't** really **fair** anyway — different athletes have access to different training facilities, coaches, equipment, etc.
- Athletes that want to compete at a **higher level** may only be able to by using performance-enhancing drugs.

Practice Questions

Q1 Where are your cruciate ligaments?

Q2 What are performance-enhancing drugs?

Exam Questions

Q1 A cricketer suffers a knee injury during a game. When examined by doctors, he's told he may need surgery.
a) Give three benefits of keyhole surgery compared to open surgery. [3 marks]
b) The damage to his knee may be so bad that it can't be repaired. How could a prosthesis help? [2 marks]

Q2 Many sports organisations have banned the use of performance-enhancing drugs.
Give two arguments in favour of such a ban. [2 marks]

Through the keyhole — who operates on a knee like this...

Wow, I didn't know sport could be so technical. Eeeee, if you hurt yourself when I were a lad, you'd be lucky to get yourself a sticky plaster... Anyway, it seems the world of sport has come on a bit since then. And you know what that means — yep, it's crucial you learn about cruciates and paramount you learn about prosthetics for your exam.

Drugs and Disease

This section is for Edexcel Unit 5 only.

Brace yourself — this section isn't as exciting as the name suggests, but it is the last. Let revision commence...

Imbalances in Some *Neurotransmitters* can *Contribute* to *Disorders*

Neurotransmitters are **chemicals** that **transmit** nerve impulses across **synapses** (see page 92). Some **disorders** are **linked** to an **imbalance** of specific neurotransmitters in the brain. Here are two examples you need to know:

Parkinson's Disease

1) Parkinson's disease is a **brain disorder** that affects the **motor skills** (the movement) of people.
2) In Parkinson's disease the **neurones** in the **parts** of the **brain** that **control movement** are **destroyed**.
3) These neurones **normally produce** the neurotransmitter **dopamine**, so **losing them** causes a **lack** of **dopamine**.
4) A lack of dopamine causes a **decrease** in the **transmission** of the **nerve impulses** involved in **movement**.
5) This leads to **symptoms** like **tremors** (shaking) and **slow movement**.
6) Scientists know that the **symptoms** are **caused by** a **lack** of **dopamine** so they've **developed drugs** (e.g. **L-dopa**, see below) to **increase** the level of **dopamine** in the brain.

Depression

1) Scientists think there's a **link** between a **low level** of the neurotransmitter **serotonin** and **depression**.
2) Serotonin transmits **nerve impulses** across synapses in the **parts** of the **brain** that **control mood**.
3) Scientists know that **depression** is **linked** to a **low level** of **serotonin** so they've **developed drugs** (antidepressants) to **increase** the level of **serotonin** in the brain.

Some *Drugs* Work by *Affecting Synaptic Transmission*

See p. 92 for more on synaptic transmission.

You need to know these two examples:

L-dopa

1) L-dopa is a drug that's used to **treat** the **symptoms** of **Parkinson's disease**.
2) Its **structure** is very **similar** to **dopamine**.
3) When L-dopa is given, it's **absorbed** into the **brain** and **converted** into **dopamine** by the enzyme **dopa-decarboxylase** (**dopamine can't be given** to treat Parkinson's disease because it **can't enter** the **brain**).
4) This **increases** the level of **dopamine** in the brain.
5) A higher level of dopamine means that **more nerve impulses** are **transmitted** across synapses in the **parts** of the **brain** that **control movement**.
6) This gives sufferers of Parkinson's disease **more control** over their **movement**.

MDMA (ecstasy)

1) MDMA **increases** the level of **serotonin** in the brain.
2) Usually, serotonin is **taken back** into a **presynaptic neurone** after triggering an action potential, to be **used again**.
3) MDMA **increases** the level of **serotonin** by **inhibiting** the **reuptake** of serotonin **into presynaptic neurones**, and by **triggering** the **release** of serotonin **from presynaptic neurones**.
4) This means that **nerve impulses** are **constantly triggered** in postsynaptic neurones in **parts** of the **brain** that **control mood**.
5) So the **effect** of MDMA is **mood elevation**.

All Eric needed was a hat and a mobile phone to increase his serotonin level.

Drugs and Disease

Info from the *Human Genome Project* is Being Used to Create *New Drugs...*

1) The **H**uman **G**enome **P**roject (**HGP**) was a 13 year long project that **identified** all of the **genes** found in **human DNA** (the human genome).
2) The **information obtained** from the HGP is **stored** in **databases**.
3) Scientists use the databases to **identify genes**, and so **proteins**, that are **involved** in **disease**.
4) Scientists are using this information to create **new drugs** that **target** the **identified proteins**, e.g. scientists have identified an **enzyme** that **helps cancer cells** to **spread** around the body — a **drug** that **inhibits** this **enzyme** is being developed.
5) The HGP has also highlighted **common genetic variations** between people.
6) It's known that **some** of these **variations** make **some drugs less effective**, e.g. some **asthma drugs** are **less effective** for people with a **particular mutation**.
7) Drug companies can use this knowledge to design **new drugs** that **are effective** in people with these **variations**.

...but this Raises *Moral* and *Ethical Issues*

1) Creating drugs for specific genetic variations will **increase research costs** for drugs companies. These new drugs will be **more expensive**, which could lead to a **two-tier health service** — only **wealthier** people could **afford** these new drugs.
2) Some people might be **refused** an **expensive drug** because their genetic make-up indicates that it **won't be that effective** for them — it may be the **only drug available** though.
3) The **information** held within a person's genome could be **used by others**, e.g. employers or insurance companies, to **unfairly discriminate** against them. For example, if a person is **unlikely** to respond to any **drug treatments** for **cancer** an insurance company might **increase** their **life insurance premium**.
4) Revealing that a drug might not work for a person could be **psychologically damaging** to them, e.g. it could be their **only hope** to treat a disease.

Practice Questions

Q1 Name a disorder that's linked to a low level of serotonin.

Q2 Describe one way that MDMA increases the level of serotonin in the brain.

Q3 What is the Human Genome Project?

Exam Questions

Q1 Parkinson's disease affects around 120 000 people in the UK.

a) Explain the role of dopamine in controlling movement. [2 marks]

b) Describe the effect that Parkinson's disease has on the brain. [4 marks]

c) Name a drug that is used to treat Parkinson's disease and explain how it works. [4 marks]

Q2 Describe how the results of the Human Genome Project are being used to create new drugs. [5 marks]

The Minnesota Donkey and Mule Association — a different kind of MDMA...

Make sure you go back over the stuff about synaptic transmission on p. 92 — it'll really help you to understand how L-dopa and MDMA work. It's not just drugs that increase your serotonin level though, chocolate does too... which is a great excuse to gobble some down — not that you really need an excuse. I think you deserve some after all this revision.

Producing Drugs Using GMOs

These pages are for Edexcel Unit 5 only.
Unsavoury characters in dark alleyways aren't the only things that produce drugs on demand...

Drugs can be *Produced* Using *Genetically Modified Organisms*

Genetically **m**odified **o**rganisms (**GMOs**) are organisms that have had their **DNA altered**. **Microorganisms**, **plants** and **animals** can all be **genetically modified** to **produce proteins** which are **used as drugs**:

Only drugs that are proteins can be produced by genetically modified organisms.

1 Genetically Modified Microorganisms

1) Here's how microorganisms are genetically engineered to produce drugs:

    1) The **gene** for the protein (drug) is **isolated** using enzymes called **restriction enzymes**.
    2) The **gene** is **copied** using **PCR** (see p. 142).
    3) **Copies** are **inserted** into **plasmids** (small circular molecules of DNA).
    4) The **plasmids** are **transferred** into **microorganisms**.
    5) The **modified microorganisms** are **grown** in large containers so that they **divide** and produce **lots** of the **useful protein**, from the inserted gene.
    6) The **protein** can then be **purified** and **used** as a drug.

Plasmids are a type of vector — vectors carry genes into an organism.

2) **Lots** of drugs are produced from **genetically modified bacteria**, for example **human insulin** (used to treat **Type 1 diabetes**) and **human blood clotting factors** (used to treat **haemophilia**).

2 Genetically Modified Plants

1) Here's how plants are genetically engineered to produce drugs:

    1) The **gene** for the protein (drug) is **inserted** into a **bacterium** (see above).
    2) The bacterium **infects** a **plant cell**.
    3) The bacterium **inserts** the **gene** into the **plant cell DNA** — the **plant cell** is now **genetically modified**.
    4) The **plant cell** is **grown** into an **adult plant** — the **whole plant** contains a **copy** of the **gene** in **every cell**.
    5) The **protein** produced from the gene can be **purified** from the **plant tissues**, or the **protein** (drug) could be **delivered** by **eating** the **plant**.

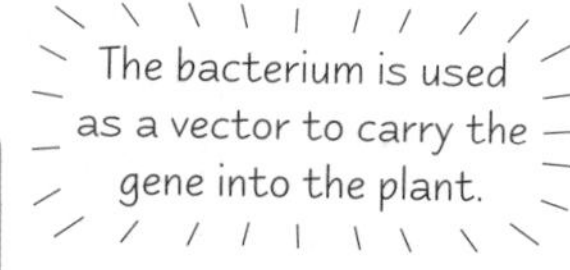

Malcolm had unwittingly eaten the Viagra plant.

2) **Some** drugs have been produced from genetically **modified** plants, for example **human insulin** and a **cholera vaccine**.

3 Genetically Modified Animals

1) Here's how animals are genetically engineered to produce drugs:

    1) The **gene** for the protein (drug) is **injected** into the **nucleus** of a **fertilised animal egg cell**.
    2) The **egg cell** is then **implanted** into an **adult animal** — it grows into a **whole animal** that contains a **copy** of the **gene** in **every cell**.
    3) The **protein** produced from the gene is normally **purified** from the **milk** of the animal.

2) Various animals have been **modified** with **human genes** to produce drugs, for example **human antithrombin** (used to treat people with a **blood clotting disorder**) has been produced from **genetically modified goats**.

Producing Drugs Using GMOs

There are **Benefits** and **Risks** Associated with Using **GMOs**

1) As well as producing drugs, GMOs are used in **agriculture** and the **food industry**. For example, genes for **herbicide resistance** can be inserted into **agricultural crops**. Herbicides can then be applied which will **kill weeds** but **not** the **herbicide-resistant crops** — the **genetically modified crop** will **thrive** without the weeds and this results in a **high yield**.
2) You need to know the **benefits** and **risks** associated with the **use** of **GMOs**:

Benefits

1) Agricultural **crops** can be modified so that they give **higher yields** or are **more nutritious**. This means these plants can be used to **reduce** the risk of **famine** and **malnutrition**.
2) Crops can also be modified to have **pest resistance**, so that **fewer pesticides** are **needed**. This **reduces costs** (making **food cheaper**) and **reduces** any **environmental problems** associated with using pesticides.
3) Industrial processes often use **enzymes**. These enzymes can be produced from genetically modified organisms in **large quantities** for less money, which **reduces costs**.
4) Some **disorders** can now be **treated** with **human proteins** from genetically engineered organisms instead of with **animal proteins**. Human proteins are **safer** and **more effective**. For example, **Type 1 diabetes** used to be treated with **cow insulin** but some people had an **allergic reaction** to it. **Human insulin**, produced from genetically modified **bacteria**, is more effective and **doesn't cause** an **allergic reaction** in humans.
5) **Vaccines** produced in plant tissues **don't need** to be **refrigerated**. This could make vaccines **available** to **more people**, e.g. in areas where **refrigeration** (usually needed for **storing** vaccines) **isn't available**.
6) **Producing drugs** using **plants** and **animals** would be **very cheap** because once the plants or animals are genetically modified they can be reproduced using **conventional farming methods**. This could make some drugs **affordable** for **more people**, especially those in poor countries.

Risks

1) Some people are **concerned** about the **transmission** of genetic material. For example, if **herbicide-resistant** crops **interbreed** with **wild plants** it could create **'superweeds'** — weeds that are **resistant** to **herbicides**, and if **drug crops** interbreed with other crops people might end up **eating drugs they don't need** (which could be harmful).
2) Some people are worried about the **long-term impacts** of using GMOs. There may be **unforeseen consequences**.
3) Some people think it's **wrong** to **genetically modify animals** purely for **human benefit**.

Practice Questions

Q1 What is a genetically modified organism?

Q2 Describe how a genetically modified plant is created.

Q3 Describe how a genetically modified animal is created.

Exam Questions

Q1 Describe how the bacterium *E. coli* is genetically modified to produce human insulin. [4 marks]

Q2 Discuss the benefits and risks associated with growing a plant that has been genetically modified to produce a hepatitis B vaccine and to be resistant to herbicides. [7 marks]

Milking a goat to get drugs — who'd have thought that was possible...

And there we go. A2 Biology. Done. Well, almost — there's still the challenge of getting all this stuff to stick in your brain. It's at times like these everyone wishes they could eat a book and absorb the information into their memory. You could try making a revision guide risotto... actually don't. Risottos can be difficult. I think I've been exposed to drugs for too long...

How to Interpret Experiment and Study Data

Science is all about getting good evidence to test your theories... so scientists need to be able to spot a badly designed experiment or study a mile off, and be able to interpret the results of an experiment or study properly. Being the cheeky little monkeys they are, your exam board will want to make sure you can do it too. Here's a quick reference section to show you how to go about interpreting data-style questions.

Here Are Some **Things** You Might be **Asked** to do...

This stuff might be familiar from AS, but you need to know it for A2 as well.

Here are three examples of the kind of data you could expect to get:

Study A

An agricultural scientist investigated the effect of three different pesticides on the number of pests in wheat fields. The number of pests was estimated in each of three fields, using ground traps, before and 1 month after application of one of the pesticides. The number of pests was also estimated in a control field where no pesticide had been applied. The table shows the results.

Pesticide	Number of pests	
	Before application	1 month after application
1	89	98
2	53	11
3	172	94
Control	70	77

Study B

Study B investigated the link between the number of bees in an area and the temperature of the area. The number of bees was estimated at ten 1-acre sites. The temperature was also recorded at each site. The results are shown in the scattergram below.

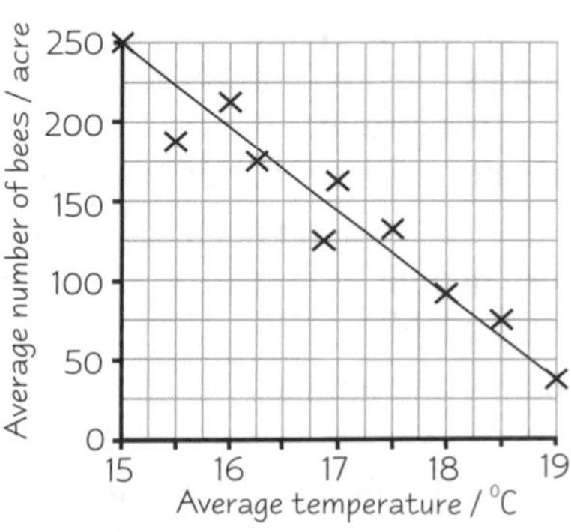

Experiment C

An experiment was conducted to investigate the effect of temperature on the rate of photosynthesis. The rate of photosynthesis in Canadian pondweed was measured at four different temperatures by measuring the volume of oxygen produced. All other variables were kept constant. The results are shown in the graph below.

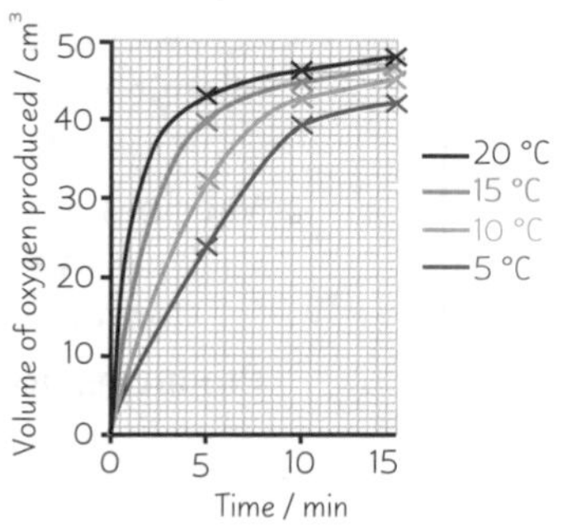

1) **Describe** and Manipulate the **Data**

You need to be able to **describe** any data you're given. The level of **detail** in your answer should be appropriate for the **number of marks** given. Loads of marks = more detail, few marks = less detail. You could also be asked to **manipulate** the data you're given (i.e. do some **calculations** on it). For the examples above:

Example — Study A

1) You could be asked to **calculate** the **percentage change** (**increase** or **decrease**) in the number of pests for each of the pesticides and the control. E.g. for pesticide 1: (98 – 89) ÷ 89 = 0.10 = **10% increase**.
2) You can then use these values to **describe** what the **data** shows — the **percentage increase** in pests in the field treated with **pesticide 1 was the same as for the control** (10% increase) (1 mark). **Pesticide 3 reduced** pest numbers by **45%**, but **pesticide 2** reduced the pest numbers the **most** (79% decrease) (1 mark).

Example — Study B

The data shows a **negative correlation** between the average number of bees and the temperature (1 mark).

Correlation describes the **relationship** between two variables — e.g. the one that's been changed and the one that's been measured. Data can show **three** types of correlation:

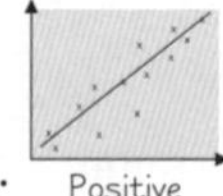
Positive

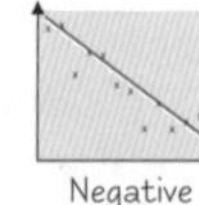
Negative

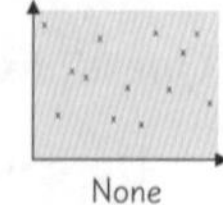
None

1) **Positive** — as one variable **increases** the other **increases**.
2) **Negative** — as one variable **increases** the other **decreases**.
3) **None** — there is **no relationship** between the two variables.

To tell if some data in a table **is correlated** — draw a **scatter diagram** of one variable against the other and **draw a line of best fit**.

Example — Experiment C

You could be asked to calculate the initial rate of photosynthesis at each temperature: The **gradient = the rate of photosynthesis**:

$$\text{Gradient} = \frac{\text{Change in Y}}{\text{Change in X}}$$

How to Interpret Experiment and Study Data

2) *Draw* or *Check* a *Conclusion*

1) Ideally, only **two** quantities would ever change in any experiment or study — everything else would be **constant**.
2) If you can keep everything else constant and the results show a correlation then you **can** conclude that the change in one variable **does cause** the change in the other.

Example — Experiment C

All other variables were **kept constant**. E.g. light intensity and CO_2 concentration **stayed the same** each time, so these **couldn't** have influenced the rate of reaction. So you **can say** that an increase in temperature up to 20 °C **causes** an increase in the rate of photosynthesis.

3) But usually all the variables **can't** be controlled, so other **factors** (that you **couldn't** keep constant) could be having an **effect**.
4) Because of this, scientists have to be very careful when **drawing conclusions**. Most results show a **link** (**correlation**) between the variables, but that **doesn't prove that a change in one causes the change in the other.**

Example — Study B

There's a **negative correlation** between the average number of bees and temperature. But you **can't** conclude that the increase in temperature **causes** the decrease in bees. **Other factors** may have been involved, e.g. there may be **less food** in some areas, there may be **more bee predators** in some areas, or **something else** you hadn't thought of could have caused the pattern...

5) The **data** should always **support** the conclusion. This may sound obvious but it's easy to **jump** to conclusions. Conclusions have to be **precise** — not make sweeping generalisations.

Example — Experiment C

A science magazine **concluded** from this data that the optimum temperature for photosynthesis is **20 °C**. The data **doesn't** support this. The rate **could** be greatest at 22 °C, or 18 °C, but you can't tell from the data because it doesn't go **higher** than 20 °C and **increases** of **5 °C** at a time were used. The rates of photosynthesis at in-between temperatures **weren't** measured.

3) *Explain* the *Evidence*

You could also be asked to **explain** the **evidence** (the data and results) — basically use your **knowledge** of the subject to explain **why** those results were obtained.

Example — Experiment C

Temperature increases the rate of photosynthesis because it **increases** the **activity** of **enzymes** involved in photosynthesis, so reactions are catalysed more quickly.

4) Comment on the *Reliability* of the *Results*

Reliable means the results can be **consistently reproduced** when an experiment or study is repeated. And if the results are reproducible they're more likely to be **true**. If the data isn't reliable for whatever reason you **can't draw** a valid **conclusion**. Here are some of the things that affect the reliability of data:

1) **Size of the data set** — For experiments, the **more repeats** you do, the **more reliable** the data. If you get the **same result** twice, it could be the correct answer. But if you get the same result **20 times**, it's much more reliable. The general rule for **studies** is the larger the **sample size**, the more **reliable** the **data** is.

E.g. Study B is quite **small** — they only studied ten 1-acre sites. The **trend** shown by the data may not appear if you studied **50 or 100 sites**, or studied them for a longer period of time.

2) **The range of values in a data set** — The **closer** all the values are to the **mean**, the **more reliable** the data set.

E.g. Study A is **repeated three more times** for pesticides 2 and 3. The percentage decrease each time is: 79%, 85%, 98% and 65% for **pesticide 2** (**mean = 82%**) and 45%, 45%, 54% and 43% for **pesticide 3** (**mean = 47%**). The data values are **closer to the mean** for **pesticide 3** than pesticide 2, so that data set is **more reliable**. The **spread** of **values about the mean** can be shown by calculating the **standard deviation** (SD).

The **smaller the SD** the **closer** the values to the **mean** and the **more reliable the data**. SDs can be shown on a graph using **error bars**. The ends of the bars show one SD **above** and one SD **below** the **mean**.

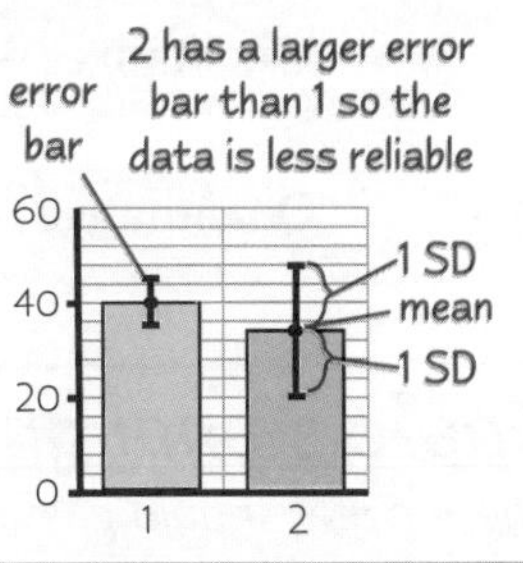

How to Interpret Experiment and Study Data

3) **Variables** — The **more variables** you **control**, the **more reliable** your data is. In an experiment you would control all the variables. In a study you try to control **as many as possible**.

The hat, trousers, shirt and tie variables had been well controlled in this study.

E.g. ideally, all the sites in Study B would have a similar **type** of land, similar **weather**, have the same **plants** growing, etc. Then you could be more sure that the one factor being **investigated** (temperature) is having an **effect** on the thing being **measured** (number of bees).

4) **Data collection** — think about all the **problems** with the **method** and see if **bias** has slipped in.

E.g. in Study A, the traps were placed on the **ground**, so pests like moths or aphids weren't included. This could have affected the results.

5) **Controls** — without controls, it's very difficult to **draw valid conclusions**. **Negative controls** are used to make sure that nothing you're doing in the experiment has an effect, **other than** what you're testing.

E.g. in Experiment C, the **negative control** would be all the equipment set up as normal but **without** the pondweed. If **no oxygen** was produced at any temperature it would show that the variation in the volume of oxygen produced when there was pondweed was due to the **effect** of temperature on the pondweed, and **not** the effect of temperature on **anything else** in the experiment.

6) **Repetition by other scientists** — for theories to become accepted as 'fact' other scientists need to **repeat** the work (see page 2). If **multiple studies** or **experiments** come to the same conclusion, then that conclusion is **more reliable**.

E.g. if a second group of scientists repeated Study B and got the same results, the results would be **more reliable**.

There Are a Few **Technical Terms** You **Need to Understand**

I'm sure you probably know these all off by heart, but it's easy to get mixed up sometimes. So here's a quick recap of some words **commonly used** when assessing and analysing experiments and studies:

1) **Variable** — A variable is a **quantity** that has the **potential to change**, e.g. weight. There are two types of variable commonly referred to in experiments:
   - **Independent variable** — the thing that's **changed** in an experiment.
   - **Dependent variable** — the thing that you **measure** in an experiment.

When drawing graphs, the dependent variable should go on the **y-axis** (the vertical axis) and the independent on the **x-axis** (the horizontal axis).

2) **Accurate** — Accurate results are those that are **really close** to the **true** answer. The true answer is **without error**, so if you can reduce error as much as possible you'll get a more accurate result. The most **accurate methods** are those that produce as **error-free** results as possible.

3) **Precise results** — These are results taken using **sensitive instruments** that measure in **small increments**, e.g. pH measured with a meter (pH 7.692) will be **more precise** than pH measured with paper (pH 8).

It's possible for results to be precise **but not** accurate, e.g. a balance that weighs to 1/1000 th of a gram will give precise results, but if it's not **calibrated** properly the results won't be accurate.

4) **Qualitative** — A **qualitative** test tells you **what's** present, e.g. an acid or an alkali.

5) **Quantitative** — A **quantitative** test tells you **how much** is present, e.g. an acid that's pH 2.46.

There's enough evidence here to conclude that data interpretation is boring...

*These pages should give you a fair idea of how to interpret data. Just use your head and remember the four things you might be asked to do — **d**escribe the **d**ata, **c**heck the **c**onclusions, **e**xplain the **e**vidence and check the **r**esults are **r**eliable.*

Answers

Section 1 — Photosynthesis and Respiration

Page 5 — Photosynthesis, Respiration and ATP

1 *Maximum of 6 marks available, from any of the 8 points below.* *In the cell, ATP is synthesised from ADP and inorganic phosphate/P_i* ***[1 mark]*** *using energy from an energy-releasing reaction, e.g. respiration* ***[1 mark]****. The energy is stored as chemical energy in the phosphate bond* ***[1 mark]****. ATP synthase catalyses this reaction* ***[1 mark]****. ATP then diffuses to the part of the cell that needs energy* ***[1 mark]****. Here, it's broken down back into ADP and inorganic phosphate/P_i* ***[1 mark]****, which is catalysed by ATPase* ***[1 mark]****. Chemical energy is released from the phosphate bond and used by the cell* ***[1 mark]****.*
Make sure you don't get the two enzymes confused — ATP **synthase** **synthesises** ATP, and ATPase breaks it down.

Page 9 — Photosynthesis

1 a) *Maximum of 1 mark available.*
The thylakoid membranes ***[1 mark]****.*
b) *Maximum of 1 mark available.*
Photosystem II ***[1 mark]****.*
c) *Maximum of 4 marks available.*
Light energy splits water ***[1 mark]****.*
H_2O ***[1 mark]*** $\rightarrow 2H^+ + \frac{1}{2}O_2$ ***[1 mark]****.*
The electrons from the water replace the electrons lost from chlorophyll ***[1 mark]****.*
The question asks you to explain the purpose of photolysis, so make sure you include why the water is split up — to replace the electrons lost from chlorophyll.
d) *Maximum of 1 mark available.*
NADP ***[1 mark]****.*

2 a) *Maximum of 6 marks available.*
Ribulose bisphosphate/RuBP and carbon dioxide/CO_2 join together to form an unstable 6-carbon compound ***[1 mark]****. This reaction is catalysed by the enzyme rubisco/ribulose bisphosphate carboxylase* ***[1 mark]****. The compound breaks down into two molecules of a 3-carbon compound called glycerate 3-phosphate/GP* ***[1 mark]****. Two molecules of glycerate 3-phosphate are then converted into two molecules of triose phosphate/TP* ***[1 mark]****. The energy for this reaction comes from ATP* ***[1 mark]*** *and the H^+ ions come from reduced NADP* ***[1 mark]****.*
b) *Maximum of 2 marks available.*
Ribulose bisphosphate is regenerated from triose phosphate/TP molecules ***[1 mark]****. ATP provides the energy to do this* ***[1 mark]****.*
This question is only worth two marks so only the main facts are needed, without the detail of the number of molecules.
c) *Maximum of 3 marks available.*
No glycerate 3-phosphate/GP would be produced ***[1 mark]****, so no triose phosphate/TP would be produced* ***[1 mark]****. This means there would be no glucose produced* ***[1 mark]****.*

Page 11 — Limiting Factors in Photosynthesis

1 *Maximum of 4 marks available.*
25 °C ***[1 mark]****. This is because photosynthesis involves enzymes* ***[1 mark]****, which become inactive at low temperatures/10 °C* ***[1 mark]*** *and denature at high temperatures/45 °C* ***[1 mark]****.*

2 a) *Maximum of 3 marks available.*
The level of GP will rise and levels of TP and RuBP will fall ***[1 mark]****. This is because there's less reduced NADP and ATP from the light-dependent reaction* ***[1 mark]****, so the conversion of GP to TP and RuBP is slow* ***[1 mark]****.*
b) *Maximum of 3 marks available.*
The levels of RuBP, GP and TP will fall ***[1 mark]****. This is because the reactions in the Calvin cycle are slower* ***[1 mark]*** *due to all the enzymes working more slowly* ***[1 mark]****.*

Page 13 — Limiting Factors in Photosynthesis

1 a) *Maximum of 4 marks available.*
By burning propane to increase air CO_2 concentration ***[1 mark]****. By adding heaters to increase temperature* ***[1 mark]****. By adding coolers to decrease temperature* ***[1 mark]****. By adding lamps to provide light at night* ***[1 mark]****.*
b) *Maximum of 2 marks available.*
Potatoes ***[1 mark]*** *because the yield showed the smallest percentage increase of 25% (850 – 680 = 170, 170 ÷ 680 × 100 = 25%)* ***[1 mark]****.*

2 *Maximum of 6 marks available, from any of the 8 points below.* *A sample of pondweed would be placed in a test tube of water* ***[1 mark]****. The test tube would be placed in a beaker containing water at a known temperature* ***[1 mark]****. The test tube would be connected to a capillary tube of water* ***[1 mark]*** *and the capillary tube connected to a syringe* ***[1 mark]****. The pondweed would be allowed to photosynthesise for a set period of time* ***[1 mark]****. Afterwards, the syringe would be used to draw the bubble of oxygen produced up the capillary tube where its length would be measured using a ruler* ***[1 mark]****. The experiment is repeated and the mean length of gas bubble is calculated* ***[1 mark]****. Then the whole experiment is repeated at several different temperatures* ***[1 mark]****.*

Page 15 — Aerobic Respiration

1 *Maximum of 6 marks available, from any of the 8 points below.* *First, the 6-carbon glucose molecule is phosphorylated* ***[1 mark]*** *by adding two phosphates from two molecules of ATP* ***[1 mark]****. This creates one molecule of 6-carbon hexose bisphosphate* ***[1 mark]*** *and two molecules of ADP* ***[1 mark]****. Then, the hexose bisphosphate is split up into two molecules of 3-carbon triose phosphate* ***[1 mark]****. Triose phosphate is oxidised (by removing hydrogen) to give two molecules of 3-carbon pyruvate* ***[1 mark]****. The hydrogen is accepted by two molecules of NAD, producing two molecules of reduced NAD* ***[1 mark]****. During oxidation four molecules of ATP are produced* ***[1 mark]****.*
When describing glycolysis make sure you get the number of molecules correct — one glucose molecule produces one molecule of hexose bisphosphate, which produces two molecules of triose phosphate. You could draw a diagram in the exam to show the reactions.

2 a) *Maximum of 3 marks available, from any of the 4 points below.* *The 3-carbon pyruvate is decarboxylated* ***[1 mark]****, then converted to acetate by the reduction of NAD* ***[1 mark]****. Acetate combines with coenzyme A (CoA) to form acetyl coenzyme A (acetyl CoA)* ***[1 mark]****. No ATP is produced* ***[1 mark]****.*
b) *Maximum of 2 marks available, from any of the 3 points below.* *The inner membrane is folded into cristae, which increase the membrane's surface area and maximise respiration* ***[1 mark]****. There are lots of ATP synthase molecules on the inner membrane to produce lots of ATP in the final stage of respiration* ***[1 mark]****. The matrix contains all the reactants and enzymes needed for the Krebs cycle to take place* ***[1 mark]****.*

Page 17 — Aerobic Respiration

1 a) *Maximum of 2 mark available.*
The transfer of electrons down the electron transport chain stops ***[1 mark]****. So there's no energy released to phosphorylate ADP/produce ATP* ***[1 mark]****.*

Answers

b) Maximum of 2 marks available.
The Krebs cycle stops **[1 mark]** because there's no oxidised NAD/FAD coming from the electron transport chain **[1 mark]**.
Part b is a bit tricky — remember that when the electron transport chain is inhibited, the reactions that depend on the products of the chain are also affected.

Page 19 — Respiration Experiments

1 a) Maximum of 1 mark available.
Because there was no proton gradient **[1 mark]**.
b) Maximum of 1 mark available.
3.7 **[1 mark]**
c) Maximum of 1 mark available.
Yes, these results support the chemiosmotic theory because they show that a proton gradient can be used by mitochondria to synthesise ATP **[1 mark]**.

2 a) Maximum of 1 mark available.
To make sure the results are only due to oxygen uptake by the woodlouse **[1 mark]**.
b) Maximum of 2 marks available.
The oxygen taken up would be replaced by carbon dioxide given out / there would be no change in air volume in the test tube **[1 mark]**. This means there would be no movement of the liquid in the manometer **[1 mark]**.
c) Maximum of 1 mark available.
Carbon dioxide/CO_2 **[1 mark]**.

Page 21 — Aerobic and Anaerobic Respiration

1 Maximum of 1 mark available.
Because lactate fermentation doesn't involve electron carriers/the electron transport chain/oxidative phosphorylation **[1 mark]**.

2 Maximum of 2 marks available.
$RQ = CO_2 \div O_2$ **[1 mark]**
So the RQ of triolein = $57 \div 80 = 0.71$ **[1 mark]**
Award 2 marks for the correct answer of 0.71, without any working.

Section 2 — Populations

Page 23 — Niches and Adaptations

1 a) Maximum of 2 marks available.
A niche is the role of a species within its habitat **[1 mark]**.
It includes both its biotic and abiotic interactions **[1 mark]**.
b) Maximum of 3 marks available.
Because every niche is unique, it can only be occupied by one species **[1 mark]**. It may look like species X and Y are occupying the same niche because they have the same biotic interactions, e.g. they feed on the same insects and are eaten by the same predator species **[1 mark]**. But there are slight differences in their niche, e.g. they feed at different times of the day **[1 mark]**.

Page 26 — Investigating Populations and Abiotic Conditions

1 a) Maximum of 1 mark available.
By taking random samples of the population **[1 mark]**.
b) Maximum of 3 marks available.
Several frame quadrats would be placed on the ground at random locations within the field **[1 mark]**. The percentage of each frame quadrat that's covered by daffodils would be recorded **[1 mark]**. The percentage cover for the whole field could then be estimated by averaging the data collected in all of the frame quadrats **[1 mark]**.

Page 29 — Investigating Populations and Analysing Data

1 a) Maximum of 5 marks available.
A group of snails would have been caught **[1 mark]**, marked in a way that wouldn't harm them, e.g. by painting a spot on their shell **[1 mark]**, then released back into the environment **[1 mark]**. After waiting a week a second sample would have been taken **[1 mark]** and the number marked in the second sample would have been counted **[1 mark]**.
b) Maximum of 2 marks available.
Total population size = $\frac{52 \times 38}{14}$ **[1 mark]**
Total population size = 141 **[1 mark]**.
Award 2 marks for correct answer of 141 without any working.

Page 31 — Variation in Population Size

1 a) Maximum of 7 marks available.
In the first three years, the population of prey increases from 5000 to 30 000. The population of predators increases slightly later (in the first five years), from 4000 to 11 000 **[1 mark]**. This is because there's more food available for the predators **[1 mark]**. The prey population then falls after year three to 3000 just before year 10 **[1 mark]**, because lots are being eaten by the large population of predators **[1 mark]**. Shortly after the prey population falls, the predator population also falls (back to 4000 by just after year 10) **[1 mark]**, because there's less food available **[1 mark]**. The same pattern is repeated in years 10-20 **[1 mark]**.
b) Maximum of 4 marks available.
The population of prey increased to around 40 000 by year 26 **[1 mark]**. This is because there were fewer predators, so fewer prey were eaten **[1 mark]**. The population then decreased after year 26 to 25 000 by year 30 **[1 mark]**. This could be because of intraspecific competition **[1 mark]**.

Page 33 — Human Populations

1 Maximum of 4 marks available.
At stage 1 of the DTM the population size is low and not increasing **[1 mark]**. At stage 5 the population is high but shrinking **[1 mark]**.
At stage 1 the population structure is made up of a lot of young people and very few older people **[1 mark]**, but at stage 5 there are few young people and a lot of older people **[1 mark]**.

Section 3 — Energy Flow and Nutrient Cycles

Page 35 — Energy Transfer and Productivity

1 a) Maximum of 4 marks available.
Because not all of the energy available from the grass is taken in by the Arctic hare **[1 mark]**. Some parts of the grass aren't eaten, so the energy isn't taken in **[1 mark]**, and some parts of the grass are indigestible, so they'll pass through the hares and come out as waste **[1 mark]**. Some energy is lost to the environment when the Arctic hare uses energy from respiration for things like movement or body heat **[1 mark]**.
b) Maximum of 2 marks available.
$(137 \div 2345) \times 100 = 5.8$ **[1 mark]**
Efficiency of energy transfer = 5.8% **[1 mark]**
Award 2 marks for correct answer of 5.8% without any working.

Answers

Page 37 — Pyramid Diagrams and Energy Transfer

1 a) *Maximum of 2 marks available.*

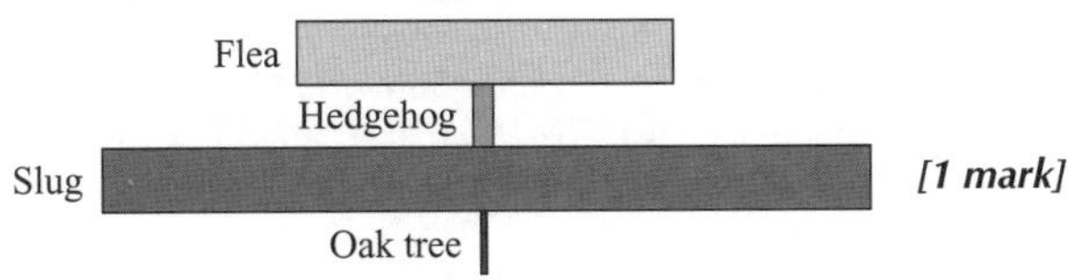

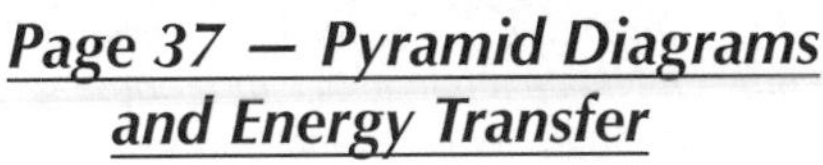

*A pyramid of numbers **[1 mark]**.*

b) *Maximum of 1 mark available.*

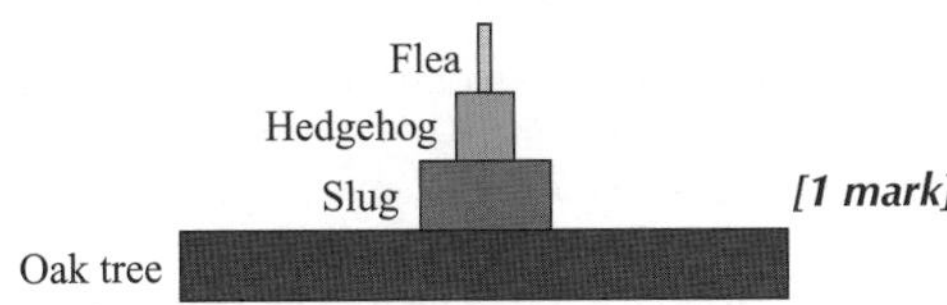

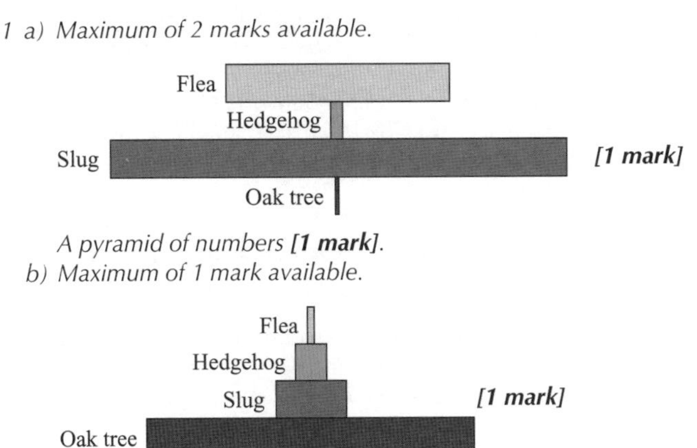

c) *Maximum of 1 mark available.*
*Pyramids of energy show the amount of energy available in each trophic level **[1 mark]**.*

Page 39 — Farming Practices and Productivity

1 *Maximum of 5 marks available.*
*Organic farmers might use biological agents **[1 mark]**. Biological agents reduce the numbers of pests, so crops lose less energy and biomass, increasing productivity **[1 mark]**. They include natural predators that eat the pest species to reduce their numbers **[1 mark]**. They could also use parasites that live in or lay their eggs on pest insects, killing the pests or reducing their ability to function **[1 mark]**. They could also introduce pathogens, which are bacteria or viruses that kill pests **[1 mark]**.*

Page 41 — The Nitrogen Cycle and Eutrophication

1 a) *Maximum of 2 marks available.*
*A — ammonification **[1 mark]***
*C — denitrification **[1 mark]***

b) *Maximum of 3 marks available.*
*Process B is nitrogen fixation **[1 mark]**. Nitrogen fixation is where nitrogen gas in the atmosphere is turned into ammonia **[1 mark]** by bacteria **[1 mark]**.*

Page 43 — The Carbon Cycle

1 a) *Maximum of 2 marks available.*
*A — Decomposition / saprobiontic nutrition **[1 mark]***
*B — Carbon compounds in decomposers **[1 mark]***

b) *Maximum of 6 marks available.*
*Carbon from CO_2 in the air and water becomes carbon compounds in plants when they photosynthesise **[1 mark]**. Carbon is then passed onto primary consumers when they eat the plants, and secondary and tertiary consumers when they eat the other consumers **[1 mark]**. When organisms die, the carbon in the dead organisms is digested by microorganisms called decomposers **[1 mark]**. Carbon is returned to the atmosphere as CO_2 because all living organisms carry out respiration, which produces CO_2 **[1 mark]**. When dead organic matter ends up in a place where there aren't any decomposers the carbon can be turned into fossil fuels over millions of years **[1 mark]**. Carbon in fossil fuels is released back into the atmosphere when they are burnt **[1 mark]**.*

Section 4 — Global Warming

Page 45 — Introduction to Global Warming

1 *Maximum of 6 marks available.*
*The diagram shows that the thickness of the pine tree rings fluctuated **[1 mark]**, but there was a trend of increasingly thicker rings from 1909 to 2009 **[1 mark]**. The thickness of each tree ring depends on the climate when the ring was formed **[1 mark]**. Warmer climates tend to give thicker rings than colder climates **[1 mark]**, which suggests that the climate where the pine tree lived became warmer over the last century **[1 mark]**. This is evidence for global warming **[1 mark]**.*

2 *Maximum of 2 marks available.*
*Scientists know the climate that different plant species live in now **[1 mark]**. When they find preserved pollen from similar plants, they know that the climate must have been similar when that pollen was produced **[1 mark]**.*

Page 47 — Causes of Global Warming

1 a) *Maximum of 4 marks available.*
*The temperature fluctuated between 1970 and 2008 **[1 mark]**, but the general trend was a steady increase from around 13.9 °C to around 14.4 °C **[1 mark]**. The atmospheric CO_2 concentration also showed a trend of increasing **[1 mark]** from around 328 ppm in 1970 to around 385 ppm in 2008 **[1 mark]**.*
You usually have to quote figures from graphs and tables in your answer to get full marks.

b) *Maximum of 2 marks available.*
*There's a positive correlation between temperature and CO_2 concentration **[1 mark]**. The increasing CO_2 concentration could be linked to the increasing temperature **[1 mark]**.*
You can't conclude from this data that it's a causal relationship because other factors may have been involved.

Page 49 — Effects of Global Warming

1 a) *Maximum of 4 marks available.*
*CO_2 concentration shows a general trend of increasing **[1 mark]** from around 338 ppm in 1980 to around 368 ppm in 2000 **[1 mark]**. The corn yield fluctuates but shows a general trend of increasing **[1 mark]** from around 105 bushels per acre in 1980 to around 135 bushels per acre in 2000 **[1 mark]**.*

b) *Maximum of 1 mark available.*
*There's a positive correlation between CO_2 concentration and corn yield / as CO_2 concentration increases so does corn yield **[1 mark]**.*
Even though there's a correlation you can't conclude that increasing CO_2 concentration is causing increases in corn yield, because there could be other factors involved, e.g. an increase in temperature or changing rainfall pattern.

c) *Maximum of 2 marks available.*
*CO_2 concentration is a limiting factor for photosynthesis **[1 mark]**, so increasing CO_2 concentration could mean crops grow faster, increasing crop yields **[1 mark]**.*

Page 51 — Effects of Global Warming

1 *Maximum of 4 marks available.*
*An increase in temperature causes an increase in enzyme activity **[1 mark]**, which speeds up metabolic reactions **[1 mark]**. Increasing the rate of metabolic reactions in a potato tuber moth will increase its rate of growth **[1 mark]**, so it will progress through its life cycle faster **[1 mark]**.*

Answers

2 *Maximum of 5 marks available.*
The student could plant some seedlings in soil trays, and measure the height of each seedling ***[1 mark]****. The student could then put the trays in incubators at different temperatures* ***[1 mark]****. All other variables would need to be kept the same for each tray* ***[1 mark]****. After a period of incubation, the student could record the change in height of each seedling* ***[1 mark]****, and then calculate the average growth rate for each tray to see how increasing temperature affects growth rate* ***[1 mark]****.*

Page 53 — Reducing Global Warming

1 *Maximum of 4 marks available.*
Biofuels are fuels produced from biomass / material that is or was recently living ***[1 mark]****. Biofuels are burnt to release energy, which produces* CO_2 ***[1 mark]****. There's no net increase in atmospheric* CO_2 *concentration because the amount of* CO_2 *produced is the same as the amount of* CO_2 *taken in when the material was growing* ***[1 mark]****. Using biofuels as an alternative to fossil fuels stops the increase in atmospheric* CO_2 *concentration caused by burning fossil fuels* ***[1 mark]****.*

2 *Maximum of 4 marks available.*
It's not actually known how emissions will change, i.e. how accurate the extrapolated CO_2 *data is* ***[1 mark]****. Scientists don't know exactly how much the extrapolated* CO_2 *changes will cause the global temperature to rise* ***[1 mark]****. The change in atmospheric* CO_2 *concentration due to natural causes isn't known* ***[1 mark]****. Scientists also don't know what attempts there will be to manage atmospheric* CO_2 *concentration, or how successful they'll be* ***[1 mark]****.*

Section 5 — Succession and Conservation

Page 55 — Succession

1 a) *Maximum of 6 marks available.*
This is an example of secondary succession, because there is already a soil layer present in the field ***[1 mark]****. The first species to grow will be the pioneer species, which in this case will be larger plants* ***[1 mark]****. These will then be replaced with shrubs and smaller trees* ***[1 mark]****. At each stage, different plants and animals that are better adapted for the improved conditions will move in, out-compete the species already there, and become the dominant species* ***[1 mark]****. As succession goes on, the ecosystem becomes more complex, so species diversity (the number and abundance of different species) increases* ***[1 mark]****. Eventually large trees will grow, forming the climax community, which is the final seral stage* ***[1 mark]****.*

b) *Maximum of 2 marks available.*
Ploughing destroys any plants that were growing ***[1 mark]****, so larger plants may start to grow, but they won't have long enough to establish themselves before the field is ploughed again* ***[1 mark]****.*

Page 57 — Conservation

1 a) *Maximum of 6 marks available.*
Conserving rainforests is important for humans as they may provide lots of things that humans need such as clothes, food or drugs ***[1 mark]****. If they're cut down, the source of these things will be lost and they won't be available in the future* ***[1 mark]****. Some people think rainforests should be conserved because it's the right thing to do — they think forests have a right to exist, and people don't have a right to cut them all down* ***[1 mark]****. The rainforests bring joy to lots of people who visit them. If they're cut down future generations won't be able to enjoy them* ***[1 mark]****. Conservation of rainforests will mean less trees are burnt. This helps to prevent climate change because burning trees releases* CO_2 *into the atmosphere, which contributes to global warming* ***[1 mark]****. Conserving rainforests will help to prevent the disruption of food chains. A decrease in one species could mean the loss of many more species in the food chain as there's less food, so more resources are lost* ***[1 mark]****.*

b) *Maximum of 2 marks available, from any of the 3 points below*
E.g. seedbanks ***[1 mark]****, captive breeding programmes* ***[1 mark]****, relocation* ***[1 mark]****.*

Page 59 — Conservation Evidence and Data

1 a) *Maximum of 4 marks available.*
The cod stock size increased from around 150 000 tonnes in 1963 to around 250 000 tonnes in 1971 ***[1 mark]****. The stock size then decreased (apart from a couple of smaller increases) to around 30 000 tonnes in 2006* ***[1 mark]****. The fishing mortality rate fluctuated but showed a trend of increasing* ***[1 mark]*** *from around 0.5 in 1963 to just below 0.8 in 2006* ***[1 mark]****.*

b) *Maximum of 2 marks available.*
There's a link between fishing mortality rate and the cod stock size ***[1 mark]****. As the fishing mortality rate increases, the cod stock size decreases/there's a negative correlation between fishing mortality rate and cod stock size* ***[1 mark]****.*

c) *Maximum of 1 mark available.*
1978/1979 ***[1 mark]***

d) *Maximum of 1 mark available.*
It could be used by governments to make decisions about cod fishing quotas (the amount of cod allowed to be removed from the sea by fishermen each year), to try to keep the cod stock above 150 000 tonnes ***[1 mark]****.*

Page 61 — Conservation of Ecosystems

1 *Maximum of 3 marks available, from any of the 4 points below. For full marks, answers must contain at least one economic, one social and one ethical reason.*
Conservation of ecosystems is important for economic reasons because ecosystems provide resources for things that are traded on a local and global scale, like clothes, drugs and food. If they're not conserved, the resources could be lost, causing large economic losses in the future ***[1 mark]****. Many ecosystems bring joy to lots of people because they're attractive to look at and people use them for activities like birdwatching and walking. If they aren't conserved the ecosystems may be lost, so future generations won't be able to use and enjoy them* ***[1 mark]****. Some people think ecosystems should be conserved because it's the right thing to do. They think organisms have a right to exist, so they shouldn't become extinct because of human activity* ***[1 mark]****. Some people also think that humans have a moral responsibility to conserve ecosystems for future human generations, so they can enjoy and use them* ***[1 mark]****.*

2 a) *Maximum of 2 marks available.*
1 mark for an explanation and 1 mark for an example.
Non-native animal species eat some native species, causing a decrease in the populations of native species ***[1 mark]****. For example, dogs, cats and black rats eat young giant tortoises and Galapagos land iguanas* ***[1 mark]*** */ pigs destroy the nests of Galapagos land iguanas and eat their eggs* ***[1 mark]*** */ goats have eaten a lot of the plant life on some of the islands* ***[1 mark]****.*

b) *Maximum of 2 marks available.*
Non-native plant species have decreased native plant populations because they compete with the native species ***[1 mark]****. For example, quinine trees are taller than some native plants. They block out light to the native plants, which then struggle to survive* ***[1 mark]****.*

Answers

c) *Maximum of 2 marks available.*
1 mark for an explanation and 1 mark for an example.
*Fishing has caused a decrease in the populations of some of the sea life around the Galapagos Islands **[1 mark]**. For example, sea cucumber and hammerhead shark populations have been reduced because of overfishing **[1 mark]** / Galapagos green turtle numbers have been reduced because of overfishing **[1 mark]** / Galapagos green turtle numbers have been reduced because they're killed accidentally when they're caught in fishing nets **[1 mark]**.*
You've been asked to explain how specific animals or plants have been affected, so you need to use named examples.

Section 6 — Inheritance

Page 63 — Inheritance

1 a) *Maximum of 3 marks available.*
*Parents' genotypes identified as RR and rr **[1 mark]**.*
*Correct genetic diagram drawn with gametes' alleles identified as R, R and r, r **[1 mark]** and gametes crossed to show Rr as the only possible genotype **[1 mark]**.*
The question specifically asks you to draw a genetic diagram, so make sure that you include one in your answer, e.g.

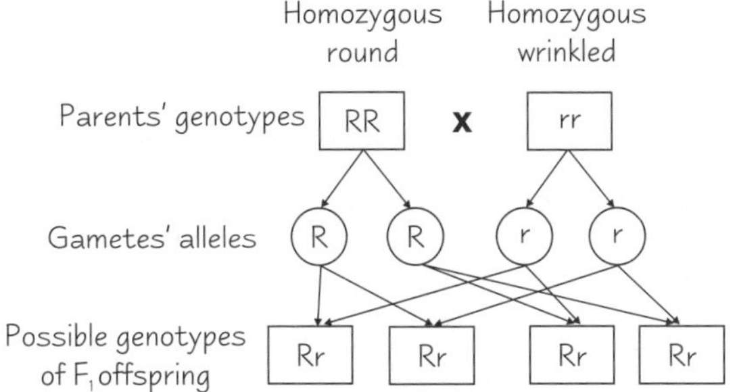

b) *Maximum of 3 marks available.*
*Gametes' alleles (produced by F_1 generation) identified as R, r and R, r **[1 mark]**. Gametes crossed to show genotypes RR, Rr and rr in a 1:2:1 ratio **[1 mark]**. RR and Rr genotypes identified as giving a round phenotype and rr as wrinkled phenotype, giving a 3:1 ratio of round : wrinkled seeds **[1 mark]**. Award three marks for a correct ratio of 3:1 for round : wrinkled seeds.*
The question doesn't ask for a genetic diagram but it can help you work out the answer, e.g.

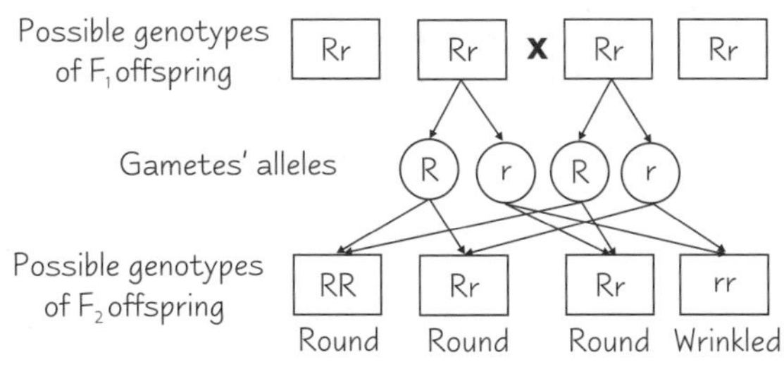

Page 65 — Inheritance

1 *Maximum of 3 marks available.*
*Men only have one copy of the X chromosome (XY) but women have two (XX) **[1 mark]**. Haemophilia A is caused by a recessive allele so females would need two copies of the allele for them to have haemophilia A **[1 mark]**. As males only have one X chromosome they only need one recessive allele to have haemophilia A, which makes them more likely to have haemophilia A than females **[1 mark]**.*

2 *Maximum of 4 marks available.*
*Genotypes of parents identified as I^AI^O and I^BI^B **[1 mark]**. Correct genetic diagram drawn with gametes' alleles identified as I^A, I^O and I^B, I^B **[1 mark]** and gametes crossed to show genotypes I^AI^B and I^BI^O in a 1:1 ratio **[1 mark]**. The probability of the couple having a child with blood group B is 0.5 (or 50%) **[1 mark]**.*
The question specifically asks you to draw a genetic diagram so make sure that you include one in your answer, e.g.

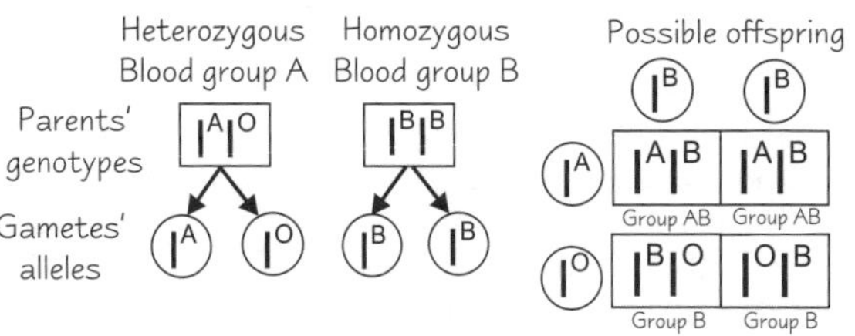

3 a) *Maximum of 1 mark available.*
*Individual 2 could have genotype AA or Aa **[1 mark]**.*
Individual 2 is an unaffected male so he must have at least one A allele (AA or Aa). But you can't say for sure if he has AA or Aa.

b) *Maximum of 2 marks available*
*Individual 6's genotype is Aa **[1 mark]**. The offspring of individuals 5 and 6 has ADA deficiency (aa), so both parents must be carriers of the recessive allele (Aa) **[1 mark]**.*
*Or, individual 3 has ADA deficiency (aa), so must have passed a recessive allele onto individual 6. Individual 6 is unaffected so must have a dominant allele as well (Aa) **[1 mark]**.*

c) *Maximum of 4 marks available*
*Parents' genotypes identified as both Aa **[1 mark]**.*
*Correct genetic diagram drawn with gametes' alleles identified as A, a and A, a **[1 mark]** and gametes crossed to show genotypes AA, Aa and aa in a 1:2:1 ratio **[1 mark]**. aa genotype identified as causing ADA deficiency, giving a probability of 0.25 (25%) **[1 mark]**.*
The question asks you to show your working so you should draw a genetic diagram, e.g.

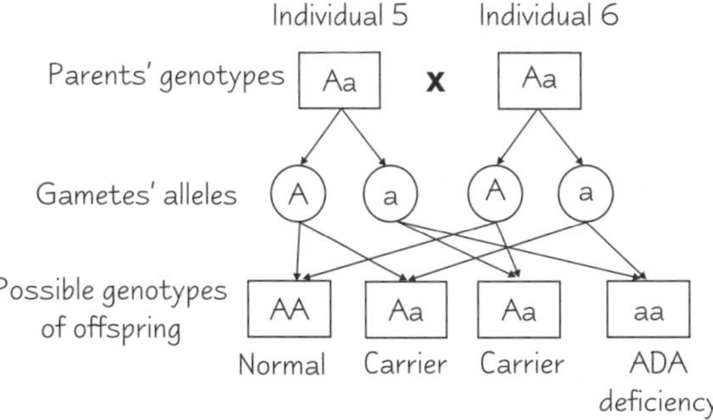

Page 67 — Phenotypic Ratios and Epistasis

1 *Maximum of 3 marks available.*
*Parents' genotypes identified as RRgg and rrGG **[1 mark]**.*
*Correct genetic diagram drawn with gametes' alleles identified as Rg and rG **[1 mark]** and gametes crossed to show RrGg as the only possible genotype of the offspring **[1 mark]**.*
The question specifically asks you to draw a genetic diagram, so make sure that you include one in your answer, e.g.

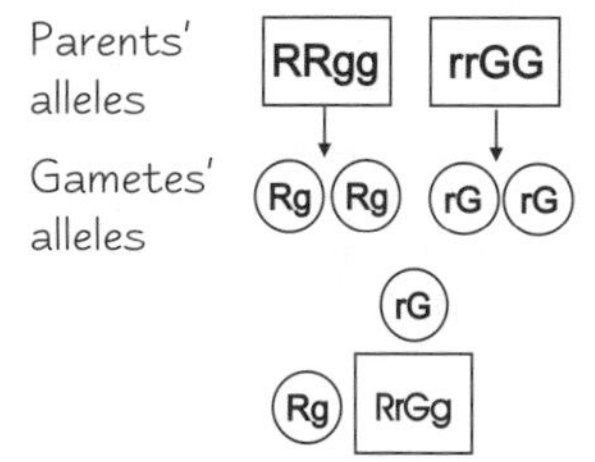

Answers

2 *Maximum of 4 marks available.*
A cross between CCGG and ccgg will produce a 9 : 3 : 4 phenotypic ratio in the F_2 generation ***[1 mark]*** *of coloured grey : coloured black : albino* ***[1 mark]****. This is because gene 1 has a recessive epistatic gene (c)* ***[1 mark]****, and two copies of the recessive epistatic gene (cc) will mask the expression of the colour gene* ***[1 mark]****.*
You don't need to draw a genetic diagram to explain the phenotypic ratio that you'd expect from this cross. You can just state the ratio and explain it using your own knowledge.

3 *Maximum of 3 marks available.*
The table shows that a cross between hhss and HHSS produces a 36 : 9 : 3 or 12 : 3 : 1 phenotypic ratio in the F_2 generation of bald : straight hair : curly hair ***[1 mark]****. This is because the hair gene has a dominant epistatic allele (H)* ***[1 mark]****, which means having at least one copy of the dominant epistatic gene (Hh or HH) will result in a bald phenotype that masks the expression of the type of hair gene* ***[1 mark]****.*

Page 69 — The Chi-Squared Test

1 a *Maximum of 4 marks available.*
(1 mark for each correct column and 1 mark for the answer).

Phenotype	Ratio	Expected Result (E)	Observed Result (O)	O – E	$O-E^2$	$\frac{(O-E^2)}{E}$
Blue with white spots	9	135	131	–4	16	0.12
Purple with white spots	3	45	52	7	49	1.09
Blue with yellow spots	3	45	48	3	9	0.2
Purple with yellow spots	1	15	9	–6	36	2.4
						3.81

b *Maximum of 2 marks available.*
The χ^2 value does support the null hypothesis ***[1 mark]*** *because it's smaller than the critical value* ***[1 mark]****.*

Page 71 — Meiosis

1 a) *Maximum of 4 marks available.*
The chromosomes condense, getting shorter and fatter ***[1 mark]****.*
Homologous chromosomes pair up ***[1 mark]****.*
The centrioles start moving to opposite ends of the cell, forming a network of protein fibres across it called the spindle ***[1 mark]****.*
The nuclear envelope breaks down ***[1 mark]****.*
The question asks you to describe the nuclear envelope, chromosomes and centrioles, so make sure you include them all to get full marks.

b) *Maximum of 2 marks available.*
A — Telophase II ***[1 mark]***
B — Anaphase II ***[1 mark]***

2 a) *Maximum of 7 marks available.*
Crossing-over of chromatids during prophase I causes genetic variation ***[1 mark]****. The non-sister chromatids twist around each other and bits of the chromatids swap over* ***[1 mark]****. This means that each of the four daughter cells contain chromatids with different combinations of alleles* ***[1 mark]****. Independent assortment of chromosomes in metaphase I produces genetic variation* ***[1 mark]****. Different combinations of maternal and paternal chromosomes go into each daughter cell, so each cell ends up with a different combination of alleles* ***[1 mark]****. Independent assortment of chromatids in metaphase II also produces genetic variation* ***[1 mark]****. Different combinations of chromatids go into each daughter cell, so each cell ends up with a different combination of alleles* ***[1 mark]****.*

b) *Maximum of 1 mark available.*
Fertilisation increases genetic variation because any egg cell can fuse with any sperm cell ***[1 mark]****.*

Section 7 — Variation, Natural Selection and Evolution

Page 73 — Variation

1 a) *Maximum of 1 mark available.*
Discontinuous variation ***[1 mark]****.*

b) *Maximum of 1 mark available.*
18.99 – 9.25 = 9.74 kg ***[1 mark]***

c) *Maximum of 2 marks available.*
Mass ***[1 mark]*** *because it shows continuous variation* ***[1 mark]****.*

2 *Maximum of 2 marks available.*
E.g. body mass ***[1 mark]*** *because large parents often have large children so it's affected by genotype, but body mass is also influenced by diet and exercise, which are environmental factors* ***[1 mark]****.*

Page 75 — Natural Selection and Genetic Drift

1 *Maximum of 2 marks available.*
Directional selection ***[1 mark]****, because the environment was changing (more coal-powered factories) and the phenotype is moving towards an extreme — darker moths* ***[1 mark]****.*

Page 77 — The Hardy-Weinberg Principle

1 *Maximum of 1 mark available.*
Frequency of the recessive allele (q) = 0.23, and p + q = 1
So the frequency of the dominant allele (p) = 1 – q
= 1 – 0.23
= 0.77 ***[1 mark]****.*

2 a) *Maximum of 3 marks available.*
Frequency of genotype CC = p^2 = 0.14 ***[1 mark]*** *so the frequency of the dominant allele C = p = $\sqrt{0.14}$ = 0.37* ***[1 mark]****.*
The frequency of the recessive allele c = q = 1 – p = 1 – 0.37 = 0.63 ***[1 mark]****. Award three marks for a correct answer of 0.63.*

b) *Maximum of 1 mark available.*
Frequency of homozygous recessive genotype cc = q^2 = 0.63^2 = 0.40 ***[1 mark]****.*

c) *Maximum of 2 marks available.*
Those that don't have a cleft chin are homozygous recessive cc = 40% ***[1 mark]****, so the percentage that do have a cleft chin, Cc or CC, is 100% – 40% = 60%* ***[1 mark]****. Award two marks for a correct answer of 60%.*
There are other ways of calculating this answer, e.g. working out the value of 2pq and adding it to p^2. It doesn't matter which way you do it as long as you get the right answer.

Page 79 — Speciation

1 a) *Maximum of 1 mark available.*
The new species could not breed with each other ***[1 mark]****.*

b) *Maximum of 3 marks available.*
Different populations of flies were isolated and fed on different foods ***[1 mark]****. This caused changes in allele frequencies between the populations* ***[1 mark]****, which made them reproductively isolated and eventually resulted in speciation* ***[1 mark]****.*

c) *Maximum of 2 marks available, from any of the 3 points below.*
Seasonal changes (become sexually active at different times) ***[1 mark]****. Mechanical changes (changes to genitalia)* ***[1 mark]****. Behavioural changes (changes in behaviour that prevent mating)* ***[1 mark]****.*

Answers

d) Maximum of 1 mark available, from any of the 5 points below or any other good point.
E.g. geographical barrier **[1 mark]**, flood **[1 mark]**, volcanic eruption **[1 mark]**, earthquake **[1 mark]**, glacier **[1 mark]**.

Page 81 — Evolution Evidence and Artificial Selection

1 Maximum of 4 marks available.
Before scientists can get their work published it must undergo peer review, which is when other scientists who work in that area read and review the work **[1 mark]**. The peer reviewer checks that the work is valid and supports the conclusion **[1 mark]**. Scientific journals also allow other scientists to repeat experiments and see if they get the same results using the same methods **[1 mark]**. If the results are replicated over and over again, the scientific community can be pretty confident that the evidence collected is reliable **[1 mark]**.

2 Maximum of 3 marks available.
Farmers could have selected a male and female with a high meat yield and bred these two together **[1 mark]**. Then they could have selected the offspring with the highest meat yields and bred them together **[1 mark]**. This process could have been continued over several generations to produce cattle with a very high meat yield **[1 mark]**.

Section 8 — Responding to the Environment

Page 84 — Nervous and Hormonal Communication

1 Maximum of 5 marks available.
Receptors detect the stimulus **[1 mark]**, e.g. light receptors/ photoreceptors in the animal's eyes detect the bright light **[1 mark]**. The receptors send impulses along neurones via the CNS to the effectors **[1 mark]**. The effectors bring about a response **[1 mark]**, e.g. the circular iris muscles contract to constrict the pupils and protect the eyes **[1 mark]**.

2 Maximum of 3 marks available, from any of the 4 points below.
The nervous system sends information as electrical impulses but the hormonal system sends information as chemical signals **[1 mark]**. Nervous responses are localised but hormonal responses may be widespread if the target cells are widespread **[1 mark]**. Nervous responses are short-lived but hormonal responses can be long-lasting **[1 mark]**. Nervous responses are faster than hormonal responses **[1 mark]**.

Page 87 — Receptors

1 Maximum of 6 marks available, from any of the 7 points below.
A tap on the arm is a mechanical stimulus that's detected by pressure receptors/mechanoreceptors called Pacinian corpuscles **[1 mark]**. The stimulus deforms the layers of connective tissue (lamellae) **[1 mark]**, which press on the sensory nerve ending **[1 mark]**. This causes deformation of stretch-mediated sodium ion channels in the neurone cell membrane **[1 mark]**.
Sodium ion channels open and sodium ions diffuse into the cell **[1 mark]**. This creates a generator potential **[1 mark]**.
If the generator potential reaches the threshold it triggers a nerve impulse/action potential **[1 mark]**.

2 Maximum of 5 marks available.
The human eye has high sensitivity because many rods join one neurone **[1 mark]**, so many weak generator potentials combine to reach the threshold and trigger an action potential **[1 mark]**.
The human eye has high acuity because cones are close together and one cone joins one neurone **[1 mark]**. When light from two points hits two cones, action potentials from each cone go to the brain **[1 mark]**. So you can distinguish two points that are close together as two separate points **[1 mark]**.

3 Maximum of 7 marks available.
Light energy bleaches rhodopsin / causes rhodopsin to break apart into retinal and opsin **[1 mark]**. This causes the sodium ion channels to close **[1 mark]**. So sodium ions are still actively transported out of the cell but they can't diffuse back in **[1 mark]**.
This means sodium ions build up on the outside of the cell, making the cell membrane hyperpolarised **[1 mark]**.
This causes the rod cell to stop releasing neurotransmitters **[1 mark]**. There's no inhibition of the bipolar neurone **[1 mark]**, so the bipolar neurone depolarises and sends an action potential to the brain via the optic nerve **[1 mark]**.

Page 89 — Nervous System — Neurones

1 a) Maximum of 1 mark available.
Stimulus **[1 mark]**.
b) Maximum of 3 marks available.
A stimulus causes sodium ion channels in the neurone cell membrane to open **[1 mark]**. Sodium ions diffuse into the cell **[1 mark]**, so the membrane becomes depolarised **[1 mark]**.
c) Maximum of 2 marks available.
The membrane was in the refractory period **[1 mark]**, so the sodium ion channels were recovering and couldn't be opened **[1 mark]**.

Page 91 — Nervous System — Neurones

1 Maximum of 2 marks available.
The refractory period makes sure that action potentials pass along as discrete impulses **[1 mark]** and it makes sure action potentials are unidirectional **[1 mark]**.

2 Maximum of 5 marks available.
Transmission of action potentials will be slower in neurones with damaged myelin sheaths **[1 mark]**. This is because myelin is an electrical insulator **[1 mark]**, so increases the speed of action potential conduction **[1 mark]**. The action potentials 'jump' between the nodes of Ranvier/between the myelin sheaths **[1 mark]**, where sodium ion channels are concentrated **[1 mark]**.
Don't panic if a question mentions something you haven't learnt about. You might not know anything about multiple sclerosis but that's fine, because you're not supposed to. All you need to know to get full marks here is how myelination affects the speed of action potential conduction.

Page 93 — Nervous System — Synaptic Transmission

1 Maximum of 5 marks available.
A — presynaptic membrane **[1 mark]**.
B — vesicle/vesicle containing neurotransmitter **[1 mark]**.
C — synaptic cleft **[1 mark]**.
D — postsynaptic receptor **[1 mark]**.
E — postsynaptic membrane **[1 mark]**.

Answers

2 Maximum of 6 marks available, from any of the 8 points below.
The action potential arriving at the presynaptic membrane stimulates voltage-gated calcium ion channels to open ***[1 mark]***, so calcium ions diffuse into the neurone ***[1 mark]***. This causes synaptic vesicles, containing acetylcholine, to fuse with the presynaptic membrane ***[1 mark]***. The vesicles release acetylcholine into the synaptic cleft ***[1 mark]***. The acetylcholine diffuses across the synaptic cleft ***[1 mark]*** and binds to cholinergic receptors on the postsynaptic membrane ***[1 mark]***. This causes sodium ion channels in the postsynaptic membrane to open ***[1 mark]*** and the influx of sodium ions triggers a new action potential to be generated at the postsynaptic membrane ***[1 mark]***.

3 Maximum of 4 marks available.
They might have weaker muscular responses than normal ***[1 mark]***. If receptors are destroyed at neuromuscular junctions then there will be fewer receptors for acetylcholine/ACh to bind to ***[1 mark]***, so fewer sodium ion channels will open ***[1 mark]***, meaning fewer muscle cells can be stimulated ***[1 mark]***.

Page 95 — Nervous System — Synaptic Transmission

1 a) Maximum of 2 marks available.
The stimulus could have been too small, so neurone A wouldn't have released enough neurotransmitter to reach the threshold in neurone X and trigger an action potential ***[1 mark]***. Neurone A could have released an inhibitory neurotransmitter, which prevented neurone X from firing an action potential ***[1 mark]***.

b) Maximum of 2 marks available.
Spatial summation could have happened if all the neurones released a small amount of neurotransmitter, which was enough to reach the threshold for neurone X ***[1 mark]***. Neurones B, C and D could have released excitatory neurotransmitters, which would overcome any inhibition of neurone X caused by an inhibitory neurotransmitter from neurone A ***[1 mark]***.

2 Maximum of 3 marks available.
Galantamine would stop acetylcholinesterase/AChE breaking down acetylcholine/ACh, so there would be more ACh in the synaptic cleft ***[1 mark]*** and it would be there for longer ***[1 mark]***. This means more nicotinic cholinergic receptors would be stimulated ***[1 mark]***.

Page 97 — Effectors — Muscle Contraction

1 Maximum of 3 marks available.
Muscles are made up of bundles of muscle fibres ***[1 mark]***. Muscle fibres contain long organelles called myofibrils ***[1 mark]***. Myofibrils contain bundles of myofilaments ***[1 mark]***.

2 a) Maximum of 3 marks available.
A = sarcomere ***[1 mark]***.
B = Z-line ***[1 mark]***.
C = H-zone ***[1 mark]***.

b) Maximum of 3 marks available.
Drawing number 3 ***[1 mark]*** because the M-line connects the middle of the myosin filaments ***[1 mark]***. The cross-section would only show myosin filaments, which are the thick filaments ***[1 mark]***.
The answer isn't drawing number 1 because all the dots in the cross-section are smaller, so the filaments shown are thin actin filaments — which aren't found at the M-line.

Page 99 — Effectors — Muscle Contraction

1 Maximum of 2 marks available.
The A-bands stay the same length during contraction ***[1 mark]***. The I-bands get shorter ***[1 mark]***.

2 Maximum of 3 marks available.
Muscles need ATP to relax because ATP provides the energy to break the actin-myosin cross bridges ***[1 mark]***. If the cross bridges can't be broken, the myosin heads will remain attached to the actin filaments ***[1 mark]***, so the actin filaments can't slide back to their relaxed position ***[1 mark]***.

3 Maximum of 3 marks available.
The muscles won't contract ***[1 mark]*** because calcium ions won't be released into the sarcoplasm, so troponin won't be removed from its binding site ***[1 mark]***. This means no actin-myosin cross bridges can be formed ***[1 mark]***.

Page 101 — Effectors — Muscle Contraction

1 Maximum of 2 marks available.
Phosphocreatine/PCr is stored inside cells ***[1 mark]***. A phosphate group is taken from phosphocreatine/PCr and added to ADP to make ATP / ATP is made by phosphorylating ADP using a phosphate group from phosphocreatine/PCr ***[1 mark]***.

2 Maximum of 5 marks available, from any of the 8 points below.
Both types of muscle have one nucleus per muscle cell/fibre ***[1 mark]***. Both types have cells/fibres that are small/about 0.2 mm long ***[1 mark]***. Neither type fatigues/gets tired quickly ***[1 mark]***. Neither type is under conscious control ***[1 mark]***. However, involuntary muscle is found in the walls of hollow internal organs like the gut, but cardiac muscle is found in the walls of the heart ***[1 mark]***. Involuntary muscle fibres are spindle-shaped but cardiac muscle fibres are cylinder-shaped with intercalated discs ***[1 mark]***. Cardiac muscle fibres are branched but involuntary muscles fibres aren't ***[1 mark]***. Cardiac muscle fibres have some cross-striations but involuntary muscle fibres have a smooth appearance ***[1 mark]***.

Page 103 — Effectors — Glands

1 Maximum of 2 marks available.
Endocrine glands secrete chemicals directly into the blood, but exocrine glands secrete into ducts ***[1 mark]***.
Endocrine glands secrete hormones, but exocrine glands usually secrete enzymes ***[1 mark]***.

2 a) Maximum of 2 marks available.
It would have no effect ***[1 mark]*** because the endocrine tissue of the pancreas doesn't secrete into the pancreatic duct ***[1 mark]***.

b) Maximum of 2 marks available.
Food might not be properly digested ***[1 mark]*** because the exocrine tissue of the pancreas normally secretes digestive enzymes into the pancreatic duct ***[1 mark]***.

3 Maximum of 4 marks available.
The first messenger is a hormone ***[1 mark]***, which carries the message from an endocrine gland to the receptor on its target tissue ***[1 mark]***. The second messenger is a signalling molecule ***[1 mark]***, which carries the message from the receptor to other parts of the cell and activates a cascade inside the cell ***[1 mark]***.

Answers

Page 105 — Responses in Animals

1 a) *Maximum of 5 marks available.*
*High blood pressure is detected by pressure receptors in the aorta called baroreceptors **[1 mark]**. Impulses are sent along sensory neurones to the medulla **[1 mark]**. Impulses are then sent to the SAN along a parasympathetic neurone **[1 mark]**. The parasympathetic neurone secretes acetylcholine, which binds to receptors on the sinoatrial node/SAN **[1 mark]**. This slows the heart rate (reducing blood pressure) **[1 mark]**.*

b) *Maximum of 2 marks available.*
*No impulses sent from the medulla would reach the SAN **[1 mark]**, so the heart rate wouldn't increase or decrease/control of the heart rate would be lost **[1 mark]**.*

Page 107 — Responses in Plants

1 *Maximum of 3 marks available.*
*Auxins are produced in the tip of shoots and they're moved around the plant, so different parts of the plant have different amounts of auxins **[1 mark]**. The uneven distribution of auxins means there's uneven growth of the plant **[1 mark]**. Auxins move to the more shaded parts of the shoots, making the cells there elongate, which makes the shoot bend towards the light **[1 mark]**.*

2 a) *Maximum of 1 mark available.*
*P_{FR} is a phytochrome molecule in a state that absorbs far-red light/light at a wavelength of 730 nm **[1 mark]**.*

b) *Maximum of 4 marks available.*
*An iris would flower in the summer/June to August **[1 mark]** because it's stimulated to flower by high levels of P_{FR}, which occurs when nights are short **[1 mark]**. Daylight contains more red light than far-red light, so more P_R is converted into P_{FR} than P_{FR} is converted into P_R **[1 mark]**. When nights are short in the summer, there's not much time for P_{FR} to be converted back into P_R, so P_{FR} builds up **[1 mark]**.*

Page 109 — Plant Hormones

1 a) *Maximum of 1 mark available.*
*Auxins **[1 mark]**.*

b) *Maximum of 2 marks available.*
*Apical dominance saves energy as it stops side shoots growing. This allows a plant in an area where there are lots of other plants to grow tall very fast, past the smaller plants, to reach the sunlight **[1 mark]**. Apical dominance also prevents side shoots of the same plant from competing with the shoot tip for light **[1 mark]**.*

2 a) *Maximum of 1 mark available.*
*Ethene **[1 mark]**.*

b) *Maximum of 1 mark available.*
*Ethene stimulates enzymes that break down cell walls, break down chlorophyll and convert starch to sugars **[1 mark]**.*

c) *Maximum of 1 mark available.*
*E.g. the tomatoes are less likely to be damaged in transport **[1 mark]**.*

Page 111 — Investigating Responses in Plants

1 a) *Maximum of 2 marks available.*
*The data shows that the plants provided with auxins grew more than those not given auxins **[1 mark]**. This is because auxins stimulate plant growth (by cell elongation) **[1 mark]**.*

b) *Maximum of 2 marks available.*
*The students should repeat the experiment to see if their results are reliable **[1 mark]**. They should also make sure all other variable conditions are the same for each group **[1 mark]**.*

c) *Maximum of 1 mark available.*
*The results suggest that auxins stimulate plant growth, so auxins could be used to increase tomato yield **[1 mark]**.*

Section 9 — Homeostasis

Page 113 — Homeostasis Basics

1 *Maximum of 3 marks available.*
*Receptors detect when a level is too high or too low **[1 mark]**, and the information's communicated via the nervous system or the hormonal system to effectors **[1 mark]**. Effectors respond to counteract the change / to bring the level back to normal **[1 mark]**.*

2 a) *Maximum of 2 marks available.*
*Statement A **[1 mark]** because body temperature continues to increase from the normal level and isn't returned **[1 mark]**.*

b) *Maximum of 2 marks available.*
*It makes metabolic reactions less efficient **[1 mark]** because the enzymes that control metabolic reactions may denature **[1 mark]**.*

Page 115 — Control of Body Temperature

1 *Maximum of 4 marks available, from any of the 8 points below. 1 mark for each method, up to a maximum of 2 marks. 1 mark for each explanation, up to a maximum of 2 marks.*
*Vasoconstriction of blood vessels **[1 mark]** reduces heat loss because less blood flows through the capillaries in the surface layers of the dermis **[1 mark]**. Erector pili muscles contract to make hairs stand on end **[1 mark]**, trapping an insulating layer of air to prevent heat loss **[1 mark]**. Muscles contract in spasms to make the body shiver **[1 mark]**, so more heat is produced from increased respiration **[1 mark]**. Adrenaline and thyroxine are released **[1 mark]**, which increase metabolism so more heat is produced **[1 mark]**.*

2 *Maximum of 2 marks available.*
*Thermoreceptors/temperature receptors in the skin detect a higher external temperature than normal **[1 mark]**.*
*The thermoreceptors/temperature receptors send impulses along sensory neurones to the hypothalamus **[1 mark]**.*

3 *Maximum of 4 marks available.*
*Snakes are ectotherms **[1 mark]**. They can't control their body temperature internally and depend on the temperature of their external environment **[1 mark]**. In cold climates, snakes will be less active **[1 mark]**, which makes it harder to catch prey, avoid predators, find a mate, etc. **[1 mark]**.*
You need to use a bit of common sense to answer this question — you know that the activity level of an ectotherm depends on the temperature of the surroundings, so in a cold environment it won't be very active. And if it can't be very active it'll have trouble surviving.

Page 117 — Control of Blood Glucose Concentration

1 *Maximum of 5 marks available, from any of the 7 points below.*
*High blood glucose concentration is detected by cells in the pancreas **[1 mark]**. Beta/β cells secrete insulin into the blood **[1 mark]**, which binds to receptors on the cell membranes of liver and muscle cells **[1 mark]**. This increases the permeability of the cell membranes to glucose, so the cells take up more glucose **[1 mark]**. Insulin also activates glycogenesis **[1 mark]** and increases the rate that cells respire glucose **[1 mark]**.*
*This lowers the concentration of glucose in the blood **[1 mark]**.*
You need to get the spelling of words like glycogenesis right in the exam or you'll miss out on marks.

Answers

2 *Maximum of 2 marks available.*
No insulin would be secreted ***[1 mark]*** *because ATP wouldn't be produced, so the potassium ion channels in the β cell plasma membrane wouldn't close / the plasma membrane of β cell wouldn't be depolarised* ***[1 mark]****.*

Page 119 — Control of Blood Glucose Concentration

1 *Maximum of 4 marks available.*
Adrenaline binds to receptors on the outside of the liver cell / liver cell membrane ***[1 mark]****, which activates an enzyme called adenylate cyclase* ***[1 mark]****. Activated adenylate cyclase converts ATP into a second messenger / cyclic AMP* ***[1 mark]****. Cyclic AMP / the second messenger activates glycogenolysis* ***[1 mark]****.*

2 *Maximum of 3 marks available.*
They have Type II diabetes ***[1 mark]****. They produce insulin, but the insulin receptors on their cell membranes don't work properly, so the cells don't take up enough glucose* ***[1 mark]****. This means their blood glucose concentration remains higher than normal* ***[1 mark]****.*

3 *Maximum of 2 marks available, from any of the 4 points below.*
It's cheaper to produce insulin using GM bacteria than to extract it from animal pancreases ***[1 mark]****. Large amounts of insulin can be made using GM bacteria, so there's enough insulin to treat everyone with Type I diabetes* ***[1 mark]****. GM bacteria make real human insulin, which is more effective and less likely to trigger an allergic response or be rejected by the immune system* ***[1 mark]****. Some people prefer insulin from GM bacteria for ethical or religious reasons* ***[1 mark]****.*

Page 121 — Control of the Menstrual Cycle

1 a) *Maximum of 5 marks available.*
Follicle-stimulating hormone/FSH stimulates a follicle to develop ***[1 mark]****. FSH also stimulates the ovary to release oestrogen* ***[1 mark]****, and oestrogen is also released by the follicle* ***[1 mark]****. Oestrogen inhibits the release of FSH from the anterior pituitary* ***[1 mark]****, so no more follicles are stimulated to develop* ***[1 mark]****.*

b) *Maximum of 3 marks available.*
High oestrogen concentration stimulates the anterior pituitary to release luteinising hormone/LH ***[1 mark]****. LH stimulates the ovary to release more oestrogen* ***[1 mark]****, which further stimulates the anterior pituitary in a positive feedback loop* ***[1 mark]****.*

2 *Maximum of 3 marks available.*
Oestrogen and progesterone inhibit follicle-stimulating hormone/ FSH release from the anterior pituitary ***[1 mark]****. Because there's no FSH, the follicle isn't stimulated to develop* ***[1 mark]****, so there is no ovulation* ***[1 mark]****.*

Section 10 — Genetics

Page 123 — DNA and RNA

1 a) *Maximum of 1 mark available.*
tRNA ***[1 mark]***

b) *Maximum of 1 mark available.*
tRNA ***[1 mark]***

2 *Maximum of 2 marks available.*
mRNA carries the genetic code from the DNA in the nucleus to the cytoplasm, where it's used to make a protein during translation ***[1 mark]****. tRNA carries the amino acids that are used to make proteins to the ribosomes during translation* ***[1 mark]****.*

Page 125 — Protein Synthesis

1 *Maximum of 2 marks available.*
The drug binds to DNA, preventing RNA polymerase from binding, so transcription can't take place and no mRNA can be made ***[1 mark]****. This means there's no mRNA for translation and so protein synthesis is inhibited* ***[1 mark]****.*

2 a) *Maximum of 2 marks available.*
$10 \times 3 = 30$ nucleotides long ***[1 mark]****. Each amino acid is coded for by three nucleotides (a codon), so the mRNA length in nucleotides is the number of amino acids multiplied by three* ***[1 mark]****.*

b) *Maximum of 6 marks available.*
The mRNA attaches itself to a ribosome and transfer RNA (tRNA) molecules carry amino acids to the ribosome ***[1 mark]****. A tRNA molecule, with an anticodon that's complementary to the first codon on the mRNA (the start codon), attaches itself to the mRNA by specific base pairing* ***[1 mark]****. A second tRNA molecule attaches itself to the next codon on the mRNA in the same way* ***[1 mark]****. The two amino acids attached to the tRNA molecules are joined by a peptide bond and the first tRNA molecule moves away, leaving its amino acid behind* ***[1 mark]****. A third tRNA molecule binds to the next codon on the mRNA and its amino acid binds to the first two and the second tRNA molecule moves away* ***[1 mark]****. This process continues, producing a chain of linked amino acids (a polypeptide chain), until there's a stop codon on the mRNA molecule, which is where translation stops* ***[1 mark]****.*

Page 127 — The Genetic Code and Nucleic Acids

1 a) *Maximum of 2 marks available.*
The mRNA sequence is 18 nucleotides long and the protein produced from it is 6 amino acids long ***[1 mark]****. $18 \div 6 = 3$, suggesting three nucleotides code for a single amino acid* ***[1 mark]****.*

b) *Maximum of 2 marks available.*
The protein produced was leucine-cysteine-glycine.
This would only be produced if the code is non-overlapping, e.g. UUGUGUGGG = UUG-UGU-GGG = leucine-cysteine-glycine ***[1 mark]****.*
If the code was overlapping the codons would be UUG-UGU-GUG-UGU, which would give the protein leucine-cysteine-valine-cysteine.
Also the protein produced is only 6 amino acids long, which is correct if the code is non-overlapping — the protein would be longer if the code overlapped ***[1 mark]****.*

2 a) *Maximum of 2 marks available. Award 2 marks if all four amino acids are correct and in the correct order. Award 1 mark if three amino acids are correct and in the correct order.*
GUG = valine
UGU = cysteine
CGC= arginine
GCA = alanine
Correct sequence = valine, cysteine, arginine, alanine.

b) *Maximum of 2 marks available. Award 2 marks if all four codons are correct and in the correct order. Award 1 mark if three codons are correct and in the correct order.*
arginine = CGC
alanine = GCA
leucine = UUG
phenylalanine = UUU
Correct sequence = CGC GCA UUG UUU.

Answers

c) Maximum of 3 marks available.
valine = GUG
arginine = CGC
alanine = GCA
mRNA sequence = GUG CGC GCA.
DNA sequence = CAC **[1 mark]** GCG **[1 mark]** CGT **[1 mark]**.

Page 129 — Regulation of Transcription and Translation

1 a) Maximum of 4 marks available.
The results of tubes 1 and 2 suggest that oestrogen affects the expression of the gene for the Chi protein **[1 mark]** because mRNA and active protein production only occur in the presence of oestrogen **[1 mark]**. When oestrogen is present it binds to the oestrogen receptor (transcription factor), forming an oestrogen-oestrogen receptor complex **[1 mark]**. This complex works as an activator, helping RNA polymerase to bind to the DNA, activating transcription and resulting in protein production in the presence of oestrogen **[1 mark]**.

b) Maximum of 3 marks available.
The mutant could have a faulty oestrogen receptor **[1 mark]**. Oestrogen might not bind to the receptor, or the oestrogen-oestrogen receptor complex might not work as an activator **[1 mark]**. This would mean even in the presence of oestrogen transcription wouldn't be activated, so no mRNA or protein would be produced **[1 mark]**.
This is quite a tricky one — drawing a diagram of how oestrogen controls transcription would help you figure out the answer.

c) Maximum of 3 marks available.
E.g. the results would be no full length mRNA and no protein produced **[1 mark]**. The siRNA and associated proteins would attach to the mRNA of the Chi protein and cut it up into smaller portions, resulting in no full length mRNA **[1 mark]**. No mRNA would be available for translation, so no protein would be produced **[1 mark]**.

Page 131 — Control of Protein Synthesis

1 Maximum of 4 marks available.
When no lactose is present, the lac repressor binds to the operator site and blocks transcription **[1 mark]**. When lactose is present, it binds to the lac repressor **[1 mark]**, changing its shape so that it can no longer bind to the operator site **[1 mark]**. RNA polymerase can now begin transcription of the structural genes, including the ones that code for β-galactosidase and lactose permease **[1 mark]**.

Page 133 — Protein Activation and Gene Mutation

1 a) Maximum of 1 mark available.
Mutations are changes to the base sequence/nucleotide sequence of DNA **[1 mark]**.

b) Maximum of 2 marks available, from any of the 5 points below.
Substitution — one base is swapped for another **[1 mark]**.
Deletion — one base is removed **[1 mark]**.
Insertion — one base is added **[1 mark]**.
Duplication — one or more bases are repeated **[1 mark]**.
Inversion — a sequence of bases is reversed **[1 mark]**.

2 a) Maximum of 1 mark available.
ATGTATTCCGGCTGT **[1 mark]**

b) Maximum of 3 marks available.
The mutation changes a triplet in the gene from TCA to TCC **[1 mark]**. But the mutated triplet still codes for serine **[1 mark]**, so the mutation would have a neutral effect on the protein that the gene codes for **[1 mark]**.

Page 135 — Mutations, Genetic Disorders and Cancer

1 a) Maximum of 1 mark available.
AGGTATGAGGCC **[1 mark]**

b) Maximum of 5 marks available.
The original gene codes for the amino acid sequence serine-tyrosine-glutamine-alanine **[1 mark]**. The mutated gene codes for the amino acid sequence arginine-tyrosine-glutamic acid-alanine **[1 mark]**. Even though there are three mutations there are only two changes to the amino acid sequence **[1 mark]**.
This is because of the degenerate nature of the DNA code, which means more than one codon can code for the same amino acid **[1 mark]**. So the substitution mutation on the last triplet doesn't alter the amino acid (GCT and GCC both code for alanine) **[1 mark]**.

c) Maximum of 2 marks available.
Acquired **[1 mark]**, because they weren't present before exposure to a mutagenic agent / they weren't present in the gametes **[1 mark]**.

d) Maximum of 3 marks available.
The mutations could result in the gene becoming an oncogene **[1 mark]**. When they are functioning normally, proto-oncogenes stimulate cell division by producing proteins that make cells divide **[1 mark]**. However, when proto-oncogenes mutate to form oncogenes the gene can become overactive. This stimulates the cells to divide uncontrollably (the rate of division increases), resulting in a tumour (cancer) **[1 mark]**.

Page 137 — Diagnosing and Treating Cancer and Genetic Disorders

1 Maximum of 25 marks available.
HINTS:
- Start off by describing what hereditary and acquired mutations are and the types of disorders they cause, e.g. genetic disorders and cancer.
- Then explain how knowing that some cancers are caused by acquired mutations affects prevention, e.g. if you know that mutagenic agents cause acquired mutations you can try to avoid them and so help prevent cancer.
- Next do the same for how knowing that some cancers are caused by acquired mutations affects diagnosis, e.g. high risk individuals can be screened more frequently to try to diagnose cancer earlier. Don't forget to include that early diagnosis increases their chances of recovery.
- Then repeat the previous two bits for hereditary cancer and genetic disorders.
- Use plenty of examples from pages 136-137 to back up your points.

Page 139 — Stem Cells

1 a) Maximum of 1 mark available.
Totipotent cells are stem cells that can mature into any cell type in an organism **[1 mark]**.

b) Maximum of 3 marks available.
The totipotent cells grew and divided at all pHs, but they grew the most at pH 4 (up to 30 g in mass) **[1 mark]**. The totipotent cells only specialised into shoot cells at pH 4 **[1 mark]**.
This suggests that the pH helps to control the specialisation of cells for this type of plant **[1 mark]**.

Answers

Page 141 — Stem Cells in Medicine

1 *Maximum of 4 marks available.*
*E.g. stem cell therapies are currently being used for some diseases affecting the blood and immune system **[1 mark]**. Bone marrow contains stem cells that can become specialised to form any type of blood cell **[1 mark]**. Bone marrow transplants can be used to replace faulty bone marrow in patients with leukaemia (a cancer of the blood or bone marrow) **[1 mark]**. The stem cells in the transplanted bone marrow divide and specialise to produce healthy blood cells **[1 mark]**.*

2 *Maximum of 2 marks available.*
*Obtaining embryonic stem cells involves the destruction of an embryo **[1 mark]**. Some people believe that embryos have a right to life and that it's wrong to destroy them **[1 mark]**.*

Section 11 — Using Gene Technology

Page 143 — Making DNA Fragments

1 *Maximum of 6 marks available.*
*The DNA sample is mixed with free nucleotides, primers and DNA polymerase **[1 mark]**. The mixture is heated to 95 °C to break the hydrogen bonds **[1 mark]**. The mixture is then cooled to between 50 – 65 °C to allow the primers to bind/anneal to the DNA **[1 mark]**. The primers bind/anneal to the DNA because they have a sequence that's complementary to the sequence at the start of the DNA fragment **[1 mark]**. The mixture is then heated to 72 °C and DNA polymerase lines up free nucleotides along each template strand, producing new strands of DNA **[1 mark]**. The cycle would be repeated over and over to produce lots of copies **[1 mark]**.*
This question asks you to describe and explain, so you need to give the reasons why each stage is done to gain full marks.

Page 145 — Common Techniques

1 *Maximum of 5 marks available.*
*A fluorescent tag is added to all the DNA fragments in the mixture so they can be viewed under UV light **[1 mark]**.*
*The DNA mixture is placed into a well in a slab of gel and covered in a buffer solution that conducts electricity **[1 mark]**.*
*An electrical current is passed through the gel and the DNA fragments move towards the positive electrode because DNA fragments are negatively charged **[1 mark]**. The DNA fragments separate according to size because the small fragments move faster and travel further through the gel **[1 mark]**. The DNA fragments are viewed as bands under UV light **[1 mark]**.*

2 a) *Maximum of 3 marks available.*
*DNA primer **[1 mark]**, free nucleotides **[1 mark]** and fluorescently-labelled modified nucleotides **[1 mark]**.*

b) *Maximum of 6 marks available.*
*The reaction mixture is added to four tubes, with a different modified nucleotide in each tube **[1 mark]**. The tubes undergo PCR to produce lots of strands of DNA of different lengths **[1 mark]**. Each strand of DNA is a different length because each one terminates at a different point depending on where the modified nucleotide was added **[1 mark]**. The DNA fragments in each tube are separated by electrophoresis and visualised under UV light **[1 mark]**. The smallest nucleotide is at the bottom of the gel and each band after this represents one more base added **[1 mark]**. So the bands can be read from the bottom of the gel to the top, forming the base sequence of the DNA fragment **[1 mark]**.*

3 *Maximum of 4 marks available.*
*The separated DNA fragments are transferred to a nylon membrane and incubated with a fluorescently labelled DNA probe **[1 mark]**. The probe is complementary to the sequence of the mutated BRCA1 gene **[1 mark]**. If the sequence is present in one of the DNA fragments, the DNA probe will hybridise to it **[1 mark]**. The membrane is then exposed to UV light and if the sequence is present in one of the DNA fragments, then that band will fluoresce **[1 mark]**.*

Page 147 — Gene Cloning

1 a) *Maximum of 2 marks available.*
*Colony A is visible/fluoresces under UV light, but Colony B isn't visible/doesn't flouresce **[1 mark]**. So only Colony A contains the fluorescent marker gene, which means it contains transformed cells **[1 mark]**.*

b) *Maximum of 3 marks available.*
*The plasmid vector DNA would have been cut open with the same restriction endonuclease that was used to isolate the DNA fragment containing the target gene **[1 mark]**. The plasmid DNA and gene (DNA fragment) would have been mixed together with DNA ligase **[1 mark]**. DNA ligase joins the sticky ends of the DNA fragment to the sticky ends of the plasmid DNA **[1 mark]**.*

Page 149 — Genetic Engineering

1 a) *Maximum of 3 marks available.*
*The drought-resistant gene could be inserted into a plasmid **[1 mark]**. The plasmid is then inserted into a bacterium **[1 mark]**, which is used as a vector to get the gene into the plant cells **[1 mark]**.*

b) *Maximum of 2 marks available.*
*The transformed wheat plants could be grown in drought-prone regions **[1 mark]**, where they would reduce the risk of famine and malnutrition **[1 mark]**.*

c) *Maximum of 1 mark available.*
*They could be concerned that the large agricultural company will have control over the recombinant DNA technology used to make the drought-resistant plants, which could force smaller companies out of business **[1 mark]**.*

Page 151 — Genetic Engineering

1 a) *Maximum of 2 marks available.*
*Colony A has grown on the agar plate containing penicillin **[1 mark]** so it contains the penicillin-resistance marker gene, which means it contains transformed cells **[1 mark]**.*

b) *Maximum of 3 marks available.*
*The plasmid vector DNA would have been cut open with the same restriction endonuclease that was used to isolate the DNA fragment containing the desired gene **[1 mark]**. The plasmid DNA and gene (DNA fragment) would have been mixed together with DNA ligase **[1 mark]**. DNA ligase joins the sugar-phosphate backbone of the two bits of DNA **[1 mark]**.*

c) *Maximum of 2 marks available.*
*It's useful for bacteria to take up plasmids because the plasmids may contain useful genes **[1 mark]** that increase their chance of survival **[1 mark]**.*

Answers

Page 153 — Genetic Engineering

1 *Maximum of 6 marks available.*
The gene for human insulin is identified and isolated using restriction enzymes ***[1 mark]****. A plasmid is cut open using the same restriction enzymes that were used to isolate the insulin gene* ***[1 mark]****. The insulin gene is inserted into the plasmid* ***[1 mark]****. The plasmid is taken up by bacteria and any transformed bacteria are identified using marker genes* ***[1 mark]****. The bacteria are grown in a fermenter and insulin is produced by the bacteria as they grow and divide* ***[1 mark]****. The insulin is extracted and purified so it can be used in humans* ***[1 mark]****.*

2 *Maximum of 7 marks available.*
The psy *and* crtl *genes are isolated using restriction enzymes* ***[1 mark]****. A plasmid is removed from the* Agrobacterium tumefaciens *bacterium and cut open using the same restriction enzymes* ***[1 mark]****. The* psy *and* crtl *genes and a marker gene are inserted into the plasmid* ***[1 mark]****. The recombinant plasmid is put back into the* A. tumefaciens *bacterium* ***[1 mark]****. Rice plant cells are incubated with the transformed* A. tumefaciens *bacteria, which infect the rice plant cells* ***[1 mark]****.* A. tumefaciens *inserts the genes into the plant cells' DNA* ***[1 mark]****. The rice plant cells are then grown on a selective medium, so only the transformed rice plants will be able to grow* ***[1 mark]****.*

Page 155 — Genetic Fingerprinting

1 a) *Maximum of 6 marks available.*
A sample of DNA is obtained, e.g. from a person's blood/saliva/skin etc. ***[1 mark]****. PCR is used to amplify multiple areas containing different sequence repeats* ***[1 mark]****, using primers that anneal to either side of a repeat so the whole repeat is amplified* ***[1 mark]****. A fluorescent tag is added to all the DNA fragments* ***[1 mark]****. The DNA mixture undergoes electrophoresis* ***[1 mark]****. The separated bands produce the genetic fingerprint, which is viewed under UV light* ***[1 mark]****.*

b) *Maximum of 2 marks available.*
Genetic fingerprint 1 is most likely to be from the child's father ***[1 mark]*** *because five out of six of the bands on his genetic fingerprint match that of the child's, compared to only one on fingerprint 2* ***[1 mark]****.*

c) *Maximum of 1 mark available, from any of the 4 points below.*
E.g. they can be used to link a person to a crime scene (forensic science) ***[1 mark]****. To prevent inbreeding between animals or plants* ***[1 mark]****. To diagnose cancer or genetic disorders* ***[1 mark]****. To investigate the genetic variability of a population* ***[1 mark]****.*

Page 157 — Sequencing Genomes and Restriction Mapping

1 *Maximum of 8 marks available.*
The genome is cut up into smaller fragments using restriction enzymes ***[1 mark]****. The individual fragments are inserted into bacterial artificial chromosomes/BACs, which are then inserted into bacteria* ***[1 mark]****. Each BAC contains a different DNA fragment, so each bacterium contains a BAC with a different DNA fragment* ***[1 mark]****. The bacteria divide, creating colonies of cloned cells that contain their specific DNA fragment* ***[1 mark]****. Together the different colonies make a complete genomic DNA library* ***[1 mark]****. DNA is extracted from each colony and cut up using restriction enzymes, producing overlapping pieces of DNA* ***[1 mark]****. Each piece of DNA is sequenced, using the chain-termination method, and the pieces are put back in order to give the full sequence from that BAC* ***[1 mark]****. Finally the DNA fragment from each different BAC is put back in order, using computers, to complete the entire genome* ***[1 mark]****.*

2 a) *Maximum of 1 mark available.*
Two ***[1 mark]****.*
Three fragments were produced in the total digest, so it must have cut one piece of DNA in two places.

b) *Maximum of 3 marks available. 1 mark for 1 fragment in the correct place, 2 marks for 2/3 fragments in the correct place and 1 mark for labelling the Sal1 restriction sites.*

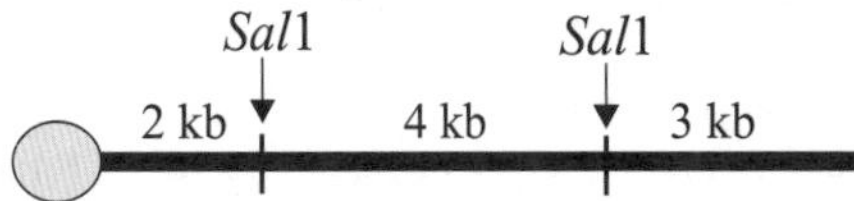

You can tell that the first cut must be after 2 kb because it's the smallest radioactive fragment in the partial digest. You can tell that the next fragment is 4 kb and not 3 kb because the other radioactive fragment in the partial digest is 6 kb long (2 kb + 4 kb). If the middle fragment was 3 kb then the radioactive fragment in the partial digest would be 5 kb long (2 + 3).

c) *Maximum of 2 marks available.*
Because Sal1 has not been left long enough to cut at all of its recognition sequences ***[1 mark]****, so there are other lengths of DNA present, i.e. 2 + 4 = 6 kb, 4 + 3 = 7 kb* ***[1 mark]****.*

Page 159 — DNA Probes in Medical Diagnosis

1 a) *Maximum of 2 marks available.*
The gene that you want to screen for is sequenced ***[1 mark]****. Multiple copies of parts of the gene are made by PCR to be used as DNA probes* ***[1 mark]****.*

b) *Maximum of 4 marks available.*
Microscopic spots of different DNA probes are attached in series to a glass slide, producing a microarray ***[1 mark]****. A sample of the person's labelled DNA is washed over the array and if any of the DNA matches any of the probes, it will stick to the array* ***[1 mark]****. The array is washed and visualised, under UV light/X-ray film* ***[1 mark]****. Any spot that shows up means that the person's DNA contains that specific gene* ***[1 mark]****.*
This question asks you to describe how many genes can be screened for at once (which is a microarray), but you could be asked how you can use DNA probes to look for a single gene too (see page 145).

2 a) *Maximum of 1 mark available.*
Because she tested negative for the mutated gene (KRAS oncogene) that the drug specifically targets ***[1 mark]****.*

b) *Maximum of 2 marks available.*
So the results of her screening can be explained to her ***[1 mark]*** *and so her treatment options can also be explained* ***[1 mark]****.*

Page 161 — Gene Therapy

1 a) *Maximum of 1 mark available.*
Gene therapy involves altering/supplementing defective genes (mutated alleles) inside cells to treat genetic disorders and cancer ***[1 mark]****.*

b) *Maximum of 1 mark available.*
Somatic gene therapy ***[1 mark]****.*

2 *Maximum of 3 marks available, from any 6 of the points below.*
E.g. the effect of the treatment may be short-lived ***[1 mark]****. The patient might have to undergo multiple treatments* ***[1 mark]****. It might be difficult to get the allele into specific body cells* ***[1 mark]****. The body may start an immune response against the vector* ***[1 mark]****. The allele may be inserted into the wrong place in the DNA, which could cause more problems* ***[1 mark]****. The allele may be overexpressed* ***[1 mark]****.*

Answers

Section 12 — Excretion

Page 163 — The Liver and Excretion

1 *Maximum of 3 marks available.*
*A — central vein **[1 mark]***
*B — sinusoid **[1 mark]***
*C — hepatocyte **[1 mark]***

2 *Maximum of 6 marks available.*
*The protein would be digested, producing amino acids **[1 mark]**. Amino acids contain nitrogen in their amino groups, but the body can't usually store nitrogenous substances, so if a lot of protein is eaten there could be an excess of amino acids that will need to be used or broken down and excreted **[1 mark]**. Excess amino acids are broken down in the liver into ammonia and organic acids in a process called deamination **[1 mark]**. Ammonia is then combined with CO_2 in the ornithine cycle to produce urea **[1 mark]**. Urea is then released into the blood and filtered out at the kidneys to produce urine **[1 mark]**. So if a large amount of protein is eaten, there may be excess amino acids that are broken down by the liver, producing a large amount of urea that's excreted in the urine **[1 mark]**.*
Don't forget to say that only excess amino acids are broken down.

Page 165 — The Kidneys and Excretion

1 a) *Maximum of 4 marks available.*
*A — nephron **[1 mark]***
*B — renal capsule / Bowman's capsule **[1 mark]***
*C — proximal convoluted tubule / PCT **[1 mark]***
*D — collecting duct **[1 mark]***

b) *Maximum of 1 mark available.*
*B (renal capsule) **[1 mark]***

c) *Maximum of 5 marks available.*
*Ultrafiltration is when substances are filtered out of the blood and enter the tubules in the kidneys **[1 mark]**. Blood enters a glomerulus, a bundle of capillaries looped inside a hollow ball called a renal capsule/Bowman's capsule **[1 mark]**. The blood in the glomerulus is under high pressure because it enters through the afferent arteriole and leaves through the smaller efferent arteriole **[1 mark]**. The high pressure forces liquid and small molecules in the blood out of the capillary and into the renal capsule **[1 mark]**. The liquid and small molecules pass through the capillary wall, the basement membrane and slits in the epithelium of the renal capsule. But larger molecules like proteins and blood cells can't pass through and stay in the blood **[1 mark]**.*

Page 167 — Controlling Water Content

1 *Maximum of 6 marks available.*
*Near the top of the ascending limb of the loop of Henle, sodium/Na^+ and chloride/Cl^- ions are actively pumped out into the medulla. This creates a low water potential in the medulla **[1 mark]**. There's now a lower water potential in the medulla than in the descending limb **[1 mark]**, so water moves out of the descending limb and into the medulla by osmosis **[1 mark]**. Near the bottom of the ascending limb sodium/Na^+ and chloride/Cl^- ions diffuse into the medulla, lowering the water potential of the medulla further **[1 mark]**. The low water potential in the medulla causes water to move out of the collecting duct by osmosis **[1 mark]**. The water in the medulla is then reabsorbed into the blood through the capillary network **[1 mark]**.*

2 *Maximum of 6 marks available.*
*Strenuous exercise causes more sweating, so more water is lost **[1 mark]**. This decreases the water content of the blood, so its water potential drops **[1 mark]**. This is detected by osmoreceptors in the hypothalamus **[1 mark]**, which stimulates the posterior pituitary gland to release more ADH **[1 mark]**.*
The answer up to this point has explained the cause of the increase in ADH in the blood. After this, the answer explains the effect on the kidney.
*The ADH increases the permeability of the walls of the distal convoluted tubule and collecting duct **[1 mark]**. This means more water is reabsorbed into the medulla and into the blood by osmosis, so a small amount of concentrated urine is produced **[1 mark]**.*

Page 169 — Kidney Failure and Detecting Hormones

1 *Maximum of 5 marks available. For full marks answers must include at least 1 advantage and 1 disadvantage.*
*Kidney transplants are cheaper in the long term than renal dialysis **[1 mark]**. Having a kidney transplant is more convenient for a person than regular dialysis sessions **[1 mark]**. A patient who has had a kidney transplant won't feel unwell between dialysis sessions **[1 mark]**. However, a transplant means the patient has to undergo a major operation, which is risky **[1 mark]**. The patient also has to take drugs to suppress the immune system, so it doesn't reject the transplant **[1 mark]**.*

2 *Maximum of 5 marks available.*
*Steroids are removed from the blood in the urine, so urine can be tested to see if a person is using steroids **[1 mark]**. It's tested using a technique called gas chromatography, where the urine is vaporised and passed through a column containing a liquid **[1 mark]**. Different substances move through the column at different speeds **[1 mark]**. The time taken for substances in the sample to pass through the column is compared to the time taken for a steroid to pass through the column **[1 mark]**. If the time taken is the same then the sample contains the steroid **[1 mark]**.*

Section 13 — Cloning and Biotechnology

Page 172 — Cloning

1 a) *Maximum of 2 marks available.*
*Reproductive cloning is used to make a complete organism that's genetically identical to another organism **[1 mark]**. Non-reproductive cloning is used to make embryonic stem cells that are genetically identical to another organism **[1 mark]**.*

b) *Maximum of 6 marks available.*
*The scientists could use nuclear transfer **[1 mark]**. They would take a body cell from an organism (organism A) and extract its nucleus **[1 mark]**. An egg cell would be taken from another organism (organism B) and its nucleus would be removed, forming an enucleated egg cell **[1 mark]**. The scientists would transfer the body cell nucleus into the enucleated egg cell **[1 mark]**. They would then stimulate the egg cell to divide **[1 mark]**. An embryo would form, which would be made up of stem cells that are genetically identical to the cells found in organism A **[1 mark]**.*
Don't forget — the technique of nuclear transfer is used in both reproductive and non-reproductive cloning.

Answers

Page 175 — Biotechnology

1 *Maximum of 8 marks available.*
*The first phase of the standard growth curve is the lag phase, when the microorganism population increases slowly **[1 mark]**. This is because the microorganisms need to make enzymes and other molecules before they can reproduce **[1 mark]**. The culture then enters the exponential phase, when the population size increases quickly **[1 mark]**. This is because there's lots of food and little competition **[1 mark]**. The next phase is the stationary phase, when the population size stays level **[1 mark]**. This is because the reproductive rate equals the death rate **[1 mark]**. The culture then enters the decline phase, when the population size begins to fall **[1 mark]**. This is because food is scarce and waste products are at toxic levels, causing microorganisms to die **[1 mark]**.*

Section 14 — The Brain and Behaviour

Page 178 — Brain Structure and Function

1 a) *Maximum of 1 mark available.*
*Hypothalamus **[1 mark]**.*
b) *Maximum of 2 marks available.*
*Control of breathing **[1 mark]**. Control of heart rate **[1 mark]**.*
c) *Maximum of 1 mark available.*
*Lack of coordinated movement / balance **[1 mark]**.*
You know that the cerebellum normally coordinates movement, so damage to it is likely to cause a lack of coordinated movement or balance.

2 a) *Maximum of 2 marks available, from any of the 3 points below.*
*An MRI scan would give information about the extent of the bleeding **[1 mark]**, the location of the bleeding **[1 mark]** and what brain functions might be affected by the bleeding **[1 mark]**.*
b) *Maximum of 1 mark available.*
*Functional magnetic resonance imaging/fMRI **[1 mark]**.*

Page 181 — Brain Development and Habituation

1 a) *Maximum of 1 mark available.*
*'Nature' means your genes **[1 mark]**.*
b) *Maximum of 2 marks available.*
*A newborn baby's brain hasn't really been affected by the environment **[1 mark]**. This means scientists can see which aspects of brain development are more likely to be due to nature than nurture **[1 mark]**.*
c) *Maximum of 2 marks available, from any of the 3 points below.*
*Twin studies **[1 mark]**. Brain damage studies **[1 mark]**. Cross-cultural studies **[1 mark]**.*
The question asks you to suggest two types of study to directly investigate the effect of nature and nurture on brain development in <u>humans</u>, so don't go writing about animal experiments.

2 a) *Maximum of 3 marks available.*
*The birds' behaviour is habituation because they showed a reduced response (they didn't fly away as much) **[1 mark]** to the unimportant stimulus of the birdwatcher **[1 mark]** after repeated exposure for an hour every day **[1 mark]**.*
b) *Maximum of 1 mark available.*
*Habituation means the birds don't waste time and energy responding to unimportant stimuli **[1 mark]**.*

Page 183 — Development of the Visual Cortex

1 a) *Maximum of 4 marks available.*
*Hubel and Wiesel stitched shut one eye of very young kittens for several months **[1 mark]**. When they unstitched the eyes, Hubel and Wiesel found that the kitten's eye that had been stitched up was blind **[1 mark]**. They also found the ocular dominance columns that were stimulated by the open eye had become bigger and had taken over the ocular dominance columns that weren't visually stimulated/for the shut eye **[1 mark]**. Hubel and Wiesel's experiments showed that the visual cortex only develops properly if both eyes are visually stimulated in the very early stages of life **[1 mark]**.*
b) *Maximum of 2 marks available.*
*Yes, their experiments give evidence for a critical 'window' in the development of the human visual system because our visual cortex is also made up of ocular dominance columns **[1 mark]**. The critical 'window' is the period of time in very early life when it's critical that you're exposed to the right visual stimuli for the visual system to develop properly **[1 mark]**.*
c) *Maximum of 2 marks available, from any of the 4 points below.*
*Animal research has led to lots of medical breakthroughs, e.g. antibiotics **[1 mark]**. Animal experiments are only done when necessary and scientists follow strict rules **[1 mark]**. Using animals is currently the only way to study how a drug affects the whole body **[1 mark]**. Some people think humans have a greater right to life than animals **[1 mark]**.*

Page 185 — Behaviour

1 *Maximum of 2 marks available.*
*Classical conditioning has occurred **[1 mark]**. The postman has learned to respond naturally to the stimulus of approaching Number 10, which wouldn't normally cause that response **[1 mark]**.*

2 *Maximum of 2 marks available.*
*A dog could be rewarded for good behaviour, e.g. it could be given a biscuit for sitting down when the trainer says, "Sit" **[1 mark]**. If the dog is repeatedly rewarded for sitting down then that behaviour will be reinforced, and the dog will learn to sit when told **[1 mark]**.*
You could answer this question by using an example of punishing a dog for bad behaviour instead.

Page 187 — Behaviour

1 a) *Maximum of 1 mark available.*
*Imprinting is where an animal learns to recognise its parents and instinctively follows them **[1 mark]**.*
b) *Maximum of 2 marks available.*
*A gosling can imprint on a human if the gosling is reared from birth/during the critical period by a human **[1 mark]**. The human will be the first moving object the gosling sees, so the gosling will imprint on the human/will follow the human **[1 mark]**.*

2 a) *Maximum of 1 mark available.*
*Social behaviour is behaviour that involves members of a group interacting with each other **[1 mark]**.*
b) i) *Maximum of 1 mark available.*
*A large group is more efficient at finding food **[1 mark]**.*
ii) *Maximum of 1 mark available, from any of the 2 points below.*
*Grooming is hygienic **[1 mark]**. Grooming helps to reinforce the social bonds within the group **[1 mark]**.*

Answers

Section 15 — Microorganisms and Immunity

Page 189 — Viral and Bacterial Infections

1 *Maximum of 7 marks available.*
*The initial symptoms of AIDS include minor infections of mucous membranes and recurring respiratory infections **[1 mark]**. These are caused by a lower than normal number of immune system cells **[1 mark]**. As AIDS progresses the number of immune system cells decreases further **[1 mark]**. Patients become susceptible to more serious infections, including chronic diarrhoea, serious bacterial infections and TB **[1 mark]**. During the late stages of AIDS, patients have a very low number of immune system cells **[1 mark]** and suffer from a range of serious infections such as toxoplasmosis and candidiasis **[1 mark]**. It's these serious infections that kill AIDS patients, not HIV itself **[1 mark]**.*

2 *Maximum of 3 marks available, from any of the 6 points below.*
*E.g. bacteria have ribosomes but viruses don't **[1 mark]**.*
*Bacteria have a cell wall but viruses don't **[1 mark]**.*
*Bacteria have cytoplasm, but viruses don't **[1 mark]**.*
*Viruses have a capsid but bacteria don't **[1 mark]**.*
*Viruses have a protein coat but bacteria don't **[1 mark]**.*
*Viruses are smaller than bacteria **[1 mark]**.*

Page 191 — Infection and The Non-Specific Immune Response

1 *Maximum of 6 marks available.*
*Immune system cells recognise foreign antigens on the surface of a pathogen and release molecules that trigger inflammation **[1 mark]**. The molecules cause vasodilation (widening of the blood vessels) around the site of infection, increasing the blood flow to it **[1 mark]**. The molecules also increase the permeability of the blood vessels **[1 mark]**. The increased blood flow brings loads of immune system cells to the site of infection **[1 mark]** and the increased permeability allows those cells to move out of the blood vessels and into the infected tissue **[1 mark]**. The immune system cells can then start to destroy the pathogen **[1 mark]**.*

2 *Maximum of 6 marks available.*
*A phagocyte recognises the antigens on a pathogen **[1 mark]**. The phagocyte engulfs the pathogen **[1 mark]**. The pathogen is now contained in a phagocytic vacuole **[1 mark]**. A lysosome fuses with the phagocytic vacuole **[1 mark]** and digestive enzymes from the lysosome break down the pathogen **[1 mark]**. The phagocyte presents the antigens to other immune system cells **[1 mark]**.*

Page 193 — The Specific Immune Response

1 a) *Maximum of 2 marks available.*
*T cells are activated when receptors on the surface of the T cells **[1 mark]** bind to complementary antigens presented to them by phagocytes **[1 mark]**.*

b) *Maximum of 3 marks available.*
*T helper cells **[1 mark]**, T killer cells **[1 mark]** and T memory cells **[1 mark]**.*

2 *Maximum of 3 marks available.*
*Antibodies agglutinate pathogens, so that phagocytes can get rid of a lot of the pathogens at once **[1 mark]**. Antibodies neutralise toxins produced by pathogens **[1 mark]**. Antibodies bind to pathogens to prevent them from binding to and infecting human cells **[1 mark]**.*
There are three marks available for this question so you need to think of three different functions.

Page 195 — Developing Immunity

1 *Maximum of 10 marks available.*
*Before the person is exposed to the antigen there are none of the right antibodies in their bloodstream **[1 mark]**. This is because the T and B cells haven't come into contact with the antigen **[1 mark]**. Shortly after the person is exposed to the antigen the concentration of the right antibody begins to rise **[1 mark]**. This is the primary response and it's slow because there aren't many B cells that can make the antibody that binds to the antigen **[1 mark]**. The concentration of antibody peaks at about 20 days after the first exposure and then it begins to fall **[1 mark]**. It begins to fall because the person's immune system is starting to overcome the infection **[1 mark]**. When the person is exposed to the same pathogen again at 60 days, the secondary response happens **[1 mark]**. The concentration of antibody rises quickly from the moment of the second exposure to a peak at about 7 days after the second exposure **[1 mark]**. This is because memory B cells that were created after the first exposure quickly divide into plasma cells **[1 mark]**. They produce the right antibody to the antigen almost immediately **[1 mark]**.*

2 *Maximum of 6 marks available.*
*HIV kills the immune systems cells that it infects **[1 mark]**. This reduces the overall number of immune system cells in the body, which reduces the chance of HIV being detected **[1 mark]**. HIV has a high rate of mutation in the genes that code for antigen proteins. The mutations change the structure of the antigens forming new strains of the virus. This is called antigenic variation **[1 mark]**. It means the antibodies produced for one strain of HIV won't recognise other strains with different antigens, so the immune system has to produce a primary response against each new strain **[1 mark]**. HIV disrupts antigen presentation in infected cells **[1 mark]**. This prevents immune system cells recognising and killing the infected cells **[1 mark]**.*

Page 197 — Antibiotics

1 a) *Maximum of 4 marks available.*
*The bacteria to be tested are spread onto the agar plate **[1 mark]**. Paper discs soaked with the antibiotics are placed apart on the plate, along with a negative control disc soaked in sterile water **[1 mark]**. The whole experiment is performed using aseptic techniques, e.g. using Bunsen burners to sterilise instruments **[1 mark]**. The plate is incubated at 25-30 °C for 24-36 hours **[1 mark]**.*

b) *Maximum of 2 marks available.*
*Erythromycin **[1 mark]**, because it has the largest inhibition zone **[1 mark]**.*

2 *Maximum of 2 marks available, 1 mark for a description of poor hygiene and 1 mark for a code of practice.*
*E.g. Hospital staff and visitors not washing their hands before and after visiting a patient **[1 mark]**. Hospital staff and visitors should be encouraged to wash their hands before and after they've been with a patient **[1 mark]**.*
*Equipment (e.g. beds or surgical instruments) and surfaces not being disinfected after they're used **[1 mark]**. Equipment and surfaces should be disinfected after they're used **[1 mark]**.*

Page 199 — Microbial Decomposition and Time of Death

1 *Maximum of 8 marks available.*
*A dead human body loses heat at a rate of approximately 1.5-2.0 °C per hour **[1 mark]**, which suggests the time of death was around 4-5 hours ago/at 17:45-18:45 **[1 mark]**. Rigor mortis has only recently started as it's limited to the upper parts of the body **[1 mark]**, which suggests that the time of death*

Answers

was around 4 to 6 hours ago ***[1 mark]****. There's a lack of visible decomposition* ***[1 mark]****, which suggests that the time of death was only a few hours ago* ***[1 mark]****. Blowfly larvae hatch from eggs approximately 24 hours after being laid* ***[1 mark]****, so no blowfly larvae on the body suggests that the time of death was less than 24 hours ago* ***[1 mark]****.*

Section 16 — Heart Rate, Ventilation and Exercise

Page 201 — Electrical Activity in the Heart

1 a) *Maximum of 1 mark available.*
The sinoatrial node acts as a pacemaker / initiates heartbeats ***[1 mark]****.*

b) *Maximum of 1 mark available.*
The bundle of His conducts the waves of electrical activity from the AVN to the Purkyne fibres in the ventricle walls ***[1 mark]****.*

2 *Maximum of 2 marks available.*
The ventricle is not contracting properly ***[1 mark]****. This could be because of muscle damage / because the AVN is not conducting impulses to the ventricles properly* ***[1 mark]****.*

Page 203 — Variations in Heart Rate and Breathing Rate

1 a) *Maximum of 5 marks available.*
The breathing rate would go up ***[1 mark]****, because carbonic acid lowers blood pH* ***[1 mark]****. This stimulates chemoreceptors in the medulla, aortic bodies and carotid bodies* ***[1 mark]****. The chemoreceptors send nerve impulses to the medulla* ***[1 mark]****. In turn, the medulla sends more frequent nerve impulses to the intercostal muscles and diaphragm* ***[1 mark]****.*
This question only asks about the breathing rate, so you won't get any extra marks for commenting on the depth of breathing or speed of gas exchange.

b) *Maximum of 1 mark available.*
cardiac output (cm^3/min) = heart rate (beats per minute) × stroke volume (cm^3) ***[1 mark]****.*

Page 205 — Investigating Ventilation

1 a) *Maximum of 2 marks available.*
Tidal volume = 1.4 – 1.0 = 0.4 dm^3 ***[1 mark]****.*
Breathing rate = 12 breaths per minute/bpm ***[1 mark]****.*

b) *Maximum of 2 marks available.*
The tidal volume would be larger ***[1 mark]*** *and the breathing rate would be faster* ***[1 mark]****.*

Page 207 — Exercise and Health

1 a) *Maximum of 1 mark available.*
The table shows that the relative risk of CHD in women increases with less physical activity ***[1 mark]****.*

b) *Maximum of 1 mark available.*
There's a correlation/link between lower levels of physical activity in women and an increased risk of CHD ***[1 mark]****.*

Page 209 — Exercise and Health

1 a) *Maximum of 3 marks available, from any of the 6 points below.*
Keyhole surgery involves a much smaller incision than open surgery so the patient loses less blood ***[1 mark]*** *and has less scarring of the skin* ***[1 mark]****. The patient usually suffers less pain after keyhole surgery than open surgery* ***[1 mark]*** *and the patient usually recovers more quickly* ***[1 mark]****. It's usually easier for the patient to return to normal activities after keyhole surgery than open surgery* ***[1 mark]*** *and their hospital stay is usually shorter* ***[1 mark]****.*

b) *Maximum of 2 marks available.*
A prosthesis could be used to replace his knee ***[1 mark]****.*
This might make it possible for him to play sport again ***[1 mark]****.*

2 *Maximum of 2 marks available.*
Banning the use of performance-enhancing drugs makes competitions fairer ***[1 mark]****. Athletes are less tempted to take drugs that can have serious health risks* ***[1 mark]****.*

Section 17 — Drugs

Page 211 — Drugs and Disease

1 a) *Maximum of 2 marks available.*
Dopamine transmits nerve impulses across synapses ***[1 mark]*** *in the parts of the brain that control movement* ***[1 mark]****.*

b) *Maximum of 4 marks available.*
In Parkinson's disease the neurones in the parts of the brain that control movement are destroyed ***[1 mark]****. These neurones normally produce the neurotransmitter dopamine* ***[1 mark]****, so losing them causes a lack of dopamine* ***[1 mark]****. This causes a decrease in the transmission of the nerve impulses involved in movement* ***[1 mark]****.*

c) *Maximum of 4 marks available.*
L-dopa is a drug that's used to treat Parkinson's disease ***[1 mark]****. L-dopa is absorbed into the brain and converted into dopamine by the enzyme dopa-decarboxylase* ***[1 mark]****. This increases the level of dopamine in the brain* ***[1 mark]****, which causes an increase in the transmission of the nerve impulses involved in movement* ***[1 mark]****.*

2 *Maximum of 5 marks available.*
Scientists use databases that store the information from the HGP to identify proteins that are involved in disease ***[1 mark]****. Scientists are using this information to create new drugs that target the identified proteins* ***[1 mark]****. The HGP has also highlighted common genetic variations between people* ***[1 mark]****. It's known that some of these variations make some drugs less effective* ***[1 mark]****. Drug companies are using this knowledge to design new drugs that are effective in people with these variations* ***[1 mark]****.*

Page 213 — Producing Drugs Using GMOs

1 *Maximum of 4 marks available.*
The human insulin gene is isolated using enzymes called restriction enzymes ***[1 mark]****. The gene is then copied using PCR and the copies of the gene are inserted into plasmids* ***[1 mark]****. The plasmids are transferred into microorganisms* ***[1 mark]****. The modified microorganisms are grown so that they divide and produce lots of human insulin* ***[1 mark]****.*

2 *Maximum of 7 marks available.*
The hepatitis B vaccine in the plant tissues won't need to be refrigerated ***[1 mark]****. This could make the vaccine available to people in areas where refrigeration isn't available* ***[1 mark]****. Herbicide resistance is a benefit because the plants will be unaffected by herbicides* ***[1 mark]****. The genetically modified plant will thrive after weeds are killed by herbicides and this will give a high yield of the vaccine* ***[1 mark]****. However, the transmission of genetic material between the genetically modified plants and wild plants could occur* ***[1 mark]****. This could create superweeds that are resistant to herbicides* ***[1 mark]****. There may also be unforeseen consequences from using the genetically modified plant* ***[1 mark]****.*
The question asked you to discuss the benefits and risks of growing the plant — make sure you write about both.

Acknowledgements

Page 43 — Data used to construct the graph of daily CO_2 concentration reproduced with kind permission from a study at Griffin Forest, Perthshire performed by the University of Edinburgh and supported by the Natural Environment Research Council.

Page 43 and 49 — Data used to construct the graph of yearly CO_2 concentration on page 43 and average yearly CO_2 concentration on page 49 reproduced with kind permission from Atmospheric CO_2 at Mauna Loa Observatory, Scripps Institution of Oceanography, NOAA Earth System Research Laboratory.

Page 44 — Data used to construct the graph of temperature change over the last 1000 years reproduced with kind permission from Climate Change 2001: The Scientific Basis, Contribution of Working Group I to the Third Assessment Report of the Intergovernmental Panel on Climate Change, SPM Figure 1. Cambridge University Press.

Page 46 — Data used to construct the graph of methane concentration © CSIRO Marine and Atmospheric Research, reproduced with permission from www.csiro.au.

Page 46 — Data used to construct the graph of CO_2 concentration reproduced with kind permission from U.S. Global Change Research Program, http://www.usgcrp.gov/usgcrp/nacc/background/scenarios/images/co2hm.gif.

Page 48 — Data used to construct the graph of wheat yield from Global scale climate-crop yield relationships and the impacts of recent warming. D. B. Lobell and C. B. Field. Environmental Research Letters 2 (2007) 014002 (7pp). IOP Publishing.

Page 49 — Data used to construct the graph of average global temperature adapted from Crown Copyright data supplied by the Met Office.

Page 49 — Data used to construct the graph of global sea temperature reproduced with kind permission from NASA Goddard Institute for Space Studies.

Page 49 — Diagram showing the distribution of subtropical plankton reproduced with kind permission from Plankton distribution changes, due to climate changes – North Sea. (February 2008). In UNEP/GRID-Arendal Maps and Graphics Library. http://maps.grida.no/go/graphic/plankton-distribution-changes-due-to-climate-changes-north-sea.

Page 49 — Data used to construct the graph of corn yield provided by the U.S. Department of Agriculture – National Agricultural Statistics Service.

Page 53 — Graph of emissions scenarios modified and based on Special Report of Working Group III of the Intergovernmental Panel on Climate Change on Emissions Scenarios (IPCC 2000).

Page 59 — Data used to construct the graph showing the stock of spawning cod in the North Sea and the rate of mortality caused by fishing since 1960 from the International Council for the Exploration of the Sea.

Page 161 — Data used to construct the graphs from S. Hacein–Bey–Abina et al. SCIENCE 302: 415–419 (2003).

Pages 177 and 178 — With thanks to Science Photo Library for permission to reproduce the photographs.

Page 206 — Data on exercise and obesity adapted by permission from Macmillan Publishers Ltd: M.Á. Martínez-González, J.A. Martínez, F.B. Hu, M.J. Gibney, J. Kearney. Physical inactivity, sendentary lifestyle and obesity in the European Union. International Journal of Obesity; 23: 1192-1201, copyright 1999.

Page 206 — Data used to construct the graph on exercise and coronary heart disease from S.G. Wannamethee, A.G. Shaper, K.G. Alberti. Physical activity, metabolic factors, and the incidence of coronary heart disease and type 2 diabetes. Archives of Internal Medicine, 2000; 160:2108-2116. Copyright © 2000 American Medical Association. All rights reserved.

Page 206 — Data used to construct the graph on exercise and type 2 diabetes from S.G. Wannamethee, A.G. Shaper, K.G. Alberti. Physical activity, metabolic factors, and the incidence of coronary heart disease and type 2 diabetes. Archives of Internal Medicine, 2000; 160:2108-2116. Copyright © 2000 American Medical Association. All rights reserved.

Page 207 — Data on exercise and osteoarthritis reproduced from U.M. Kujala, J. Kapiro, S. Sarna. Osteoarthritis of weight bearing joints of lower limbs in former elite male athletes. BMJ 2004; 308:231-234. Data reproduced with permission from BMJ Publishing Group Ltd.

Page 207 — Data on exercise and the immune system from L. Spence, W.J. Brown, D.B. Pyne, M.D. Nissen, T.P. Sloots, J.G. McCormack, A.S. Locke, P.A. Fricker. Incidence, etiology and sypmtomatology of upper respiratory illness in elite athletes. Med Sci Sports Exerc 2007; 39:577-586.

Page 207 — Data in the exam question from T.Y. Li, J.S. Rana, J.E. Manson, W.C. Willett, M.J. Stampfer, G.A. Colditz, K.M. Rexrode, F.B. Hu. Obesity as compared with physical activity in predicting risk of coronary heart disease in women. Circulation 2006; 113:499-506.

Index

A

abiotic conditions/factors 22, 23
abiotic stress 106
abundance of organisms 22, 24
accuracy 216
acetyl coenzyme A (acetyl CoA) 15, 16
acetylcholine (ACh) 92
acquired immune deficiency syndrome (AIDS) 188
acquired mutations 135, 136
actin 97-99
action potentials 89, 90
active immunity 194
adaptations 23
adenosine triphosphate (ATP) 4-9, 14, 16-18, 20
ADH (antidiuretic hormone) 167
adrenal glands 103, 104
adrenaline 103, 104, 118
aerobic respiration 4, 14-17, 20, 21
age-sex pyramids 33
Agrobacterium tumefaciens 152
AIDS (acquired immune deficiency syndrome) 188
alcoholic fermentation 20
allele frequency 74, 76-78
alleles 62
anabolic steroids 169
anaerobic respiration 4, 20, 21
animal models 182
antagonistic muscle pairs 96
antibiotics 196, 197
antibodies 192-194
anticodons 123
antidiuretic hormone (ADH) 167
antigen-presenting cells 191
antigenic variation 195
antigens 190-195
apical dominance 108, 110
apoptosis (programmed cell death) 131
artificial immunity 194
artificial selection 81
asepsis 175
ATP (adenosine triphosphate) 4-9, 14-18, 20
ATP-phosphocreatine system 100
autonomic nervous system 83, 104, 114
autotrophs 4
auxins 106-111
AVN (atrioventricular node) 200

B

B cells 192-194
B effector cells 192
BACs (bacterial artificial chromosomes) 156
bacterial structure 189
bacteriocidal antibiotics 196
bacteriophages 146, 150
bacteriostatic antibiotics 196
baroreceptors 104
base triplets 122, 123, 126
batch culture 175
beating trays 27
behaviour 181, 184-187
biceps muscles 96
biofuels 52
biological species concept 79
biotechnology 173-175
biotic conditions/factors 22, 23, 30
bleaching of rhodopsin 87
blood glucose concentration 116-119
body plans 130, 131
body temperature 114, 115, 198
brain development 179, 180, 182
brain structure 176
breathing rate 202, 204, 205
brine shrimp hatch rate 51

C

Calvin cycle 6, 8, 9
cAMP (cyclic AMP) 103, 118, 132
cancer 135-137, 158-160
carbon cycle 42, 52, 198
cardiac muscle 101, 200
cardiac output 203
cardiovascular control centres 203
carriers 64
carrying capacity 30
cell signalling 82
central nervous system (CNS) 82, 83, 96
cerebellum 176
cerebrum 176
chain-termination method 144
chemical mediators 105
chemiosmosis 7, 17, 18
chemoreceptors 104, 202, 203
chi-squared test 68, 69
chlorophyll 6, 7
chloroplasts 6, 9
chromatids 70
classical conditioning 184
classifying species 79
climax communities 54, 55
cloning organisms 170-172
closed cultures 174
CNS (central nervous system) 82, 83, 96
codominance 62, 63
codons 122, 123
coenzymes 5
colour blindness 64
commercial uses of plant hormones 109
communication systems 82-84
communities 22
complementary base pairing 122
complementary DNA (cDNA) 143
conclusions (checking and drawing) 215
cones 86
conflicting evidence 58
conservation 55-61
continuous culture 175
continuous variation 72
control of heart rate 104, 203
controls 3, 216
correlations 214
coronary heart disease (CHD) 206
countercurrent multiplier mechanism 166
cristae 14
critical values 69
critical windows 182, 183
crossing-over of chromatids 71
cruciate ligaments 208
CT (computed tomography) scanners 177
cyclic AMP (cAMP) 103, 118, 132
cystic fibrosis 65, 76, 133

D

data collection 216
decarboxylation 5
decline phase 172
decomposers 34, 42
decomposition 199
degrees of freedom 69
dehydrogenation 5
Demographic Transition Model (DTM) 32
dendrochronology 44
dependent variables 216
depression 210
describing data 214
diabetes 118, 119, 206
dialysis 168
differential reproductive success 74
dihybrid inheritance 66
directional selection 74
discontinuous variation 72
distribution of organisms 22, 24, 29
DNA 122, 123
- fragments 142, 143
- ligase 146, 150
- microarrays 158
- polymerase 142, 144
- probes 145, 158
- profiling 154, 155
- sequencing 144

dominant alleles 62
dopamine 210
dopamine receptor D4 186
drawing conclusions 215
drugs
- action at synapses 95, 208
- from genetically modified organisms 212, 213

E

ECG (electrocardiogram) 200, 201
ecosystems 22, 34
- conservation of 60, 61

ectotherms 115
effectors 82, 96-103

Index

elbow joints 96
electrical activity in the heart 200, 201
electrocardiograms (ECG) 200, 201
electron carriers 7, 17
electron transport chain 7, 17
electrophoresis 144
endocrine glands 102
endocrine system 84
endotherms 115
energy transfers 34, 35, 37
entomology 198
epistasis 66, 67
escape reflexes 184
ethical issues
  of genetic engineering 149, 153
  of stem cell use 141
  of using animals in medical research 183
  with doing fieldwork 28
  with cloning humans 170
  with the Human Genome Project 211
eutrophication 40, 41
evading the immune system 195
evidence 2, 3, 215
evolution 74, 75, 80, 81
excretion 162-165
exercise 202-209
exocrine glands 102
exons 124
exponential phase 172
extensor muscles 96

F

farming practices and productivity 38, 39
fast twitch muscle fibres 100
fermentation vessels 174
fight or flight response 104
first messengers 103
flexor muscles 96
fMRI (functional magnetic resonance imaging) scanners 178
food chains 34
food webs 34
forensic entomology 198
frame quadrats 24
frequency of organisms 24
FSH (follicle-stimulating hormone) 120, 121

G

GALP (glyceraldehyde 3-phosphate) 8
Galapagos islands 61
gametes 70
genes 62, 122
gene cloning 146, 147
gene pool 74
gene sequencing 144, 157
gene therapy 160, 161
generator potentials 85
genetic fingerprinting 154, 155
genetic bottlenecks 75
genetic code 122, 126
genetic counselling 159
genetic diagrams 62, 63, 65, 66
  pedigree diagrams 65
  Punnett squares 63
genetic disorders 135, 137, 160
genetic drift 75
genetic engineering 148-153
genetic variation 71
genetically modified organisms 148, 150, 210, 211
genome sequencing 156
genotypes 62, 156
geographical isolation 78
geotropism 106, 107
germ line gene therapy 160
gibberellins 106, 108-110
glands 84, 102, 103
global warming 44-53
glucagon 116, 117
gluconeogenesis 116
glycerate 3-phosphate (GP) 8, 11
glycogenesis 116
glycogenolysis 116
glycolysis 14, 20
Golden Rice 148, 152
GP (glycerate 3-phosphate) 8, 11
greenhouse gases 46
gross productivity 34
growth factors 106, 107
growth hormones 106-109
gut and skin flora 190

H

habitat 22
habituation 181, 184
HAIs (hospital acquired infections) 197
Hardy-Weinberg principle 76, 77
heart rate 104, 202, 203
hereditary mutations 135, 137
heterotrophs 4
heterozygotes 62
hexose bisphosphate 14
histamine 105
HIV (human immunodeficiency virus) 188, 195
homeobox sequences 130, 156
homeostasis 112, 113
homologous pairs 70
homozygotes 62
hormonal communication 84
hormones 84, 102, 103
hospital acquired infections (HAIs) 197
how science works 2, 3
Hubel and Wiesel 182
Human Genome Project 211
human populations 32, 33
hydrolysis 5, 143
hypothalamus 114, 176
hypothermia 113
hypotheses 2

I

IAA (indoleacetic acid) 107
immobilised enzymes 173
immune response 190-193
immunity 194
imprinting 186
in vitro cloning 146, 147
in vivo cloning 146, 147
independent assortment 71
independent variables 216
infectious diseases 188
inflammation 190
inheritance 62-65
innate behaviour 184, 186
insight learning 185
insulin 116-119, 148, 152
  production of 148, 152
intensive farming 38
interferons 191
interpreting data 214-216
  on auxins 111
  on breathing rate and tidal volume 205
  on conservation issues 58
  on eutrophication 41
  on exercise and health 206, 207
  on gene expression 129
  on gene therapy 161
  on global warming 48, 49
  on limiting factors in photosynthesis 12
  on natural selection 77
  on nucleic acids 127
  on the distribution of organisms 29
  on tissue culture 139
interspecific competition 30
intraspecific competition 30
introns 124
involuntary muscle 101
islets of Langerhans 102, 116

J

joints 96

K

keyhole surgery 208
kidneys 164-169
kilobase (kb) 144
kinetic responses (kineses) 105, 184
knee joint replacement 208
Krebs cycle 16

Index

L

L-dopa 210
lab experiments 3
lac operon 130
lactate fermentation 20
lactic acid (lactate) 20
lag phase 172
latent learning 185
leaching 40
leaf loss in deciduous plants 108
learned behaviour 181, 184-186
LH (luteinising hormone) 120, 121
ligaments 96
ligation 146, 150
light-dependent reaction 6, 7
light-independent reaction 6, 8, 9
limiting factors for population size 31
limiting factors in photosynthesis
    10, 12, 13
link reaction 15
linkage 62, 71
liver 162, 163
locus 62
loop of Henle 165, 166
lysosomes 191
lysozyme 190

M

making decisions 3
  about conservation 59
mark-release-recapture 28
marker genes 146, 151, 152
MDMA 210
medulla 104, 176, 202, 203
meiosis 70, 71
memory cells 194
menstrual cycle 120, 121
metaphase 70
microbial decomposition 198
mitochondria 14, 15
mitosis 192
monogenic characteristics 72
monohybrid inheritance 62
motor neurones 82, 88
MRI (magnetic resonance imaging) 177
mRNA 123
mucosal surfaces 190
multiple alleles 64
muscle contraction 96-101, 198
mutagenic agents 134
mutations 74, 78, 132-137
Mycobacterium tuberculosis 188, 195
myelinated neurones 91
myofibrils 96, 97
myofilaments 97
myosin 97-99

N

natural immunity 194
natural selection 74, 78, 81
nature-nurture debate 179
negative feedback 112, 113
nephrons 164
nervous communication 82, 83
nervous system 82, 83, 88-95
net productivity 34, 35
neuromuscular junctions 93
neurones 82, 88-91
neurotransmitters 92-95
  imbalances in 210
niches 22
nitrogen cycle 40
nodes of Ranvier 91
non-reproductive cloning 170
non-specific immune response 190, 191
nuclear transfer 170
nucleic acids 126, 127
null hypothesis 68, 69

O

obesity 206
ocular dominance columns 182
oestrogen 120, 121, 128
oestrous cycle 120, 121
oncogenes 135
operant conditioning 185
opsin 87
oxidation 5, 14
oxidative phosphorylation 16, 17

P

Pacinian corpuscles 85
pancreas 102, 116
parasympathetic nervous system 83
Parkinson's disease 210
passive immunity 194
pathogens 188
PCR (polymerase chain reaction) 142
peer review 2, 80
percentage cover 24
performance-enhancing drugs 209
PGD (preimplantation genetic diagnosis)
    137
PGH (preimplantation genetic
    haplotyping) 155
phagocytes 188, 191
phagocytosis 191
phenotypes 62
phenotype frequency 78
phenotypic ratios 66, 67
phosphorylation 5, 14
photolysis 5-7
photophosphorylation 5-7
photoreceptors 86, 107
  in animals 86
  in plants 107
photosynthesis 4, 6-9, 10-13, 34
photosystems 6, 7
phototropism 106, 107
phylogenetic species concept 79
phytochromes 107
pioneer species 54
pifall traps 27
pituitary gland 120, 121, 176
plant hormones 108, 109
plant responses 106, 107
plasma cells 192-194
plasmids 146, 150-152, 160, 189
point quadrats 25
pollen in peat bogs 45
polygenic characteristics 72
polymerase chain reaction (PCR) 142
pondweed 13
pooters 27
populations 22, 24, 25, 27,
    30-33, 74
population growth curves 33
population sizes 30, 31
positive feedback 113
potential difference 85
precision 216
predation 31
predictions 2
pregnancy tests 169
preimplantation genetic diagnosis (PGD)
    137
preimplantation genetic haplotyping
    (PGH) 155
preservation of ecosystems 60
primary metabolites 172
primary productivity 35
primary immune response 194
primers 142, 144, 154
producers 34
productivity 35, 36, 38, 39
progesterone 120, 121
programmed cell death (apoptosis) 131
prophase 70
prostaglandins 105
prostheses 208
protein activation 132
protein synthesis 124, 125
  genetic control of 130, 131
proteomics 80
proto-oncogenes 135
Punnett squares 63
pupil dilation and constriction 83
pyramid diagrams 36

Q

quadrats 24, 25
qualitative 216
quantitative 216

Index

R

random samples 24
receptors 82, 85-87, 128, 192
recessive alleles 62
recognition sequences in DNA 143
recombinant DNA 146, 148, 150
reduction 5
redox reactions 5
reflexes 83
reforestation 52
refractory periods 89, 90
relay neurones 82, 88
reliability 215
reproductive cloning 170
reproductive isolation 78
respiration 5, 14-21, 198
  aerobic 4, 14-21, 100
  anaerobic 4, 20, 21, 100
respiratory centres 202
respiratory loss in food chains 34
respiratory quotient (RQ) 21
respirometer 19
resting potential of cells 85
resting potential of neurones 88
restriction endonuclease enzymes 143
restriction mapping 157
retinal 87
reverse transcriptase 143
rhodopsin 87
ribulose bisphosphate (RuBP) 8, 9, 11
ribulose bisphosphate carboxylase (rubisco) 8, 11
rigor mortis 198
risk assessments 28
RNA 122, 123
RNA polymerase 124
rods 86, 87
rooting hormones 109
RQ (respiratory quotient) 21
rubisco (ribulose bisphosphate carboxylase) 8, 11
RuBP (ribulose bisphosphate) 8, 9, 11

S

saltatory conduction 91
SAN (sinoatrial node) 104, 200
saprobiontic nutrition 42
sarcomeres 97
Schwann cells 91
scientific conferences 80
scientific journals 2, 80
screening for diseases 158, 159
second messengers 103, 118
secondary metabolites 172
seedling growth rates 51
selective reabsorption 164, 165
sensitivity of photoreceptors 86
sensory neurones 82, 88
sequencing genomes 156
seral stages 54
serotonin 210
severe combined immunodeficiency (SCID) 140
sex-linkage 64
shivering 114
sickle-cell anaemia 63, 158
simple responses 105
sinoatrial node (SAN) 104, 200
siRNA 128
skeletal muscle 96, 100, 101
skin 190
sliding filament theory 98
slow twitch muscle fibres 100
smooth muscle 101
social behaviours 187
sodium-potassium pumps 88, 89
somatic gene therapy 160
spatial summation 94
speciation 78
species 74, 79
  classification of 79
specific base pairing 122, 124, 125
specific immune response 192, 193
spirometers 204, 205
splicing 124
stabilising selection 74
standard deviation (SD) 215
standard growth curve 174
start codons 125, 126
stationary phase 172
stem cells 119, 138-141
stem elongation 108, 110
steroids 169
sticky ends 143, 150
stimuli 82, 85
stomach acid 190
stop codons 125, 126
stroke volume 203
studies 3, 214-216
substrate-level phosphorylation 16
succession (ecology) 54, 55
succession (in dead bodies) 199
summation 94
survival curves 33
sweating 114
sympathetic nervous system 83, 104
synapses 92-95

T

T cells 192, 194
tactic responses (taxes) 105, 184
target cells of hormones 84
TB (tuberculosis) 188
telophase 70
temperate woodland 60
temperature records 44
temporal summation 94
tendons 96
theories 2
thermoreceptors 114
thermoregulation 114
threshold level 85
thyroid hormone 131
tidal volume 204, 205
time of death (TOD) 198, 199
tissue culture 139, 171
totipotent stem cells 138
TP (triose phosphate) 8, 9, 11
transcription 124
  regulation of 128, 130, 131
transcription factors 128, 130, 131
transducers 85
transects 25
transformed cells 146, 150
transformed organisms 148, 150-152
transgenic organisms 150
translation 125, 128
triceps muscles 96
triose phosphate (TP) 8, 9, 11
tRNA 123
trophic levels 34
tropisms 106, 107
tropomyosin 98, 99
troponin 98, 99
tuberculosis (TB) 188
tumour suppressor genes 135

U

ultrafiltration 164
urea 164, 165
urine 164-167

V

vaccines 194
variables 216
variation 72, 73
vasoconstriction 114
vasodilation 114, 190
vectors 146, 150, 160
vegetative propagation 172
ventilation centres 202
ventilation rate 202, 204
virus structure 189
visual acuity of photoreceptors 86
visual cortex 182
voluntary muscle 101

W

water content of urine 166, 167
wear and tear of joints 207
Wiesel and Hubel 182

X

X-linked disorders 64
xenotransplantation 152